Cell Stress Proteins

PROTEIN REVIEWS

Editorial Board:

EDITOR-IN-CHIEF: M. ZOUHAIR ATASSI, *Baylor College of Medicine, Houston, Texas*

EDITORIAL BOARD: LAWRENCE J. BERLINER, *University of Denver, Denver, Colorado*
ROWEN JUI-YOA CHANG, *University of Texas, Houston, Texas*
HANS JÖRNVALL, *Karolinska Institutet, Stockholm, Sweden*
GEORGE L. KENYON, *University of Michigan, Ann Arbor, Michigan*
BRIGITTE WITTMAN-LIEBOLD, *Wittman Institute of Technology and Analysis, Tetlow, Germany*

Recent Volumes in this Series

VIRAL MEMBRANE PROTEINS: STRUCTURE, FUNCTION, AND DRUG DESIGN
    Edited by Wolfgang B. Fischer

THE p53 TUMOR SUPPRESSOR PATHWAY AND CANCER
    Edited by Gerard P. Zambetti

PROTEOMICS AND PROTEIN PROTEIN INTERACTIONS: BIOLOGY, CHEMISTRY, BIOINFORMATICS, AND DRUG DESIGN
    Edited by Gabriel Waksman

PROTEIN MISFOLDING, AGGREGATION AND CONFORMATIONAL DISEASES
PART A: PROTEIN AGGREGATION AND CONFORMATIONAL DISEASES
    Edited by Vladimir N. Uversky and Anthony L. Fink

PROTEIN INTERACTIONS: BIOPHYSICAL APPROACHES FOR THE STUDY OF COMPLEX REVERSIBLE SYSTEMS
    Edited by Peter Schuck

PROTEIN MISFOLDING, AGGREGATION, AND CONFORMATIONAL DISEASES
PART B: MOLECULAR MECHANISMS OF CONFORMATIONAL DISEASES
    Edited by Vladimir N. Uversky and Anthony L. Fink

CELL STRESS PROTEINS
    Edited by Stuart K. Calderwood

A Continuation Order Plan is available for this series. A continuation order will bring delivery of each new volume immediately upon publication. Volumes are billed only upon actual shipment. For further information please contact the publisher.

Contents

Foreword ix

Acknowledgement xi

List of Contributors xiii

1. Introduction: Heat Shock Proteins—From *Drosophila* Stress Proteins to Mediators of Human Disease 1
   Stuart K. Calderwood

Part I. Stress Response and Molecular Chaperones   5

2. Biology of the Heat Shock Response and Stress Conditioning 7
   George A. Perdrizet, Michael J. Rewinski, Emily J. Noonan, and Lawrence E. Hightower

3. Bacterial Stress Sensors 36
   Wolfgang Schumann

4. Unfolded Protein Response: Contributions to Development and Disease 57
   Nan Liao and Linda M. Hendershot

Part II. Molecular Mechanisms of Stress Protein Expression   89

5. Genetic Models of HSF Function 91
   András Orosz and Ivor J. Benjamin

6. HSF1 and HSP Gene Regulation 122
   Richard Voellmy

Contents

Part III. Cellular Stress Proteins   141

7. Small Heat Shock Proteins in Physiological and Stress-Related Processes.. 143
   Diana Orejuela, Anne Bergeron, Geneviève Morrow, and Robert M. Tanguay

8. Large Mammalian hsp70 Family Proteins, hsp110 and grp170, and Their Roles in Biology and Cancer Therapy........................... 178
   Xiang-Yang Wang, Douglas P. Easton, and John R. Subjeck

Part IV. Molecular Chaperones and Protein Folding   207

9. Regulation of Hsp70 Function: Hsp40 Co-Chaperones and Nucleotide Exchange Factors... 209
   Robert T. Youker and Jeffrey L. Brodsky

10. Protein Disassembly by Hsp40–Hsp70 228
    Samuel J. Landry

11. Mammalian HSP40/DnaJ Chaperone Proteins in Cytosol.................. 255
    Kazutoyo Terada and Masataka Mori

Part V. Role of Molecular Chaperones in Cell Regulation   279

12. FKBP Co-Chaperones in Steroid Receptor Complexes..................... 281
    Joyce Cheung-Flynn, Sean P. Place, Marc B. Cox, Viravan Prapapanich, and David F. Smith

13. Up and Down: Regulation of the Stress Response by the Co-Chaperone Ubiquitin Ligase CHIP... 313
    Shu-Bing Qian and Cam Patterson

14. Role of Cdc37 in Protein Kinase Folding..................................... 326
    Atin K. Mandal, Devi M. Nair, and Avrom J. Caplan

Part VI. Intracellular and Extracellular Stress Proteins in Human Disease   339

15. Targeting Hsp90 in Cancer and Neurodegenerative Disease.............. 341
    Len Neckers and Percy Ivy

16. gp96 and Tumor Immunity: A Simple Matter of Cross-Presentation Antigens?.. 364
    Christopher V. Nicchitta

Cell Stress Proteins

Edited by

STUART K. CALDERWOOD
Department of Radiation Oncology, Harvard Medical School, Beth Israel Deaconess Medical Center, Boston, Massachusetts

Stuart K. Calderwood
Department of Radiation Oncology, Harvard Medical School,
Beth Israel Deaconess Medical Center, Boston,
Massachusetts

Library of Congress Control Number: 2006932956

ISBN-10: 0-387-39714-0 e-ISBN-10: 0-387-39717-5
ISBN-13: 978-0-387-39714-6 e-ISBN-13: 978-0-387-39717-7

Printed on acid-free paper

© 2007 Springer Science+Business Media, LLC

All rights reserved. This work may not be translated or copied in whole or in part without the written permission of the publisher (Springer Science+Business Media, LLC, 233 Spring Street, New York, NY 10013, USA), except for brief excerpts in connection with reviews or scholarly analysis. Use in connection with any form of information storage and retrieval, electronic adaptation, computer software, or by similar or dissimilar methodology now known or hereafter developed is forbidden.
The use in this publication of trade names, trademarks, service marks, and similar terms, even if they are not identified as such, is not to be taken as an expression of opinion as to whether or not they are subject to proprietary rights.

9 8 7 6 5 4 3 2 1

springer.com

17. Immunoregulatory Activities of Extracellular Stress Proteins 377
    A. Graham Pockley and Munitta Muthana

18. Heat Shock Proteins and Neurodegenerative Diseases 396
    Ian R. Brown

19. Heat Shock Proteins in the Progression of Cancer 422
    Stuart K. Calderwood, Abdul Khalique, and Daniel R. Ciocca

Index 451

Foreword

Stress proteins such as the heat shock proteins (Hsp) and glucose-regulated proteins (Grp) are front-line molecules in responses to cellular insult and play key roles in the viability of single cell organisms exposed to environmental stresses. However, the discovery of the roles of Hsp and Grp as molecular chaperones indicates much wider functions in the physiology of cells and organisms. It is now clear that some stress proteins are expressed constitutively and are key mediators of housekeeping protein folding in the day-to-day existence of the cell. The maturation of enzymes, transcription factors, and cell surface receptors relies on these functions of the stress proteins. In addition, the ability of stress proteins to manipulate the structures of target proteins has lent them cell regulatory properties over and above their role in folding the proteome and they play key roles in controlling signal transduction, cell death pathways and transcription. Recently, novel extracellular roles for the Hsp have also emerged as it has become apparent that the Hsp can escape from the cytoplasm of cells and play a significant extracellular role in signaling to neighbor cells and in immunosurveillance. As might be expected with such key molecules, dysregulation of Hsp expression over time can lead to disastrous results in terms of the health of the organism. Under-expression of the Hsp is associated with advanced aging and neurodegeneration. Elevated expression is associated with malignant progression. We have aimed in this volume to indicate advances in each of these aspects of stress protein research, with chapters ranging from basic studies of the role of Hsp in protein folding to reviews examining the breakdown of stress protein regulation in disease.

Stuart K. Calderwood
Harvard Medical School
July 25, 2006

Acknowledgement

We acknowledge the outstanding contribution of Mary Parkman in the Department of Radiation Oncology at the Beth Israel Deaconess medical center, Boston who played a major role in coordinating the review and assembly of this volume.

List of Contributors

Ivor J. Benjamin, MD, PhD
Department of Internal Medicine,
Division of Cardiology
University of Utah Health Sciences
  Center
Salt Lake City, UT, USA.

Anne Bergeron, PhD
Laboratoire de génetique céllulaire et
  développementale
Département de médecine et CREFSIP
Université Laval, Quebec, Canada.

Jeffrey L. Brodsky, PhD
Department of Biological Sciences
University of Pittsburgh
Pittsburgh, PA, USA.

Ian R. Brown, PhD
Center for the Neurobiology of Stress
University of Toronto at Scarborough
Toronto, Ontario, Canada.

Stuart K. Calderwood, PhD
Department of Radiation Oncology
Harvard Medical School
Boston, MA, USA.

Avrom Caplan, PhD
Department of Pharmacology and
  Biological Chemistry
Mount Sinai School of Medicine
New York, NY, USA.

Joyce Cheung-Flynn, PhD
Department of Biochemistry and
  Molecular Biology
Mayo Clinic
Scottsdale, AZ, USA.

Daniel R. Ciocca, MD
Oncology Laboratory
Institute of Experimental Medicine
  and Biology of Cuyo
Regional Center for Scientific and
  Technological Research
Mendoza, Argentina.

Marc B. Cox, PhD
Department of Biochemistry and
  Molecular Biology
Mayo Clinic
Scottsdale, AZ, USA.

Douglas P. Easton, PhD
Department of Biology
State University of New York College
  at Buffalo
Buffalo, NY, USA.

Linda M. Hendershot, PhD
Department of Molecular
  Pharmacology
St. Jude Children's Research
  Hospital
Memphis, TN, USA.

List of Contributors

Lawrence E. Hightower, PhD
Dept. of Molecular & Cell
    Biology
University of Connecticut
Storrs, CT, USA.

Percy Ivy, PhD
Urologic Oncology Branch
    Center for Cancer Research
National Cancer Institute
Bethesda, MD, USA.

Md. Abdul Khalique, PhD
Department of Radiation Oncology
Beth Israel Deaconess Medical
    Center
Harvard Medical School
Boston, MA, USA.

Samuel J. Landry, PhD
Department of Biochemistry
Tulane University Health Sciences
    Center
New Orleans, LA, USA.

Nan Liao, PhD
Department of Molecular
    Pharmacology
St. Jude Children's Research
    Hospital,
Memphis, TN, USA.

Atin K. Mandal
Department of Pharmacology and
    Biological Chemistry
Mount Sinai School of Medicine
New York, NY, USA.

Masataka Mori, PhD
Department of Molecular
    Genetics
Graduate School of Medical
    Sciences
Kumamoto University
Kumamoto, Japan.

Geneviève Morrow, PhD
Laboratoire de génétique cellulaire et
    développementale
Département de médecine et CREFSIP
Université Laval, Quebec, Canada.

Munitta Muthana, PhD
Department of Immunobiology
Sheffield, University
Sheffield, UK.

Devi M. Nair
Department of Pharmacology and
    Biological Chemistry
Mount Sinai School of Medicine
New York, NY, USA.

Len Neckers, PhD
Urologic Oncology Branch
Center for Cancer Research
National Cancer Institute
Bethesda, MD, USA.

Christopher V. Nicchitta, PhD
Department of Cell Biology
Duke University Medical Center
Durham, NC, USA.

Emily J. Noonan, PhD
Dept. of Molecular & Cell Biology
University of Connecticut
Storrs, CT, USA.

Diana Orejuela, PhD
Laboratoire de génétique cellulaire et
    développementale
Département de médecine et CREFSIP
Université Laval, Quebec, Canada.

András Orosz, PhD
Department of Internal
    Medicine, Division of Cardiology
University of Utah Health Sciences
    Center
Salt Lake City, UT, USA.

Cam Patterson
Carolina Cardiovascular Biology Center
School of Medicine
University of North Carolina
Chapel Hill, NC, USA.

George A. Perdrizet, MD
Division of Trauma
Hartford Hosptital
Hartford, CT, USA.

Sean P. Place, PhD
Department of Biochemistry and Molecular Biology
Mayo Clinic
Scottsdale, AZ, USA.

A. Graham Pockley, PhD
Department of Immunobiology
Sheffield, University
Sheffield, UK.

Viravan Prapapanich, PhD
Department of Biochemistry and Molecular Biology
Mayo Clinic
Scottsdale, Arizona, USA.

Shu-Bing Qian, PhD
Carolina Cardiovascular Biology Center
School of Medicine
University of North Carolina
Chapel Hill, NC, USA.

Michael J. Rewinski, MS
Department of Surgery
Hartford Hospital
Hartford, CT, USA.

Wolfgang Schumann, PhD
Institute of Genetics
University of Bayreuth
Bayreuth, Germany.

David F. Smith, PhD
Department of Biochemistry and Molecular Biology
Mayo Clinic
Scottsdale, AZ, USA.

John R. Subjeck, PhD
Department of Cell Stress Biology
Roswell Park Cancer Institute
Buffalo, NY, USA.

Robert M. Tanguay, PhD
Laboratoire de génétique cellulaire et développementale
Département de médecine et
CREFSIP
Université Laval, Quebec, Canada.

Kazutoyo Terada, PhD
Department of Molecular Genetics
Graduate School of Medical Sciences
Kumamoto University
Kumamoto, Japan.

Richard Voellmy, PhD
HSF Pharmaceuticals S.A.
1009 Pully, Switzerland

Xiang-Yang Wang, PhD
Department of Cell Stress Biology
Roswell Park Cancer Institute,
Buffalo, NY, USA.

Robert T. Youker, PhD
Department of Biological Sciences
University of Pittsburgh,
Pittsburgh, PA, USA.

1
Introduction: Heat Shock Proteins—From *Drosophila* Stress Proteins to Mediators of Human Disease

STUART K. CALDERWOOD

Division of Molecular and Cellular Radiation Oncology, Beth Israel Deaconess Medical Center, Harvard Medical School, Boston, MA 02215; Department of Medicine, Boston University School of Medicine, MA 02118

Nowadays heat shock proteins (HSP) seem to be everywhere and can apparently do anything. But it was not always so. For many years *HSP* genes were academic *arcana*, curiosities apparently confined to the salivary glands of fruit flies. Their study was initiated by the discovery of a new gene expression pattern, through a happy accident involving the overheating of a *Drosophila* salivary gland preparation on a microscope stage (Ritossa, 1962). This was first reported as, "A new puffing pattern induced by temperature shock and DNP in *Drosophila*" in 1962 (Ritossa, 1962). However, it was to take another 10–15 years before the first *Drosophila* HSP mRNA was isolated (Ashburner, 1982). Around this time (1978) the HSP "went global" and were discovered in mammalian tissue culture cells, in *E. coli*, in yeast, and in plants (Kelley and Schlesinger, 1978; Lemeaux et al., 1978; Bouche et al., 1979; Miller et al., 1979; Barnett et al., 1980; Hightower and White, 1981). The heat shock field emerged as a major study area in experimental biology at the 1982 meeting *Heat Shock: From Bacteria To Man*, held at the Cold Spring Harbor Laboratory (Ashburner, 1982). At this time, however, the functions of the HSP remained mysterious and the details of regulation of *hsp* gene expression were only beginning to emerge. All that was known was that the proteins appeared to possess "homeostatic activity" and were (as they are to this day) associated with resistance to heat shock and other stresses (Chapter 2). However, with the intensive international effort and the wealth of experimental systems available in the early 1980s, the concept began to emerge that the HSP belonged to a new kind of proteins which function to modify the structures of other proteins. Most notably the functions of the *E. coli* genes *DNA-K* (HSP70), *DNA-J* (HSP40), and *GroEL* (HSP60) were determined genetically in study of λ-phage replication and biochemical analysis of the clathrin coated pits at the membranes of mammalian cells hinted at a new function for HSP70 (Schlossman et al., 1984; Georgopolis and Welch, 1993). The HSP were apparently required to fold the proteins involved in λ-page replication and to disassemble the proteins involved in the huge lattice structures of the clathrin coats with the aid of ATP hydrolysis. HSP70 was described as an "unfolding ATPase," a protein that could use the energy derived from

ATP and an intrinsic ATPase activity, to influence quaternary interactions between proteins (Schlossman et al., 1984). Heat shock proteins thus became known as "molecular chaperones" due to their role in associating with other proteins and among other things, discouraging promiscuous interactions, and this name has been retained to the present time (Georgopolis and Welch, 1993). The molecular and biochemical mechanisms involved in "molecular chaperoning" by members of the "small HSP family," HSP70 and HSP110 are described in Chapters 7–11. In addition we will discuss a gene family which is an offshoot of these HSP families with specialized roles in protein trafficking in the endoplasmic reticulum; and the properties of these "glucose regulated proteins, or GRP, are described in Chapter 4. We have not included chapters on the elegant structural studies of GroEL/HSP60 HSP family, as we felt that this would require a volume all of its own.

During this period 1980–1990, huge strides were also made in elucidating the processes underlying transcriptional regulation of HSP genes. In eukaryotes, a *cis*-acting element (the heat shock element, or HSE) that conferred heat shock regulation on genes was discovered first in *Drosophila*, and then strikingly similar sequences were found in yeast, avian, and mammalian cells (Pelham and Bienz, 1982) (Chapter 5). Discovery of the HSE sequence was then instrumental in the isolation of the transcription factors (heat shock factor or HSF) that could respond to heat shock, bind to the HSE, and activate HSP gene transcription (Sorger and Pelham, 1988). An excellent review describes the early studies on the refinement of the canonical HSE sequence and the early studies on HSF regulation (Wu, 1995) and chapters are included that describe current investigations on the cellular (Chapter 6) and genetic (Chapter 5) analysis of HSF. Although HSP structure and function is remarkably conserved between prokaryotes and eukaryotes, regulation of expression involves almost entirely remote systems and prokaryotic HSP regulons are controlled through novel stress responsive σ-factors and a signaling cascade entirely different from that in eukaryotes (see Chapter 3).

In recent years the HSP have been assigned a more empowered role than that of just chaperones and are now envisaged as central regulators of cell metabolism. This new concept was heralded by the finding that HSP90 is required for glucocorticoid receptor activity and that in the absence of HSP90, GR fails to mature to a transcriptionally active form (Picard et al., 1990). It has since been shown that HSP90 carries out these functions in combination with HSP70 as well as a host of cofactors (co-chaperones), in over 100 molecules most of which are signal transduction molecules that must be maintained in a form poised for activation by extracellular or intracellular signals (Pratt and Toft, 2003). Signaling through HSP90 is discussed in Chapter 12 by Cheung-Flynn et al. and in Chapter 14 by Avram Caplan. As an alternative, particularly during stress, molecular chaperones can also mark their substrates for a more destructive fate. HSP70 and HSP90 have been shown to bind the ubiquitin E3 ligase CHIP and thus their associated client proteins are tagged with a polyubiquitin chain and delivered to the proteasome for destruction (Chapter 13). The HSP thus function at the cross roads between protein function and destruction.

Of course, proteins with such essential roles in the cell are fine regulated in terms of intracellular expression. Powerful mechanisms exist for both up- and down-regulation of HSP expression (Chapters 5, 6). However, when regulation breaks down and HSP levels increase or decrease beyond their prescribed levels, a range of pathologies have been shown to develop. Low intracellular levels of HSP are involved in the etiology of protein aggregation diseases characterized by the inclusion of large protein aggregates in compromised or dying cells, features characteristic of a range of neurodegenerative diseases (Chapters 15, 18). When HSP levels increase due to the coopting of the HSP transcriptional pathway by oncogenic signaling proteins, tumor growth is enhanced and a wide range of tumor types contain aberrantly high HSP levels (Chapter 19). In addition, when cell membranes become permeable in cells dying of necrosis, HSP can be released from cells along with their cargo of chaperoned proteins and peptides, and these molecules can be taken up by cells of the immune system and lead to immune interactions of the immune system with the cells releasing HSP-antigen complexes (Chapters 16, 17).

A long road has thus been traveled since the *Drosophila* stress genes were stumbled upon, leading to the current status of HSP as key physiological intermediates with roles in protein folding and cell regulation in all cellular organisms. We aim here to give an overview of the current status of some of the key areas in the basic study of heat shock proteins as well as their role in biology and medicine. With such a large subject it is impossible to be entirely inclusive and we apologize to those of our colleagues whose contributions may have been left unmentioned and for the exclusion of some topics which might seem more significant to other people in the HSP field.

References

Ashburner, M. (1982) Cold Spring Harbor laboratory Publications. Cold Spring Harbor Laboratory, Cold Spring Harbor.

Barnett, T. M., Altschuler, C. N., McDaniel, C. N., and Mascarentes, J. P. (1980). Heat shock induced proteins in plant cells. *Dev Genet* 1:331–40.

Bouche, G., Amalric, F., Caizergues-Ferrer, M., and Zalta, J. P. (1979) Effects of heat shock on gene expression and subcellular protein distribution in Chinese hamster ovary cells. *Nucleic Acids Res* 7:1739–47.

Georgopolis, C., and Welch, W. J. (1993) Role of the major heat shock proteins as molecular chaperones. *Ann Rev Cell Biol* 9:601–34.

Hightower, L. E., and White, F. P. (1981) Cellular responses to stress: comparison of a family of 71–73-kilodalton proteins rapidly synthesized in rat tissue slices and canavanine-treated cells in culture. *J Cell Physiol* 108:261–75.

Kelley, P. M., and Schlesinger, M. J. (1978) The effect of amino acid analogues and heat shock on gene expression in chicken embryo fibroblasts. *Cell* 15:1277–86.

Lemeaux, P. G., Herendeen, S. L., Bloch, P. L., and Neihardt, F. C. (1978) Transient rates of synthesis of individual polypeptides in *E. coli* following temperature shifts. *Cell* 13:427–34.

Miller, M. J., Xuong, N. -H., and Geiduschek, E. P. (1979) A response of protein synthesis to temperature shift in the yeast *Saccharomyces cerevisiae*. *Proc Natl Acad Sci USA* 76:1117–21.

Pelham, H. R., and Bienz, M. (1982) A synthetic heat-shock promoter element confers heat-inducibility on the herpes simplex virus thymidine kinase gene. *EMBO J* 1:1473–7.

Picard, D., Khursheed, B., Garabedian, M. J., Fortin, M. G., Lindquist, S., Yamamoto, K. R. (1990) Reduced levels of hsp90 compromise steroid receptor action *in vivo*. *Nature* 348:166–8.

Pratt, W. B., and Toft, D. O. (2003) Regulation of signaling protein function and trafficking by the hsp90/hsp70-based chaperone machinery. *Exp Biol Med (Maywood)* 228:111–33.

Ritossa, F. (1962) A new puffing pattern induced by temperature shock and DNP in *Drosophila*. *Experientia* 18:571–3.

Schlossman, D. M., Schmid, S. L., Braell, W. A., and Rothman, J. E. (1984) An enzyme that removes clathrin coats: purification of an uncoating ATPase. *J Cell Biol* 99:723–33.

Sorger, P. K., and Pelham, H. R. (1988) Yeast heat shock factor is an essential DNA-binding protein that exhibits temperature-dependent phosphorylation. *Cell* 54:855–64.

Wu, C. (1995) Heat shock transcription factors: structure and regulation. *Ann Rev Cell Dev Biol* 11:441–69.

I
Stress Response and Molecular Chaperones

2
Biology of the Heat Shock Response and Stress Conditioning

GEORGE A. PERDRIZET,[1] MICHAEL J. REWINSKI,[1] EMILY J. NOONAN,[2] AND LAWRENCE E. HIGHTOWER[2]

[1] Departments of Surgery and Trauma, Hartford Hospital and University of Connecticut School of Medicine, Hartford
[2] Department of Molecular & Cell Biology, University of Connecticut, Storrs, CT

The heat shock or stress response has been studied mainly as a cellular response. Most of the data come from bacterial cells, eukaryotic microorganisms (yeast primarily), and cultured animal cells. Often these cultured cells are tumor cell lines, i.e., cells that are functionally eukaryotic microorganisms as a consequence of genetic changes that change their social behavior and proliferative control. These systems have provided useful information about stress protein function and their roles in the defensive cellular state of cytoprotection. However, a full understanding of stress response biology in complex multicellular organisms requires different thinking and different models. This conclusion stems from the paradigm that the basic unit of function in animals and plants is not the individual cell but the tissue. Therefore stress response biology in these complex biological systems is primarily about tissue-level protection. Ultimately we would like to know how these responses are deployed in humans and how these inducible defenses may be used to prevent tissue damage from disease and from surgical intervention.

It is now abundantly clear that cell stress proteins function both inside and outside of cells. The exobiology of these proteins has been the subject of a recent monograph entirely devoted to this topic and interested readers are referred to this book (Henderson and Pockley, 2005). Here, we will describe four animal and tissue models that have provided information on tissue-level responses to stress: (1) the Sonoran Desert topminnow *Poeciliopsis*, (2) in vitro cultures of secretory epithelium of the winter flounder, (3) ex vivo studies of the rat vascular endothelium, and (4) a rat kidney model for acute ischemia, either as kidney procurement, cold storage, and transplantation or as in situ renal ischemia. Studies using these model systems have been published separately (Hightower et al., 2000; Norris and Hightower, 2000; House et al., 2001), but this is the first article in which the stress biology of all of these systems has been described together.

1. *Poeciliopsis* as a Model Organism

The idea that selection for enhanced production of heat shock proteins (HSPs) may occur in organisms living in a thermally stressful environment is a natural outgrowth of the demonstration that HSPs play a central role in at least some forms of thermotolerance. Species of *Poeciliopsis* from the river systems of northwestern Mexico are an ideal system to test this hypothesis. These live-bearing topminnows are adapted to a variety of habitats, including relatively cool mountain headwaters, small streams that shrink to isolated pools during the dry season, and broad rivers. In the desert environment, exposure to near-lethal heat (>40°C) occurs routinely at certain times of the year, while in other seasons extreme cold is encountered. Even on a daily basis, rapid changes in temperature can occur, such as a 22°C change over a period of 3 hours (Bulger and Schultz, 1979). As poikilotherms, these fish encounter temperature changes unbuffered by homeothermic mechanisms. Individual species are found in habitats that differ in the degree of thermal stress encountered. Survival of acute heat and cold stress differs among species and has been correlated with the thermal characteristics of different habitats (Bulger and Schultz, 1979, 1982). Local extinctions and fragmentation in populations of desert species as a result of seasonal changes in water flow have also contributed to the evolution of these fish (Vrijenhoek, 1989). In contrast, representatives of this genus from southern Mexico inhabit a tropical environment characterized by high rainfall and low seasonality. These fish provide a backdrop against which adaptations to the desert environment can be studied.

As an experimental model, the biological properties of these small aquarium fish offer some specific advantages. Adults become sexually mature at 2.5–3 months of age and have a life span of about 2 years. Young are born at 8–12-day intervals, at a size of 7–8 mm, usually in broods of 8–20 but occasionally as many as 30 fish. These fish are live-bearers and pregnant females carry active sperm for up to 6 months and can continue reproducing without being re-mated. In additional to heterosexual populations, unisexual hybrid populations exist both in nature and as laboratory creations. These hybridogenetic all-female fish have a hemiclonal reproductive mechanism in which only the maternal genome is inherited clonally (no reassortment or recombination with the paternal genome) and the paternal genome is replaced each generation.

2. Heat Shock Response of *Poeciliopsis*

Conservation of the heat shock response across diverse taxa is seen in both the mechanism of heat-inducibility of HSP synthesis, and in the conservation of individual HSPs at the level of function, protein sequence, and/or nucleotide sequence. These common characteristics were used in an investigation of the heat shock response in one species *P. lucida* (White et al., 1994). Identification of the

HSPs of *P. lucida* was made possible through the use of heterologous antibodies and cDNA probes, as well as comparison of HSP induction profiles to those from other organisms.

In many respects, the heat shock response of *Poeciliopsis* cells was typical of that seen in other systems. Induction of Hsp70 synthesis is a threshold phenomenon which generally corresponds to the upper range of temperatures organisms experience in the wild (diIorio, 1994). This is due to differences in the temperature-sensing mechanism within particular cell types (Corces et al., 1981). Hsp70 was first detected in gill tissue of fish that had been given a 33°C heat shock, whereas Hsp30 was first detected in fish incubated at 37°C (diIorio, 1994). When the thermal preferences of *P. lucida* were determined in a temperature-gradient tank (Fielding, 1992), fish were frequently seen at temperatures that induce synthesis of Hsp70, and only rarely seen at temperatures that induce Hsp30 synthesis. At temperatures $\geq 37°C$ a much more pronounced induction of Hsp70 occurs, along with the induction of Hsp30. Small increases in Hsp70 may be used by these fish to cope with slightly elevated, but commonly encountered, temperatures which do not acutely affect survival, thus allowing them to thrive in thermally unstable habitats. Hsp30, along with higher levels of Hsp70, may be important in surviving less frequent, severe thermal stress.

Poeciliopsis as well as other Poeciliidae are thought to have arisen in the tropics and then radiated northward (Rosen and Bailey, 1963). The distribution of Hsp70 isoforms in desert and tropical species of *Poeciliopsis* is consistent with this hypothesis. As ancestral fish adapted to the warm, thermally stable tropical environment moved northward and colonized desert streams, they would have encountered fluctuating temperatures, including cooler temperatures than those experienced in the tropics. In the desert environment, frequent exposure to temperatures that induce Hsp70 synthesis would be followed by rapid return to normal temperature. Inducible Hsp70 is thought to be deleterious when expressed under non-stress conditions (Feder et al., 1992; Krebs and Feder, 1997a,b). This idea needs to be qualified to accommodate the fact that Hsp70 contributes to the defensive inducible state of cytoprotection. This altered state of cellular physiology may persist for three or four days in homeotherms, long after the triggering stress has gone, and is clearly an advantageous, evolutionarily highly conserved mechanism when expressed transiently. It may be that chronic maintenance of the cytoprotected state with its upregulation of stress proteins and downregulation of proteins involved in differentiated tissue function is the problem.

In Poeciliopsis, adaptation to the desert environment may have involved selection for isoforms that can be rapidly turned over. Thus, the preponderance of isoform 3 in desert species may be linked to its short half-life during recovery from stress (Hightower et al., 1999). In this context it is interesting to consider the human HSP70B' isoform, that has a relatively short half-life and is degraded by a proteasomal pathway (Noonan et al., 2006). Rodents do not have an HSP70B' isoform. This raises an interesting question. Did a common ancestor of both humans and fish have an ancestral gene encoding a stress-inducible, rapidly turning over HSP70 or is HSP70B′ a recent evolutionary addition to the human HSP70 repertoire? If such an isoform is the product of an ancient gene, why was it retained in humans but lost in rodents?

3. Thermotolerant State

Thermotolerance has been defined as the ability of a cell or organism to survive a normally lethal heat stress. This ability can be acquired in a number of ways, including alterations in growth status (Elliott et al., 1996) and long-term acclimation to increased temperature in both cultured cells (Laszlo and Li, 1985) and *Drosophila* (Cavicchi et al., 1995), but has been most widely studied as "heat hardening." This type of thermotolerance is induced by exposure to elevated, sublethal temperature. Following a recovery period, the conditioned cells (or organisms) exhibit a transient ability to survive a heat shock which kills the majority of unconditioned cells. The establishment and decay of the thermotolerant state has been correlated in both cultured cells and at the organismal level with changes in the levels of HSPs. Cell lines that over-express Hsp70 (Parsell and Lindquist, 1993; Li and Nussenzweig, 1996) or HSP27 (Landry et al., 1989) become thermotolerant, while cells in which the accumulation of HSPs is blocked become thermosensitive (Li and Nussenzweig, 1996). However, multiple mechanisms to achieve the thermotolerant state exist: some of these are independent of changes in HSP level (Hall, 1983; Easton et al., 1987; Borrelli et al., 1996) and may involve changes in the levels of naturally occurring "chemical chaperones" (Welch and Brown, 1996). Even in cultured cells, induction of thermotolerance by stressors other than heat shock, or by overexpression of single HSPs, results in the protection of different subsets of the cellular structures and processes that are adversely affected by heat shock (e.g., protein synthesis, rRNA transcription and processing, mRNA splicing, and microfilament and nuclear integrity (Arrigo and Landry, 1994; Corell et al., 1994; Li and Nussenzweig, 1996). Protection of any one of these targets leads to an increase in survival, but these "partially protected" states are probably not equivalent to the thermotolerant state induced by heat shock, in which multiple targets are protected.

At the organismal level, increased thermal resistance is an even more complex phenomenon. Measurements of differences in thermal resistance among species have often employed critical thermal maxima, that are determined by acute exposure to borderline lethal temperatures. This procedure generally precludes expression of any inducible responses during the heat stress. The inhibition of protein synthesis at high temperature blocks the accumulation of HSPs until well into the recovery period. This type of intrinsic thermal tolerance is a heritable trait, but it is unclear what genetic loci are involved. The use of ecologically relevant thermal stress regimens has also revealed differences in thermal resistance among closely related species (Bosch et al., 1988; Sanders et al., 1991), among different populations within a species (Bulger and Schultz, 1982; Hoffmann and Parsons, 1991), and among individuals within a population (Norris et al., 1995; Krebs and

Feder, 1997a,b). Some of these studies (Bosch et al., 1988; Sanders et al., 1991) correlated increased HSP expression with the ability to survive in thermally stressful environments. A positive correlation between the amount of Hsp70 synthesized and thermal resistance was also seen in individuals from outbred populations of a live-bearing fish, *Poeciliopsis gracilis* (Norris et al., 1995). In addition, *Poeciliopsis* hybrids have been used to obtain quantitative evidence that both constitutive Hsc70 and inducible Hsp70 contribute to acquired thermotolerance (diIorio et al., 1996). However, lines of *Drosophila melanogaster* that exhibited both increased Hsp70 expression and higher acquired thermal resistance also showed decreased survival of larvae to adulthood in the absence of stress (Krebs and Feder, 1997a,b). Strains that accumulate Hsp70 to higher levels after heat shock may also accumulate higher levels of Hsp70 during early development, when synthesis is induced in the absence of stress, and this may have deleterious consequences. Thus, it is possible that variation in Hsp70 levels in populations is maintained in part by trade-offs between beneficial and deleterious effects. For a more complete analysis of the literature on the heat shock response in organismal biology, readers are referred to a review by Feder and Hofmann (Feder and Hofmann, 1999).

4. Special Features of *Poeciliopsis* Heat Shock Proteins

The most novel findings from our work on *Poeciliopsis* HSPs was the discovery in these fishes of two subfamilies of small HSPs that previously were known only in separate species, and the discovery that both HSP70-dependent and independent mechanisms of acquired thermotolerance exist within one population of a tropical species of these fishes *P. gracilis*, and that the former kind of thermotolerance correlates with the presence of a specific isoform 3 of HSP70. Hsp70 isoform 3 appears to be the most common allele in *Poeciliopsis*, where it is found in six of the eight species analyzed. This isoform is also synthesized in the confamilial species *Gambusia affinis* collected from two sites in Nevada.

We have tested the correlations between acquired thermotolerance and Hsp70 abundance for fish containing isoform 3, the most frequently encountered Hsp70 isoform, and for those not containing this isoform (Hightower et al., 1999). For fish having isoform 3, we concluded that there is a strong positive association between survival and amount of Hsp70 ($N = 21, r = 0.58, Z = 2.590, p = 0.01$). This correlation is linear up to about 5 arbitrary units of total Hsp70, an amount which may be necessary for these fish to acquire maximum thermotolerance. Fish with more than twice this amount of Hsp70 were not more thermotolerant, and they appeared to have reached an upper limit to thermal resistance. A remarkably similar relationship between survival of heat stress and level of expression of Hsp70 was reported for clones of Rat-1 cells carrying a human Hsp70 gene (Li and Nussenzweig, 1996). A strong positive correlation between Hsp70 levels and survival was also found for isofemale lines derived from a single population of *Drosophila melanogaster* (Krebs and Feder, 1997a,b). For fish containing isoform complements found only in the tropics and with no isoform 3 influence, there is little or no association between survival and total amount of Hsp70 ($N = 26$, $r = 0.28$, $Z = 1.4$ and $p = 0.16$). These correlation plots are consistent with the interpretation that isoform complements containing isoform 3 contribute to acquired thermotolerance, whereas those containing only isoform 1 and both 0 and 1 do not. It is quite possible that another Hsp, such as Hsp30, or perhaps an Hsp-independent thermotolerance mechanism compensates for the lack of the Hsp70 isoform contribution to survival at 41°C in the fish carrying the 0 and 0, 1 isoform patterns. The results of this analysis raise the possibility that the Hsp70 isoforms may be qualitatively different in their ability to contribute to thermotolerance and that the mechanism of thermotolerance that includes a contribution by Hsp70 isoform 3 may provide an advantage in the desert.

The small heat shock proteins (sHSPs) are a diverse group of stress-inducible proteins characterized minimally by a molecular mass of 15–30 kDa and a conserved region of approximately 90 amino acid residues in the C-terminal region of the protein. This conserved region is also found in the α-crystallins, major proteins of the vertebrate eye lens. Together these proteins make up the α-crystallin/sHSP superfamily (de Jong et al., 1993). Two distinct sHSPs, Hsp27 and Hsp30, have been characterized in *Poeciliopsis lucida*. Both Hsp27 and Hsp30 are more similar to homologous small HSPs from other organisms than to each other, and they share regions of identity across the entire protein sequence with their respective homologs (Norris et al., 1997). cDNA clones for *P. lucida* Hsp27 and Hsp30 were sequenced and evolutionary analysis was performed using the derived protein sequences. Hsp27 is most similar to a group of mammalian and avian sHSPs, with which it shares induction patterns that differ from those of Hsp70, stress-inducible phosphorylation (Arrigo and Landry, 1994), and sequence similarity. The *P. lucida* Hsp30 sequence is most similar to that of *Xenopus* and salmon Hsp30s.

Poeciliopsis Hsp27, like its human counterpart (Landry et al., 1992), is phosphorylated at two of three possible sites following heat shock. Increased phosphorylation of human Hsp27 has been demonstrated following exposure to stressors other than heat shock (i.e., arsenite and hydrogen peroxide), after stimulation by mitogens and differentiation-inducing factors, and upon exposure to inflammatory cytokines (Arrigo and Landry, 1994). The conservation of phosphorylation sites between the human and *Poeciliopsis* Hsp27 sequences makes it likely that *Poeciliopsis* Hsp27 also plays a role in signal transduction to the actin cytoskeleton. In contrast, the ability of mammalian Hsp27 to function as a molecular chaperone in vitro is independent of phosphorylation. Recombinant murine Hsp27 prevents aggregation of unfolded proteins and assists refolding regardless of phosphorylation state, i.e., the recombinant protein was phosphorylated in vitro with purified MAPKAP kinase-2 (Knauf et al., 1994). Thus, the lack of phosphorylation of Hsp30 would not preclude a role for it as a molecular chaperone. It may be that in *Poeciliopsis*, the two sHSPs play complementary roles, in contrast to the more multifunctional role of mammalian Hsp27.

5. Winter Flounder Renal Epithelial Transport In Vitro

Secretion of small organic molecules and ions across the renal epithelium requires an intact, differentiated monolayer of epithelial cells complete with tight junctions and apical membrane domains with microvilli. Therefore, protection of net secretory function can only be tested at the tissue level. Dissociation of winter flounder renal tubules yields a population of cells highly enriched for secretory epithelial cells which reform a functional secretory epithelium when plated on native collagen. After 12 to 14 days incubation at 22°C, the monolayers on collagen pads were mounted in Ussing chambers in which transepithelial electrical characteristics and unidirectional [^{35}S]sulfate fluxes were measured. Sublethal heating and recovery (27°C for 6 h followed by 1.5 h at 22°C), i.e., stress conditioning, resulted in a 30% increase in sulfate transport. Cycloheximide or actinomycin D prevented the enhancing and protective effects of stress conditioning and blocked the induction of heat shock proteins. A challenge severe heat shock (32°C for 1.5 h followed by 1.5 h at 22°C) reduced transport by about 30%, essentially to control levels (Brown et al., 1992).

Zinc ions can also be used to stress condition the epithelium with similar results. Preincubation of primary epithelium in 100 mM $ZnCl_2$ for 6 h followed by a 1.5-h recovery in zinc-free medium enhanced net sulfate flux and protected transport from a severe heat shock. Cycloheximide prevented the induction of heat shock proteins in response to treatment with zinc ions and prevented the acquisition of protection. Induction of cytoprotection by zinc was not specific for sulfate transport since sodium-dependent glucose transport was also protected (Renfro et al., 1993).

Essentially all hypotheses on the mechanism of cytoprotection have assumed that the protection allows cells to return to near-normal physiological functions by stopping damage to macromolecules and/or facilitating their repair. In our studies it was shown that cytoprotection is characterized by both the presence of stress proteins and increased renal secretory capacity well above control levels. The actual protection of transport is not due to a lack of damage to or repair of transporters, but rather to extra capacity, which is inactivated during the challenge stress, but only back to control levels. It is possible that higher amounts of molecular chaperones in the stress conditioned epithelium allow the assembly of more transporters by a direct chaperoning function. An interesting morphological effect was observed using scanning electron microscopy. Microvilli, in which sulfate transporters are located, disappear after a severe heat treatment, presumably due to effects on cytoskeletal elements; however, these structures remain after thermal challenge of cytoprotected epithelium.

6. Rat Ex Vivo Vascular Endothelium Model

Vascular endothelium also has functions that require an intact, differentiated monolayer of cells, e.g., attachment and transendothelial migration of activated leukocytes from the blood stream into tissues in the early stages of inflammation.

How does stress conditioning affect this tissue-level function (House et al., 2001)? To answer this question rats were subjected to either heating to 42C for 15 min followed by a two-day recovery period or to injection IP with stannous chloride (0.15 mg/kg) and exposure for 16 hours. The rats were then prepared for intravital microscopy by exposing mesentery tissue pulled through a midsagittal abdominal incision. The microcirculation of the exposed mesentery tissue was observed using a Nikon UM3 metallurgic microscope adapted for intravital microscopy. Microcirculatory events were recorded using a video camera and video cassette recorder. Stress conditioning rats with either heat or stannous chloride blocks extravasation of neutrophils across venules in response to a proinflammatory stimulus. Since white blood cell flux decreased significantly in response to the pro-inflammatory peptide formyl-methionyl-leucyl-phenylalanine (FMLP) in both conditioned and placebo animals, we concluded that the initial low affinity interactions between lymphocytes and endothelium were not blocked. However, firm attachment, measured in a leukocyte-endothelial adhesion assay, was blocked in conditioned animals. Hsp70 was detected by Western blotting of extracts of aortas from heat shocked and stannous chloride-treated rats but not in aortas from placebo rats. Our working hypothesis is that vascular endothelial cells and/or neutrophils are in a cytoprotected state in which they do not respond to signals that would normally up-regulate cellular adhesion molecules involved in the firm attachment of neutrophils to the vascular endothelium.

7. Discussion of the Winter Flounder Renal Epithelial Transport Model and the Rat Ex Vivo Vascular Endothelium Model

Our studies indicate that differentiated functions specific to a particular tissue are altered in thermotolerant (cytoprotected) animals, functions such as transepithelial transport in renal epithelium and attachment and transmigration of leukocytes across vascular endothelium in response to mediators of inflammation. One venue of inflammatory responses in vertebrates is wound healing. Inflammation of a wound is essential for proper healing but prolonged inflammation and excessive destruction of cells in the wound interfere with healing. We propose that heat shock proteins are induced relatively early in wound responses and cytoprotection begins to develop, a process which requires about six-eight hours. Ian Brown and collegues have documented the accumulation of Hsp70 mRNA in surgical wounds in rat brain tissue (Brown et al., 1989). Cytoprotection would serve as a brake on inflammation, protecting cells from oxidative and heat damage in the inflamed wound and contributing to the throttling down of the inflammatory response, in part by shut-down of signal transduction pathways as suggested below. Barbara Polla was among the first to suggest that inflammation is a major venue for the heat shock response and that it may serve as a brake on inflammatory responses (Polla, 1988). Blood vessels are now returning to center stage in studies of cytoprotection in intact animals and tissues. We say "returning" because the studies of Fredric White done 25 years ago (White, 1980a,b) showed that cells associated with the brain microvasculature are among the most stress-responsive cells in explants and in heat shocked rats.

Previous studies have shown that cytoprotected cells are unresponsive to inducers of proliferation and apoptosis. We now add a pro-inflammatory mediator to this list. Recent studies suggest that a major reason for the unresponsiveness of cytoprotected cells is that products of stress-inducible genes block signal transduction pathways. For example, Sherman and coworkers showed that Hsp70 prevents the activation of JNK and p38 kinases and inhibits heat-induced apoptosis in human tumor cell lines (Gabai et al., 1997). The mechanism involves increased rate of inactivation of stress kinase JNK (Volloch et al., 2000). Wong and coworkers obtained data suggesting that heat induction of the inhibitory protein I-kB inhibits the activation of the pro-inflammatory transcription factor NF-kB (Wong et al., 1999). Calderwood and colleagues found another anti-inflammatory effect of the heat shock response: transcription factor Hsf1 acts as a transcriptional repressor of genes encoding several pro-inflammatory cytokines including IL1β and TNFα (Xie et al., 1999). A transient period of unresponsiveness appears to be an important general characteristic of the cytoprotected state of cell physiology.

8. Rat Kidney Models of Acute Ischemia

Modern advances in medical care have created stressful environments to which patients are exposed on a routine basis. Many of these iatrogenic stressors are short lived and associated with infrequent, but major complications. The North American Symptomatic Carotid Endarterectomy Trial (Committee 1991) has determined that the benefit of carotid endarterectomy (CEA) is dependent upon the complication rate associated with surgery, and should be less than 6%, the combined rates of stroke and death. The combined mortality and stroke rates following CEA were estimated to range between 5% and 11% for all Medicare patients in 1991. This data does not reflect less severe degrees of neurologic dysfunction experienced by this patient population (Committee 1991). Rare but devastating complications greatly detract from the overall benefit associated with advanced, but nevertheless traumatic, medical/surgical procedures. Major complications associated with *successful* invasive cardiovascular surgical procedures include neurologic damages following coronary artery bypass or renal failure and paraplegia following thoracoabdominal aortic aneurysm repair (Cunningham, 1998; Gharagozloo et al., 1998; Hogue et al., 1999). Less dramatic, but impacting a far greater number of individuals, are the more subtle injuries suffered by those subjected to modern diagnostic and therapeutic methods. For example, the medical literature reports the incidence of contrast-induced nephropathy in diabetics requiring angiographic evaluations to be 43%, acute pulmonary dysfunction following major joint replacement in the elderly to be 30%, and new cognitive deficits following exposure to cardiopulmonary bypass support to be 53%. These medical procedures and associated iatrogenic injuries are extremely common events (Weisberg et al., 1994; Lane et al., 1997; Newman et al., 2001). We suggest that the collateral damage currently being suffered by patients, as unwanted side effects of modern diagnostic and therapeutic methods, is a widespread phenomenon. A common theme within these examples is that damage to complex tissues results from obligatory but anticipated interventions which can be prepared for. Efforts to minimize complications associated with invasive medical care have almost exclusively relied upon risk stratification coupled with patient selection or technical manipulations at the time of, or even after, exposure to the threatening procedure. The few preparatory interventions utilized are limited to minimizing effects that modifiable risk factors may have on outcomes, such as hydrating the diabetic patient prior to exposure to intravascular contrast agents. Currently there exists a paucity of techniques designed to enhance the intrinsic resistance of cells, tissues and organs to iatrogenic threats. The primary objective of this communication is to describe a technique, *stress conditioning*, whereby the cell-stress response (a.k.a. heat-shock response, HSR) is utilized as a potent preventative agent applied prior to common invasive medical procedures. Stress conditioning, by exposure of tissues to a short-lived, sublethal stressor followed by a period of recovery, can awaken the cytoprotective potential of the HSR found within all cells. Preoperative stress-conditioning protocols should attenuate damages and reduce complications associated with a myriad of modern medical and surgical interventions.

The stress-conditioning hypothesis was first tested in a translational model of organ preservation (Perdrizet et al., 1989). The formulation of this hypothesis was based upon scientific principles developed within the fields of heat shock biology and therapeutic hyperthermia. Both disciplines had made a similar observation; prior exposure to sublethal hyperthermia followed by an intervening period of recovery, would result in a marked increase in resistance to lethal doses of hyperthermia. Heat shock biologists focused their work on the regulation of heat shock gene expression, first reported by Ferruccio Ritossa in Italy (Ritossa, 1963). Scientists studying hyperthermia focused their efforts on methods to enhance the efficacy of modern anticancer therapies. Both groups clearly described the stress-conditioning phenomenon that is the subject of this work (Crile, 1963; Ohtsuka and Laszlo, 1992). Finally, the historical works of Hans Selye place these more recent scientific observations on a solid historical foundation upon which the stress-conditioning hypothesis is firmly based (Selye, 1946). Since the initial report of stress conditioning in organ preservation, numerous laboratory and clinical examples have been published which continue to develop the theme of stress conditioning, underscoring the adaptive power of the HSR (Perdrizet, 1997; DeMaio, 1999; Jaattela, 1999). A random sampling of recent, supportive literature is presented in Table 1. Furthermore, recent growth within the fields of thermal biology and molecular genetics have provided valuable insight into mechanisms for the potent cytoprotection that is observed to follow exposure to a sublethal stressor (Table 2). Stress conditioning is based upon the fundamental and universal tenets of the HSR, reviewed elsewhere (Cotto and Morimoto, 1999; Feder and Hofmann, 1999). A testable hypothesis is that stress conditioning in the form of exposure of

TABLE 1. Stress Conditioning: Recent Medical Literature

Organ system	Stress conditioner	Stress challenger	Species	Citation
Liver	Heat shock	Ischemia—reperfusion	Rodent	(Kume et al., 1996)
Intestinal epithelial cell line IEC-18	HSP70	Oxidant and thermal injury	Rodent	(Buress et al., 1996)
Brain—astrocytes	Heat shock	In vitro reperfusion	Rodent	(Takuma et al., 1996)
Lung—endothielial cells	Heat shock	Oxidation by Hydrogen peroxide	Bovine, rodent	(Wang et al., 1996)
Liver	HSP70	Thioacetamide	Rodent	(Fujimori et al., 1997)
Eye—retinal edema	HSP70	Ischemia	Rodent	(Yu et al., 2001)
Heart—myocytes	Small HSPs	Ischemia—reperfusion	Rodent	(Martin et al., 1997)
Heart—myocardial function	Heat shock	Ischemia	Rodent	(Amrani et al., 1998)
Heart—myocardial function	Ethanol pretreatment	Acute ischemia	Rodent	(McDonough, 1999)
Intestinal—colon	Heat shock	Acetic acid induced colitis	Murine	(Otani et al., 1997)
Intestinal—ileum	Heat shock	Acute inflammation	Rodent	(Stojadinovic et al., 1997)
Kidney—proximal renal tubule cells	Cadmium induction of HSP	Zinc stress	Rodent	(Liu et al., 1996)
Kidney—renal tubule cells	Liposomal Hsp72 or heat	Simulated ischemia	Rodent	(Meldrum et al., 2003)
Eye—retinal ganglion cells	Heat shock and hypoxia	Anoxia and excitotoxicity	Rodent	(Caprioli et al., 1996)
Fibroblast	HSPs	MPP+	Rodent	(Freyaldenhoven and Ali, 1996)
Skeletal muscle	Heat shock	Acute ischemia	Rodent	(Lepore et al., 2000)
Skin—myocutaneous flap	Heat shock	Acute ischemia	Rodent	(Harder et al., 2005)
Lung—bronchial epithelial cell line BEAS-2B	Heat shock	NO toxicity	Human	(Wong et al., 1997)
Lung—epithelial cell	Allergic inflammation-HSP27	H_2SO_4 cytotoxicity	Human	(Hastie et al., 1997)
Lung—endothelial cells	HSR	LPS-mediated apoptosis	Ovine	(Wong et al., 1996)
Pancreas	Heat shock	Cerulein-induced pancreatitis	Rodent	(Wagner et al., 1996)
Lung carcinoma A549 cells	Transfection of chimeric Hsp27	Heat shock	Human	(Borrelli et al., 2002)
Whole body	Heat shock	Total-body gamma irradiation	Murine	(Patil et al., 1996)
Heart & Liver	Heat shock	Hemorrhage/trauma	Rodent	(Mizushima et al., 2000)

(cont.)

TABLE 1. (*Continued*)

Organ system	Stress conditioner	Stress challenger	Species	Citation
Liver	Geranylgeranyl—acetone plus heat	Acute ischemia	Rodent	(Fan et al., 2005)
Liver	Heat shock	Acute ischemia	Rodent	(Yamagami et al., 2003)
Liver	Heat shock	Acute ischemia	Rodent	(Uchinami et al., 2002)
Whole body	Heat shock	Burn wound	Rodent	(Meyer et al., 2000)
Vascular endothelium	Heat shock	Hydrogen peroxide oxidation	Human	(Gill et al., 1998)
Intestine—Caco-2 cell line	Arsenite	LPS/cytokines	Human	(Swank et al., 1998)

TABLE 2. Stress Conditioning: Potential Mechanisms of Action

Model	Class	Citation
Yeast	Chaperone rescue of aggregated proteins	(Glover and Lindquist, 1998)
Rabbit	Anti-inflammatory	(Franci et al., 1996)
Human—HSP70 transgenic mouse	Cytoprotective—mitochondrial preservation	(Jayakumar et al., 2001)
Rodent mesenteric vasculature	Preservation of microcirculation	(Chen et al., 1997)
Rabbit spinal cord	Prevents PCD	(Sakurai et al., 1997)
Human HeLa cell line	Modulates iCalcium flux in ER	(Lievremont et al., 1997)
Renal epithelial cells	Prevents oxidative stress, iCa disturbance, cell death	(Liu et al., 1997)
Rodent epithelial cells	Heat shock preserves PKC	(Meldrum et al., 2001)
Human embryonic renal epithelial cells	Hsp90 supports clc-2 Chloride channel	(Hinzpeter et al., 2006)
C2c12 myocyte	Hsp72 mutant cells loose protection	(Voss et al., 2005)
Neonatal rodent cardiomyocytes	Smac/DIABLO release from mitochondria is reduced by heat shock	(Jiang et al., 2005)
Rodent myocardium	Mn-SOD activation	(Yamashita et al., 1998)
Rodent myocardium	Hsp72 myocardial ischemic protection	(Shinohara et al., 2004)
Rodent myocardium	Hsp32 (HO-1) ischemic protection	(Lu et al., 2002)
Rodent fibroblast	Suppression of PCD	(Guenal et al., 1997)
Rodent renal cells	Hsf1 knock out	(Yan et al., 2005)
Fibroblast—embryonic	Hsf1 knockout	(Luft et al., 2001)
J774—human macrophages	Blocks peroxynitrite cytotoxicity	(Szabo et al., 1996)
Murine stem cells	HSP27	(Wu and Welsh, 1996)
Human endothelial cells	Neutrophil-mediated necrosis	(Wang et al., 1995)

an individual or their tissues to a sublethal stressor (e.g., A-heat shock) followed by a critical recovery period will provide transient resistance to subsequent exposure to lethal stressors (A-hyperthermia, or B-ischemia/reperfusion, or C-etc.). This transient state of heightened resistance to injury is termed the protected phenotype. The protected phenotype is characterized by cytoprotection at the cellular level, and an anti-inflammatory state at the tissue level.

9. Materials and Methods

The following translational models were selected to provide simple, unequivocal endpoints with which to test the stress-conditioning hypothesis. All models address relevant, common complications that arise from the application of standard medical/surgical procedures. All experiments were performed within the guidelines outlined by the *Guide for the Care and Use of Laboratory Animals*, published by the Institute of Laboratory Animal Resources, Commission on Life Sciences, (National Research Council, 1996) and in compliance with the Animal Care and Use Committee at our institution.

9.1. Rodent Renal Transplant

Whole-organ procurement and preservation are stressful events. With modern cold-storage techniques, the maximal safe cold storage time of the rodent kidney is limited to 24 hours. Currently there is no record of successful cold storage of a rat kidney for 48 hours. We tested the hypothesis that stress conditioning (heat shock and recovery) will improve the ability of the rat kidney to withstand the stresses of conventional organ procurement and preservation.

Ten-week-old LBN donor rats were anesthetized (60–80 mg chloral hydrate, IP), hydrated (2 cc isotonic saline, IV), positioned with an intraperitoneal temperature probe into a standard tissue culture incubator pre-equilibrated to a temperature of 45°C until a core body temperature of 42.5°C was reached for 5 minutes. The animals were then returned to their cages and allowed to recover at room temperature for 6–8 hours before organ retrieval and cold storage. The total time during which core-body temperature equaled 42.5°C was 15 minutes. Group 1 (sham HS, n = 12) represents controls that did not receive heat shock. Group 2 (HS, n = 5) received heat shock and an optimal normothermic recovery of 6–8 hours. After recovery, organs were harvested and placed in cold storage (4°C) for 48 hours. The 48-hour time point was chosen as being a uniformly lethal storage time. Following storage, kidneys were heterotopically transplanted into syngeneic recipients. Seven days after transplantation a bilateral native nephrectomy was performed. Animals alive 50 days after native nephrectomy were considered survivors.

9.2. Warm Ischemia-Reperfusion of the In Situ Rodent Kidney

Obligatory episodes of acute ischemia and reperfusion of many organs and tissues frequently accompanies major surgical procedures. We wished to determine whether heat shock pretreatment could protect the in situ rodent kidney against a severe episode of warm ischemia. We studied three outcomes of acute renal injury (survival, serum creatinine levels, and renal vascular resistance) following 60 minutes of warm in situ ischemia.

9.2.1. Survival and Functional Study

Male Sprague-Dawley rats (200–300 g) were divided into two groups: Group I represents controls that did not receive heat shock pretreatment (sham HS, n = 28); and Group 2 did receive heat shock pretreatment (HS, n = 10). All animals were anesthetized (ketamine/acepromazine), hydrated, and positioned on aluminum trays floated in a water bath. Group 2, HS animals were exposed to a 45°C water bath for 20–25 minutes, until a core body temperature of 42.5°C was reached and maintained for an average of 10 minutes (core-body temperature range 42–42.9°C for 8–11 minutes). Group 1, sham HS animals received the same anesthetics but were exposed to a 37°C water bath for 20–25 minutes. Renal ischemia was induced surgically following systemic anticoagulation (heparin, 0.25 U/g, IV) and exposure of the kidneys through a midline laparotomy incision. The kidneys were dissected from the surrounding perinephric fat and a microvascular clamp was applied to the left renal artery for 60 minutes. Renal artery occlusion and reperfusion were confirmed visually. The contralateral kidney was removed and stored in liquid nitrogen for subsequent protein analysis. Core body temperature was maintained at 35–37°C with a heating pad and lamps throughout the ischemic period. The warm ischemia time of 60 minutes was chosen to yield a 50% survival rate. Three animals were exposed to heat shock alone (no ischemia) to determine the effect heat shock has on renal function. Serum creatinine measurements were performed at 0, 2, 4, and 7 days following renal ischemia. Survival was determined on day 14 following renal ischemia.

9.2.2. Renal Vascular Resistance

Male Sprague-Dawley rats (200–300 g) were divided into three groups: Group I, control (CON, no pretreatment and no ischemia, n = 13); Group 2, sham heat shock pretreatment (sham HS, n = 15); and Group 3, heat-shock pretreatment (HS, n = 8). All animals were anesthetized (pentobarbital, 5 mg/100 g body wt, IP), hydrated, and heparinized prior to induction of 60 minutes of warm ischemia as described above. Following ischemia, a polyethylene catheter (PE-50) with an in-line pressure transducer (Tektronix) was placed in the left renal artery through an infra-renal aortotomy incision. Each kidney was then perfused with phosphate buffered isotonic saline (27°C) at a constant flow rate (0.76 mL/min) and perfusion pressures were recorded (mm Hg) after an equilibration period 10 minutes. Data was later converted to renal vascular resistance (mmHg/mL/g). Renal vascular resistance typically increases following acute renal ischemia, reflecting loss of microvascular integrity. Following determination of renal vascular resistance, representative kidneys (n = 2) from each group were injected with a silicone rubber compound (Microfil, Flow Tek, Inc., Boulder, CO) according to manufacturers guidelines and then transected through the long axis to visually depict the patency of the renal vasculature.

9.3. Sponge Matrix Heterograft

A significant contribution to injury following episodes of acute ischemia-reperfusion is thought to result from the induction of an acute inflammatory reaction that occurs during reperfusion (Barone and Feuerstein, 1999). Evidence from the hyperthermia literature suggests that immune responses are depressed following exposure to hyperthermia both in vitro and in vivo (Skeen et al., 1986; Maridonneau-Parini et al., 1988). We wish to test the hypothesis that heat shock pretreatment will attenuated an acute inflammatory reaction in vivo. The sponge matrix model was selected to act as a foreign body into which inflammatory cells would readily migrate and from which these cells could be easily eluted.

Male Sprague-Dawley rats (200–300 g) were divided into two groups: Group I, control, no pretreatment (CON, n = 2); and Group 2, heat shock pretreatment (HS, n = 4). Heat shock pretreatment was performed by whole-body immersion, to the neckline, in a water bath pre-equilibrated to 45°C following the administration of general anesthesia (phenobarbital, 5 mg/100 g body wt, IP). Control animals received the same anesthesia and immersion into a normothermic water bath (37°C). Immediately following the sham or heat shock pretreatments, all animals had the subcutaneous implantation of a single sterilized polyurethane foam sponge (10 × 10 × 10 mm) saturated with Hanks' balanced salt solution (HBSS) through a dorsal midline incision. Sponges had been previously processed according to methods described elsewhere (Ascher et al., 1983). Animals were recovered and allowed free access to food and water for 24 hours, at which time they were euthanized. Sponges were recovered, immediately weighed, and placed in iced HBSS. Each sponge was eluted three times with 5-cc volumes of iced HBSS to remove infiltrating cells. Total number of infiltrating cells was determined in duplicate and pooled data presented as mean number of cells per milligram of freshly excised sponge.

9.4. Protein Gel Electrophoresis

9.4.1. Western Blot Analysis

Rat kidneys/tissues were homogenized in 40 mM Tris-HCl containing 10% glycerol and 2% SDS. Proteins isolated from tissue homogenates were separated using a 12% SDS-polyacrylamide gels. Assays to detect the inducible isoform of HSP70 (iHSP70) were performed by standard western blotting techniques with a monoclonal antibody specific for iHSP70 (SPA810, StressGen Biotechnologies, Victoria, BC, Canada). Bound primary antibody was detected by a secondary antibody conjugated with horseradish peroxidase and visualized using an enhanced chemiluminescence system (Amersham Pharmacia, Piscataway, NJ).

TABLE 3. Stress Conditioning Protects Rodent Kidneys Against 48 Hours of Cold Storage

Group	N	Survival (%)
1. Sham HS	12	0 (0)
2. Heat shock	5	4 (80)*

* $p < 0.002$, by Fisher's exact test.

10. Results

10.1. Rodent Renal Transplant (Perdrizet, Heffron et al., 1989)

Survivors were observed only in the group of animals that had received the heat shock pretreated grafts, (Group 2, Table 3). Group 1, which represents state-of-the-art cold storage at that time, had no survivors. Associated with these functional data was the observation made at 15–30 minutes after revascularization of the kidney graft, that there was immediate reperfusion and urine formation in the heat shock pretreated grafts only. Figure 1 represents photographs of representative kidneys from Group 1, sham HS (left) and Group 2, heat shock pretreated (right), respectively, 20 minutes following reperfusion in the recipient animal. The renal graft protected by the heat shock appears virtually identical to the adjacent native graft, that had not been exposed to cold storage. The control graft appears dark and cyanotic and did not function secondary to vascular thrombosis. This work was originally presented at the Resident's Forum of the Thirtieth Annual Meeting of the Society of University Surgeons, Feb. 11–13, 1988, San Antonio, TX, and published as an extended abstract (Perdrizet et al., 1989).

10.2. Warm Ischemia-Reperfusion of the Rodent Kidney

10.2.1. Survival and Functional Study

Survival and serum creatinine values were significantly lower in the heat shock pretreated animals (Group 2) than in the sham HS animals (Group 1) not receiving

Stress Conditioning Protects Rodent Kidneys Against 48 Hours of Cold Storage

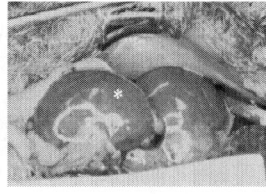

Sham-Heat Shock   Heat Shock

FIGURE 1. Fifteen minutes following reperfusion, the sham-HS kidney (*right organ in left photo) is dark and cyanotic compared to the HS kidney (*left organ in right photo), which is pink and well perfused. Reproduced with permission (Perdrizet, 1997). [See Color Plate I]

TABLE 4. Stress Conditioning Protects Rodent Kidneys Against 60 Minutes of Warm Ischemia

Group	N	Survival (%)	Serum creatinine (mg/dL), d 2, 4, 7
1. Sham HS	28	18 (64)	$5.0 \pm 0.3, 4.0 \pm 0.6, 1.9 \pm 0.4$
2. Heat shock	10	9 (90)*	$2.5 \pm 0.8, 1.8 \pm 1.0, 1.1 \pm 0.1$**

* $p < 0.04$ by chi square.
** $p = 0.002$ by repeated-measures analysis.

any pretreatment, following sixty minutes of warm ischemia and reperfusion (Table 4). The renal function of two animals that received heat shock treatment and no ischemia remained normal on days 2, 4, and 7, data not shown. This work was originally presented at the Resident's Forum of the Thirty Second Annual Meeting of the Society of University Surgeons, Feb.10, 1990, Los Angeles, CA, and published as an extended abstract (Chatson et al., 1990).

10.2.2. Renal Vascular Resistance

Renal vascular resistance increased significantly from baseline, 45.7 ± 10.7 mm Hg/mL/g to 60.7 ± 14.2 mm Hg/mL/g following 60 minutes of warm in situ ischemia. Heat shock pretreatment 6–8 hours prior to warm ischemia reduced the vascular resistance from 60.7 ± 14.2 to 37.6 ± 10.2 mm Hg/mL/g, reflecting attenuation of ischemic damage and preservation of microvasculature (Table 5). Preservation of renal vasculature, within the organ protected by heat shock, is demonstrated by representative silicone rubber casts (Fig. 2). A nonischemic, baseline kidney is shown on the left for comparison and demonstrates a clear corticomedullary junction and segmental vessels containing the silicone agent. The kidney in the middle is representative of the changes that occur in the sham HS organ following 60 minutes of ischemia, in which loss of corticomedullary junction, disruption of segmental renal vessels, and areas of hemorrhagic necrosis are seen. The kidney on the right represents the HS pretreated organ following 60 minutes of warm ischemia, with preservation of the corticomedullary junction and segmental vessels seen in a pattern similar to that of the baseline organ, suggesting preservation of renal vasculature. This work was previously presented at the 52nd Annual Sessions of the Owen H. Wangensteen Surgical Forum, Clinical Congress, American College of Surgeons, Oct 12–17, 1997, Chicago, IL, and published as an extended abstract (Garcia et al., 1997).

TABLE 5. Stress Conditioning Preserves Renal Vascular Resistance in Rodent Kidneys After 60 Minutes of Warm Ischemia

Group	N	Vascular resistance (mm Hg/mL/g)
1. No ischemia	13	45.7 ± 10.7
2. Sham HS	15	60.7 ± 14.2
3. Heat shock	8	37.6 ± 10.2*

* $p < 0.001$ for Group 2 vs. Group 3 by Student's t-test.

Stress Conditioning Protects Rodent Kidneys Against 60 Minutes of Warm Ischemia

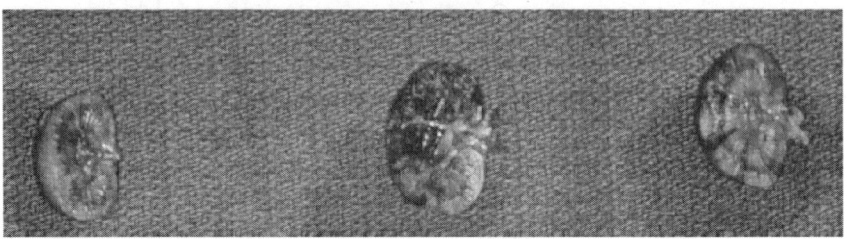

No Ischemia   Sham Heat Shock   Heat Shock
                    Plus Ischemia   Plus Ischemia

FIGURE 2. Rodent kidneys at baseline (left) have a clearly visible corticomedullary junction* and patent renal vessel (arrow). Loss of this junction and disruption of the renal vessel followed warm ischemia in the sham-HS organ (center) but is much less in the HS organ (right).

10.3. Sponge Matrix Heterograft

Sponges placed in animals pretreated with heat shock (HS, Group 2) contained approximately ten times less infiltrating cells twenty-four hours after implantation, compared to sponges removed from the non-heated control animals (CON, Group 1) (Table 6). Heat shock pretreatment dramatically attenuates acute inflammatory reactions in vivo.

10.4. Protein Gel Electrophoresis

Analysis of protein lysates demonstrated enhanced expression of the inducible isoform of the 70kDa heat shock protein (iHSP70) in pulmonary (P), hepatic (H), renal (R), and aortic (A) tissues taken from animals pretreated with heat shock, (lanes 1–4 and 9–12, Fig. 3) and not in the same tissues taken from the non-heated, control animal (lanes 5–8, Fig. 3). The elevated iHSP70 content of heat shock pretreated tissues is temporally associated with the cytoprotected and anti-inflammatory states described above.

TABLE 6. Stress Conditioning Attenuates Cell Infiltration of Sponge Heterografts in the Rodent

Group	N	Infiltrating cells ($\times 10^3$/mg sponge)
1. No heat shock	2	31.5 ± 10.1
2. Heat shock	4	4.5 ± 2.4*

*$p = 0.003$ by Student's t-test.

FIGURE 3. Renal, hepatic, cardiac, and aortic tissues (R, H, C, A, respectively) show increased iHSP72 following HS, lanes 1–4 and 9–12, but not in the same tissues taken from a sham-HS animal, lanes 5–8.

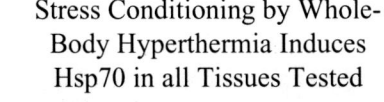
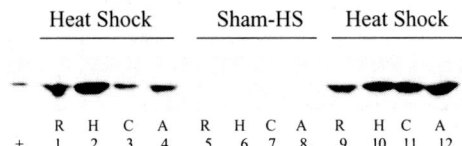

11. Discussion of Rat Kidney Models of Acute Ischemia

The need to prevent tissue injury associated with iatrogenic interventions is expected to increase in the future due to the aging demographics of modern society. With the increase in the number of older adults, there will be an increase in the number of invasive medical procedures and their associated complications. The recently published practice guidelines for coronary bypass surgery from the American College of Cardiology and the American Heart Association listed the top ten target areas for perioperative cardiac surgery management (Eagle, 1999). The first two goals are to reduce type I and II brain injuries. The fourth targeted area is the reduction of systemic consequences (*complications*) of cardiopulmonary bypass. It is important that the medical community begin to develop preventative strategies that will minimize the collateral damage associated with these needed interventions.

Medical bias presumes that debility will follow exposure to sub-lethal stressors when, in fact, just the opposite occurs at all levels of biology from single celled organisms to complex animals and plants. We conclude that stress conditioning, by heat shock pretreatment and recovery, is both cytoprotective and anti-inflammatory in the rodent. During the past decade the phenomenon of stress conditioning has been repeatedly confirmed by ourselves and others (see Table 1). It is the universal nature of the cell-stress response that will ultimately translate into wide clinical utility. The cellular-stress response is driven by a fundamental class of genes, the heat shock genes, which are universally present and have been highly conserved through hundreds of millions of years of evolution. The timely activation of these conserved genes can make the difference between a life or death outcome. These findings provide an opportunity to practice preventive medicine at the molecular level. As of yet, no human clinical trials have taken advantage of the prior induction of the HSR to provide tissue protection. The late Hans Selye clearly described the fundamental behavior of the stress response as being predictable and triphasic in nature (Selye, 1936, 1946). We now have insight into these phenotypic behaviors at the molecular level. This should provide investigators with clues to control this complex metabolic response.

Until the time when stress-conditioning protocols are available for general clinical use, it should be recognized that the cellular response to stress colors the clinical syndromes of daily medical and surgical practice. Currently there exists unrecognizable (and unmeasurable) alterations in the intrinsic resistance to stress that occurs in persons exposed to stressors, whether they are natural (disease) or man-made (iatrogenic, environmental). Modern examples of the triphasic nature of the stress response can be found in the current medical literature. For example, in the trauma/critical care disciplines, the phenomenon of damage control laparotomy has greatly improved the outcome from formerly lethal, traumatic events. The tenets of damage control laparotomy are remarkably similar to those of stress conditioning. When definitive surgical repair (a stressor) is delayed in the setting of overwhelming injury (potentially lethal) an improvement in survival is observed (Morris et al., 1996). Resistance to the second injury (iatrogenic, surgery) takes time to develop relative to the first injury (trauma). The ability of a sublethal trauma to condition animals to resist lethal trauma is an old observation that has never been directly related to the clinical situation (Nobel, 1943). A second example in which stress-conditioning principles are active is in the setting of acute myocardial infarction with or without preceding angina. Episodes of preinfarction angina are associated with improved preservation of cardiac function and improved clinical outcome compared to myocardial infarction not preceded by angina (Kloner et al., 1998; Tomai et al., 1999). As predicted by other cellular stress response phenomena, the beneficial effects of pre-infarction angina are time dependent. Angina is most protective during the 24 hours preceding the myocardial infarction. It is now recognized that this clinical observation in part, represents a natural example of ischemic preconditioning which is also known to activate heat shock gene expression.

The cell-stress response is a complex metabolic response. The precise mechanisms behind the cytoprotection and anti-inflammation characteristic of this response are being investigated (Table 2). Recent work on cytokine gene-transcription has shed light on a molecular mechanism whereby induction of the HSR leads to inhibition of IL-1β and TNF-α production following LPS stimulation of the human monocyte in vitro (Xie et al., 1999). The inhibition of cytokine gene-expression is the result of heat shock transcription factor (HSF) activation and binding to heat shock sequence elements (HSE) (located within the promoter regions upstream of the IL-1β and TNF-α structural genes) at the time of heat shock. The inhibition of acute inflammation is beneficial in the setting of ischemia/reperfusion injury, but left unchecked, could become problematic, as in critical illness following trauma or major surgery. We anticipate a future need to turn the HSR "off" as well as "on" in a myriad of medically relevant conditions.

Finally, if the widespread application of stress-conditioning principles is to occur, methods to assay for and identify the existence of the protected phenotype will need to be developed. This will allow clinicians to identify when an individual's resistance or tolerance to stressful events will be maximal and should allow the scheduling of noxious medical interventions accordingly. If the principles of stress conditioning are to be integrated into the mainstream of clinical medicine, safe, effective and clinically convenient inducers of the cell stress response must be developed (Morimoto and Santoro, 1998; House et al., 2001). The application of whole-body hyperthermia to the average clinical setting is impractical and undesirable. Ideally, the development of pharmaceutical inducers of the cell-stress response will occur in the near future.

In summary, there is a large and growing clinical need to make invasive medical interventions safer. We suggest that the risks associated with these interventions can be reduced by the timely induction of the cell-stress response through the application of the principle of stress conditioning. The potent ability of the cell-stress response to protect against common iatrogenic injuries can be integrated into existing clinical protocols. The phenomenon of stress conditioning has been repeatedly confirmed in laboratory models, including human cell culture, and suggests that its potential for application will be as universal as the cellular response to stress itself.

12. Concluding Remarks

The Hsp70 gene family in humans has a complex evolutionary history shaped by multiple gene duplications, divergence, and deletion. Analysis of the protein sequences of Hsp70 family members across a wide range of organisms including bacteria, yeast, *Drosophila*, Xenopus, plants, and mammals indicates that over 75% of the respective sequence length has been conserved throughout evolution (Rensing and Maier, 1994). While these findings indicate the biological value of Hsp70 chaperones among different lineages, the diversity of genes and protein products within the Hsp70 families of different organisms has become increasingly evident. Using 2-dimensional gel electrophoresis, many Hsp70 isoforms have been detected which are species or cell/tissue-type specific (Allen et al., 1988; White et al., 1994; Gutierrez and Guerriero, 1995; Norris et al., 1995; Tavaria et al., 1996; Manzerra et al., 1997; Place and Hofmann, 2005). Sequence analysis of Hsp70s in closely related species indicates a rapid turnover and differential sorting of genes within the Hsp70 family despite preserving the function of these proteins as a whole (Boorstein et al., 1994; Aoki et al., 2002; Martin and Burg, 2002). For example, Hsp70B' is an inducible Hsp70 gene presumably conserved in the mammalian lineage, with homologs in *Saguinus oedipus* (cottontop tamarin), *Sus scrofa* (pig), *Bos taurus* (cow), and *Homo sapiens* (human). Understanding the relationship of Hsp70B' to other members of the human Hsp70 family may help us understand the specific functional role, if any, of this protein in the identified organisms. Variation in thermal response is most likely the result of species-specific differences in the presence of heat-inducible Hsp70s, as has been noted in vertebrates (Yamashita et al., 2004). The presence of different isoforms within species can give rise to slight phenotypic differences, and ultimately play a role in the development of molecular pathologies unique to a given tissue or organism. Therefore, careful selection of model systems is critical in studying chaperone regulated pathologies.

Acknowledgments. G.P. wishes to acknowledge the Research Committee and Animal Care & Use Committee, University of Chicago, Chicago, IL, and Hartford Hospital, Hartford Connecticut.
Hartford Hospital Research Administration for funding.

Renal transplants and photography were performed by the late Mr. Francis Buckingham, Chicago, IL. L.E.H. Wishes to acknowledge and thank the members of the Hightower Laboratory who contributed to the *Poeciliopsis* studies, particularly Carol Norris, Mary Brown, and Philip DiIorio. The contributions of Jack and Mary Schultz are gratefully acknowledged as well. For the winter flounder work, my collaborator J. Larry Renfro is acknowledged; and for the ex vivo vascular endothelium studies, Peter Guidon, Jr., and Steve House are gratefully acknowledged.

References

Allen, R. L., O'Brien, D. A., et al. (1988) Expression of heat shock proteins by isolated mouse spermatogenic cells. *Mol Cell Biol* 8(8):3260–6.

Amrani, M., Latif, N., et al. (1998) Relative induction of heat shock protein in coronary endothelial cells and cardiomyocytes: Implications for myocardial protection. *J Thorac Cardiovasc Surg* 115(1):200–9.

Aoki, K., Kragler, F., et al. (2002) A subclass of plant heat shock cognate 70 chaperones carries a motif that facilitates trafficking through plasmodesmata. *Proc Natl Acad Sci U S A* **99**(25):16342–7.

Arrigo, A.-P., and Landry, J. (1994) Expression and function of the low-molecular-weight heat shock proteins. In Morimoto, R. I. Tissiéres, A., and Georgopoulos, C (eds.): *The Biology of Heat Shock Proteins and Molecular Chaperones*. Cold Spring Harbor Laboratory Press, Plainville, NY, pp. 335–74.

Ascher, N. L., Chen, S., et al. (1983) Maturation of cytotoxic T cells within sponge matrix allografts *J Immunol* 131(2):617–21.

Barone, F. C., and Feuerstein, G. Z. (1999) Inflammatory mediators and stroke: New opportunities for novel therapeutics. *J Cereb Blood Flow Metab* 19(8):819–34.

Boorstein, W. R., Ziegelhoffer, T., et al. (1994) Molecular evolution of the hsp70 multigene family, *J Mol Evol* 38:1–17.

Borrelli, M. J., Bernock, L. J., et al. (2002) Stress protection by a fluorescent Hsp27 chimera that is independent of nuclear translocation or multimeric dissociation. *Cell Stress Chaperones* 7(3):281–96.

Borrelli, M. J., Stafford, D. M., et al. (1996) Thermotolerance expression in mitotic CHO cells without increased translation of heat shock proteins. *J Cell Physiol* 169:420–8.

Bosch, T. C. G., Gellner, K. L., et al. (1988) Thermotolerance and synthesis of heat shock proteins: These responses are present in *Hydra attenuata* but absent in *Hydra oligactis*. *Proc Natl Acad Sci U S A* 85:7927–31.

Brown, I. R., Rush, S. J., et al. (1989) Induction of a heat shock gene at the site of tissue injury in the rat brain. *Neuron* 2:1559–64.

Brown, M., Upender, R., et al. (1992) Thermoprotection of a functional epithelium: Heat stress effects on transepithelial transport by flounder renal tubule in primary monolayer culture. *Proc Natl Acad Sci U S A* 89:3246–50.

Bulger, A. J., and Schultz, R. J. (1979) Heterosis and interclonal variation in thermal tolerance in unisexual fishes. *Evolution* 33:848–59.

Bulger, A. J., and Schultz, R. J. (1982) Origin of thermal adaptations in northern versus southern populations of a unisexual hybrid fish. *Evolution* 36:1041–50.

Buress, G. C., Musch, M. W., et al. (1996) Induction of heat-shock-protein-70 protects intestinal epithelial IEC-18 cells from oxidant and thermal injury. *Am J Physiol* 39:C429–36.

Caprioli, J., Kitano, S., et al. (1996) Hyperthermia and hypoxia increase tolerance of retinal ganglion cells to anoxia and excitotoxicity. *Invest Ophthalmol Vis Sci* 37(12): 2376–81.

Cavicchi, S., Guerra, D., et al. (1995) Chromosomal analysis of heat-shock resistance in *Drosophila melanogaster* evolving at different temperatures in the laboratory. *Evolution* 49:676–84.

Chatson, G., Perdrizet, G., et al. (1990) Heat shock protects kidneys against warm ischemic injury. *Curr Surg* 47(6):420–3.

Chen, G., Kelly, C., et al. (1997) Induction of heat shock protein 72 kDa expression is associated with attenuation of ischaemia-reperfusion induced microvascular injury. *J Surg Res* 69(2):435–9.

Committee, N. A. S. C. E. T. N. S. (1991) North American Symptomatic Carotid Endarterectomy Trial: Methods, patients characteristics, and progress. *Stroke* 22:711–20.

Corces, V., Pellicer, A., et al. (1981) Integration, transcription, and control of a *Drosophila* heat shock gene in mouse cells. *Proc Natl Acad Sci U S A* 78:7038–42.

Corell, R. A., Riordan, J. A., et al. (1994) Chemical induction of stress proteins does not induce splicing thermotolerance under conditions producing survival thermotolerance. *Exp Cell Res* 211:189–96.

Cotto, J. J., and Morimoto, R. I. (1999) Stress-induced activation of the heat-shock response: Cell and molecular biology of heat-shock factors. *Biochem Soc Symp* 64:105–18.

Crile, G., Jr. (1963) The effects of heat and radiation on cancers implanted on the feet of mice. *Cancer Res* 23:372–80.

Cunningham, J. N., Jr. (1998) Spinal cord ischemia. Introduction. *Semin Thorac Cardiovasc Surg* 10(1):3–5.

de Jong, W. W., Leunissen, J. A. M., et al. (1993) Evolution of the α-crystallin/small heat-shock protein family. *Mol Biol Evol* 10:103–26.

DeMaio, A. (1999) Heat shock proteins: Facts, thoughts, and dreams. *Shock* 12:323–5.

diIorio, P. J. (1994) Heat shock proteins and acquired thermotolerance in species and hemiclones of the livebearing fish, *Poeciliopsis*. *Ecology and Evolutionary Biology*. University of Connecticut, Storrs, CT.

diIorio, P. J., K. Holsinger, et al. (1996) Quantitative evidence that both Hsc70 and Hsp70 contribute to thermal adaptation in hybrids of the livebearing fishes *Poeciliopsis*. *Cell Stress Chap* 1(2):139–47.

Eagle, K. A., et al. (1999) ACC/AHA Guidelines for coronary artery bypass surgery: A report of the American College of Cardiology/American Heart Association Task Force on practice guidelines. *J Am Col Card* 34:1247–62.

Easton, D. P., Rutledge, P. S., et al. (1987) Heat shock protein induction and induced thermal tolerance are independent in adult salamanders. *J Exp Zool* 241:263–7.

Elliott, B., Haltiwanger, R. S., et al. (1996) Synergy between trehalose and Hsp104 for thermotolerance in *Saccharomyces cerevisiae*. *Genetics* 144:923–3.

Fan, N., Yang, G. S., et al. (2005) Oral administration of geranylgeranylacetone plus local somatothermal stimulation: A simple, effective, safe and operable preconditioning combination for conferring tolerance against ischemia-reperfusion injury in rat livers. *World J Gastroenterol* 11(36):5725–31.

Feder, J. H., Rossi, J. M., et al. (1992) The consequences of expressing hsp70 in *Drosophila* cells at normal temperatures. *Genes Dev* 6:1402–13.

Feder, M. E., and Hofmann, G. E. (1999) Heat-shock proteins, molecular chaperones, and the stress response: Evolutionary and ecological physiology. *Annu Rev Physiol* 61:243–82.

Fielding, E. (1992) *Metabolic and Behavioral Responses to Temperature in the All-Female Hybrid Topminnow, Poeciliopsis monacha-lucida.* University of Connecticut, Storrs, CT.

Franci, O., Amici, A., et al. (1996) Influence of thermal and dietary stress on immune response of rabbits. *J Anim Sci* 74(7):1523–9.

Freyaldenhoven, T. E., and Ali S. F. (1996) Heat shock proteins protect cultured fibroblasts from the cytotoxic effects of MPP+. *Brain Res* 735(1):42–9.

Fujimori, S., Otaka, M., et al. (1997) Induction of a 72-kDa heat shock protein and cytoprotection against thioacetamide-induced liver injury in rats. *Dig Dis Sci* 42(9):1987–94.

Gabai, V., Meriin, A., et al. (1997) Hsp70 prevents activation of stress kinases. *J Biol Chem* 272(29):18033–7.

Garcia, J. C., Perdrizet, G. A., et al. (1997) Shock protects rodent kidneys from acute ischemia/reperfusion injury. *Surg Forum* 48:381–2.

Gharagozloo, F., Neville, R. F. Jr., et al. (1998) Spinal cord protection during surgical procedures on the descending thoracic and thoracoabdominal aorta: A critical overview. *Semin Thorac Cardiovasc Surg* 10(1):73–86.

Gill, R. R., Gbur, C. J. Jr., et al. (1998) Heat shock provides delayed protection against oxidative injury in cultured human umbilical vein endothelial cells. *J Mol Cell Cardiol* 30(12):2739–49.

Glover, J. R., and Lindquist, S. (1998) Hsp104, Hsp70, and Hsp40: A novel chaperone system that rescues previously aggregated proteins. *Cell* 94(1):73–82.

Guenal, I., Sidoti-de Fraisse, C., et al. (1997) Bcl-2 and Hsp27 act at different levels to suppress programmed cell death. *Oncogene* 15(3):347–60.

Gutierrez, J. A., and Guerriero, V. Jr. (1995) Chemical modifications of a recombinant bovine stress-inducible 70 kDa heat-shock protein (Hsp70) mimics Hsp70 isoforms from tissues. *Biochem J* 305(Pt 1):197–203.

Hall, B. G. (1983) Yeast thermotolerance does not require protein synthesis. *J Bact* 156:1363–5.

Harder, Y., Amon, M., et al. (2005) Heat shock preconditioning reduces ischemic tissue necrosis by heat shock protein (HSP)-32-mediated improvement of the microcirculation rather than induction of ischemic tolerance. *Ann Surg* 242(6):869–78, see also discussion 878–9.

Hastie, A. T., Everts, K. B., et al. (1997) HSP27 elevated in mild allergic inflammation protects airway epithelium from H2SO4 effects. *Am J Physiol* 273(2 Pt 1):L401–9.

Henderson, B., and Pockley, A. G. (2005) *Molecular Chaperones and Cell Signalling.* Cambridge University Press, Cambridge.

Hightower, L. E., Brown, et al. (2000) Tissue-level cytoprotection. *Cell Stress Chaperones* 5(5):412–4.

Hightower, L. E., Norris, C. E., et al. (1999) Heat shock responses of closely related species of tropical and desert fish. *Am Zool* 39:877–88.

Hinzpeter, A., Lipecka, J., et al. (2006) Association between Hsp90 and the ClC-2 chloride channel upregulates channel function. *Am J Physiol Cell Physiol* 290(1):C45–56.

Hoffmann, A. A., and Parsons, P. A. (1991) *Evolutionary Genetics and Environmental Stress.* Oxford University Press, New York.

Hogue, C. W., Jr., Sundt, T. M. III, et al. (1999) Neurological complications of cardiac surgery: The need for new paradigms in prevention and treatment. *Semin Thorac Cardiovasc Surg* 11(2):105–15.

House, S. D., Guidon, P. T. Jr., et al. (2001) Effects of heat shock, stannous chloride, and gallium nitrate on the rat inflammatory response. *Cell Stress Chap* 6(2):164–71.

Jaattela, M. (1999) Heat shock proteins as cellular lifeguards. *Ann Med* 31(4):261–71.

Jayakumar, J., Suzuki, K., et al. (2001) Heat shock protein 70 gene transfection protects mitochondrial and ventricular function against ischemia-reperfusion injury. *Circulation* 104(12 Suppl. 1):I303–7.

Jiang, B., Xiao, W., et al. (2005) Heat shock pretreatment inhibited the release of Smac/DIABLO from mitochondria and apoptosis induced by hydrogen peroxide in cardiomyocytes and C2C12 myogenic cells. *Cell Stress Chap* 10(3):252–62.

Kloner, R. A., Shook, T., et al. (1998) Prospective temporal analysis of the onset of pre-infarction angina versus outcome: An ancillary study in TIMI-9B. *Circulation* 97(11): 1042–5.

Knauf, U., Jakob, U., et al. (1994) Stress- and mitogen-induced phosphorylation of the small heat shock protein Hsp25 by MAPKAP kinase 2 is not essential for chaperone properties and cellular thermoresistance. *EMBO J* 13:54–60.

Krebs, R. A., and Feder, M. E. (1997a) Deleterious consequences of Hsp70 overexpression in *Drosophila melanogaster* larvae. *Cell Stress Chap* 2:60–71.

Krebs, R. A., and Feder, M. E. (1997b) Natural variation in the expression of the heat-shock protein Hsp70 in a population of *Drosophila melanogaster* and its correlation with tolerance of ecologically relevant thermal stress. *Evolution* 51:173–9.

Kume, M., Yamamoto, Y., et al. (1996) Ischemic preconditioning of the liver in rats: Implications of heat shock protein induction to increase tolerance of ischemia-reperfusion injury. *J Lab Clin Med* 128(3):251–8.

Landry, J., Chrétien, P., et al. (1989) Heat shock resistance conferred by expression of the human hsp27 gene in rodent cells. *J Cell Biol* 109:7–15.

Landry, J., Lambert, H., et al. (1992) Human hsp27 is phosphorylated at serines 78 and 82 by heat shock and mitogen-activated kinases that recognize the same amino acid motif as S6 kinase. *J Biol Chem* 267:794–803.

Lane, G. J., Hozack, W. J., et al. (1997) Simultaneous bilateral versus unilateral total knee arthroplasty. Outcomes analysis. *Clin Orthop Relat Res* (345):106–12.

Laszlo, A., and Li, G. C. (1985) Heat-resistant variants of Chinese hamster fibroblasts altered in expression of heat shock protein. *Proc Natl Acad Sci U S A* 82:8029–33.

Lepore, D. A., Hurley, J. V., et al. (2000) Prior heat stress improves survival of ischemic-reperfused skeletal muscle *in vivo*. *Muscle Nerve* 23(12):1847–55.

Li, G. C., and Nussenzweig, A. (1996) Thermotolerance and heat shock proteins. In Feige, U., Morimoto, R. I., Yahara, I., and Polla B. S. (eds.): *Stress-Inducible Cellular Responses*. Birkhäuser, Boston, pp. 425–50.

Lievremont, J. P., Rizzuto, R., et al. (1997) BiP, a major chaperone protein of the endoplasmic reticulum lumen, plays a direct and important role in the storage of the rapidly exchanging pool of Ca2+. *J Biol Chem* 272(49):30873–9.

Liu, H., Bowes, R. C. III, et al. (1997) Endoplasmic reticulum chaperones GRP78 and calreticulin prevent oxidative stress, Ca2+ disturbances, and cell death in renal epithelial cells. *J Biol Chem* 272(35):21751–9.

Liu, J., Squibb, K. S., et al. (1996) Cytotoxicity, zinc protection, and stress protein induction in rat proximal tubule cells exposed to cadmium chloride in primary cell culture. *Ren Fail* 18(6):867–82.

Lu, R., Peng, J., et al. (2002) Heme oxygenase-1 pathway is involved in delayed protection induced by heat stress against cardiac ischemia-reperfusion injury. *Int J Cardiol* 82(2):133–40.

Luft, J. C., Benjamin, I. J., et al. (2001) Heat shock factor 1-mediated thermotolerance prevents cell death and results in G2/M cell cycle arrest. *Cell Stress Chap* 6(4): 326–36.

Manzerra, P., Rush, S. J., et al. (1997) Tissue-specific differences in heat shock protein hsc70 and hsp70 in the control and hyperthermic rabbit. *J Cell Physiol* 170(2):130–7.

Maridonneau-Parini, I., Clerc, J., et al. (1988) Heat shock inhibits NADPH oxidase in human neutrophils. *Biochem Biophys Res Commun* 154(1):179–86.

Martin, A. P., and Burg, T. M. (2002) Perils of paralogy: Using HSP70 genes for inferring organismal phylogenies. *Syst Biol* 51(4):570–87.

Martin, J. L., Mestril, R., et al. (1997) Small heat shock proteins and protection against ischemic injury in cardiac myocytes. *Circulation* 96(12):4343–8.

McDonough, K. H. (1999) The role of alcohol in the oxidant antioxidant balance in heart. *Front Biosci* 4:D601–6.

Meldrum, K. K., Burnett, A. L., et al. (2003) Liposomal delivery of heat shock protein 72 into renal tubular cells blocks nuclear factor-kappaB activation, tumor necrosis factor-alpha production, and subsequent ischemia-induced apoptosis. *Circ Res* 92(3):293–9.

Meldrum, K. K., Meldrum, D. R., et al. (2001) Heat shock prevents simulated ischemia-induced apoptosis in renal tubular cells via a PKC-dependent mechanism. *Am J Physiol Regul Integr Comp Physiol* 281(1):R359–64.

Meyer, T. N., da Silva, A. L., et al. (2000) Heat shock response reduces mortality after severe experimental burns. *Burns* 26(3):233–8.

Mizushima, Y., Wang, P., et al. (2000) Preinduction of heat shock proteins protects cardiac and hepatic functions following trauma and hemorrhage. *Am J Physiol Regul Integr Comp Physiol* 278(2):R352–9.

Morimoto, R. I., and Santoro, M. G. (1998) Stress-inducible responses and heat shock proteins: New pharmacologic targets for cytoprotection. *Nat Biotechnol* 16(9):833–8.

Morris, J. A., Jr., Eddy, V. A., et al. (1996) The trauma celiotomy: The evolving concepts of damage control. *Curr Probl Surg* 33(8):611–700.

Newman, M. F., Kirchner, J. L., et al. (2001) Longitudinal assessment of neurocognitive function after coronary-artery bypass surgery. *N Engl J Med* 344(6):395–402.

Nobel, R. L. (1943) The development of resistance by rats and guinea pigs to amounts of trauma usually fatal. *Am J Physiol* 138:346–51.

Noonan, E. J., Place, R. F., et al. (2006) Cell number-dependent regulation of Hsp70B' expression: Evidence of an extracellular regulator. *J Cell Physiol* 210(1):201–211.

Norris, C. E., Brown, M. A., et al. (1997) Low-molecular-weight heat shock proteins in a desert fish (*Poeciliopsis lucida*): homologs of human Hsp27 and *Xenopus* Hsp30. *Mol Biol Evol* 14(10):1050–61.

Norris, C. E., diIorio, P. J., et al. (1995) Variation in heat shock proteins within tropical and desert species of Poeciliid fishes. *Mol Biol Evol* 12(6):1048–62.

Norris, C. E., and Hightower, L. (2000) The heat shock response of tropical and desert fish (genus *Poeciliopsis*). In Storey, K.B., and Storey, J. (eds.): *Environmental Stressors and Gene Responses*. Elsevier Science, Amsterdam.

Ohtsuka, K., and Laszlo, A. (1992) The relationship between hsp 70 localization and heat resistance. *Exp Cell Res* 202(2):507–18.

Otani, S., Otaka, M., et al. (1997) Effect of preinduction of heat shock proteins on acetic acid-induced colitis in rats. *Dig Dis Sci* 42(4):833–46.

Parsell, D. A., and Lindquist, S. (1993) The function of heat-shock proteins in stress tolerance: Degradation and reactivation of damaged proteins. *Ann Rev Genet* 27:437–96.

Patil, M. S., Kaklij, G. S., et al. (1996) Radioprotective effect of whole-body hyperthermia on mice exposed to lethal doses of total-body gamma irradiation. *Indian J Exp Biol* 34(9):842–4.

Perdrizet, G. A. (1997) *Heat Shock Response and Organ Preservation: Models of Stress Conditioining.* Georgetown, TX.

Perdrizet, G. A., Heffron, T. G., et al. (1989) Stress conditioning: A novel approach to organ preservation. *Curr Surg* 46(1):23–6.

Place, S. P., and Hofmann, G. E. (2005) Comparison of Hsc70 orthologs from polar and temperate notothenioid fishes: Differences in prevention of aggregation and refolding of denatured proteins. *Am J Physiol Regul Integr Comp Physiol* 288(5):R1195–202.

Polla, B. S. (1988) A role for heat shock proteins in inflammation. *Immunol Today* 9(5): 134–7.

Renfro, J., Brown, M., et al. (1993) Relationship of thermal and chemical tolerance to transepithelial transport by cultured flounder renal epithelium. *J Pharmacol Exp Ther* 265(2):992–1000.

Rensing, S. A., and Maier, U. G. (1994) Phylogenetic analysis of the stress-70 protein family. *J Mol Evol* 39(1):80–6.

Ritossa, F. (1963) New puffs induced by temperature shock, DNP and salicilate in salivary chromosomes of D. melanogaster. *Drosoph Inf Serv* 37:122–3.

Rosen, D. E., and Bailey, R. M. (1963) *The Poeciliid Fishes (Cyprinodontiformes); Their Structure, Zoogeography, and Systematics.* New York.

Sakurai, M., Aoki, M., et al. (1997) Selective motor neuron death and heat shock protein induction after spinal cord ischemia in rabbits. *J Thorac Cardiovasc Surg* 113(1):159–64.

Sanders, B. M., Hope, C., et al. (1991) Characterization of the stress protein response in two species of *Collisella* limpets with different temperature tolerances. *Physiol Zool* 64:1471–89.

Selye, H. (1936) A syndrome produced by diverse nocuous agents. *J Neuropsychiatry Clin Neurosci* 10(2):230–1.

Selye, H. (1946) The general adaptation syndrome and the diseases of adaptation. *J Clin Endocrinol* 6(2):117–230.

Shinohara, T., Takahashi, N., et al. (2004) Estrogen inhibits hyperthermia-induced expression of heat-shock protein 72 and cardioprotection against ischemia/reperfusion injury in female rat heart. *J Mol Cell Cardiol* 37(5):1053–61.

Skeen M. J., McLaren, J. R., Olkowski, Z. L. (1986) Influences of hyperthermia on immunological functions. In Angheleri L. J., Roberts J. (eds.): *Hyperthermia in Cancer Treatment.* Vol 1. Boca Raton: CRC Press, pp. 94–105.

Stojadinovic, A., Kiang, J., et al. (1997) Induction of the heat shock response prevents tissue injury during acute inflammation of the rat ileum. *Crit Care Med* 25(2):309–17.

Swank, G. M., Lu, Q., et al. (1998) Effect of acute-phase and heat-shock stress on apoptosis in intestinal epithelial cells (Caco-2). *Crit Care Med* 26(7):1213–7.

Szabo, C., Wong, H. R., et al. (1996) Pre-exposure to heat shock inhibits peroxynitrite-induced activation of poly(ADP) ribosyltransferase and protects against peroxynitrite cytotoxicity in J774 macrophages. *Eur J Pharmacol* 315(2):221–6.

Takuma, K., Matsuda, T., et al. (1996) Heat shock protects cultured rat astrocytes in a model of reperfusion injury. *Brain Res* 735(2):265–70.

Tavaria, M., Gabriele, T., et al. (1996) A hitchhiker's guide to the human Hsp70 family. *Cell Stress Chap* 1:23–28.

Tomai, F., Crea, F., et al. (1999) Preinfarction angina and myocardial preconditioning. *Cardiologia* 44(11):963–7.

Uchinami, H., Yamamoto, Y., et al. (2002) Effect of heat shock preconditioning on NF-kappaB/I-kappaB pathway during I/R injury of the rat liver. *Am J Physiol Gastrointest Liver Physiol* 282(6):G962–71.

Volloch, V., Gabai, V., et al. (2000) Hsp72 can protect cells from heat-induced apoptosis by accelerating the inactivation of stress kinase JNK. *Cell Stress Chap* 5(2):139–47.

Voss, M. R., Gupta, S., et al. (2005) Effect of mutation of amino acids 246–251 (KRKHKK) in HSP72 on protein synthesis and recovery from hypoxic injury. *Am J Physiol Heart Circ Physiol* 289(6):H2519–25.

Vrijenhoek, R. C. (1989) Genotypic diversity and coexistence among sexual and clonal lineages of *Poeciliopsis*. In Otte, D., and Endler, J. (eds.): *Speciation and Its Consequences*. Sinauer Press, Sunderland, MA, pp. 386–400.

Wagner, A. C. C., Weber, H., et al. (1996) Hyperthermia induces heat shock protein expression and protection against cerulein-induced pancreatitis in rats. *Gastroenterology* 111:1333–42.

Wang, J. H., Redmond, H. P., et al. (1995) Induction of heat shock protein 72 prevents neutrophil-mediated human endothelial cell necrosis. *Arch Surg* 130(12):1260–5.

Wang, Y. R., Xiao, X. Z., et al. (1996) Heat shock pretreatment prevents hydrogen peroxide injury of pulmonary endothelial cells and macrophages in culture. *Shock* 6(2):134–41.

Weisberg, L. S., Kurnik, P. B., et al. (1994) Risk of radiocontrast nephropathy in patients with and without diabetes mellitus. *Kidney Int* 45(1):259–65.

Welch, W. J., and Brown, C. R. (1996) Influence of molecular and chemical chaperones on protein folding. *Cell Stress Chap* 1:109–15.

White, C. N., Hightower, L. E., et al. (1994) Variation in heat-shock proteins among species of desert fishes (Poeciliidae, *Poeciliopsis*). *Mol Biol Evol* 11:106–19.

White, F. P. (1980a) Differences in protein synthesized *in vivo* and *in vitro* by cells associated with the cerebral microvasculature: A protein synthesized in response to trauma. *Neuroscience* 5:1793–9.

White, F. P. (1980b) The synthesis and possible transport of specific proteins by cells associated with brain capillaries. *J Neurochem* 35:88–94.

Wong, H., Ryan, M., et al. (1999) Heat shock activates the I-κBα promoter and increases I-κBα mRNA expression. *Cell Stress Chap* 4(1):1–7.

Wong, H. R., Mannix, R. J., et al. (1996) The heat-shock response attenuates lipopolysaccharide-mediated apoptosis in cultured sheep pulmonary artery endothelial cells. *Am J Respir Cell Mol Biol* 15(6):745–51.

Wong, H. R., Ryan, M., Menendez (1997) Heat shock protein induction protects human respiratory epithelium against nitric oxide-mediated cytotoxicity. *Shock* 8:213–218.

Wu, W., and Welsh, M. J. (1996) Expression of the 25-kDa heat-shock protein (HSP27) correlates with resistance to the toxicity of cadmium chloride, mercuric chloride, cis-platinum(II)-diammine dichloride, or sodium arsenite in mouse embryonic stem cells transfected with sense or antisense HSP27 cDNA. *Toxicol Appl Pharmacol* 141(1):330–9.

Xie, Y., Cahill, C. M., et al. (1999) Heat shock proteins and regulation of cytokine expression. *Infect Dis Obstet Gynecol* 7(1–2):26–30.

Yamagami, K., Enders, G., et al. (2003) Heat-shock preconditioning protects fatty livers in genetically obese Zucker rats from microvascular perfusion failure after ischemia reperfusion. *Transpl Int* 16(8):456–63.

Yamashita, M., Hirayoshi, K., et al. (2004) Characterization of multiple members of the HSP70 family in platyfish culture cells: Molecular evolution of stress protein HSP70 in vertebrates. *Gene* 336(2):207–18.

Yamashita, N., Hoshida, S., et al. (1998) Whole-body hyperthermia provides biphasic cardioprotection against ischemia/reperfusion injury in the rat. *Circulation* 98(14): 1414–21.

Yan, L. J., Rajasekaran, N. S., et al. (2005) Mouse HSF1 disruption perturbs redox state and increases mitochondrial oxidative stress in kidney. *Antioxid Redox Signal* 7(3–4): 465–71.

Yu, Q., Kent, C. R., et al. (2001) Retinal uptake of intravitreally injected Hsc/Hsp70 and its effect on susceptibility to light damage. *Mol Vis* 7:48–56.

3
Bacterial Stress Sensors

WOLFGANG SCHUMANN
Institute of Genetics, University of Bayreuth, D-95440 Bayreuth, Germany

1. Introduction

Bacterial cells have limited abilities to modify and choose their dynamic environment. They utilize information processing systems to monitor their surroundings constantly for important changes. Among the appropriate responses to environmental changes are alterations in physiology, development, virulence, and location. In most species, highly sophisticated global regulatory networks modulate the expression of genes. These effects are mediated in large part through the activation or repression of mRNA transcript initiation by DNA-binding proteins, σ-factors, and corresponding signal transduction systems. This adaptive response is based on appropriate genetic programmes allowing them to respond rapidly and effectively to environmental changes that impair growth or even threaten their life. Cellular homeostasis is achieved by a multitude of sensors and transcriptional regulators, which are able to sense and respond to changes in temperature (heat and cold shock), external pH (alkaline and acid shock), reactive oxygen species (hydrogen peroxide and superoxide), osmolarity (hyper- and hypoosmotic shock), and nutrient availability to mention the most important ones. These changes are often called stress factors, and stresses can come at a sudden (catastrophic stress) or grow and grow (pervasive stress). Each stress factor leads to the induction of a subset of genes, the stress genes coding for stress proteins. It should be mentioned that challenge to any stress factor will not only result in induction of genes, but also in repression or even turn off of a subset of genes, but the underlying mechanisms are largely unknown. While some genes are induced by only one single stress factor, others respond to several. The former are termed specific stress genes and the latter general stress genes.

The focus of this chapter will be on the stress sensors and the signal transduction pathways leading either to the transient or constitutive induction of the appropriate stress genes. I will first discuss the stress response pathway including all components followed by a detailed description of the most prominent and well-studied stress response(s). There are numerous review articles dealing with stress responses in bacteria, both in general and dealing with specific stresses. For obtaining an overview, I would like to recommend the book by Storz and Hengge-Aronis (2000) and the review article by Wick and Egli (2004). Review articles dealing with specific stress factors will be cited below.

2. The Stress Response Pathway and Stress Sensors

Adaptation of the different kind of physical and chemical stresses is genetically regulated and involves several distinct steps. The first step leads to the generation of a signal. Here, either the stress factor itself acts as a signal (reactive oxygen species) or leads to the generation of a signal either inside (denatured proteins) or outside of the cytoplasm (perturbations of the cytoplasmic membrane). This signal is registered by a sensor which is either a transcriptional regulator (alternative sigma factor, activator or repressor protein) or the transcript itself leading to a conformational change (ROSE element, *rpoH* transcript). This, in turn, leads either to the transient or constitutive enhanced expression of a subset of stress genes involved in adaptation to the appropriate stress factor. When adaptation has occurred expression of the stress genes will be reduced, and in most cases the expression level will return to a level two- to three-fold higher than the basal level provided that stress factor is still present (e.g., heat shock). While some stress factors can persist for a long time (heat and cold), others can be destroyed by appropriate enzymes (reactive oxygen species).

Three classes of stress sensors have been described so far: DNA, mRNA, and proteins. By which principles do these macromolecules sense stress? In most cases, alternative conformations (secondary structures, folding) of these macromolecules are crucial. In the case of DNA, curvature and DNA-binding proteins play a crucial role where the bent is temperature-dependent. Some mRNA molecules are also able to respond to temperature. Here, translation initiation is impaired at low temperatures where a stem-loop structure sequesters the Shine–Dalgarno sequence partly or fully largely preventing binding of the small ribosomal subunit. If the temperature increases, the secondary structure starts to unfold allowing binding of the 30S subunit. Three examples will be presented, mRNAs containing the ROSE (repression of heat shock gene expression) element, the *rpoH* and *prfA* mRNAs coding for small heat shock proteins (HSPs), the alternative sigma factor σ^{32}, and the transcriptional activator PrfA, respectively (for recent review, see Narberhaus et al., 2002b, 2006).

In most cases, proteins sense stress factors. In the case of heat and cold shock, molecular chaperones, proteases, sensor kinases, and transcriptional regulators are able to sense changes in temperature. Molecular chaperones and proteases both sense denatured proteins which appear after a sudden increase in temperature, while one sensor kinase has been described sensing changes in the cytoplasmic membrane occurring immediately after a temperature downshock (Mansilla and De Mendoza, 2005). In the case of oxidative stress, three different sensor proteins have been described which become activated through the formation of disulfide bonds.

3. Heat Shock and High-Temperature Response Sensors

Two fundamental different responses toward sudden increases in temperature have to be distinguished, the heat shock response and the high-temperature response. The former responds to temperature increments, is transient, and results in the expression of so-called heat shock genes coding for HSPs. On the contrary, the latter is a response towards the absolute temperature, is constitutive and results in the expression of high-temperature genes coding for high-temperature proteins (HTPs). Many pathogenic bacteria to monitor infection of their mammalian host use this response. The heat shock response regulates the transient high-level of expression of the heat shock genes which allow the cells to quickly adapt to the stressful situation (about 10 min in *E. coli* and *B. subtilis*). In contrast, the high-temperature response allows cells to express a subset of genes at a high level as long as they are exposed to the high temperature. It follows that in the former case, the heat shock should generate a signal, which will disappear leading to a shut-off of expression of the heat shock genes, while in the latter case the absolute temperature itself is the signal. DNA, mRNA and proteins, can sense increases in temperature. In the case of the high-temperature response, it results in conformational changes, which persist as long as the bacteria are exposed to that temperature.

3.1. Sensors of the High-Temperature Response

The paradigm for DNA acting as a sensor of high-temperature is the *virF* gene of the human enteropathogen *Shigella* coding for the primary regulator of the invasion function. The first event following the shift from the outside environment to its mammalian host is the expression of the *virF* gene, and the VirF transcriptional activator triggers expression of several operons encoding invasion functions (Durand et al., 2000; Prosseda et al., 2002; Tobe et al., 1993). The activation of *virF* occurs at temperatures above 32°C (Colonna et al., 1995; Prosseda et al., 1998). Below 32°C, the DNA-binding protein H-NS is responsible for repressing *virF* expression by interaction with two sites within the *virF* promoter region where both H-NS complexes are assumed to make contact with each other. This is aided by a bend in the promoter region halfway between the two H-NS sites bringing the two complexes into close proximity. At low temperatures, the *virF* promoter occurs in a closed conformation and involves a curvature strong enough to maintain the two H-NS sites sufficiently close to favour formation of a stable contact between them occluding the access of RNA polymerase. While the temperature raises, the bent weakens, and at 32°C, the curved structure collapses destroying the contact between the two H-NS sites and allowing binding of the RNA polymerase (Prosseda et al., 2004).

Well-studied examples for mRNA molecules acting as heat sensors are the *rpoH* transcript coding for the alternative sigma factor σ^{32}, mRNAs containing the ROSE element, and the *prfA* mRNA. The *rpoH* gene is constitutively expressed, but translation of the *rpoH* transcript occurs in the temperature-dependent manner.

While translation at ≤37°C is low, it starts to increase at higher temperatures. This is due to a secondary structure which sequesters the Shine–Dalgarno sequence, making it almost inaccessible to the ribosomes. High temperature will destabilize the secondary structure, and the ribosomes will bind to initiate translation (Morita et al., 1999). The ROSE element, an ~100-bp RNA sequence located between the transcriptional and translational start sites, has been identified within the 5′ untranslated region (UTR) of *Bradyrhizobium japonicum* and other *Rhizobiae* (Nocker et al., 2001a,b). This *cis*-acting sequence mainly regulates genes that encode small HSPs. As described for *rpoH*, a secondary structure sequesters the SD sequence at temperatures of ≤30°C and largely prevents binding of the small ribosomal subunit at low temperatures. If the temperature increases above 30°C, the secondary structure will be destabilized allowing binding of the ribosomes and translation to start. The *prfA* transcript is present in *Listeria monocytogenes* and codes for a transcriptional activator of virulence genes (Johansson et al., 2002). At temperatures ≤30°C, the 5′ UTR preceding the open reading frame forms a secondary structure, which masks the SD sequence. When *L. monocytogenes* infects its mammalian host, it is exposed to 37°C destabilizing the secondary structure and resulting in translation of the *prfA* coding region. In all three cases described, translation occurs as long as the bacterial cells are exposed to the high temperature.

Examples for proteins acting as sensors of high temperature are the TlpA and the RheA repressors. TlpA is a *Salmonella typhimurium* virulence plasmid-encoded protein acting as a sequence-specific DNA-binding autoregulator (Hurme et al., 1996). This repressor protein contains an N-terminal DNA-binding region and a long coiled-coil domain. The coiled-coil motif enables TlpA to sense temperature shifts directly based on monomer-to-coiled-coil equilibrium (Hurme et al., 1997). At temperatures below 37°C, the TlpA repressor interacts with its operators to prevent transcription of virulence genes and to autoregulate its own expression. When *S. typhimurium* infects a mammalian host, the 37°C environment leads to a conformational change of TlpA causing its dissociation from the DNA. The RheA repressor is encoded by *Streptomyces albus* and regulates expression of HSP18, a small HSP. Dichroism circular spectroscopy revealed a reversible change of the RheA conformation in relation with the temperature (Servant et al., 2000). In both cases, as described before, transcription continues unimpaired as long as cells stay at the high temperature.

3.2. Sensors of the Heat Shock Response

The heat shock response is a ubiquitous phenomenon that enables cells to survive a variety of environmental stresses. The HSPs help to process misfolded, damaged, or aggregated polypeptide chains and support protein maturation and trafficking (Morimoto et al., 1994). HSPs do so by functioning as molecular chaperones and ATP-dependent proteases (Feder and Hofmann, 1999; Morimoto et al., 1994). A sudden increase in temperature results in the production of partially or totally unfolded proteins generally called non-native proteins. Due to hydrophobic amino acid residues normally buried within proteins and now exposed on the outside of non-native polypeptide chains, proteins can form aggregates, and large aggregates can kill the cell. Therefore, cells have to prevent formation of protein aggregates and do this by using two different strategies. HSPs are involved in protein folding and refolding of non-native proteins or their degradation. Many important molecular chaperones belong to the first class of HSPs and can be divided into three classes. **Folder chaperones** bind non-native proteins in a 1:1 stoichiometry can allow their refolding. Well-studied examples are the DnaK-DnaJ-GrpE team, the GroEL-GroES team and the ClpA, ClpC, ClpX and ClpY chaperones (Mogk and Bukau, 2004). **Holder chaperones** form large oligomeric structures, bind non-native proteins and pass them over to folder chaperones for refolding. They act as chaperone buffer when the folder chaperones are overloaded (Narberhaus, 2002a). **Disaggregating chaperones**, together with the DnaK-DnaJ-GrpE team, can act on protein aggregates to disaggregate and finally refold single polypeptide chains. ClpB is here a well-known example (Weibezahn et al., 2005). The second major class of HSPs are ATP-dependent proteases. While some such as Lon recognize any denatured protein, others such as the Clp proteases or the membrane-anchored FtsH protease act on specific target proteins. All ATP-dependent proteases first bind there substrates sometimes using an adaptor protein such as ClpS, then use ATP-hydrolysis to unfold the protein and feed it into the proteolytic chamber of the protease subunit (Dougan et al., 2002).

Both molecular chaperones and ATP-dependent proteases can act as heat shock sensors. The underlying principle is titration by denatured proteins appearing immediately after a temperature upshock. DnaK has been identified in two cases. The first case is σ^{32} of *E. coli,* which has an half-life of less than 1 minute at 30°C. This short half-life is caused by the concerted action of the DnaK team and the FtsH protease. At low temperatures, the DnaK chaperone machine converts σ^{32} into a form recognized by FtsH that will subsequently degrade the destabilized sigma factor (Herman et al., 1995; Tatsuta et al., 1998; Tomoyasu et al., 1995, 1998). After a heat shock, DnaK is titrated by non-native proteins, and σ^{32} will escape binding to DnaK as long as this chaperone is involved in dealing with denatured proteins. Enhanced expression of the σ^{32}-dependent heat shock genes leads to the removal of all non-native proteins within the cytoplasm. The more DnaK chaperone is unoccupied, the more σ^{32} will be converted to a form accessible to degradation by the FtsH protease, leading to a gradual turn-off of the heat shock response. The second example is the HspR repressor of *Streptomyces coelicolor.* The *hspR* gene is the last gene of the tetracistronic *dnaK* operon and binds to three inverted repeats located in front of this operon (Bucca et al., 1995). The repressor protein is not able to bind to these repeats by itself, but needs DnaK as co-repressor, which is titrated by non-native proteins causing relief of repression (Bucca et al., 2000).

GroEL is the second chaperone identified to be involved in regulation of the activity of another repressor called HrcA. This repressor protein regulates negatively expression of the *groESL* operon and sometimes in addition the *dnaK* operon in more than 100 bacterial species (Schumann, 2003). At low temperatures, HrcA binds to its operators to allow low level of transcription. After a heat shock, HrcA dissociates from its operators triggered by a thus far unknown mechanism and rebinds later. It is assumed that both newly synthesized HrcA and HrcA dissociated from its operator are unable to bind and rebind, respectively, to the DNA. They need to interact with GroEL allowing them to fold into their active conformation (Mogk et al., 1997). While under physiological conditions, sufficient amounts of GroEL are present to allow folding of HrcA, this chaperone is titrated immediately after a heat shock by the non-native proteins appearing in the cytoplasm. The more of these denatured proteins are removed from the cytoplasm, the more HrcA molecules can bind to GroEL to become folded into their active conformation resulting in a turn off of the heat shock response. Therefore, GroEL acts indirectly as sensor of heat stress.

Another example deals with the transcriptional repressor protein CtsR studied in *B. subtilis* and present in several Gram-positive species (Schumann et al., 2002). The homodimeric CtsR repressor, which binds to a highly conserved heptanucleotide direct repeat, located upstream of *clpP*, *clpE* and the *clpC* operon forming the CtsR regulon. At 37°C, a basal steady-state level of CtsR is maintained, and after a heat shock, the repressor protein is rapidly degraded by the ClpCP protease (Krüger et al., 2001). All genes of the tetracistronic *clpC* operon consisting of *ctsR*, *mcsA*, *mcsB* and *clpC* are involved in the regulation of the activity of CtsR. While ClpC interacts with ClpP to form an ATP-dependent protease, *mcsB* codes for a tyrosine kinase and McsA stimulates the kinase activity of McsB (Kirstein et al., 2005). In the absence of a heat shock, ClpC interacts with McsB to inhibit its kinase activity. Following a sudden increase in temperature, ClpC is titrated by the non-native proteins, and McsA binds to McsB to stimulate its kinase activity. Subsequently, McsB phosphorylates McsA, CtsR and itself, turning CtsR~P into a substrate for the ClpCP protease (Kirstein et al., 2005). Here, the ClpC chaperone acts as a sensor of denatured proteins.

In the last example to be described here, the DegS protease of *E. coli* can detect partially unfolded outer membrane proteins carrying the specific signature at their C-terminal end (Ades, 2004; Alba and Gross, 2004). Besides σ^{32}, σ^E is another alternative sigma factor involved in the heat shock response when non-native proteins appear within the periplasm such as denatured outer membrane proteins. σ^E is kept inactive by binding to an anti-sigma factor, RseA, which is anchored in the cytoplasmic membrane and consists of three functional domains. While the C-terminal domain is extracytoplasmic, the central transmembrane domain anchors the protein in the membrane, and the N-terminal domain exposed into the cytoplasm sequesters σ^E. Activation of σ^E needs proteolytic cleavage of RseA by three different proteases acting consecutively. If denatured outer membrane proteins carrying the C-terminal YQF-COOH or YYF-COOH motif appear, e.g., due to overexpression, this signature will be recognized by the PDZ domain of the DegS protease. This protease is anchored in the inner membrane and the active site faces the periplasm. In the absence of stress, the PDZ domain inhibits its proteolytic activity. If the PDZ domain moves away from the proteolytic site, DegS will cleave the C-terminal domain of RseA at a specific site. This in turn leads to the activation of the second protease, RseP, which will also cleave at a specific site within the transmembrane domain. This causes release of the N-terminal part of RseA with two alanine residues exposed at its new C-terminal end into the cytoplasm with σ^E still bound to it. The last step involves the complete degradation of RseA. First, the adaptor protein SspB will recognize the truncated RseA and pass it over to the ClpXP protease (Flynn et al., 2004). The ClpX ATPase subunits will unfold the RseA protein and feed it into the proteolytic ClpP chamber. This will finally cause release of σ^E able to interact with the core RNA polymerase, and the genes of the σ^E regulon will transcribed.

4. Cold Shock Sensors

A sudden decrease in temperature induces the cold shock response in bacteria. This physical stress factor influences several vital parameters such as solute diffusion rates, enzyme kinetics, and membrane fluidity and affects conformation, flexibility, and topology of macromolecules such as DNA, RNA, and proteins. Two major problems arise from exposing a bacterial cell to a sudden decrease in temperature to which bacteria have to adapt: (i) the fluidity of the inner membrane is reduced impairing the free movement of integral membrane proteins. (ii) The decrease in temperature induces secondary structures in mRNA molecules thereby impairing translation. Bacteria have evolved strategies to cope with both problems (Eriksson et al., 2002; Gualerzi et al., 2003; Weber and Marahiel, 2003).

The cold shock response of exponentially growing cells can roughly be separated into three different phases. Phase I represents the initial but transient cold shock response (called **acclimation phase** = lag period of cell growth) that immediately follows cold exposure and, depending on the bacterial species, may take up to several hours during which a profound reduction of the growth rate as well as reprogramming of the translation machinery occurs. When the mesophile *E. coli* is transferred from 37°C to a temperature below 20°C, cell growth stops for an acclimation period of one to several hours. During phase I, homeoviscous adaptation occurs and a dramatic reprogramming of gene expression takes place leading to a transient increase in the rate of synthesis of a small set of cold shock proteins (CSPs) to overcome the deleterious effects of the cold shock, whereas that of most of the other genes are repressed. Many CSPs facilitate translation by adapting ribosomes to the lower temperature and preventing the formation of RNA secondary structures. During phase II, the recovery phase, cells start to grow significantly faster. These cells are considered to be cold-adapted and later enter phase III, the stationary phase.

To restore the fluidity of the inner membrane, the amount of double-bonds within the fatty acids of the phospholipids is enhanced. This can occur by one of two strategies: incorporation of unsaturated fatty acids during *de novo* synthesis of phospholipids or, alternatively, introduction of double-bonds into preexisting phospholipids. *E. coli* uses the first strategy. It codes for an enzyme called FabF (gene: *fabF*) which is synthesized constitutively, but which is inactive at temperatures above 30°C and which gains activity after a cold shock. This enzyme accepts palmitoleyl-ACP as substrate, leading to increased amounts of unsaturated *cis*-vaccenic acid in membrane phospholipids (Garwin et al., 1980). Here, the enzyme β-ketoacyl-[acyl-carrier-protein] synthase II encoded by *fabF* acts as cold sensor.

The second strategy is used by *B. subtilis* and cyanobacteria. They code for one (*B. subtilis*) or more enzymes called desaturases which introduce double bonds into preexisting saturated fatty acids. The *B. subtilis* desaturase Des (gene: *des*) (Aguilar et al., 1998) is absent at temperatures above 30°C and transiently induced after a cold shock. Induction of the *des* gene is regulated by a two-component signal transduction system consisting of the sensor kinase DesK (gene: *desK*) and the response regulator DesR (gene: *desR*) (Aguilar et al., 2001). DesK is an integral membrane protein that senses the decrease in fluidity of the cytoplasmic membrane after a sudden temperature downshift. This induces autophosphorylation at its invariant histidine residue and subsequent transfer of the phosphor group to the invariant aspartate residue of the response regulator. The active DesR protein then binds to a DNA site upstream of the *des* promoter to induce transcription of the *des* gene. The desaturase introduces double-bonds into the fatty acids of the phospholipids. When the fluidity of the membrane has been restored the activity of the DesK protein changes from a kinase to a phosphatase and DesR will be dephosphorylated. This leads to turning off of the expression of the *des* gene. If *B. subtilis* cells stay for an extended time at the low temperature, the *des* gene will be expressed at a level sufficient to ensure the fluidity of the membrane.

Besides restoring the membrane fluidity, bacteria have to deal with secondary structures in their mRNAs induced by the cold shock. This is accomplished by the synthesis of CSPs acting as RNA chaperones such as CspA and RNA helicases (CsdA in *E. coli*, CrhC in cyanobacteria) which bind to RNA molecules to either prevent formation of secondary structures or to remove them. What do we know about the regulation of these cold-inducible genes and the sensor(s) involved? From studies carried out in *E. coli* we know that there are two levels of regulation, one post-transcriptional regulation based on the modulation of transcript stability and one based on translation control (Gualerzi et al., 2003).

The *cspA* and some additional cold shock genes contain a long 5′ UTR of more than 100 nucleotides, which renders the transcript extremely unstable ($t_{1/2} \sim 12$ s) (Goldenberg et al., 1996). Thirty minutes after transfer of *E. coli* cells from 37°C to 15°C, its half-life increased to ≥ 70 min. It is assumed that the presence of a stem-loop like the hp2 element might be responsible for the RNaseE-dependent degradation at 37°C (Diwa et al., 2000). Toward the end of the acclimation phase, a new decay pathway will become effective to prevent the dangerous accumulation of unnecessary cold-shock transcripts. Here, alternative secondary structures present within the 5′ UTR of many cold-inducible genes of *E. coli* functions as a cold sensor.

The second layer of regulation is exerted at the level of translation initiation. In *E. coli*, there is an obvious cold-shock translation bias becoming affective during the acclimation phase. This means that the translation apparatus of cold-shocked cells translates at 15°C cold-shock mRNA at a faster rate and to a higher level than non-cold-shock mRNA. Both *cis*- and *trans*-elements contribute to the cold-shock translation bias. As to the nature of the *cis*-elements, they most probably are constituted by elements of secondary/tertiary structure than by sequence motifs. In the case of *cspA* such an element has been located in the upstream half of the 5′ UTR. As to the nature of the *trans*-acting factors, IF3 was identified as being the most important element capable of conferring translational selectivity in the cold in favour of cold-shock mRNAs (Giuliodori et al., 2004). In addition, the RNA chaperone CspA stimulates mRNA translation by favouring unstructured mRNA conformation with the ribosomes (Brandi et al., 1999; Jiang et al., 1997). Here, IF3 can be assumed to act as a cold sensor.

5. Sensors of Water Stress

Osmosensing and osmoregulation are important survival mechanisms for all cells. The cytoplasm of a bacterial cell typically contains 300–400 g/L of macromolecules (DNA, RNA, proteins), which occupy 20–30% of the cellular volume (Zimmerman and Trach, 1991). The primary contributors to cytoplasmic osmolality (defined as the osmotic pressure of a solution at a particular temperature, expressed as moles of solute per kg of solvent), which has been determined to be 100–200 g/L, are low molecular weight solutes also termed osmosolutes, where most of them are ionic. The osmolyte concentration gradient across the bacterial cytoplasmic membrane results in an osmotic pressure difference of about 4 atm for *E. coli* and 20–30 in Gram-positive bacteria. An increase (hyperosmotic upshift) or a decrease (hypoosmotic downshift) in the external osmolality causes water to flow across the membrane resulting in concentrating or diluting the cytoplasm. Alterations in the water content result in changes of the turgor, volume, viscosity, membrane tension, and possibly in membrane potential and ion gradient. Furthermore, changes of the thermodynamic activity of cytoplasmic water results in altered interactions among ions and macromolecules, e.g., hydrogen bonding, hydration, electrostatic interactions, and macromolecular crowding. When the osmolality in the medium drops, water will flow into the cell, which may cause leakage and even lysis of the cell. When the medium osmolality increases, water will exit the cell and plasmolysis will occur. To prevent lysis or plasmolysis, cells respond to osmotic up- and downshifts by rapidly accumulating or releasing low molecular weight osmolytes. Osmoregulatory channels and transporters sense and respond to osmotic stress via different mechanisms (Poolman et al., 2004; Wood, 1999; Wood et al., 2001).

Upon osmotic upshift, water will flow out of the cell, the turgor will decrease and the cells may plasmolyse. To prevent lysis and plasmolysis, bacteria adjust their intracellular osmolyte concentrations by taken up particular zwitterionic organic solvents such as glycine betaine or ectoine as osmoprotectants generally referred to as compatible solutes (Wood, 1999). These compounds can be accumulated to molar levels and stabilize native protein structures. To accumulate compatible solutes upon osmotic upshift, bacteria use ATP-binding cassettes or ion-motive force driven transporters. If the turgor becomes to high, they activate mechanosensitive channels (MSCs) allowing excretion of compatible solutes. These MSCs excrete solutes with little discrimination, except for size. MscL with a huge pore size (30–40 Å) appears to be a final resort to release high pressures.

Osmoregulation has been extensively studied in *E. coli* and in *S. typhimurium* and is coordinated into a particular sequence of reactions resulting in cellular adaptation to the unfavourable osmotic condition. The primary event after an osmotic upshock is the efflux of water by diffusion across the lipid bilayer and through aquaporins mediating rapid and large water fluxes in both directions. This will lead to physical and structural changes in the cell triggering the osmotic response. The first response in *E. coli* is the massive uptake of K^+, which is mediated by the Trk (high K_M for K^+, estimated of 0.3 to 3 mM) and Kdp (low K_M for K^+ with 2 µM) systems and occurs within seconds after the osmotic shift (Bossemeyer et al., 1989; Meury and Kohiyama, 1992; Voelkner et al., 1993). Within a few minutes, glutamate begins to accumulate acting as the counterion of K^+ (Dinnbier et al., 1988; McLaggan et al., 1994). The accumulation occurs via an increase in biosynthesis and a decrease of glutamate utilization. Changes in the turgor contribute to the regulation in activity of both the Trk and the Kdp transporters where the latter is regulated via a two-component signal transduction system. Signal perception induces autophosphorylation of the KdpD sensor kinase which subsequently transfers the phosphor group to the response regulator KdpE . KdpE~P then activates transcription of the *kdp* operon. Besides the turgor, KdpD somehow senses the availability of K^+. The K^+ uptake system is encoded by the *kdpFABC* genes, where KdpA is believed to span the cytoplasmic membrane 10 times and to translocate K^+ in a process involving two sequentially occupied K^+ binding sites (Buurman et al., 1995). The catalytic subunit KdpB accepts phosphate from ATP during the K^+ transport cycle, while the functions of KdpC and KdpF are unknown. Because high cytoplasmic K^+ concentrations have negative effects on protein functions and DNA-protein interactions, the initial increase in cellular K^+ is followed by the accumulation of compatible solutes. This allows the cells to discharge large amounts of the initially acquired K^+ through specific and nonspecific efflux (Kef) systems (Douglas et al., 1991).

Cytoplasmic membrane-based proteins are implicated in the earliest bacterial responses to osmotic up- and downshifts. Their membrane localization focuses the search for the stimulus they detect. The sensory domains can remain associated with the periplasmic and cytoplasmic faces of the membrane. Does the sensor protein detect changes in transmembrane solvent gradients or in absolute osmolality? Are these changes detected directly or are solvent-induced changes in the phospholipids of the membrane monitored? Does the sensor protein detect membrane changes that are mechanical and/or chemical in origin? The homologous proteins TrkG and TrkH, both predicted to span the membrane 12 times, are believed to be the earliest sensors of and the first respondents to osmotic upshifts in *E. coli*.

6. Sensors of Reactive Oxygen Species

Molecular oxygen (O_2) serves as an important role in fundamental cellular functions, including the process of aerobic respiration. Reactive oxygen species (ROS), including superoxide radicals ($O_2\cdot^-$) and hydrogen peroxide (H_2O_2), are continuously generated as by-products of respiration. Reaction of H_2O_2 with free iron generates hydroxyl radicals ($\cdot OH$) in the Fenton reaction. These hydroxyl radicals are highly reactive and react at virtually diffusion-limited rates with most biomolecules (Imlay, 2002, 2003). ROS are also produced by the host immune system as the first line of defence against invading bacteria. Macrophages and neutrophiles release high concentrations of hydrogen peroxide and hypochlorous acid (HOCl) in an attempt to kill the bacteria (Storz and Imlay, 1999). Proteins are the major target of oxidative damage because of their high cellular abundance (Davies, 2005). Iron–sulfur cluster containing proteins are especially vulnerable to ROS, and elevated levels of ROS are furthermore associated with the oxidation of methionines to methionine sulfoxide, which is reversible (Grimaud et al., 2001; Weissbach et al., 2002), and oxidation of thiolate groups in cysteines. Another irreversible modification is amino acid carbonylation (Fredriksson et al., 2005; Requena et al., 2001; Reverter-Branchat et al., 2004).

Living cells have developed numerous mechanisms to withstand oxidative stress. While in *E. coli* many enzymes have been identified that protect against oxidative damage (Kiley and Storz, 2004), no mechanisms to degrade ROS are known. Microorganisms are able to detect ROS within seconds. Upon sensing ROS, the expression of specific genes is induced whose products protect cells from the otherwise fatal consequences of oxidative stress. Induction of these genes involves transcription factors that are specifically activated by oxidative stress. I will emphasize recent insights into the mechanism of peroxide sensing for three families of proteins (OxyR, PerR, and OhrR; (Mongkolsuk and Helmann, 2002) and of nitric oxide and superoxide sensing by SoxR all of them acting as transcription activators.

The *E. coli* OxyR was the first peroxide-sensing transcription factor to be characterized in detail sensing oxidants directly (Zheng and Storz, 2000). While reduced OxyR binds to two adjacent major grooves separated by one helical turn, in its oxidized form, it binds four adjacent major groove regions and activates transcription by recruitment of the RNA polymerase holoenzyme, which subsequently initiates transcription of its 22 target genes (Choi et al., 2001; Tao et al., 1993; Toledano et al., 1994). In total, treatment of *E. coli* cells with H_2O_2 induces about 140 genes most of them in an OxyR-independent way. The initial reaction of OxyR with H_2O_2 is postulated to occur at Cys-199, leading to the formation of an unstable Cys-sulphenic acid (Cys-SOH) intermediate (Zheng et al., 1998). Next, the oxidized Cys-199 reacts with Cys-208 to form an intramolecular disulfide bond. Since these two cysteine residues are separated by ≈ 17 Å in the reduced state, formation of the disulfide bond is accompanied by a refolding of a central domain in the OxyR monomer (Choi et al., 2001). Aerobically growing *E. coli* cells maintain cytoplasmic H_2O_2 concentrations near 20 nM, primarily because of the peroxidase activity of Ahp (Seaver and Imlay, 2001) and needs as little as 100 nM intracellular H_2O_2 to become activated (Gonzalez-Flecha and Demple, 1997). Reduction occurs with a half-time of 5 min through the activity of glutaredoxin 1 and glutathione (Zheng et al., 1998). Though OxyR has been identified primarily as an H_2O_2 sensor, it is also able to respond to disulfide stress resulting as a consequence from defects in the systems that function to maintain an intracellular reducing environment (Aslund et al., 1999). Transcription regulators such as OxyR are widely distributed in most Gram-negative and some Gram-positive bacteria, and the Cys-199 and Cys-208 residues are absolutely conserved, suggesting a common redox-sensing mechanism (Zheng and Storz, 2000).

The PerR family of metallo-regulatory proteins includes several small, dimeric DNA-binding proteins that respond to metal ions such as Fur (iron sensor), Zur (zinc sensor) and PerR (peroxide sensor) (Bsat et al., 1998; Escolar et al., 1999; Gaballa and Helmann, 1998). PerR was identified as the major regulator of the inducible peroxide stress response in *B. subtilis* and acts as the prototype of related peroxide-sensing repressors present in both Gram-positive and Gram-negative bacteria (Herbig and Helmann, 2001). PerR contains two metal binding sites per monomer: one site binds Zn(II) and may play a structural role, while the second site binds a regulatory metal. In vivo, either Fe(II) or Mn(II) function as co-repressors, but both metals influence induction after addition of H_2O_2 quite differently. While in Mn(II)-supplemented medium, induction of the genes of the PerR regulon are tightly repressed, peroxide leads to an induction in medium with added iron. Two different models can account for peroxide sensing by PerR. In model one, H_2O_2 reacts directly with the regulatory metal ion leading to its dissociation and concomitant derepression. In the second model, H_2O_2 reacts with one or more cysteine residues, leading to the formation of a disulfide bond and concomitant loss of metal ion. Since all PerR homologues have a highly conserved CxxC motif near their C-terminus, and these cysteine residues are critical for repressor function, these observations favour the second model.

The OhrR (organic hydroperoxide resistance) family of proteins is found in a wide variety of bacteria. In *B. subtilis*, OhrR binds to a pair of inverted repeat sequences overlapping promoter sites thereby blocking transcription initiation (Fuangthong et al., 2001, 2002). Peroxide sensing requires oxidation of Cys-15 and formation of a Cys-SOH (sulphenic acid) and concomitant dissociation from its operators (Fuangthong et al., 2002).

The second identified transcription factor, SoxR, is responsive to nitric oxide and superoxide (Ding and Demple, 2000). Like OxyR, SoxR is expressed constitutively in an inactive state. SoxR is a homodimer in solution with each subunit containing an [2Fe-2S] cluster that functions as a redox-responsive switch. Oxidation or direct nitrosylation of both clusters activates SoxR to enhance transcription of *soxS* up to 100-fold. The SoxS protein in turn activates expression of 37 genes of the SoxRS regulon (Pomposiello and Demple, 2001; Touati, 2000). It should be mentioned that superoxide modulates the expression of at least 112 *E. coli* genes.

7. Sensors of External pH Stress

Each bacterial species in nature has an optimal pH and drastic changes in extracellular pH values trigger a pH stress response that results in overexpression of certain genes and suppression of others. The promoters of those genes respond to the pH changes (Storz and Hengge-Aronis, 2000), but the mechanisms of pH control of pH-regulated genes are poorly understood. Bacteria display an amazing capacity to survive and grow in life-threatening stresses in a variety of pathogenic and natural situations among them the bacterial response to extremes of pH (Bearson et al., 1997; Cotter and Hill, 2003; Foster, 2004; Hall et al., 1996). As an example, enteric bacteria manage to grow in minimal medium ranging between pH 5 and 8.5 representing more than a 3,000-fold range in H^+-ion concentration. These bacteria can even survive from pH 4 to pH 9 (a 100,000-fold range) for extended periods of time. By which mechanisms, bacterial cells sense pH and subsequently alter transcription? A mechanism commonly employed by bacteria involves two-component signal transduction systems, where several examples have been described for acid stress.

7.1. Acid Stress

Acid stress can be described as the combined biological effect of low pH and weak (organic) acids present in the environment. Weak acids include volatile fatty acids like butyrate and proprionate, which, in their uncharged, protonated form can diffuse across the cell membrane and dissociate inside the cell, lowering the internal pH. The lower the external pH, the more undissociated weak acid will be available to cross the cytoplasmic membrane. Several two-component signal transduction systems have been described where the sensor kinase somehow monitors a drop in the external pH value and subsequently activates a response regulator.

The acid stress response has been studied in great detail using *E. coli* as a genetic model organism (Foster, 2004) and to some extent in several other bacterial species (Cotter and Hill, 2003; Hall et al., 1996). Four different acid-resistance (AR) systems have been described. The first system becomes apparent when *E. coli* cells are grown to stationary phase in LB medium buffered to pH 5.5. When these cells are diluted into minimal media at pH 2.5, they will survive while cells grown at pH 8 will be killed. The alternative sigma factor σ^S and CRP (cAMP receptor protein) are required to develop acid tolerance (Castanie-Cornet et al., 1999). The other three acid-resistance systems are developed by cells growing in glucose-containing LB medium supplemented either with glutamate (second system), arginine (third system) or lysine (fourth system). At least 11 regulatory proteins are known to affect induction of the AR2 system, and the transcriptional activator GadE plays a central role. It binds to a 20-bp *gad* box located upstream of the transcriptional start sites of *gadA* and *gadB* (both coding for glutamate decarboxylase) (Castanie-Cornet and Foster, 2001; Ma et al., 2003). Activation of *gadE* needs the two-component signal transduction system EvgS (sensor kinase) and EvgA (response regulator), where EvgA activates transcription of *gadE* (Ma et al., 2004). AR3, coding for an arginine decarboxylase (*adiA*) and *adiC* (arginine/agmatine antiporter), is activated by CysB under anaerobic conditions at low pH values in complex media, and CysB is assumed to be the sensor of these factors (Shi and Bennett, 1994).

One virulence system that responds to pH is the PhoP-PhoQ regulon of *Salmonella typhimurium* which is capable of growth within macrophage phagolysosomes (Finlay and Falkow, 1989). PhoQ is thought to be the environmental sensor of the phagocytic habitat, and induction of the *pag* gene is maximal when the pH of the phagosome drops below 5.0 (Alpuche Aranda et al., 1992). PhoQ senses both Mg^{2+} and pH and appears to be responsible for the low pH induction of only a few of the many acid shock proteins (Bearson et al., 1998). One of the key steps in the activation of genes involved in the invasion into host epithelial cells in the intestine is the synthesis of the first global activator, VirF in *Shigella* and HilA in *Salmonella*. The expression level of *Shigella sonnei virF* is controlled in response to pH which is accomplished by the two-component regulatory system *cpxR-cpxA* (Nakayama and Watanabe, 1995, 1998). In the case of *Salmonella*, the *hilA* gene is activated by the response regulator SirA, which is phosphorylated by the sensor kinase BarA (Altier et al., 2000; Johnson et al., 1991). The virulence regulator ToxR of *Vibrio cholerae* is a transmembrane sensor protein which uses its N-terminal cytoplasmic domain to bind to promoter regions of virulence factor genes while its C-terminus is exposed in the periplasm where it senses environmental stimuli (Miller et al., 1987). ToxR regulates a series of genes, defined as the ToxR regulon (Parsot and Mekalanos, 1991), and its induction occurs during growth at pH 6.5 but not at pH 8.4 (DiRita, 1992). The last example is the ChvG sensor protein and ChvI response regulator which play a role in coordinating the expression of acid-inducible genes in *A. tumefaciens* (Li et al., 2002).

What do all these sensor kinases sense? Do they all sense the same signal? Is the signal an increased extracellular proton concentration, changes in the outer membrane or alterations within the sensor module leading to a conformational change leading to autophosphorylation?

7.2. Alkaline Stress

Far less is known about base-inducible genes and their regulation. At high pH, the Na^+/H^+ antiporter helps maintain internal pH and protects cells from excess sodium (Gerchman et al., 1993). *E. coli* codes for two antiporters, NhaA and NhaB, which specifically exchange Na^+ or Li^+ for H^+ (Schuldiner and Padan, 1993). The expression of *nhaA* is positively regulated by *nhaR* (Rahav-Manor et al., 1992), and is induced by Na^+, where induction is increased with pH (Padan et al., 1989). NhaA detects external pH directly (Rahav-Manor et al., 1992). Replacement of His-226 with arginine confers a lowered pH range of activity and loss of activity above pH 7.5 (Gerchman et al., 1993).

The F_0F_1 ATPase operon seems to be among the alkali-inducible genes in both Gram-negative and Gram-positive species. In *E. coli*, the ATP synthase imports protons at high pH to counteract the alkaline stress response on cytoplasmic pH on one hand and prefers to minimize proton export on the other hand (Maurer et al., 2005). High pH represses synthesis of flagella, which expend the proton motive force. The underlying regulatory mechanism including the sensor remains elusive. In *Corynebacterium glutamicum*, the F_0F_1 ATPase operon is strongly induced by alkaline pH, too, and expression of the operon under these conditions seems to be controlled by σ^H (Barriuso-Iglesias et al., 2006).

In *B. subtilis*, about 80 genes are base-inducible at least fourfold following a pH upshock from 7.4 to 8.9, where about 60 of these genes belong to the σ^W regulon (Wiegert et al., 2001). The alternative sigma factor σ^W is sequestered by the anti-sigma factor RsiW anchored in the cytoplasmic membrane. RsiW consists of three functional domains where the N-terminal domain is exposed in the cytoplasm and involved in binding σ^W, the central domain anchors the protein in the membrane, and the C-terminal domain is located on the outside of the cell (Schöbel et al., 2004). Release of σ^W after a sudden pH upshock requires three different proteases acting successively. The first protease termed PrsW cleaves the RsiW within its extracytoplasmic domain (Ellermeier and Losick, 2006; Heinrich and Wiegert, 2006). This site-1 cleavage activates the RasP protease which cleaves within the transmembrane domain (Schöbel et al., 2004). This site-2 cleavage reaction causes release of the N-terminal domain into the cytoplasm with σ^W still bound. Next, the truncated RsiW is attacked by the ATP-dependent ClpXP protease which degrades the protein into peptides (site-3 cleavage) finally causing release of σ^W into the cytoplasm where it will interact with the RNA polymerase core enzyme to transcribe the about 60 genes of the σ^W regulon (Zellmeier et al., 2006). Since this system consists of five components namely the sigma, the anti-sigma factor, and three different proteases, we suggest calling it a five-component signal transduction system. Which protein acts as a sensor of alkali stress? Three possibilities are envisaged: the C-terminal part of RsiW, the first protease PrsW or a third so far unknown protein interacting either with RsiW, the site-1 protease, or both.

References

Ades, S. E. (2004) Control of the alternative sigma factor σ^E in *Escherichia coli*. *Curr Opin Microbiol* 7:157–62.

Aguilar, P. S., Cronan, J. E., Jr., and De Mendoza, D. (1998) A *Bacillus subtilis* gene induced by cold shock encodes a membrane phospholipid desaturase. *J Bacteriol* 180:2194–200.

Aguilar, P. S., Hernandez-Arriaga, A. M., Cybulski, L. E., Erazo, A. C., and De Mendoza, D. (2001) Molecular basis of thermosensing: A two-component signal transduction thermometer in *Bacillus subtilis*. *EMBO J* 20:1681–91.

Alba, B. M., and Gross, C. A. (2004) Regulation of the *Escherichia coli* σ^E-dependent envelope stress response. *Mol Microbiol* 52:613–9.

Alpuche Aranda, C. M., Swanson, J. A., Loomis, W. P., and Miller, S. I. (1992) *Salmonella typhimurium* activates virulence gene transcription within acidified macrophage phagosomes. *Proc Natl Acad Sci U S A* 89:10079–83.

Altier, C., Suyemoto, M., Ruiz, A. I., Burnham, K. D., and Maurer, R. (2000) Characterization of two novel regulatory genes affecting *Salmonella* invasion gene expression. *Mol Microbiol* 35:635–46.

Aslund, F., Zheng, M., Beckwith, J., and Storz, G. (1999) Regulation of the OxyR transcription factor by hydrogen peroxide and the cellular thiol-disulfide status. *Proc Natl Acad Sci U S A* 96:6161–5.

Barriuso-Iglesias, M., Barreiro, C., Flechoso, F., and Martin, J. F. (2006) Transcriptional analysis of the F0F1 ATPase operon of *Corynebacterium glutamicum* ATCC 13032 reveals strong induction by alkaline pH. *Microbiology* 152:11–21.

Bearson, S., Bearson, B., and Foster, J. W. (1997) Acid stress responses in enterobacteria. *FEMS Microbiol Lett* 147:173–80.

Bearson, B. L., Wilson, L., and Foster, J. W. (1998) A low PH-inducible, PhoPQ-dependent acid tolerance response protects *Salmonella typhimurium* against inorganic acid stress. *J Bacteriol* 180:2409–17.

Bossemeyer, D., Borchard, A., Dosch, D. C., Helmer, G. C., Epstein, W., Booth, I. R., et al. (1989) K^+-transport protein TrkA of *Escherichia coli* is a peripheral membrane protein that requires other *trk* gene products for attachment to the cytoplasmic membrane. *J Biol Chem* 264:16403–10.

Brandi, A., Spurio, R., Gualerzi, C. O., and Pon, C. L. (1999) Massive presence of the *Escherichia coli* 'major cold-shock protein' CspA under non-stress conditions. *EMBO J* 18:1653–9.

Bsat, N., Herbig, A., Casillas-Martinez, L., Setlow, P., and Helmann, J. D. (1998) *Bacillus subtilis* contains multiple Fur homologues: Identification of the iron uptake (Fur) and peroxide regulon (PerR) repressors. *Mol Microbiol* 29:189–98.

Bucca, G., Ferina, G., Puglia, A. M., and Smith, C. P. (1995) The *dnaK* operon of *Streptomyces coelicolor* encodes a novel heat-shock protein which binds to the promoter region of the operon. *Mol Microbiol* 17:663–74.

Bucca, G., Brassington, A. M. E., Schönfeld, H. -J., and Smith, C. P. (2000) The HspR regulon of *Streptomyces coelicolor*: A role for the DnaK chaperone as a transcriptional co-repressor. *Mol Microbiol* 38:1093–103.

Buurman, E. T., Kim, K. T., and Epstein, W. (1995) Genetic evidence for two sequentially occupied K^+ binding sites in the Kdp transport ATPase. *J Biol Chem* 270:6678–85.

Castanie-Cornet, M. P., and Foster, J. W. (2001) *Escherichia coli* acid resistance: cAMP receptor protein and a 20 bp *cis*-acting sequence control pH and stationary phase expression of the *gadA* and *gadBC* glutamate decarboxylase genes. *Microbiology* 147:709–15.

Castanie-Cornet, M. P., Penfound, T. A., Smith, D., Elliott, J. F., and Foster, J. W. (1999) Control of acid resistance in *Escherichia coli*. *J Bacteriol* 181:3525–35.

Choi, H., Kim, S., Mukhopadhyay, P., Cho, S., Woo, J., Storz, G., et al. (2001) Structural basis of the redox switch in the OxyR transcription factor. *Cell* 105:103–13.

Colonna, B., Casalino, M., Fradiani, P. A., Zagaglia, C., Naitza, S., Leoni, L., et al. (1995) H-NS regulation of virulence gene expression in enteroinvasive *Escherichia coli* harboring the virulence plasmid integrated into the host chromosome. *J Bacteriol* 177:4703–12.

Cotter, P. D., and Hill, C. (2003) Surviving the acid test: Responses of Gram-positive bacteria to low pH. *Microbiol Mol Biol Rev* 67:429–53.

Davies, M. J. (2005) The oxidative environment and protein damage. *Biochim Biophys Acta* 1703:93–109.

Ding, H., and Demple, B. (2000) Direct nitric oxide signal transduction via nitrosylation of iron-sulfur centers in the SoxR transcription activator. *Proc Natl Acad Sci U S A* 97:5146–50.

Dinnbier, U., Limpinsel, E., Schmid, R., and Bakker, E. P. (1988) Transient accumulation of potassium glutamate and its replacement by trehalose during adaptation of growing cells of *Escherichia coli* K-12 to elevated sodium chloride concentrations. *Arch Microbiol* 150:348–57.

DiRita, V. J. (1992) Co-ordinate expression of virulence genes by ToxR in *Vibrio cholerae*. *Mol Microbiol* 6:451–8.

Diwa, A., Bricker, A. L., Jain, C., and Belasco, J. G. (2000) An evolutionarily conserved RNA stem-loop functions as a sensor that directs feedback regulation of RNase E gene expression. *Genes Dev* 14:1249–60.

Dougan, D. A., Reid, B. G., Horwich, A. L., and Bukau, B. (2002) ClpS, a substrate modulator of the ClpAP machine. *Mol Cell* 9:673–83.

Douglas, R. M., Roberts, J. A., Munro, A. W., Ritchie, G. Y., Lamb, A. J., and Booth, I. R. (1991) The distribution of homologues of the *Escherichia coli* KefC K(+)-efflux system in other bacterial species. *J Gen Microbiol* 137:1999–2005.

Durand, J. M., Dagberg, B., Uhlin, B. E., and Bjork, G. R. (2000) Transfer RNA modification, temperature and DNA superhelicity have a common target in the regulatory network of the virulence of *Shigella flexneri*: The expression of the *virF* gene. *Mol Microbiol* 35:924–35.

Ellermeier, C. D., and Losick, R. (2006) Evidence for a novel protease governing regulated intramembrane proteolysis and resistance to antimicrobial peptides in *Bacillus subtilis*. *Genes Dev* 18:2292–2301.

Eriksson, S., Hurme, R., and Rhen, M. (2002) Low-temperature sensors in bacteria. *Philos Trans R Soc Lond B* 357:887–93.

Escolar, L., Perez-Martin, J., and de, L. V. (1999) Opening the iron box: Transcriptional metalloregulation by the Fur protein. *J Bacteriol* 181:6223–9.

Feder, M. E., and Hofmann, G. E. (1999) Heat-shock proteins, molecular chaperones, and the stress response: Evolutionary and ecological physiology. *Annu Rev Physiol* 61:243–82.

Finlay, B. B., and Falkow, S. (1989) *Salmonella* as an intracellular parasite. *Mol Microbiol* 3:1833–41.

Flynn, J. M., Levchenko, I., Sauer, R. T., and Baker, T. A. (2004) Modulating substrate choice: The SspB adaptor delivers a regulator of the extracytoplasmic-stress response to the AAA+ protease ClpXP for degradation. *Genes Dev* 18:2292–301.

Foster, J. W. (2004) *Escherichia coli* acid resistance: Tales of an amateur acidophile. *Nat Rev Microbiol* 2:898–907.

Fredriksson, A., Ballesteros, M., Dukan, S., and Nystrom, T. (2005) Defense against protein carbonylation by DnaK/DnaJ and proteases of the heat shock regulon. *J Bacteriol* 187:4207–13.

Fuangthong, M., Atichartpongkul, S., Mongkolsuk, S., and Helmann, J. D. (2001) OhrR is a repressor of *ohrA*, a key organic hydroperoxide resistance determinant in *Bacillus subtilis*. *J Bacteriol* 183:4134–41.

Fuangthong, M., Herbig, A. F., Bsat, N., and Helmann, J. D. (2002) Regulation of the *Bacillus subtilis fur* and *perR* genes by PerR: Not all members of the PerR regulon are peroxide inducible. *J Bacteriol* 184:3276–86.

Gaballa, A., and Helmann, J. D. (1998) Identification of a zinc-specific metalloregulatory protein, Zur, controlling zinc transport operons in *Bacillus subtilis*. *J Bacteriol* 180:5815–21.

Garwin, J. L., Klages, A. L., and Cronan, J. E., Jr. (1980) β-Ketoacyl-acyl carrier synthase II of *Escherichia coli*. Evidence for function in the thermal regulation of fatty acid synthesis. *J Biol Chem* 255:3263–56.

Gerchman, Y., Olami, Y., Rimon, A., Taglicht, D., Schuldiner, S., and Padan, E. (1993) Histidine-226 is part of the pH sensor of NhaA, a Na^+/H^+ antiporter in *Escherichia coli*. *Proc Natl Acad Sci U S A* 90:1212–6.

Giuliodori, A. M., Brandi, A., Gualerzi, C. O., and Pon, C. L. (2004) Preferential translation of cold-shock mRNAs during cold adaptation. *RNA* 10:265–76.

Goldenberg, D., Azar, I., and Oppenheim, A. B. (1996) Differential mRNA stability of the *cspA* gene in the cold-shock response of *Escherichia coli*. *Mol Microbiol* 19:241–8.

Gonzalez-Flecha, B., and Demple, B. (1997) Homeostatic regulation of intracellular hydrogen peroxide concentration in aerobically growing *Escherichia coli*. *J Bacteriol* 179:382–8.

Grimaud, R., Ezraty, B., Mitchell, J. K., Lafitte, D., Briand, C., Derrick, P. J., et al. (2001) Repair of oxidized proteins. Identification of a new methionine sulfoxide reductase. *J Biol Chem* 276:48915–20.

Gualerzi, C. O., Giuliodori, A. M., and Pon, C. L. (2003) Transcriptional and posttranscriptional control of cold-shock genes. *J Mol Biol* 331:527–39.

Hall, H. K., Karem, K. L., and Foster, J. W. (1996) Molecular responses of microbes to environmental pH stress. *Adv Microb Physiol* 36:229–72.

Heinrich, J., and Wiegert, T. (2006) YpdC determines site-1 degradation in regulated intramembrane proteolysis of the RsiW anti-sigma factor of *Bacillus subtilis*. *Mol Microbiol* 62:566–79.

Herbig, A. F., and Helmann, J. D. (2001) Roles of metal ions and hydrogen peroxide in modulating the interaction of the *Bacillus subtilis* PerR peroxide regulon repressor with operator DNA. *Mol Microbiol* 41:849–59.

Herman, C., Thévenet, D., D' Ari, R., and Bouloc, P. (1995) Degradation of σ^{32}, the heat shock regulator in *Escherichia coli*, is governed by HflB. *Proc Natl Acad Sci U S A* 92:3516–20.

Hurme, R., Berndt, K. D., Namok, E., and Rhen, M. (1996) DNA binding exerted by a bacterial gene regulator with an extensive coiled-coil domain. *J Biol Chem* 271:12626–31.

Hurme, R., Berndt, K. D., Normark, S. J., and Rhen, M. (1997) A proteinaceous gene regulatory thermometer in *Salmonella*. *Cell* 90:55–64.

Imlay, J. A. (2002) How oxygen damages microbes: Oxygen tolerance and obligate anaerobiosis. *Adv Microb Physiol* 46:111–53.

Imlay, J. A. (2003) Pathways of oxidative damage. *Annu Rev Microbiol* 57:395–418.

Jiang, W., Hou, Y., and Inouye, M. (1997) CspA, the major cold-shock protein of *Escherichia coli*, is an RNA chaperone. *J Biol Chem* 272:196–202.

Johnson, K., Charles, I., Dougan, G., Pickard, D., O'Gaora, P., Costa, G., et al. (1991) The role of a stress-response protein in *Salmonella typhimurium* virulence. *Mol Microbiol* 5:401–7.

Johansson, J., Mandin, P., Renzoni, A., Chiaruttini, C., Springer, M., and Cossart, P. (2002) An RNA thermosensor controls expression of virulence genes in *Listeria monocytogenes*. *Cell* 110:551–61.

Kiley, P. J., and Storz, G. (2004) Exploiting thiol modifications. *PLoS Biol* 2:e400.

Kirstein, J., Zühlke, D., Gerth, U., Turgay, K., and Hecker, M. (2005) A tyrosine kinase and its activator control the activity of the CtsR heat shock repressor in *Bacillus subtilis*. *EMBO J* 24:3435–45.

Krüger, E., Zühlke, D., Witt, E., Ludwig, H., and Hecker, M. (2001) Clp-mediated proteolysis in Gram-positive bacteria is autoregulated by the stability of a repressor. *EMBO J* 20:852–63.

Li, L., Jia, Y., Hou, Q., Charles, T. C., Nester, E. W., and Pan, S. Q. (2002) A global pH sensor: *Agrobacterium* sensor protein ChvG regulates acid-inducible genes on its two chromosomes and Ti plasmid. *Proc Natl Acad Sci U S A* 99:12369–74.

Ma, Z., Gong, S., Richard, H., Tucker, D. L., Conway, T., and Foster, J. W. (2003) GadE (YhiE) activates glutamate decarboxylase-dependent acid resistance in *Escherichia coli* K-12. *Mol Microbiol* 49:1309–20.

Ma, Z., Masuda, N., and Foster, J. W. (2004) Characterization of EvgAS-YdeO-GadE branched regulatory circuit governing glutamate-dependent acid resistance in *Escherichia coli*. *J Bacteriol* 186:7378–89.

Mansilla, M. C., and De Mendoza, D. (2005) The *Bacillus subtilis* desaturase: A model to understand phospholipid modification and temperature sensing. *Arch Microbiol* 183:229–35.

Maurer, L. M., Yohannes, E., Bondurant, S. S., Radmacher, M., and Slonczewski, J. L. (2005) pH regulates genes for flagellar motility, catabolism, and oxidative stress in *Escherichia coli* K-12. *J Bacteriol* 187:304–19.

McLaggan, D., Naprstek, J., Buurman, E. T., and Epstein, W. (1994) Interdependence of K+ and glutamate accumulation during osmotic adaptation of *Escherichia coli*. *J Biol Chem* 269:1911–7.

Meury, J., and Kohiyama, M. (1992) Potassium ions and changes in bacterial DNA supercoiling under osmotic stress. *FEMS Microbiol Lett* 99:159–64.

Miller, V. L., Taylor, R. K., and Mekalanos, J. J. (1987) Cholera toxin transcriptional activator ToxR is a transmembrane DNA binding protein. *Cell* 48:271–9.

Mogk, A., and Bukau, B. (2004) Molecular chaperones: Structure of a protein disaggregase. *Curr Biol* 14:R78–80.

Mogk, A., Homuth, G., Scholz, C., Kim, L., Schmid, F. X., and Schumann, W. (1997) The GroE chaperonin machine is a major modulator of the CIRCE heat shock regulon of *Bacillus subtilis*. *EMBO J* 16:4579–90.

Mongkolsuk, S., and Helmann, J. D. (2002) Regulation of inducible peroxide stress responses. *Mol Microbiol* 45:9–15.

Morimoto, R. I., Tissières, A., and Georgopoulos, C. (1994) *The biology of heat shock proteins and molecular chaperones*. Cold Spring Harbor Laboratory Press, Cold Spring Harbor.

Morita, M. T., Tanaka, Y., Kodama, T. S., Kyogoku, Y., Yanagi, H., and Yura, T. (1999) Translational induction of heat shock transcription factor σ^{32}: Evidence for a built-in RNA thermosensor. *Genes Dev* 13:655–65.

Nakayama, S., and Watanabe, H. (1995) Involvement of *cpxA*, a sensor of a two-component regulatory system, in the pH-dependent regulation of expression of *Shigella sonnei virF* gene. *J Bacteriol* 177:5062–9.

Nakayama, S., and Watanabe, H. (1998) Identification of *cpxR* as a positive regulator essential for expression of the *Shigella sonnei virF* gene. *J Bacteriol* 180:3522–8.

Narberhaus, F. (2002a) Alpha-crystallin-type heat shock proteins: Socializing minichaperones in the context of a multichaperone network. *Microbiol Mol Biol Rev* 66:64–93.

Narberhaus, F. (2002b) mRNA-mediated detection of environmental conditions. *Arch Microbiol* 178:404–10.

Narberhaus, F., Waldminghaus, T., and Chowdhury, S. (2006) RNA thermometers. *FEMS Microbiol Rev* 30:3–16.

Nocker, A., Hausherr, T., Balsiger, S., Krstulovic, N. P., Hennecke, H., and Narberhaus, F. (2001a) A mRNA-based thermosensor controls expression of rhizobial heat shock genes. *Nucleic Acids Res* 29:4800–7.

Nocker, A., Krstulovic, N. P., Perret, X., and Narberhaus, F. (2001b) ROSE elements occur in disparate rhizobia and are functionally interchangeable between species. *Arch Microbiol* 176:44–51.

Padan, E., Maisler, N., Taglicht, D., Karpel, R., and Schuldiner, S. (1989) Deletion of *ant* in *Escherichia coli* reveals its function in adaptation to high salinity and an alternative NA^+/H^+ antiporters system(s). *J Biol Chem* 264: 20097–302.

Parsot, C., and Mekalanos, J. J. (1991) Expression of the *Vibrio cholerae* gene encoding aldehyde dehydrogenase is under control of ToxR, the cholera toxin transcriptional activator. *J Bacteriol* 173:2842–51.

Pomposiello, P. J., and Demple, B. (2001) Redox-operated genetic switches: The SoxR and OxyR transcription factors. *Trends Biotechnol* 19:109–114.

Poolman, B., Spitzer, J. J., and Wood, J. M. (2004) Bacterial osmosensing: Roles of membrane structure and electrostatics in lipid–protein and protein–protein interactions. *Biochim Biophys Acta* 1666:88–104.

Prosseda, G., Fradiani, P. A., Di, L. M., Falconi, M., Micheli, G., Casalino, M., et al. (1998) A role for H-NS in the regulation of the *virF* gene of *Shigella* and enteroinvasive *Escherichia coli*. *Res Microbiol* 149:15–25.

Prosseda, G., Falconi, M., Nicoletti, M., Casalino, M., Micheli, G., and Colonna, B. (2002) Histone-like proteins and the *Shigella* invasivity regulon. *Res Microbiol* 153:461–8.

Prosseda, G., Falconi, M., Giangrossi, M., Gualerzi, C. O., Micheli, G., and Colonna, B. (2004) The *virF* promoter in *Shigella*: More than just a curved DNA stretch. *Mol Microbiol* 51:523–37.

Rahav-Manor, O., Carmel, O., Karpel, R., Taglicht, D., Glaser, G., Schuldiner, S., et al. (1992) NhaR, a protein homologous to a family of bacterial regulatory proteins (LysR), regulates *nhaA*, the sodium proton antiporter gene in *Escherichia coli*. *J Biol Chem* 267:10433–8.

Requena, J. R., Chao, C. C., Levine, R. L., and Stadtman, E. R. (2001) Glutamic and aminoadipic semialdehydes are the main carbonyl products of metal-catalyzed oxidation of proteins. *Proc Natl Acad Sci U S A* 98:69–74.

Reverter-Branchat, G., Cabiscol, E., Tamarit, J., and Ros, J. (2004) Oxidative damage to specific proteins in replicative and chronological-aged *Saccharomyces cerevisiae*: Common targets and prevention by calorie restriction. *J Biol Chem* 279:31983–9.

Schöbel, S., Zellmeier, S., Schumann, W., and Wiegert, T. (2004) The *Bacillus subtilis* σ^W anti-sigma factor RsiW is degraded by intramembrane proteolysis through YluC. *Mol Microbiol* 52:1091–105.

Schuldiner, S., and Padan, E. (1993) Molecular analysis of the role of Na+/H+ antiporters in bacterial cell physiology. *Int Rev Cytol* 137C:229–66.

Schumann, W. (2003) The *Bacillus subtilis* heat shock stimulon. *Cell Stress Chap* 8:207–17.

Schumann, W., Hecker, M., and Msadek, T. (2002) Regulation and function of heat-inducible genes in *Bacillus subtilis*. In Sonenshein, A. L., Hoch, J. A., and Losick, R. (eds.): *Bacillus subtilis and Its Closest Relatives: From Genes to Cells*. American Society for Microbiology, Washington, D. C. pp. 359–68.

Seaver, L. C., and Imlay, J. A. (2001) Alkyl hydroperoxide reductase is the primary scavenger of endogenous hydrogen peroxide in *Escherichia coli*. *J Bacteriol* 183:7173–81.

Servant, P., Grandvalet, C., and Mazodier, P. (2000) The RheA repressor is the thermosensor of the HSP18 heat shock response in *Streptomyces albus*. *Proc Natl Acad Sci U S A* 97:3538–43.

Shi, X., and Bennett, G. N. (1994) Effects of *rpoA* and *cysB* mutations on acid induction of biodegradative arginine decarboxylase in *Escherichia coli*. *J Bacteriol* 176:7017–23.

Storz, G., and Imlay, J. A. (1999) Oxidative stress. *Curr Opin Microbiol* 2:188–194.

Storz, G., and Hengge-Aronis, R. (2000) *Bacterial Stress Responses*. American Society for Microbiology, Washington, DC.

Tao, K., Fujita, N., and Ishihama, A. (1993) Involvement of the RNA polymresae α subunit C-terminal region in co-operative interaction and transcriptional activation with OxyR protein. *Mol Microbiol* 7:859–64.

Tatsuta, T., Tomoyasu, T., Bukau, B., Kitagawa, M., Mori, H., Karata, K., et al. (1998) Heat shock regulation in the *ftsH* null mutant of *Escherichia coli*: Dissection of stability and activity control mechanisms of σ^{32} in vivo. *Mol Microbiol* 30:583–94.

Tobe, T., Yoshikawa, M., Mizuno, T., and Sasakawa, C. (1993) Transcriptional control of the invasion regulatory gene *virB* of *Shigella flexneri*: Activation by *virF* and repression by H-NS. *J Bacteriol* 175:6142–9.

Toledano, M. B., Kullik, I., Trinh, F., Baird, P. T., Schneider, T. D., and Storz, G. (1994) Redox-dependent shift of OxyR-DNA contacts along an extended DNA-binding site: A mechanism for differential promoter selection. *Cell* 78:897–909.

Tomoyasu, T., Gamer, J., Bukau, B., Kanemori, M., Mori, H., Rutman, A. J., et al. (1995) *Escherichia coli* FtsH is a membrane-bound, ATP-dependent protease which degrades the heat-shock transcription factor σ^{32}. *EMBO J* 14:2551–60.

Tomoyasu, T., Ogura, T., Tatsuta, T., and Bukau, B. (1998) Levels of DnaK and DnaJ provide tight control of heat shock gene expression and protein repair in *Escherichia coli*. *Mol Microbiol* 30:567–82.

Touati, D. (2000) Sensing and protecting against superoxide stress in *Escherichia coli*–How many ways are there to trigger *soxRS* response? *Redox Rep* 5:287–93.

Voelkner, P., Puppe, W., and Altendorf, K. (1993) Characterization of the KdpD protein, the sensor kinase of the K^+-translocating Kdp system of *Escherichia coli*. *Eur J Biochem* 217:1019–26.

Weber, M. H. W., and Marahiel, M. A. (2003) Bacterial cold shock responses. *Sci Prog* 86:9–75.

Weibezahn, J., Schlieker, C., Tessarz, P., Mogk, A., and Bukau, B. (2005) Novel insights into the mechanism of chaperone-assisted protein disaggregation. *Biol Chem* 386:739–44.

Weissbach, H., Etienne, F., Hoshi, T., Heinemann, S. H., Lowther, W. T., Matthews, B., et al. (2002) Peptide methionine sulfoxide reductase: Structure, mechanism of action, and biological function. *Arch Biochem Biophys* 397:172–8.

Wick, L. M., and Egli, T. (2004) Molecular components of physiological stress responses in *Escherichia coli*. *Adv Biochem Eng Biotechnol* 89:1–45.

Wiegert, T., Homuth, G., Versteeg, S., and Schumann, W. (2001) Alkaline shock induces the *Bacillus subtilis* σ^W regulon. *Mol Microbiol* 41:59–71.

Wood, J. M. (1999) Osmosensing by bacteria: Signals and membrane-based sensors. *Microbiol Mol Biol Rev* 63:230–62.

Wood, J. M., Bremer, E., Csonka, L. N., Kraemer, R., Poolman, B., Van der Heide, T., et al. (2001) Osmosensing and osmoregulatory compatible solute accumulation by bacteria. *Comp Biochem Physiol A Mol Integr Physiol* 130:437–60.

Zellmeier, S., Schumann, W., and Wiegert, T. (2006) Involvement of Clp protease activity in modulating the *Bacillus subtilis* σ^W stress response. *Mol Microbial* 61:1569–82.

Zheng, M., Åslund, F., and Storz, G. (1998) Activation of the OxyR transcription factor by reversible disulfide bond formation. *Science* 279:1718–21.

Zheng, M., and Storz, G. (2000) Redox sensing by prokaryotic transcription factors. *Biochem Pharmacol* 59:1–6.

Zimmerman, S. B., and Trach, S. O. (1991) Estimation of macromolecule concentrations and excluded volume effects for the cytoplasm of *Escherichia coli*. *J Mol Biol* 222:599–620.

4
Unfolded Protein Response: Contributions to Development and Disease

NAN LIAO[1] AND LINDA M. HENDERSHOT[1,2]
[1]Department of Molecular Sciences, University of Tennessee, Health Science Center, Memphis, TN 38163 and
[2]Department of Tumor Cell Biology, St. Jude Children's Research Hospital, Memphis, TN 38105

Abstract: The unfolded protein response (UPR) is a multifaceted signal transduction pathway that is activated in all eukaryotic organisms in response to changes in the environment of the endoplasmic reticulum (ER) that adversely affect protein folding and assembly in the secretory pathway. The response is generally thought to protect cells from the transient alterations that can occur in the ER environment and serves to restore homeostatis in this organelle. Under extreme or prolonged stress, apoptotic pathways can be activated to destroy the cell. Recent studies reveal that in addition to protecting cells from adverse physiological conditions, the UPR plays an essential role in the normal development and functioning of some tissues and can be a major contributor to the pathology of some diseases.

1. Characteristics of the UPR

In 1977, GRP78 and GRP94 were first observed as transformation-related proteins that were later demonstrated to be up-regulated by glucose deprivation and rapid metabolism (Pouyssegur et al., 1977; Shiu et al., 1977). Later, these two proteins were independently identified as ER resident chaperones, which monitor and aid the folding and assembly of secreted and membrane-bound proteins (Hendershot et al., 1988; Lee, 1992).

The concept of an unfolded protein response (UPR) pathway was originally described in early 1990s when mammalian cells were found to respond to severe ER conditions, such as altered pH, low levels of glucose or oxygen, or alterations in the oxidizing state of the ER by up-regulating the resident ER molecular chaperones. All these agents perturb the normal folding and maturation of the secretory pathway proteins that are synthesized in this organelle, and indeed the entire UPR pathway can be activated by simply expressing a mutant protein that cannot fold properly in the ER (Kozutsumi et al., 1988). While this signal transduction pathway is primarily designed to safeguard the ER, its effects are extended to other organelles, and in extreme cases it protects the organism by terminating cells experiencing chronic ER stress (Ma and Hendershot, 2004; Xu et al., 2005).

A number of changes in the normal ER environment can affect nascent protein folding. For instance, lowering the ER pH alters side chain charges on polypeptides; low glucose interferes with both the glycosylation of nascent proteins and inhibits energy production; and decreased oxygen causes the ER environment to become more reducing, which blocks the formation of disulfide bonds. When these post-translational modifications are inhibited, the nascent peptides cannot achieve their correct mature conformation and the incompletely folded proteins accumulate in the ER due to quality control programs that prevent them from further transport along the secretory pathway (Ellgaard and Helenius, 2003). A number of pharmacological agents are routinely used to activate the UPR for experimental purposes (Lee, 1992). For example, thapsigargin, an inhibitor of the ER Ca^{2+} ATPase, depletes ER calcium, and tunicamycin prevents the addition of N-linked glycans to nascent ER proteins. DTT and 2-β mercaptoethanol alter the oxidizing environment of the ER and interfere with formation of disulfide bonds that normally stabilize the folding and assembly of secretory pathway proteins, and 2-deoxyglucose (2-DG) mimics low glucose conditions by competing with glucose and prevents both N-linked glycosylation and energy production, which is required for protein folding (Helenius, 1994; Kornfeld and Kornfeld, 1985). Finally, Brefeldin A, an inhibitor of intracellular protein transport, causes proteins to accumulate in the ER and activate the UPR.

The UPR is thought to function primarily as a cytoprotective response that protects cells from transient but frequent changes in their environment (Fig. 1). The hallmark of the response is the up-regulation of ER chaperones, which bind to unfolded regions on proteins and act to prevent their aggregation and promote proper refolding if stress conditions are alleviated. This aspect of the response is conserved in all organisms (Lee, 1987). A second feature of UPR activation is a transient inhibition of protein synthesis (Brostrom et al., 1996), which serves to limit the load of unfolded proteins. Interestingly, the block in protein translation is not specific to ER proteins and occurs in all metazoans. A third component of the UPR is the expansion of the degradative capacity of the cell, which allows for the disposal of unfolded proteins (Brodsky et al., 1999; Kostova and Wolf, 2003) and is found in all eukaryotic organisms. ER-associated degradation (ERAD) is a highly selective mechanism used for both regulated degradation of certain functional ER-resident proteins and for the removal of aberrant and unassembled proteins from the ER. Another long recognized aspect of the UPR is the arrest of cells in the G1 phase of cell cycle (Lee et al., 1986), which serves to protect the organism by inhibiting the proliferation of cells that are experiencing stress. Finally, if ER stress persists or is particularly severe, apoptotic pathways are initiated to eliminate the stressed cell. The first three functions contribute to preventing the aggregation of unfolded proteins by both limiting the load of unfolded proteins in the ER and increasing the concentration of molecular chaperones to deal with existing proteins. They constitute the cytoprotective functions of the UPR, whereas the arrest of cells in the G1 phase is considered to be both protective to the cell and the organism.

4. Unfolded Protein Response: Contributions to Development and Disease

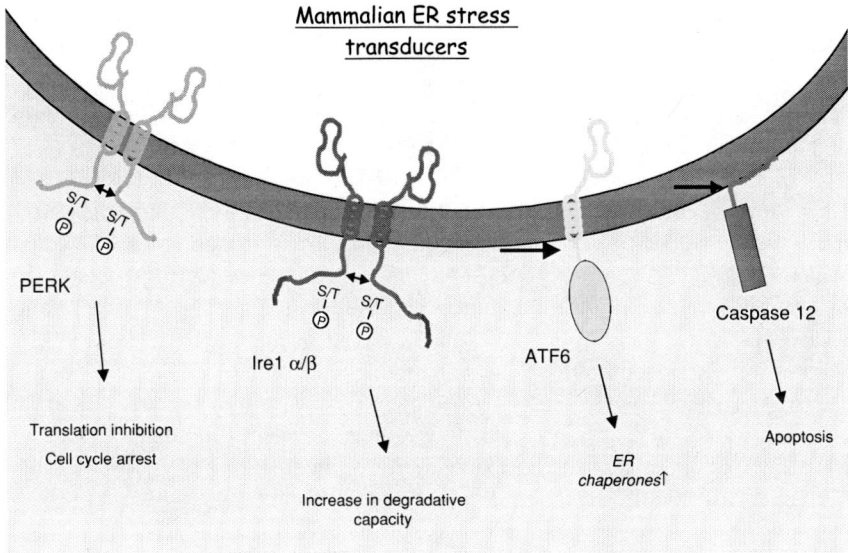

FIGURE 1. ER localized UPR transducers and the downstream responses they control. Four ER localized proteins have been identified that monitor evidence of stress conditions in the ER lumen and upon activation regulate the downstream responses. These include PERK (blue), an eIF-2α kinase that is responsible for the transient inhibition of protein synthesis and cell cycle arrest. There are two Ire1 homologues (green), Ire1α which is ubiquitously expressed and Ire1β which is expressed in the gut epithelium. Activation of Ire1 leads to cleavage of XBP-1, which in turn regulates components of the degradative machinery. The third transducer is ATF6 (yellow), which up-regulates ER chaperones, and the fourth stress sensor is procaspase 12 (pink), which activates apoptotic programs.

Activation of apoptosis signals the point where the cytoprotective aspects of the response shift to cytodestructive ones in order to protect the organism. Thus, the entire response can be considered a protective one that is first active at the cellular level and later shifts to the organismal level. Where this threshold is set is not currently well understood and appears to vary between cell types.

2. Components of the UPR Pathway

The UPR response occurs in all known eukaryotes. It was first delineated in yeast which has a single ER stress transducer that activates the entire pathway. This component is conserved in higher eukaryotes but is greatly expanded.

2.1. The Yeast UPR

To identify the components of the UPR pathway, a 22-bp cis-acting element from the yeast BiP promoter (UPRE) was identified that was sufficient to confer ER stress-induced regulation of a heterologous transcript (Mori et al., 1992). This element was used in a genetic screen that led to the identification of Ire1 (Cox et al., 1993; Mori et al., 1993), which was originally discovered as inositol-requiring gene 1 (Ire1). Ire1 possesses an N-terminal ER targeting sequence, a luminal "stress-sensing" domain, a transmembrane domain, a cytosolically disposed kinase domain, and a C-terminal domain that possesses endonuclease activity (Sidrauski and Walter, 1997). Ire1 is required for yeast survival during ER stress but not during normal cell growth. The target of Ire1's endonuclease activity is the *HAC1* transcript, which encodes a basic leucine zipper transcription factor (Cox and Walter, 1996). *HAC1* mRNA is constitutively expressed but not translated due to the presence of a 252-nucleotide intron that inhibits translation. Activated Ire1p cleaves the *HAC1* transcript at either end of the intron and Rlg1p, a tRNA ligase, re-ligates the transcript in a spliceosome-independent reaction (Sidrauski et al., 1996). The spliced *HAC1* mRNA is efficiently translated to produce Hac1p, a transcription factor that translocates to the nucleus where it binds to UPREs in ER chaperone promoters and up-regulates these and numerous other UPR targets (Travers et al., 2000).

2.2. The UPR Pathway in Mammalian Cells

Mammals have preserved the basic components of the yeast UPR and greatly expanded it (Figs. 1, 2). Two Ire1 homologues exist in mammalian cells: IRE1α and IRE1β, which possess the same features as yeast Ire1p (Tirasophon et al., 1998; Wang et al., 1998b). The only known target of Ire1's endonuclease activity is the X-box protein 1 (XBP-1) transcript (Calfon et al., 2002; Yoshida et al., 2001), which in the unspliced form encodes a protein with a DNA binding domain but no transactivation domain. The excision of 26 bases from the XBP-1 transcript by activated Ire1 changes the reading frame of the C-terminus of XBP-1, so that the spliced form of XBP-1 now encodes both a DNA binding domain and a transactivation domain (Calfon et al., 2002; Yoshida et al., 2001). The remodeled XBP-1(S) regulates components of the ERAD pathway, like EDEM (Yoshida et al., 2003), co-factors of the ER chaperone BiP, including ERdj3 and ERdj4 (Lee et al., 2003), an inhibitor of PERK's kinase activity, p58IPK, which serves in part to restore protein translation (Yan et al., 2002), and components of lipid synthesis that play a role in the expansion of ER membranes during the differentiation of some secretory tissues (Sriburi et al., 2004).

Unlike yeast, a transient inhibition of protein synthesis occurs during the UPR, which is achieved by phosphorylating the eukaryotic translation initiation factor-2α (eIF-2α) through the activation of PKR-like endoplasmic reticulum (ER) kinase (PERK) (Harding et al., 1999; Shi et al., 1998). This leads to the loss of cyclin D1 from cells (Brewer and Diehl, 2000), causing a G1 arrest that prevents the proliferation of cells experiencing ER stress. Paradoxically, the translation inhibition specifically allows expression of activating transcription factor 4 (ATF4) by suppressing the usage of several small open reading frames at the 5' end of the transcript that overlap with the true ATF4 start site (Harding et al., 2000). ATF4 regulates a number of genes including those involved in metabolism and energy

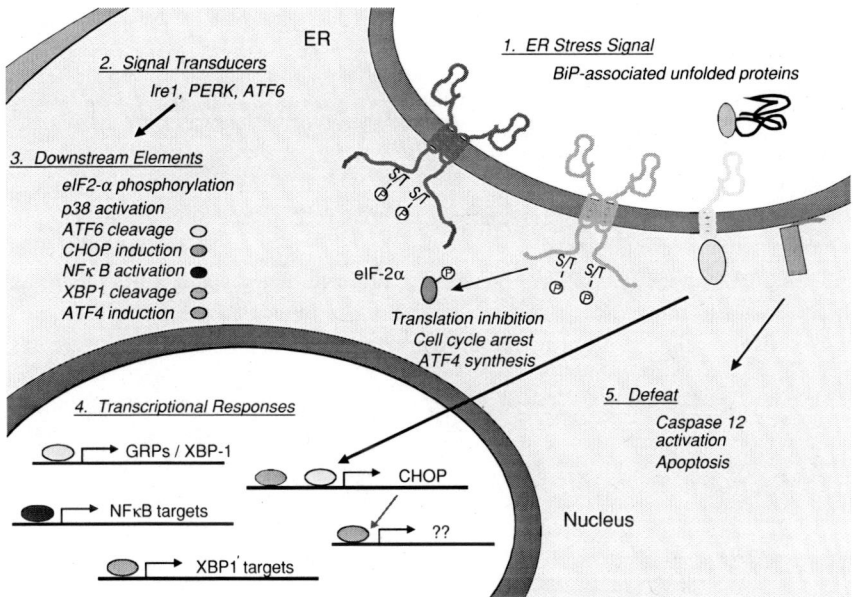

FIGURE 2. Components of the mammalian ER stress response. The accumulation of unfolded proteins in the ER leads to the activation of three ER membrane proteins (Ire1 (green), PERK (blue), and ATF6 (yellow) that act as signal transducers. These three arms of the response are largely cytoprotective and serve to regulate downstream targets which are ultimately responsible for the up-regulation of ER chaperones, inhibition of translation, cell cycle arrest, a number of other transcriptional responses including those downstream of ATF6 (yellow oval), AFT4 (green oval) NFκB (navy oval), and XBP-1 (blue oval). If the stress is not resolved, caspase 12 is activated to initiate apoptosis.

production (Harding et al., 2003; Siu et al., 2002). In addition, ATF4 plays a role in reversing the translation arrest by transactivating GADD34 (Ma and Hendershot, 2003; Novoa et al., 2001), the regulatory subunit of the PP1 phosphatase that dephosphorylates eIF-2α. This serves to reverse the translation arrest and contributes to apoptosis by inducing the C/EBP-homologous protein (CHOP) (McCullough et al., 2001; Zinszner et al., 1998). PERK is also required for NF-κB activation (Jiang et al., 2003), which up-regulates anti-apoptotic proteins like Bcl2 during ER stress. Thus, PERK sits at the crossroad of both survival and death signals.

Using a one-hybrid screen to identify proteins that bound to the ER stress regulated element (ERSE) found in ER chaperone promoters, Mori's group identified ATF6, a transcription factor that is synthesized as an ER-localized transmembrane protein, with a luminal stress sensing domain and a cytosolically oriented transcription factor domain (Haze et al., 1999). In response to ER stress, ATF6 is transported to the Golgi, where it is cleaved by the Golgi-localized S1P and S2P proteases (Ye et al., 2000), releasing the cytosolic transcription-factor domain. ATF6 up-regulates ER chaperones such as BiP, GRP94, and calreticulin, and folding enzymes like peptidyl-prolyl isomerase B (PPIase B) and protein disulfide isomerase (PDI) (Yoshida et al., 1998). ATF6 also serves to up-regulate the XBP-1 transcript (Yoshida et al., 2001).

If ER stress conditions are not resolved, apoptotic pathways are activated. Procaspase-12 is localized to the cytosolic face of the ER membrane and is activated in response to ER stress (Nakagawa et al., 2000) by IRE1-dependent mechanisms involving calpain activation (Nakagawa and Yuan, 2000). Although the human caspase-12 gene contains several missense mutations and clearly cannot have a role in UPR-induced apoptosis, recent data indicate that caspase-4 in humans is homologous to murine caspase-12 and is activated in an ER-stress-specific manner, indicating that it might be the human caspase-12 orthologue (Hitomi et al., 2004).

3. Mechanisms of UPR Activation

3.1. Conventional Means of UPR Activation

The pharmacological agents that are used to induce the UPR share the ability to dramatically affect protein folding in the ER (Lee, 1992). Indeed, an early study demonstrated that over-expression of an unfolded secretory pathway protein was sufficient to activate the UPR in mammalian cells (Kozutsumi et al., 1988). Somewhat surprisingly, further studies revealed that not all unfolded proteins were able to initiate the response. The PiZ mutant of α1-antitrypsin, which binds to calnexin and is retained in the ER, did not activate the response (Graham et al., 1990). On the other hand, unfolded hemagglutinin from influenza virus (strain A/Japan/305/57, H2N2) (Kozutsumi et al., 1988), hemagglutinin-neuraminidase (HN) from simian virus 5 (Watowich et al., 1991), and unassembled immunoglobulin μ-heavy chains (Lenny and Green, 1991) all bind to BiP and activate the UPR. A series of deletion mutants were made in the extracellular domain of HN, and in all cases, loss of BiP binding correlated with an inability of the unfolded HN protein to activate the UPR (Watowich et al., 1991). Together these studies suggested that since only BiP binding proteins appeared to activate the response, some aspect of BiP might be monitored to detect adverse changes in the folding environment of the ER. This possibility was supported by studies in which overexpression of BiP inhibited UPR activation (Dorner et al., 1992) in a Chinese hamster ovary cell line, whereas overexpression of ERp72, PDI, or calreticulin did not (Dorner et al., 1990; Llewellyn and Roderick, 1998).

Once the proximal UPR-transducers were identified, a number of studies revealed that BiP directly regulated the UPR by controlling the activation status of all three transducers. BiP binds to both Ire1 and PERK under normal physiological conditions, which keeps them in a monomeric, inactive state. Treatment of cells with either thapsigargin or DTT induces a rapid loss of BiP from the luminal domains of both proteins; leading to their oligomerization and activation (Bertolotti et al., 2000). Similar results were obtained with yeast Ire1p (Kimata et al., 2003;

Okamura et al., 2000). The latter study suggested that BiP bound to the signal transducer as though it was an unfolded protein, an observation that is supported by the finding that BiP can be released from the mammalian kinases in vitro with ATP (Bertolotti et al., 2000). Most recently, efforts to map the BiP binding site on yeast Ire1p revealed that deletion of the BiP-binding site did not lead to a constitutively active kinase, but did increase its sensitivity to some stressors (Kimata et al., 2004). This suggests that BiP is not the only determinant controlling Ire1 activity, but that it may act as an adjustor for sensitivity to various stress signals. ATF6 also associates with BiP in the absence of ER stress, which in this case serves to retain ATF6 in the ER. Activation of the stress response leads to BiP release and transport of ATF6 to the Golgi for proteolytic liberation (Shen et al., 2002). Although BiP is rapidly released from ATF6 in response to ER stress, it is not constantly cycling on and off this transducer under normal physiological conditions (Shen et al., 2005). Thus, it is not entirely clear at present how unfolded proteins are able to "draw" BiP off the various UPR transducers. It is possible that as stress conditions are alleviated and the pool of free BiP increases, it also plays a role in shutting down the response. Although this possibility has not been directly tested, a recent study found that BiP is not readily translated early in the stress response even though BiP transcripts begin to increase at this time (Ma and Hendershot, 2003), suggesting that regulation of BiP levels may be important in controlling UPR activation throughout the response.

3.2. Alternative Mechanisms for Activating the UPR

While it is clear that the accumulation of unfolded proteins signals UPR activation in eukaryotic cells, two recent studies demonstrate that either there are alternative ways to activate the UPR or that other changes can unexpectedly affect protein folding in the ER. Multiple studies have shown that the macrophages populating atherosclerotic lesions accumulate cholesterol, which ultimately induces apoptosis in these cells and further exacerbates the progression of the disease. It had been postulated that changes in plasma membrane cholesterol triggered apoptosis in these cells. However, cholesterol also traffics to multiple internal membranes, including the ER, where it was shown to deplete ER calcium, thereby activating the UPR and inducing apoptosis (Feng et al., 2003). Separate studies to understand the mechanism of neurodegeneration in a lysosomal storage disease found that the accumulation of GM1-ganglioside in the ER led to UPR-mediated apoptosis in the brain and spinal cord neurons (Tessitore et al., 2004). While the accumulation of a glycolipid would not be expected to bind to BiP or to directly affect protein folding in this organelle, similar to the studies on ER loading of cholesterol, excess GM1-ganglioside-induced calcium loss from the ER and altered normal protein folding. Thus, the aberrant localization or accumulation of non-protein molecules in the ER can lead to UPR activation by altering ER calcium and secondarily affecting protein folding. Finally, studies have shown that high levels of circulating free fatty acids are toxic to pancreatic islet cells and play a role in the development of type 2 diabetes (McGarry and Dobbins, 1999). Recently, the β-cells were demonstrated to succumb to apoptosis via UPR activation (Kharroubi et al., 2004), although it is presently unclear as to whether this is mediated by calcium release from the ER or involves a distinct mechanism.

3.3. Activating the UPR During Plasma Cell Differentiation

Plasma cells are highly specialized, terminally differentiated secretory cells that produce tremendous quantities of a single product, the antibody molecule. In differentiating from a quiescent B cell, the plasma cell must undergo a dramatic architectural metamorphosis, which includes augmenting the secretory organelles and the proteins that populate them, up-regulating their energy and translation potential, and increasing all aspects of the quality control system. This transformation is accomplished in part by activating the UPR. Although the expression of Ig heavy chains is required for optimal production of XBP-1(S), μH-chain-deficient B cells exhibit low, but readily detectable, amounts of XBP-1(S) when stimulated to undergo differentiation with LPS (Iwakoshi et al., 2003). Moreover, the activation of Ire1 and induction of XBP-1(S) expression precedes the massive increase in Ig production in several B cell lymphoma models and in normal splenic B cells (Gass et al., 2002; Iwakoshi et al., 2003; van Anken et al., 2003). Thus, the nature of the signals that can elicit the UPR in physiologic settings is not yet fully understood.

4. Balance Between Cytoprotective and Proapoptotic Aspects of the Response

In mammalian cells, the UPR has a wide range of effects on regulating cell survival and apoptosis (Fig. 3). In keeping with its function as a cytoprotective response, a number of UPR targets are anti-apoptotic factors. The translation arrest that occurs downstream of PERK leads to the loss of IκB, thereby activating NFκB (Jiang et al., 2003), which in turn up-regulates anti-apoptotic proteins like Bcl2. Other studies suggest that a pool of BiP, which is up-regulated by the ATF6 branch of the UPR, can relocalize to the cytosol where it binds caspase-7 and caspase-12 and prevents their activation (Rao et al., 2002; Reddy et al., 2003). Glycogen synthase kinase-3β (GSK3β) is activated by NFκB and can contribute to cell survival by phosphorylating p53 and accelerating its degradation (Qu et al., 2004).

On the other hand, several proapoptotic genes are also activated by the UPR pathway. CHOP, which is downstream of the PERK and ATF6 pathways, is a proapoptotic gene that down-regulates anti-apoptotic proteins like Bcl2 and increases free oxygen species, causing mitochondrial membrane damage and cytochrome c release, which induces apoptosis (McCullough et al., 2001). The increases in cytosolic calcium that occur during UPR activation lead to up-regulation of the proapoptotic protein BAD, leading to increases in cytochrome c release and activation of APAF1 (Wang et al., 1999). The higher cytosolic calcium levels lead to calpain activation, which can induce procaspase-12 cleavage (Nakagawa and Yuan, 2000) and activation of the caspase-9 cascade (Morishima et al., 2002).

4. Unfolded Protein Response: Contributions to Development and Disease

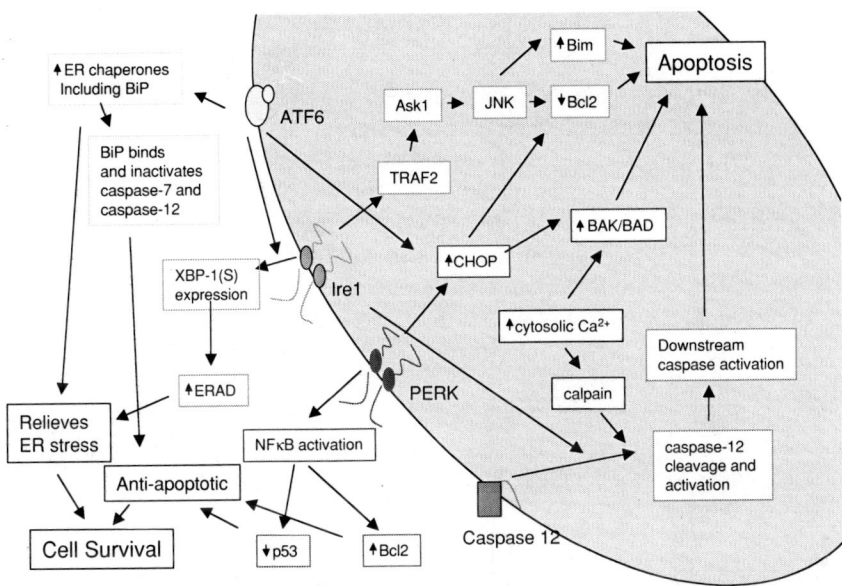

FIGURE 3. The UPR includes both antiapoptotic and proapoptotic elements. The left part of the figure (white) shows antiapoptotic components of the UPR which serve to protect cells undergoing ER stress from apoptosis and contribute to cell survival. They include BiP which is up-regulated by ATF6 (yellow), XBP-1(S), which is regulated by both ATF6 (yellow) and Ire1 (green), and NFκB, which is activated by loss of IκB via PERK kinase (blue) activation. Together they contribute to cell survival either by directly or indirectly inhibiting the apoptotic pathways, or by increasing the ERAD machinery for disposal of unfolded or misfolded proteins to relieve ER stress. The right part of the figure (grey) shows proapoptotic components of UPR that trigger apoptotic programs. They include TRAF2/Ask1, which is activated by Ire1 (green) and serves to activate JNK. In addition, CHOP, which is up-regulated by PERK (blue) and ATF6 (yellow), caspase-12 (red), which is cleaved by Ire1 and activates downstream caspases, and finally increases in cytosolic calcium during UPR activation all contribute to the destruction of chronically stressed cells.

Finally, during ER stress Ire1 can recruit TRAF2 and induce procaspase-12 clustering and activation (Yoneda et al., 2001). In addition, Ask1 can be recruited to Ire1/TRAF2 complex and activated, leading to JNK activation. JNK in turn induces the proapoptotic protein Bim (Lei and Davis, 2003; Putcha et al., 2003) and inhibits the anti-apoptotic protein Bcl2 (Yamamoto et al., 1999). In cell culture experiments, fibroblasts and some tumor lines survive for long periods of time before they stop fighting and induce apoptosis in response to ER stress, while others, like splenic B cells and some neuronal cells, are very sensitive to UPR activation. Thus, the balance between cell survival and apoptosis may vary in a cell type dependent fashion. More studies are needed to understand this very important aspect of the UPR.

5. Role in Normal Development and Differentiation

While the UPR has largely been thought to serve as a means to protect cells from transient adverse changes in their environment or to destroy cells that are chronically or irreversibly stressed, several recent studies demonstrate that the UPR is also activated during normal development and can play an essential role in maintaining the homeostasis of some organs. Two transgenic mice have been created that express a UPR-inducible reporter gene. The first of these rely on excision of the 26 bases of the XBP-1 stem-loop structure to restore the correct reading frame to GFP (Iwawaki et al., 2004). In this mouse, there was no detectable fluorescence in any organ during embryonic development or early postnatal stages, whereas the day 14 pancreas strongly expressed GFP demonstrating UPR activation in this tissue. In addition, intense expression of GFP was detected in the skeletal muscles of older mice but not in young mice. The second transgenic mouse expresses a LacZ gene that is regulated by the rat GRP78/BiP promoter (Mao et al., 2004). Examination of the developing embryo revealed easily detectable β-gal expression in the heart and somites with lower levels observed in neural tubes. This is consistent with activation of at least the ATF-6 arm during embryogenesis. It is not clear if the lack of detectable GFP in these organs during embryogenesis was due to decreased sensitivity of this reporter gene or if there are differences in activation of the Ire1 pathway versus the ATF6 pathway in developing mice. Finally, the creation of mice that are nullizygous for various components of the UPR demonstrates an essential role for UPR activation in the normal functioning and differentiation of various tissues as described below.

5.1. Contribution of the UPR to Pancreatic Homeostasis

PERK was originally identified in rat pancreatic islet cells as a novel eIF-2α kinase (PEK) (Shi et al., 1998). Although ubiquitously expressed, northern blot analyses revealed that its expression was highest in pancreatic cells. Moreover, pancreatic tissue exhibits high constitutive activity for both PERK/PEK and Ire1 (Harding et al., 2001; Iwawaki et al., 2004). Several studies demonstrate that this activation plays an essential role in regulating protein synthesis in β islet cells in response to the sporadic demands for high levels of insulin synthesis. A transgenic mouse was created in which the normal eIF-2α gene was replaced with an eIF-2α mutant that cannot be phosphorylated by PERK and other eIF-2α kinases (Scheuner et al., 2001). Mice homozygous for the targeted knock-in of this point mutation of $eIF2\alpha^{S51\rightarrow A}$ are severely deficient in glucose metabolism. They die soon after birth due to an inability to regulate protein synthesis levels in response to alterations in blood glucose levels, which leads to runaway protein synthesis, defective gluconeogenesis, and increased apoptosis in pancreatic β cells (Scheuner et al., 2001). The production of a PERK null mouse further demonstrated the role of the UPR in pancreatic homeostasis (Harding et al., 2001). Although both the exocrine and endocrine pancreas develop normally in these mice in utero, postnatally the exocrine cells begin to show evidence of an extended ER, abnormal activation of Ire1, and increased cell death, leading to progressive diabetes (Harding et al., 2001). The critical role of this pathway in maintaining pancreatic function was further revealed when the heterozygous eIF-2α^{S51A} mice were fed a high fat diet (Scheuner et al., 2005). These mice became obese and showed profound glucose intolerance due to reduced insulin secretion. This was accompanied by abnormal distension of the ER lumen, defective trafficking of proinsulin, and a reduced number of insulin granules in β cells. Finally, studies on the p58IPK null mice revealed that runaway PERK activation lead to gradual hyperglycemia due to increased apoptosis of β islet cells (Ladiges et al., 2005), suggesting that this component of the UPR also plays a role in regulating normal pancreatic function. Together these studies suggest that when increased demands are put on the pancreas to produce high levels of insulin to control blood glucose levels, it is essential to control protein synthesis via UPR activation in order protect the cell. This concept is supported by the identification of human mutations in PERK that are linked to familial diabetes (Biason-Lauber et al., 2002; Delepine et al., 2000; Harding and Ron, 2002).

5.2. A Role for the UPR in Plasma Cell Differentiation

The differentiation of resting B cells to plasma cells that secrete large quantities of antibody molecules is accompanied by a dramatic architectural remodeling of the cell. The ER undergoes an expansion into an elaborate network that extends throughout the cytoplasm (Shohat et al., 1973; Wiest et al., 1990), as well as enlargement of the Golgi complex (Wiest et al., 1990), and increases in mitochondria to supply the increased energy demands. An essential connection between the UPR and plasma cell development was revealed when XBP-1 null ES cells were used to reconstitute recombination activating gene-2 (Rag-2)-deficient mice. The chimeric animals produced normal numbers of mature B cells in all compartments, but they were unable to differentiate into plasma cells (Reimold et al., 2001). Enforced expression of XBP-1(S) in these animals was sufficient to restore differentiation potential to their B cells (Iwakoshi et al., 2003), and in a B cell line it was shown to increase the ER, Golgi, mitochondria, and lysosomes (Shaffer et al., 2004). A separate study found that enforced expression of XBP-1(S) in fibroblasts induced PtdCho production, elevated the overall mass of membrane phospholipids, and expanded the rough ER (Sriburi et al., 2004). Organelle expansion in both types of cells overexpressing XBP-1(S) was accompanied by elevated expression of many genes encoding secretory pathway components. These include proteins that target and translocate nascent polypeptides into the ER, chaperones and their co-factors that promote protein folding and assembly, oxidoreductases that regulate oxidative protein folding, glycosylation enzymes, regulators of vesicular trafficking, and proteins implicated in ERAD (Shaffer et al., 2004).

In addition to activation of the Ire1-XBP-1 branch of the UPR, ATF6 was shown to undergo proteolytic cleavage during LPS-induced differentiation of the CH12

B cell lymphoma (Gass et al., 2002). While a role for ATF6 in developing plasma cells has not been established, it is reasonable to suggest that this factor contributes to the induction of *XBP-1* mRNA (Calfon et al., 2002;Yoshida et al., 2001), as well as to genes encoding ER chaperones and folding enzymes (Yoshida et al., 2000). PERK, on the other hand, does not appear to be activated in the normal course of terminal B cell differentiation, although it can be activated in response to ER stress-inducing agents (Gass et al., 2002). This implies that the physiological UPR may be uniquely customized according to the specific needs of distinct cell types.

5.3. Contribution of the UPR to the Normal Physiology of Other Tissues

A growing body of data suggests that UPR activation may play a role in other tissues. Apparently PERK plays a critical role in the normal physiology of osteoblasts, which secrete the Type I collagen that constitutes the matrix of compact bone (Biason-Lauber et al., 2002; Zhang et al., 2002). In addition to diabetes, PERK null mice exhibit skeletal dysplasias at birth and show evidence of postnatal growth retardation (Zhang et al., 2002). The skeletal defects include deficient mineralization, osteoporosis, and abnormal compact bone development. The rough ER of the major secretory cells that comprise the skeletal system show evidence of distortion, which is likely to contribute to the skeletal dysplasia in these mice (Zhang et al., 2002). The PERK target CHOP, was originally identified due to its ability to heterodimerize with classic C/EBPs and negatively regulate their activity (Ron and Habener, 1992) as well as to transactivate novel genes (Ubeda et al., 1996). As such, CHOP interferes with adipocyte differentiation (Batchvarova et al., 1995) and accelerates osteoblastogenesis in vitro (Pereira et al., 2004). When CHOP null mice were examined, they were found to have a decreased rate of bone formation with normal numbers of osteoblasts, indicating a defect in osteoblastic function (Pereira et al., 2005). Thus, unlike other systems where CHOP induces apoptosis, in osteoblasts it is essential to their normal function.

A slightly less clear-cut link exists between the UPR and the development of the liver, another highly secretory tissue. Targeted disruption of the XBP-1 gene results in embryonic lethality due in part to failure of the hepatocytes to differentiate (Reimold et al., 2000). This phenotype was observed before investigators realized that the XBP-1 transcript encoded both the unspliced and spliced forms of the XBP-1 protein. While it is very likely that the levels of protein synthesis that occur in hepatocytes are high enough to activate the UPR, this possibility remains to be formally tested. Very recently the unspliced form of XBP-1, which encodes a protein that possesses a DNA binding domain but lacks the transactivation domain, was shown to act as a dominant negative protein and repress the transcription of some genes (Yoshida et al., 2006). Thus, it is currently unclear which XBP-1 function is responsible for the block in liver development.

6. UPR and Disease

Although the UPR can be considered a cytoprotective response that evolved to protect cells from the transient and repeated alterations that can occur in their environment, it is increasingly clear that UPR activation is associated with a number of disease states. Interestingly diseases can arise from both cytoprotective and cytodestructive components of the UPR, and in some pathological states UPR activation initially protects the tissue and later contributes to the disease state.

6.1. Cytoprotective Responses that Can Contribute to Disease State

6.1.1. Viral Infections

A common feature of many viruses is to corrupt the cellular machinery in order to produce large numbers of viral particles. In the case of enveloped viruses, this requires the synthesis of membrane proteins that are synthesized in the ER. Data exist for a number of viruses to show evidence of UPR activation, and as would be expected from their incredible diversity, they use the UPR in very different ways. In some cases activation of the UPR assists in viral production and therefore contributes to the spread of virus. In other cases, activation of the UPR by viral infection induces apoptotic pathways that contribute to disease pathology. Finally, in other cases the virus uses a combination of UPR responses to increase its production and produce cytotoxic effects.

UPR activation has been shown to contribute to viral production by a variety of means. The Ire1 activation that occurs during infection of cells with Hepatitis B virus leads to production of XBP-1(S), which in turns activates the HBV S promoter (Huang et al., 2005). On the other hand infection of hepatocytes with hepatitis C virus blocks Ire1 activation, thereby suppressing ERAD components but maintaining the induction of ER chaperones, leading to increased production of viral envelope proteins (Tardif et al., 2004). Studies on human cytomegalovirus (Isler et al., 2005), Herpes simplex virus (Cheng et al., 2005), and African swine fever virus (Netherton et al., 2004) suggest they are also able to modify the UPR in order to maintain the protective components of the UPR without decreasing protein synthesis or activating apoptotic components respectively.

6.1.2. Cancer

6.1.2.1. Tumor Establishment and Survival

Once genotoxic alterations occur that are essential to the development of a cancer cell, the tumor encounters an inadequate environment that can become growth limiting. These conditions lead to the activation of cytoprotective responses including the hypoxia–induced response (Graeber et al., 1996; Hockel and Vaupel, 2001a,b) and the UPR (Feldman et al., 2005; Ma and Hendershot, 2004). Primary hepatocellular carcinomas show evidence of XBP-1 and ATF6 activation (Shuda et al., 2003), and downstream targets like CHOP, BiP, GRP94, and GRP170 (also known as ORP150) have been reported to be up-regulated in breast tumors (Fernandez et al., 2000), hepatocellular carcinomas (Shuda et al., 2003), gastric tumors (Song et al., 2001), and esophageal adenocarcinomas (Chen et al., 2002). One study reported that BiP was overexpressed more frequently in the higher-grade, estrogen-receptor-negative tumors than in lower-grade, estrogen-receptor-positive tumors (Fernandez et al., 2000), suggesting that UPR activation is correlated with a clinically more aggressive phenotype. In support of this, a fibrosarcoma cell line that was engineered to have lower levels of BiP formed tumors initially but they were rapidly resolved (Jamora et al., 1996). In two more recent studies, researchers found that transformed PERK null MEFs produced much smaller tumors that exhibited higher levels of apoptosis in hypoxic areas than wild-type MEFs (Bi et al., 2005), and that XBP-1 null MEFs and XBP-1–knockdown cells did not form tumors in mice, even though their growth rate and secretion of VEGF were similar to wild-type cells in culture (Romero-Ramirez et al., 2004). Since both PERK and XBP-1 activation are specific to the UPR pathway, this evidence suggests that not only is the UPR activated but that it can play an essential role in tumorigenesis.

Many studies have focused on the signal transduction pathway activated in response to hypoxic conditions and demonstrated that hypoxia-inducible factor1α (Hif1α) plays a critical role in tumor growth and angiogenesis (Pugh and Ratcliffe, 2003; Semenza et al., 2000). Several recent studies provide evidence to suggest that the UPR and Hif1α pathways are related to each other and may interact to regulate downstream targets. For instance, cells cultured under hypoxic conditions activated PERK (Koumenis et al., 2002), leading to eIF2α phosphorylation and expression of both ATF4 and GADD34 (Blais et al., 2004), suggesting that this branch of UPR is fully activated by hypoxia. Hypoxia can also activate NFκB through a phosphorylation-mediated degradation of IκB (Koong et al., 1994), while ER stress activates NFκB through a PERK dependent loss of IκB (Jiang et al., 2003). Since PERK can be activated by both pathways, it represents a possible point of synergy in the activation of this important anti-apoptotic protein. CHOP, a proapoptotic factor, is also up-regulated by PERK. It is presently unclear if these represent opposing effects on cell survival or if the dismantling of apoptotic machinery that is a common feature to the transformation process interferes with the CHOP effect in tumor cells. Indeed, a number of the proapoptotic components that are activated during ER stress converge on the caspase 3/9 pathway, which is often mutated or disabled (Soung et al., 2004) in tumor cells, and the down-regulation of Bcl2 by CHOP maybe countered by the NFκB's activation of antiapoptic factors like c-IAP1 and c-IAP2 (Wang et al., 1998a) or its suppression of targets like PTEN (Vasudevan et al., 2004).

There are also some targets that are regulated differently by the two pathways. For example, the proapoptotic tumor suppressor p53 is stabilized in response to hypoxia through a reduction in the level of MDM2, which targets p53 for proteasomal degradation (Alarcon et al., 1999). However, during ER stress, glycogen synthase kinase-3β (GSK3β) is activated and phosphorylates p53, which relocalizes p 53 to the cytoplasm and accelerates its degradation (Qu et al., 2004). It is not currently clear which of these two pathways is dominant in cancer cells that retain p53 function and have activated both pathways.

6.1.2.2. Effect of UPR Activation on Angiogenesis

Tumors respond to their inadequate vascularization by secreting the proangiogenic factor, VEGF (Ferrara and Davis-Smyth, 1997). Recent studies demonstrate that in addition to its well characterized induction via Hif1α transactivation (Poellinger and Johnson, 2004; Semenza, 2001), VEGF is also up-regulated by the UPR through an ATF4-dependent pathway (Roybal et al., 2004). Furthermore, its processing in the ER and secretion is controlled by GRP170, an ER chaperone that is up-regulated during both ER stress (Lin et al., 1993) and hypoxia (Ikeda et al., 1997; Tamatani et al., 2001). GRP170 increases the resistance of cells to hypoxia (Ozawa et al., 1999), and tumors cells that were manipulated to express high levels of GRP170 showed enhanced secretion of VEGF, whereas decreasing the amount of GRP170 led to ER retention of VEGF (Ozawa et al., 2001). When these cells were used in xenograft studies, the tumors expressing low levels of GRP170 grew very poorly in animals, whereas cells that overexpressed GRP170 produced larger tumors than the parental line (Ozawa et al., 2001). This suggests that UPR activation could play an essential role in tumorigenesis by synergizing with hypoxia activated pathways in promoting angiogenesis.

6.1.2.3. UPR Alters Sensitivity of Tumors to Chemotherapeutic Agents

In addition to a role in promoting tumor growth, there are a number of studies to suggest that UPR activation in tumors might alter their sensitivity to chemotherapy. The in vitro treatment of cultured cells with drugs that activate the UPR increases their resistance to topoisomerase II (topo II) poisons (Hughes et al., 1989), which is thought to occur via a UPR-mediated decrease in topo II levels (Gosky and Chatterjee, 2003; Shen et al., 1989; Yun et al., 1995). Other studies have shown that the increased levels of BiP that occur during UPR activation can result in the relocalization of BiP to the cytosol where it binds caspase7 and caspase12 and prevents their activation in response to some chemotherapeutic agents (Rao et al., 2002; Reddy et al., 2003). Activation of the UPR has also been shown to up-regulate P-glycoprotein (Ledoux et al., 2003), which induces multiple drug resistance. However, UPR activation can also increase the sensitivity of some tumors to DNA cross-linking agents, like cisplatin (Chatterjee et al., 1997; Yamada et al., 1999). A recent study reported that cisplatin is a potent UPR activator and induces a calpain-dependent activation of caspase-12 (Mandic et al., 2003). Changes in the level of any of the components of the DNA repair system in response to UPR activation could also contribute to the increased sensitivity to these agents, although there are currently no data to support this hypothesis. Together these studies suggest that it may be important to couple information on UPR activation in tumors with treatment choices.

6.2. UPR Scenarios that First Protect and then Destroy Tissue

6.2.1. Ischemia

Ischemia or decreased blood flow to tissues that occurs in response to arterial occlusion or cardiac arrest can result in hypoxia and hypoglycemia, which in turn activates the UPR (Kumar et al., 2001; Paschen, 2004). Reperfusion of the affected tissue triggers oxidative stress leading to production of nitric oxide (Montie et al., 2005; Oyadomari et al., 2001), which poisons the ER calcium ATPase and depletes ER calicum stores (Doutheil et al., 2000; Kohno et al., 1997), and of reactive oxygen species, which can directly modify secretory pathway proteins (Hayashi et al., 2003), affecting their normal folding in the ER. And finally, the acidic environment that is established during ischemia further contributes to ER stress (Aoyama et al., 2005). Studies exist which suggest UPR activation can be protective to the tissue as well as ones that demonstrate it contributes to the disease pathology (Kumar et al., 2001; Montie et al., 2005; Paschen, 2004).

In vivo and in vitro studies demonstrated UPR activation in glomerular nephritis. The induction of ER chaperones by phospholipase A(2) diminished cytotoxicity, and PERK null fibroblasts were more susceptible to both complement- and ischemia-reperfusion-mediated death (Cybulsky et al., 2005), suggesting that controlling the load of misfolded proteins in the ER and protecting them from aggregation is important in early stages of ischemia in this tissue. In keeping with this, pretreatment of cardiac myocytes with tunicamycin increased their survival after ischemic shock (Zhang et al., 2004), and Purkinge cells from GRP170/ORP150 transgenic mice showed increased resistance to ischemia (Kitao et al., 2004). In addition, anti-sense constructs that reduced GRP94 levels lead to increased apoptosis in neuronal cells, whereas overexpression of GRP94 protected neurons both in vitro and in vivo from ischemia/reperfusion (Bando et al., 2003). Similarly, overexpression of PDI in neuronal cells protected them from apoptosis in response to nitric oxide or staurosporine (Ko et al., 2002). The importance of up-regulating ER chaperones during ER stress in the brain might be underscored by the presence of an astrocyte-specific member of the ATF6 family, OASIS (Kondo et al., 2005), which regulates chaperone levels, and the remarkable resistance of astrocytes to short term ischemia as compared to neurons (Benavides et al., 2005). Finally, siRNA was used to reduce stannicalcin 2 levels, a UPR inducible protein that contributes to ER homeostatis and is rapidly induced by ischemia. This led to increased apoptosis in response to ER stress, whereas overexpression of stannincalcin 2 protected cells (Ito et al., 2004).

However, other studies have shown a correlation between PERK activation, CHOP induction, and caspase-12 expression and apoptosis in response to ischemia. The use of antisense oligonucleotides to decrease either CHOP (Benavides et al., 2005) or caspase-12 (Aoyama et al., 2005) expression dramatically reduced astrocyte cell death, and CHOP null mice have less tissue loss after stroke (Aoyama et al., 2005), suggesting that these UPR targets play a critical role in mediating cell death in response to ischemia. Thus, it would appear that during initial phases of ischemia the up-regulation of ER chaperones and decreased protein synthesis play a major role in protecting affected tissues. However if the stress continues, the activation of proapoptotic elements of the UPR can ultimately destroy cells affected by ischemia.

6.2.2. Diabetes

The pancreas undergoes dramatic changes in the synthesis of secretory pathway proteins (in particular insulin) in response to fluctuations in blood glucose levels. This increased load of proteins in the ER activates the UPR and most importantly induces PERK phosphorylation leading to transient decreases in translation. Failure to activate PERK can lead to diabetes. Wolcott–Rallison syndrome, which is characterized by infantile diabetes due to pancreatic hypoplasia, is caused by mutations in the PERK gene (Delepine et al., 2000). Similarly, PERK null mice are normal at birth but begin to develop diabetes as they age due to loss of β cell mass (Harding et al., 2001). β cells isolated from these mice show runaway insulin synthesis and secretion, suggesting that controlling of the rate of protein synthesis may be essential to protecting the β cells. In keeping with this, Wolfram syndrome, another hereditary form of diabetes, is caused by mutations in a gene that has homology to Hrd3, which is a UPR-inducible component of the ER degradation machinery in *C. elegans* (Inoue et al., 1998; Strom et al., 1998). Together these data suggest that the dramatic increases in insulin production that occur in the ER of pancreatic islet cells must be modulated to protect the islet cell and that activation of the UPR, specifically PERK, plays a major role in their survival.

However, it is increasingly clear in a number of model systems and in human disease that prolonged, rather than intermittent, UPR activation in pancreatic islet cells can lead to cell death. In this case, CHOP appears to be a major mediator of apoptosis. The production of inflammatory cytokines by pancreatic cells lead to nitric oxide production, which down-regulates the ER calcium ATPase leading to UPR activation and apoptosis (Cardozo et al., 2005; Oyadomari et al., 2001). Cells from CHOP null mice are dramatically more resistant to nitric oxide-induced apoptosis (Oyadomari et al., 2001). Similarly the accumulation of mutant insulin in the ER of Akira mice induces the UPR and leads to loss of β cell mass due to apoptosis (Oyadomari et al., 2002). β cell loss is prevented and development of diabetes is delayed by crossing these mice with the CHOP null mouse (Araki et al., 2003; Oyadomari et al., 2002). Full CHOP induction requires the ATF3 transcription factor (Jiang et al., 2004). Studies show that transgenic mice expressing ATF3 in β cells develop abnormal islets and β cell insufficiency, whereas ATF3 null islets are partially protected from cytokine and nitric oxide induced apoptosis (Hartman et al., 2004). Finally, enforced expression of Bcl2 in human β islet cultures protects them from death by pharmacological inducers of the UPR and by nitric oxide (Contreras et al., 2003). Loss of Bcl2 during UPR activation is regulated by CHOP (McCullough et al., 2001). While CHOP appears to play a major role in contributing to the development of diabetes by inducing islet cell death, it should be noted that both the PERK null (Harding et al., 2000) and eIF2-α mutant (Scheuner et al., 2001) mice are unable to induce CHOP and still show enhanced apoptosis of islet cells, suggesting that other pathways can also contribute. Thus, it appears that the unusual demands placed on pancreatic cells in response to continuous changes in blood glucose levels require PERK activation to control protein synthesis levels and protect the islet cells. However, prolonged or continuous pressure on the ER leads to a more malevolent UPR that results in cell death. Loss of the UPR in the first case leads to diabetes, while in the latter case activation of the UPR does as well!

6.3. Cytodestructive Responses that Contribute to Disease Pathology

6.3.1. Viral Infections

In addition to the UPR's role in contributing to viral severity by enhancing virus production, it is clear that for many viruses that activation of the UPR contributes to the death of infected cells, which plays a major role in the disease pathology. For instance, neuropathogenic retroviruses like Molony murine leukemia virus-ts1 (Kim et al., 2004; Liu et al., 2004) and FrCas^E (Dimcheff et al., 2003), simian virus 5 (Sun et al., 2004), hantaviruses (Li et al., 2005), and hepatotoxic hepatitis C virus (Benali-Furet et al., 2005; Ciccaglione et al., 2005; Waris et al., 2002) induce apoptosis via activation or upregulation of UPR components including caspase-12/4, CHOP, and JNK. In the case of hepatitis C virus, the HCV core triggers ER calcium depletion leading to UPR activation. This in turn induces CHOP and initiates a caspase-3-mediated apoptotic cascade (Benali-Furet et al., 2005). Thus, at least for HCV, data suggest that UPR activation both contributes to the viral load by increasing the production of envelop proteins and to the disease pathology by destroying infected cells.

6.3.2. Neurodegenerative Diseases

Neuronal death underlies the pathology of a number of disorders, including Alzheimer's, Parkinson's, and Huntington's diseases, as well as polyglutamine expansion disease, stroke, lysosomal storage diseases, amyotrophic lateral sclerosis, and prion diseases (Mattson, 2000). Evidence now exists to suggest that the UPR is activated in these diseases and that it contributes to neuronal cell death (Cutler et al., 2002; Nishitoh et al., 2002; Paschen and Mengesdorf, 2005; Ryu et al., 2002; Silva et al., 2005). We will focus on just two of these here.

6.3.2.1. Lysosomal Storage Disease

Lysosomal storage disease represents a diverse group of inherited metabolic disorders in which various lysosomal enzymes are deficient. Although the enzymes are expressed in all cells, a rapid, progressive degeneration of the nervous system leading to death at an early age represents the major pathology of these diseases. Recent studies of several different lysosomal diseases have linked UPR activation to apoptosis of affected neurons. Sandoff disease is caused by a deficiency in hexosaminidase A leading to a massive accumulation of the GM2 ganglioside in the brain. Somewhat surprisingly, the accumulation of this glycolipid results in an inhibition in calcium uptake into the ER, which triggers the UPR and induces apoptosis (Pelled et al., 2003). Very similar results were obtained when a mouse model for β-galactosidase deficiency was examined. The deficiency leads to the accumulation of GM1 gangliosides in the brain, which was shown to activate the UPR via changes in ER calcium and to induce caspase-12 cleavage and JNK activation leading to apoptosis (Tessitore et al., 2004). In keeping with this theme, UPR activation has also been linked to disease pathology in Gaucher's disease, which is caused by an inherited deficiency in glucocerbrosidase. In both animal models and human disease, the resultant accumulation of GlcCer in the ER causes enhanced calcium release via the ryanodine receptor, which in turn activates the UPR (Korkotian et al., 1999; Pelled et al., 2005). Although a number of other lysosomal storage diseases exist that affect the degradation of secretory pathway proteins and which could presumably lead to UPR activation, it is not presently known if the UPR also plays a role in these diseases.

6.3.2.2. Alzheimer's Disease

The ER appears to be a prime contributor to the cellular changes that result in neuronal dysfunction and death in Alzheimer's disease (AD). The β-amyloid precursor protein (APP) is synthesized in the ER where it binds to the molecular chaperone BiP (Yang et al., 1998). Aberrant proteolytic cleavage of APP in the ER results in the generation of a fragment (Aβ) that is highly prone to aggregation (Yang et al., 1998). This fragment is produced by two successive cleavages by the enzymes β-secretase and γ-secretase. Mutations in APP and in one of the subunits of γ-secretase are responsible for several forms of hereditary AD (Mattson et al., 2001). In addition to forming aggregates that damage the ER, the Aβ fragment causes abnormalities in ER calcium homoeostasis and activates the UPR (Hoozemans et al., 2005; Kudo et al., 2002; Mattson et al., 2001). Evidently, some components of UPR activation are protective during the early stages of the disease, including the up-regulation of Herp (Chan et al., 2004), which is thought to play a role in the degradation of ER proteins. However, long-term activation of the UPR is responsible for inducing apoptosis in affected neurons, in part via CHOP (Milhavet et al., 2002), ASK1 (Kadowaki et al., 2005), and caspase 4 activation (Hitomi et al., 2004). In cases of AD that are caused by mutation of presenilin-1 (PS-1), there is some evidence to link expression of this protein to UPR activation (Chan et al., 2002). However, this point is somewhat controversial (Sato et al., 2000), and mice with a PS-1 knock-in mutation do not show evidence of abnormal UPR activation (Siman et al., 2001).

7. Manipulating the UPR as a Possible Means of Disease Intervention

The finding that UPR activation plays a detrimental role in so many diverse diseases makes it an attractive target for therapeutic intervention. A recent study conducted a screen for small molecules that would protect a rat cell line from ER stress-induced apoptosis. They identified a molecule that they are calling salubrinal, which worked at reasonably low concentrations and was more effective than a pan-caspase inhibitor in blocking cell death (Boyce et al., 2005). This drug was shown to induce eIF-2α phosphorylation by selectively inhibiting both the GADD34 and CReP phosphatase complexes that dephosphorylate eIF-2α under ER stress and normal conditions respectively. Salubrinal also blocked eIF-2α dephosphorylation mediated by a herpes simplex virus protein and inhibited viral replication in Vero cells. This suggests that medically useful agents might be developed that interfere with UPR activation and bodes well for future small molecule screens.

References

Alarcon, R., Koumenis, C., Geyer, R. K., Maki, C. G., and Giaccia, A. J. (1999) Hypoxia induces p53 accumulation through MDM2 down-regulation and inhibition of E6-mediated degradation. *Cancer Res* 59:6046–51.

Aoyama, K., Burns, D. M., Suh, S. W., Garnier, P., Matsumori, Y., Shiina, H., et al. (2005) Acidosis causes endoplasmic reticulum stress and caspase-12-mediated astrocyte death. *J Cereb Blood Flow Metab* 25:358–70.

Araki, E., Oyadomari, S., and Mori, M. (2003) Impact of endoplasmic reticulum stress pathway on pancreatic beta-cells and diabetes mellitus. *Exp Biol Med* 228:1213–7.

Bando, Y., Katayama, T., Kasai, K., Taniguchi, M., Tamatani, M., and Tohyama, M. (2003) GRP94 (94 kDa glucose-regulated protein) suppresses ischemic neuronal cell death against ischemia/reperfusion injury. *Eur J Neurosci* 18:829–40.

Batchvarova, N., Wang, X.-Z., and Ron, D. (1995) Inhibition of adipogenesis by the stress-induced protein CHOP (Gadd153). *EMBO J* 14:4654–61.

Benali-Furet, N. L., Chami, M., Houel, L., De, G. F., Vernejoul, F., Lagorce, D., Buscail, L., Bartenschlager, R., Ichas, F., Rizzuto, R., and Paterlini-Brechot, P. (2005) Hepatitis C virus core triggers apoptosis in liver cells by inducing ER stress and ER calcium depletion. *Oncogene* 24:4921–33.

Benavides, A., Pastor, D., Santos, P., Tranque, P., and Calvo, S. (2005) CHOP plays a pivotal role in the astrocyte death induced by oxygen and glucose deprivation. *Glia* 52:261–75.

Bertolotti, A., Zhang, Y., Hendershot, L. M., Harding, H. P., and Ron, D. (2000). Dynamic interaction of BiP and ER stress transducers in the unfolded-protein response. *Nat Cell Biol* 2:326–32.

Bi, M., Naczki, C., Koritzinsky, M., Fels, D., Blais, J., Hu, N., Harding, H., Novoa, I., Varia, M., Raleigh, J., Scheuner, D., Kaufman, R. J., Bell, J., Ron, D., Wouters, B. G., and Koumenis, C. (2005) ER stress-regulated translation increases tolerance to extreme hypoxia and promotes tumor growth. *EMBO J* 24:3470–81.

Biason-Lauber, A., Lang-Muritano, M., Vaccaro, T., and Schoenle, E. J. (2002) Loss of kinase activity in a patient with Wolcott–Rallison syndrome caused by a novel mutation in the EIF2AK3 gene. *Diabetes* 51:2301–5.

Blais, J. D., Filipenko, V., Bi, M., Harding, H. P., Ron, D., Koumenis, C., Wouters, B. G., and Bell, J. C. (2004) Activating transcription factor 4 is translationally regulated by hypoxic stress. *Mol Cell Biol* 24:7469–82.

Boyce, M., Bryant, K. F., Jousse, C., Long, K., Harding, H. P., Scheuner, D., Kaufman, R. J., Ma, D., Coen, D. M., Ron, D., and Yuan, J. (2005) A selective inhibitor of eIF2alpha dephosphorylation protects cells from ER stress. *Science* 307:935–9.

Brewer, J. W., and Diehl, J. A. (2000) PERK mediates cell-cycle exit during the mammalian unfolded protein response. *Proc Natl Acad Sci U S A* 97:12625–30.

Brodsky, J. L., Werner, E. D., Dubas, M. E., Goeckeler, J. L., Kruse, K. B., and McCracken, A. A. (1999) The requirement for molecular chaperones during endoplasmic reticulum-associated protein degradation demonstrates that protein export and import are mechanistically distinct. *J Biol Chem* 274:3453–60.

Brostrom, C. O., Prostko, C. R., Kaufman, R. J., and Brostrom, M. A. (1996) Inhibition of translational initiation by activators of the glucose-regulated stress protein and heat shock protein stress response systems. Role of the interferon-inducible double-stranded RNA-activated eukaryotic initiation factor 2alpha kinase. *J Biol Chem* 271:24995–5002.

Calfon, M., Zeng, H., Urano, F., Till, J. H., Hubbard, S. R., Harding, H. P., Clask, S.G., and Ron, D. (2002) IRE1 couples endoplasmic reticulum load to secretory capacity by processing the *XBP-1* mRNA. *Nature* 415:92–6.

Cardozo, A. K., Ortis, F., Storling, J., Feng, Y. M., Rasschaert, J., Tonnesen, M., Van Eylen, F., Mandrup-Poulsen, T., Herchuelz, A., and Eizirik, D. L. (2005) Cytokines downregulate the sarcoendoplasmic reticulum pump Ca2+ ATPase 2b and deplete endoplasmic reticulum Ca2+, leading to induction of endoplasmic reticulum stress in pancreatic beta-cells. *Diabetes* 54:452–61.

Chan, S. L., Culmsee, C., Haughey, N., Klapper, W., and Mattson, M. P. (2002) Presenilin-1 mutations sensitize neurons to DNA damage-induced death by a mechanism involving perturbed calcium homeostasis and activation of calpains and caspase-12. *Neurobiol Dis* 11:2–19.

Chan, S. L., Fu, W., Zhang, P., Cheng, A., Lee, J., Kokame, K., and Mattson, M. P. (2004) Herp stabilizes neuronal Ca2+ homeostasis and mitochondrial function during endoplasmic reticulum stress. *J Biol Chem* 279:28733–43.

Chatterjee, S., Hirota, H., Belfi, C. A., Berger, S. J., and Berger, N. A. (1997) Hypersensitivity to DNA cross-linking agents associated with up-regulation of glucose-regulated stress protein GRP78. *Cancer Res* 57:5112–6.

Chen, X., Ding, Y., Liu, C. G., Mikhail, S., and Yang, C. S. (2002) Overexpression of glucose-regulated protein 94 (Grp94) in esophageal adenocarcinomas of a rat surgical model and humans. *Carcinogenesis* 23:123–30.

Cheng, G., Feng, Z., and He, B. (2005) Herpes simplex virus 1 infection activates the endoplasmic reticulum resident kinase PERK and mediates eIF-2alpha dephosphorylation by the gamma(1)34.5 protein. *J Virol* 79:1379–88.

Ciccaglione, A. R., Costantino, A., Tritarelli, E., Marcantonio, C., Equestre, M., Marziliano, N., and Rapicetta, M. (2005) Activation of endoplasmic reticulum stress response by hepatitis C virus proteins. *Arch Virol* 150:1339–56.

Contreras, J. L., Smyth, C. A., Bilbao, G., Eckstein, C., Young, C. J., Thompson, J. A., Curiel, D. T., and Eckhoff, D. E. (2003) Coupling endoplasmic reticulum stress to cell death program in isolated human pancreatic islets: Effects of gene transfer of Bcl-2. *Transpl Int* 16:537–42.

Cox, J. S., Shamu, C. E., and Walter, P. (1993) Transcriptional induction of genes encoding endoplasmic reticulum resident proteins requires a transmembrane protein kinase. *Cell* 73:1197–206.

Cox, J. S. and Walter, P. (1996) A novel mechanism for regulating activity of a transcription factor that controls the unfolded protein response. *Cell* 87:391–404.

Cutler, R. G., Pedersen, W. A., Camandola, S., Rothstein, J. D., and Mattson, M. P. (2002) Evidence that accumulation of ceramides and cholesterol esters mediates oxidative stress-induced death of motor neurons in amyotrophic lateral sclerosis. *Ann Neurol* 52:448–57.

Cybulsky, A. V., Takano, T., Papillon, J., and Bijian, K. (2005) Role of the endoplasmic reticulum unfolded protein response in glomerular epithelial cell injury. *J Biol Chem* 280:24396–403.

Delepine, M., Nicolino, M., Barrett, T., Golamaully, M., Lathrop, G. M., and Julier, C. (2000) EIF2AK3, encoding translation initiation factor 2-alpha kinase 3, is mutated in patients with Wolcott–Rallison syndrome. *Nat Genet* 25:406–9.

Dimcheff, D. E., Askovic, S., Baker, A. H., Johnson-Fowler, C., and Portis, J. L. (2003) Endoplasmic reticulum stress is a determinant of retrovirus-induced spongiform neurodegeneration. *J Virol* 77:12617–29.

Dorner, A. J., Wasley, L. C., and Kaufman, R. J. (1992) Overexpression of GRP78 mitigates stress induction of glucose regulated proteins and blocks secretion of selective proteins in Chinese hamster ovary cells. *EMBO J* 11:1563–71.

Dorner, A. J., Wasley, L. C., Raney, P., Haugejorden, S., Green, M., and Kaufman, R. J. (1990) The stress response in Chinese hamster ovary cells. Regulation of ERp72 and protein disulfide isomerase expression and secretion. *J Biol Chem* 265:22029–34.

Doutheil, J., Althausen, S., Treiman, M., and Paschen, W. (2000) Effect of nitric oxide on endoplasmic reticulum calcium homeostasis, protein synthesis and energy metabolism. *Cell Calcium* 27:107–15.

Ellgaard, L., and Helenius, A. (2003) Quality control in the endoplasmic reticulum. *Nat Rev Mol Cell Biol* 4:181–91.

Feldman, D. E., Chauhan, V., and Koong, A. C. (2005) The unfolded protein response: A novel component of the hypoxic stress response in tumors. *Mol Cancer Res* 3:597–605.

Feng, B., Yao, P. M., Li, Y., Devlin, C. M., Zhang, D., Harding, H. P., Sweeney, M., Rong, J. X., Kuriakose, G., Fisher, E. A., Marks, A. R., Ron, D., Tabas, I. (2004) The endoplasmic reticulum is the site of cholesterol-induced cytotoxicity in macrophages. *Nature Cell Biol.* 5:781–92.

Fernandez, P. M., Tabbara, S. O., Jacobs, L. K., Manning, F. C., Tsangaris, T. N., Schwartz, A. M., Kennedy, K. A., and Patierno, S. R. (2000) Overexpression of the glucose-regulated stress gene GRP78 in malignant but not benign human breast lesions. *Breast Cancer Res Treat* 59:15–26.

Ferrara, N., and Davis-Smyth, T. (1997) The biology of vascular endothelial growth factor. *Endocr Rev* 18:4–25.

Gass, J. N., Gifford, N. M., and Brewer, J. W. (2002) Activation of an unfolded protein response during differentiation of antibody-secreting B cells. *J Biol Chem* 277:49047–54.

Gosky, D., and Chatterjee, S. (2003) Down-regulation of topoisomerase II alpha is caused by up-regulation of GRP78. *Biochem Biophys Res Commun* 300:327–32.

Graeber, T. G., Osmanian, C., Jacks, T., Housman, D. E., Koch, C. J., Lowe, S. W., and Giaccia, A. J. (1996) Hypoxia-mediated selection of cells with diminished apoptotic potential in solid tumours. *Nature* 379:88–91.

Graham, K. S., Le, A., and Sifers, R. N. (1990) Accumulation of the insoluble PiZ variant of human alpha 1-antitrypsin within the hepatic endoplasmic reticulum does not elevate the steady-state level of grp78/BiP. *J Biol Chem* 265:20463–8.

Harding, H. P., Novoa, I., Zhang, Y., Zeng, H., Wek, R., Schapira, M., and Ron, D. (2000) Regulated translation initiation controls stress-induced gene expression in mammalian cells. *Mol Cell* 6:1099–108.

Harding, H. P., and Ron, D. (2002) Endoplasmic reticulum stress and the development of diabetes: A review. *Diabetes* 51:S455–61.

Harding, H. P., Zeng, H., Zhang, Y., Jungries, R., Chung, P., Plesken, H., Sabatini, D. D., and Ron, D. (2001) Diabetes mellitus and exocrine pancreatic dysfunction in $perk^{-/-}$ mice reveals a role for translational control in secretory cell survival. *Mol Cell* 7:1153–63.

Harding, H. P., Zhang, Y., and Ron, D. (1999) Protein translation and folding are coupled by an endoplasmic-reticulum-resident kinase. *Nature* 397:271–4.

Harding, H. P., Zhang, Y., Zeng, H., Novoa, I., Lu, P. D., Calfon, M., Sadri, N., Yun, C., Popko, B., Paules, R., Stojdl, D. F., Bell, J. C., Hettmann, T., Leiden, J. M., and Ron, D. (2003) An integrated stress response regulates amino acid metabolism and resistance to oxidative stress. *Mol Cell* 11:619–33.

Hartman, M. G., Lu, D., Kim, M. L., Kociba, G. J., Shukri, T., Buteau, J., Wang, X., Frankel, W. L., Guttridge, D., Prentki, M., Grey, S. T., Ron, D., and Hai, T. (2004) Role for activating transcription factor 3 in stress-induced beta-cell apoptosis. *Mol Cell Biol* 24:5721–32.

Hayashi, T., Saito, A., Okuno, S., Ferrand-Drake, M., Dodd, R. L., Nishi, T., Maier, C. M., Kinouchi, H., and Chan, P. H. (2003) Oxidative damage to the endoplasmic reticulum is implicated in ischemic neuronal cell death. *J Cereb Blood Flow Metab* 23:1117–28.

Haze, K., Yoshida, H., Yanagi, H., Yura, T., and Mori, K. (1999) Mammalian transcription factor ATF6 is synthesized as a transmembrane protein and activated by proteolysis in response to endoplasmic reticulum stress. *Mol Biol Cell* 10:3787–99.

Helenius, A. (1994) How N-linked oligosaccharides affect glycoprotein folding in the endoplasmic reticulum. *Mol Biol Cell* 5:253–65.

Hendershot, L. M., Ting, J., and Lee, A. S. (1988) Identity of the immunoglobulin heavy-chain-binding protein with the 78,000-dalton glucose-regulated protein and the role of post-translational modifications in its binding function. *Mol Cell Biol* 8:4250–6.

Hitomi, J., Katayama, T., Eguchi, Y., Kudo, T., Taniguchi, M., Koyama, Y., Manabe, T., Yamagishi, S., Bando, Y., Imaizumi, K., Tsujimoto, Y., and Tohyama, M. (2004) Involvement of caspase-4 in endoplasmic reticulum stress-induced apoptosis and A{beta}-induced cell death. *J Cell Biol* 165:347–56.

Hockel, M., and Vaupel, P. (2001a) Biological consequences of tumor hypoxia. *Semin Oncol* 28:36–41.

Hockel, M., and Vaupel, P. (2001b) Tumor hypoxia: Definitions and current clinical, biologic, and molecular aspects. *J Natl Cancer Inst* 93:266–76.

Hoozemans, J. J., Veerhuis, R., Van Haastert, E. S., Rozemuller, J. M., Baas, F., Eikelenboom, P., and Scheper, W. (2005) The unfolded protein response is activated in Alzheimer's disease. *Acta Neuropathol* 110:165–72.

Huang, Z. M., Tan, T., Yoshida, H., Mori, K., Ma, Y., and Yen, T. S. (2005) Activation of hepatitis B virus S promoter by a cell type-restricted IRE1-dependent pathway induced by endoplasmic reticulum stress. *Mol Cell Biol* 25:7522–33.

Hughes, C. S., Shen, J. W., and Subjeck, J. R. (1989) Resistance to etoposide induced by three glucose-regulated stresses in Chinese hamster ovary cells. *Cancer Res* 49:4452–4.

Ikeda, J., Kaneda, S., Kuwabara, K., Ogawa, S., Kobayashi, T., Matsumoto, M., Yura, T., and Yanagi, H. (1997) Cloning and expression of cDNA encoding the human 150 kDa oxygen-regulated protein, ORP150. *Biochem Biophys Res Commun* 230:94–9.

Inoue, H., Tanizawa, Y., Wasson, J., Behn, P., Kalidas, K., Bernal-Mizrachi, E., Mueckler, M., Marshall, H., Donis-Keller, H., Crock, P., Rogers, D., Mikuni, M., Kumashiro, H., Higashi, K., Sobue, G., Oka, Y., and Permutt, M. A. (1998) A gene encoding a transmembrane protein is mutated in patients with diabetes mellitus and optic atrophy (Wolfram syndrome). *Nat Genet* 20:143–8.

Isler, J. A., Skalet, A. H., and Alwine, J. C. (2005) Human cytomegalovirus infection activates and regulates the unfolded protein response. *J Virol* 79:6890–9.

Ito, D., Walker, J. R., Thompson, C. S., Moroz, I., Lin, W., Veselits, M. L., Hakim, A. M., Fienberg, A. A., and Thinakaran, G. (2004) Characterization of stanniocalcin 2, a novel target of the mammalian unfolded protein response with cytoprotective properties. *Mol Cell Biol* 24:9456–69.

Iwakoshi, N. N., Lee, A. H., Vallabhajosyula, P., Otipoby, K. L., Rajewsky, K., and Glimcher, L. H. (2003) Plasma cell differentiation and the unfolded protein response intersect at the transcription factor XBP-1. *Nat Immunol* 4:321–9.

Iwawaki, T., Akai, R., Kohno, K., and Miura, M. (2004) A transgenic mouse model for monitoring endoplasmic reticulum stress. *Nat Med* 10:98–102.

Jamora, C., Dennert, G., and Lee, A. S. (1996) Inhibition of tumor progression by suppression of stress protein GRP78/BiP induction in fibrosarcoma B/C10ME. *Proc Natl Acad Sci U S A* 93:7690–4.

Jiang, H. Y., Wek, S. A., McGrath, B. C., Lu, D., Hai, T., Harding, H. P., Wang, X., Ron, D., Cavener, D. R., and Wek, R. C. (2004) Activating transcription factor 3 is integral to the eukaryotic initiation factor 2 kinase stress response. *Mol Cell Biol* 24:1365–77.

Jiang, H. Y., Wek, S. A., McGrath, B. C., Scheuner, D., Kaufman, R. J., Cavener, D. R., and Wek, R. C. (2003) Phosphorylation of the alpha subunit of eukaryotic initiation factor 2 is required for activation of NF-kappaB in response to diverse cellular stresses. *Mol Cell Biol* 23:5651–63.

Kadowaki, H., Nishitoh, H., Urano, F., Sadamitsu, C., Matsuzawa, A., Takeda, K., Masutani, H., Yodoi, J., Urano, Y., Nagano, T., and Ichijo, H. (2005) Amyloid beta induces neuronal cell death through ROS-mediated ASK1 activation. *Cell Death Differ* 12:19–24.

Kharroubi, I., Ladriere, L., Cardozo, A. K., Dogusan, Z., Cnop, M., and Eizirik, D. L. (2004) Free fatty acids and cytokines induce pancreatic beta-cell apoptosis by different mechanisms: Role of nuclear factor-kappaB and endoplasmic reticulum stress. *Endocrinology* 145:5087–96.

Kim, H. T., Waters, K., Stoica, G., Qiang, W., Liu, N., Scofield, V. L., and Wong, P. K. (2004) Activation of endoplasmic reticulum stress signaling pathway is associated with neuronal degeneration in MoMuLV-ts1-induced spongiform encephalomyelopathy. *Lab Invest* 84:816–27.

Kimata, Y., Kimata, Y. I., Shimizu, Y., Abe, H., Farcasanu, I. C., Takeuchi, M., Rose, M. D., and Kohno, K. (2003) Genetic evidence for a role of BiP/Kar2 that regulates Ire1 in response to accumulation of unfolded proteins. *Mol Biol Cell* 14:2559–69.

Kimata, Y., Oikawa, D., Shimizu, Y., Ishiwata-Kimata, Y., and Kohno, K. (2004) A role for BiP as an adjustor for the endoplasmic reticulum stress-sensing protein Ire1. *J Cell Biol* 167:445–56.

Kitao, Y., Hashimoto, K., Matsuyama, T., Iso, H., Tamatani, T., Hori, O., Stern, D. M., Kano, M., Ozawa, K., and Ogawa, S. (2004) ORP150/HSP12A regulates Purkinje cell survival: A role for endoplasmic reticulum stress in cerebellar development. *J Neurosci* 24:1486–96.

Ko, H. S., Uehara, T., and Nomura, Y. (2002) Role of ubiquilin associated with protein-disulfide isomerase in the endoplasmic reticulum in stress-induced apoptotic cell death. *J Biol Chem* 277:35386–92.

Kohno, K., Higuchi, T., Ohta, S., Kohno, K., Kumon, Y., and Sakaki, S. (1997) Neuroprotective nitric oxide synthase inhibitor reduces intracellular calcium accumulation following transient global ischemia in the gerbil. *Neurosci Lett* 224:17–20.

Kondo, S., Murakami, T., Tatsumi, K., Ogata, M., Kanemoto, S., Otori, K., Iseki, K., Wanaka, A., and Imaizumi, K. (2005) OASIS, a CREB/ATF-family member, modulates UPR signalling in astrocytes. *Nat Cell Biol* 7:186–194.

Koong, A. C., Chen, E. Y., and Giaccia, A. J. (1994) Hypoxia causes the activation of nuclear factor kappa B through the phosphorylation of I kappa B alpha on tyrosine residues. *Cancer Res* 54:1425–30.

Korkotian, E., Schwarz, A., Pelled, D., Schwarzmann, G., Segal, M., and Futerman, A. H. (1999) Elevation of intracellular glucosylceramide levels results in an increase in endoplasmic reticulum density and in functional calcium stores in cultured neurons. *J Biol Chem* 274:21673–8.

Kornfeld, R., and Kornfeld, S. (1985) Assembly of asparagine-linked oligosaccharides. *Annu Rev Biochem* 54:631–64.

Kostova, Z., and Wolf, D. H. (2003) For whom the bell tolls: Protein quality control of the endoplasmic reticulum and the ubiquitin-proteasome connection. *EMBO J* 22:2309–17.

Koumenis, C., Naczki, C., Koritzinsky, M., Rastani, S., Diehl, A., Sonenberg, N., Koromilas, A., and Wouters, B. G. (2002) Regulation of protein synthesis by hypoxia via activation of the endoplasmic reticulum kinase PERK and phosphorylation of the translation initiation factor eIF2alpha. *Mol Cell Biol* 22:7405–16.

Kozutsumi, Y., Segal, M., Normington, K., Gething, M. J., and Sambrook, J. (1988) The presence of malfolded proteins in the endoplasmic reticulum signals the induction of glucose-regulated proteins. *Nature* 332:462–4.

Kudo, T., Katayama, T., Imaizumi, K., Yasuda, Y., Yatera, M., Okochi, M., Tohyama, M., and Takeda, M. (2002) The unfolded protein response is involved in the pathology of Alzheimer's disease. *Ann N Y Acad Sci* 977:349–55.

Kumar, R., Azam, S., Sullivan, J. M., Owen, C., Cavener, D. R., Zhang, P., Ron, D., Harding, H. P., Chen, J. J., Han, A., White, B. C., Krause, G. S., and DeGracia, D. J. (2001) Brain ischemia and reperfusion activates the eukaryotic initiation factor 2alpha kinase, PERK. *J Neurochem* 77:1418–21.

Ladiges, W. C., Knoblaugh, S. E., Morton, J. F., Korth, M. J., Sopher, B. L., Baskin, C. R., MacAuley, A., Goodman, A.G., LeBoeuf, R.C., and Katze, M.G. (2005) Pancreatic beta-cell failure and diabetes in mice with a deletion mutation of the endoplasmic reticulum molecular chaperone gene P58IPK. *Diabetes* 54:1074–81.

Ledoux, S., Yang, R., Friedlander, G., and Laouari, D. (2003) Glucose depletion enhances P-glycoprotein expression in hepatoma cells: Role of endoplasmic reticulum stress response. *Cancer Res* 63:7284–90.

Lee, A. H., Iwakoshi, N. N., and Glimcher, L. H. (2003) XBP-1 regulates a subset of endoplasmic reticulum resident chaperone genes in the unfolded protein response. *Mol Cell Biol* 23:7448–59.

Lee, A. S. (1987) Coordinated regulation of a set of genes by glucose and calcium ionophores in mammalian cells. *Trends Biochem Sci* 12:20–3.

Lee, A. S. (1992) Mammalian stress response: Induction of the glucose-regulated protein family. *Curr Opin Cell Biol* 4:267–73.

Lee, A. S., Wells, S., Kim, K. S., and Scheffler, I. E. (1986) Enhanced synthesis of the glucose/calcium-regulated proteins in a hamster cell mutant deficient in transfer of oligosaccharide core to polypeptides. *J Cell Physiol* 129:277–82.

Lei, K., and Davis, R. J. (2003) JNK phosphorylation of Bim-related members of the Bcl2 family induces Bax-dependent apoptosis. *Proc Natl Acad Sci U S A* 100:2432–7.

Lenny, N., and Green, M. (1991) Regulation of endoplasmic reticulum stress proteins in COS cells transfected with immunoglobulin mu heavy chain cDNA. *J Biol Chem* 266:20532–7.

Li, X. D., Lankinen, H., Putkuri, N., Vapalahti, O., and Vaheri, A. (2005) Tula hantavirus triggers pro-apoptotic signals of ER stress in Vero E6 cells. *Virology* 333:180–9.

Lin, H. Y., Masso-Welch, P., Di, Y. P., Cai, J. W., Shen, J. W., and Subjeck, J. R. (1993) The 170-kDa glucose-regulated stress protein is an endoplasmic reticulum protein that binds immunoglobulin. *Mol Biol Cell* 4:1109–9.

Liu, N., Kuang, X., Kim, H. T., Stoica, G., Qiang, W., Scofield, V. L., and Wong, P. K. (2004) Possible involvement of both endoplasmic reticulum- and mitochondria-dependent pathways in MoMuLV-ts1-induced apoptosis in astrocytes. *J Neurovirol* 10:189–98.

Llewellyn, D. H., and Roderick, H. L. (1998) Overexpression of calreticulin fails to abolish its induction by perturbation of normal ER function. *Biochem Cell Biol* 76:875–80.

Ma, Y., and Hendershot, L. M. (2003) Delineation of the negative feedback regulatory loop that controls protein translation during ER stress. *J Biol Chem* 278:34864–73.

Ma, Y., and Hendershot, L. M. (2004) The role of the unfolded protein response in tumour development: Friend or foe? *Nat Rev Cancer* 4:966–77.

Mandic, A., Hansson, J., Linder, S., and Shoshan, M. C. (2003) Cisplatin induces endoplasmic reticulum stress and nucleus-independent apoptotic signaling. *J Biol Chem* 278:9100–6.

Mao, C., Dong, D., Little, E., Luo, S., and Lee, A. S. (2004) Transgenic mouse model for monitoring endoplasmic reticulum stress *in vivo*. *Nat Med* 10:1013–4.

Mattson, M. P. (2000) Apoptosis in neurodegenerative disorders. *Nat Rev Mol Cell Biol* 1:120–9.

Mattson, M. P., Gary, D. S., Chan, S. L., and Duan, W. (2001) Perturbed endoplasmic reticulum function, synaptic apoptosis and the pathogenesis of Alzheimer's disease. *Biochem Soc Symp* 151–62.

McCullough, K. D., Martindale, J. L., Klotz, L. O., Aw, T. Y., and Holbrook, N. J. (2001) Gadd153 sensitizes cells to endoplasmic reticulum stress by down-regulating Bcl2 and perturbing the cellular redox state. *Mol Cell Biol* 21:1249–59.

McGarry, J. D., and Dobbins, R. L. (1999) Fatty acids, lipotoxicity and insulin secretion. *Diabetologia* 42:128–38.

Milhavet, O., Martindale, J. L., Camandola, S., Chan, S. L., Gary, D. S., Cheng, A., Holbrook, N J., and Mattson, M. P. (2002) Involvement of Gadd153 in the pathogenic action of presenilin-1 mutations. *J Neurochem* 83:673–81.

Montie, H. L., Haezebrouck, A. J., Gutwald, J. C., and DeGracia, D. J. (2005) PERK is activated differentially in peripheral organs following cardiac arrest and resuscitation. *Resuscitation* 66:379–89.

Mori, K., Ma, W., Gething, M. J., and Sambrook, J. (1993) A transmembrane protein with a cdc2+/CDC28-related kinase activity is required for signalling from the ER to the nucleus. *Cell* 74:743–56.

Mori, K., Sant, A., Kohno, K., Normington, K., Gething, M. J., and Sambrook, J. F. (1992) A 22 bp cis-acting element is necessary and sufficient for the induction of the yeast KAR2 (BiP) gene by unfolded proteins. *EMBO J* 11:2583–93.

Morishima, N., Nakanishi, K., Takenouchi, H., Shibata, T., and Yasuhiko, Y. (2002) An endoplasmic reticulum stress-specific caspase cascade in apoptosis. Cytochrome c-independent activation of caspase-9 by caspase-12. *J Biol Chem* 277:34287–94.

Nakagawa, T., and Yuan, J. (2000) Cross-talk between two cysteine protease families. Activation of caspase-12 by calpain in apoptosis. *J Cell Biol* 150:887–94.

Nakagawa, T., Zhu, H., Morishima, N., Li, E., Xu, J., Yankner, B. A., and Yuan, J. (2000) Caspase-12 mediates endoplasmic-reticulum-specific apoptosis and cytotoxicity by amyloid-beta. *Nature* 403:98–103.

Netherton, C. L., Parsley, J. C., and Wileman, T. (2004) African swine fever virus inhibits induction of the stress-induced proapoptotic transcription factor CHOP/GADD153. *J Virol* 78:10825–8.

Nishitoh, H., Matsuzawa, A., Tobiume, K., Saegusa, K., Takeda, K., Inoue, K., Hori, S., Kakizuka, A., and Ichijo, H. (2002) ASK1 is essential for endoplasmic reticulum stress-induced neuronal cell death triggered by expanded polyglutamine repeats. *Genes Dev* 16:1345–55.

Novoa, I., Zeng, H., Harding, H. P., and Ron, D. (2001) Feedback inhibition of the unfolded protein response by GADD34-mediated dephosphorylation of eIF2alpha. *J Cell Biol* 153:1011–22.

Okamura, K., Kimata, Y., Higashio, H., Tsuru, A., and Kohno, K. (2000) Dissociation of Kar2p/BiP from an ER sensory molecule, Ire1p, triggers the unfolded protein response in yeast. *Biochem Biophys Res Commun* 279:445–50.

Oyadomari, S., Koizumi, A., Takeda, K., Gotoh, T., Akira, S., Araki, E., and Mori, M. (2002) Targeted disruption of the Chop gene delays endoplasmic reticulum stress-mediated diabetes. *J Clin Invest* 109:525–32.

Oyadomari, S., Takeda, K., Takiguchi, M., Gotoh, T., Matsumoto, M., Wada, I., Akira, S., Araki, E., and Mori, M. (2001) Nitric oxide-induced apoptosis in pancreatic beta cells is mediated by the endoplasmic reticulum stress pathway. *Proc Natl Acad Sci U S A* 98:10845–50.

Ozawa, K., Kuwabara, K., Tamatani, M., Takatsuji, K., Tsukamoto, Y., Kaneda, S., Yanagi, H., Stern, D. M., Eguchi, Y., Tsujimoto, Y., Ogawa, S., and Tohyama, M. (1999) 150-kDa oxygen-regulated protein (ORP150) suppresses hypoxia-induced apoptotic cell death. *J Biol Chem* 274:6397–404.

Ozawa, K., Tsukamoto, Y., Hori, O., Kitao, Y., Yanagi, H., Stern, D. M., and Ogawa, S. (2001) Regulation of tumor angiogenesis by oxygen-regulated protein 150, an inducible endoplasmic reticulum chaperone. *Cancer Res* 61:4206–13.

Paschen, W. (2004) Endoplasmic reticulum dysfunction in brain pathology: Critical role of protein synthesis. *Curr Neurovasc Res* 1:173–81.

Paschen, W., and Mengesdorf, T. (2005) Endoplasmic reticulum stress response and neurodegeneration. *Cell Calcium* 38:409–15.

Pelled, D., Lloyd-Evans, E., Riebeling, C., Jeyakumar, M., Platt, F. M., and Futerman, A. H. (2003). Inhibition of calcium uptake via the sarco/endoplasmic reticulum Ca2+-ATPase in a mouse model of Sandhoff disease and prevention by treatment with N-butyldeoxynojirimycin. *J Biol Chem* 278:29496–501.

Pelled, D., Trajkovic-Bodennec, S., Lloyd-Evans, E., Sidransky, E., Schiffmann, R., and Futerman, A. H. (2005) Enhanced calcium release in the acute neuronopathic form of Gaucher disease. *Neurobiol Dis* 18:83–8.

Pereira, R. C., Delany, A. M., and Canalis, E. (2004). CCAAT/enhancer binding protein homologous protein (DDIT3) induces osteoblastic cell differentiation. *Endocrinology* 145:1952–60.

Pereira, R. C., Stadmeyer, L., Marciniak, S. J., Ron, D., and Canalis, E. (2005) C/EBP homologous protein is necessary for normal osteoblastic function. *J Cell Biochem* 97:633–40.

Poellinger, L., and Johnson, R. S. (2004) HIF-1 and hypoxic response: The plot thickens. *Curr Opin Genet Dev* 14:81–5.

Pouyssegur, J., Shiu, R. P., and Pastan, I. (1977) Induction of two transformation-sensitive membrane polypeptides in normal fibroblasts by a block in glycoprotein synthesis or glucose deprivation. *Cell* 11:941–7.

Pugh, C. W., and Ratcliffe, P. J. (2003) Regulation of angiogenesis by hypoxia: Role of the HIF system. *Nat Med* 9:677–84.

Putcha, G. V., Le, S., Frank, S., Besirli, C. G., Clark, K., Chu, B., Alix, S., Youle, R. J., LaMarche, A., Maroney, A. C., and Johnson, E. M., Jr. (2003) JNK-mediated BIM phosphorylation potentiates BAX-dependent apoptosis. *Neuron* 38:899–914.

Qu, L., Huang, S., Baltzis, D., Rivas-Estilla, A. M., Pluquet, O., Hatzoglou, M., Koumenis, C., Taya, Y., Yoshimura, A., and Koromilas, A. (2004) Endoplasmic reticulum stress induces p53 cytoplasmic localization and prevents p53-dependent apoptosis by a pathway involving glycogen synthase kinase-3beta. *Genes Dev* 18:261–77.

Rao, R. V., Peel, A., Logvinova, A., Del Rio, G., Hermel, E., Yokota, T., Goldsmith, P. C., Ellerby, L. M., Ellerby, H. M., and Bredesen, D. E. (2002) Coupling endoplasmic reticulum stress to the cell death program: Role of the ER chaperone GRP78. *FEBS Lett* 514:122–8.

Reddy, R. K., Mao, C., Baumeister, P., Austin, R. C., Kaufman, R. J., and Lee, A. S. (2003) Endoplasmic reticulum chaperone protein GRP78 protects cells from apoptosis induced by topoisomerase inhibitors: Role of ATP binding site in suppression of caspase-7 activation. *J Biol Chem* 278:20915–24.

Reimold, A. M., Etkin, A., Clauss, I., Perkins, A., Friend, D. S., Zhang, J., Horton, H. F., Scott, A., Orkin, S. H., Byrne, M. C., Grusby, M. J., and Glimcher, L. H. (2000) An essential role in liver development for transcription factor XBP-1. *Genes Dev* 14: 152–7.

Reimold, A. M., Iwakoshi, N. N., Manis, J., Vallabhajosyula, P., Szomolanyi-Tsuda, E., Gravallese, E. M., Friend, D., Grusby, M. J., Alt, F., and Glimcher, L. H. (2001) Plasma cell differentiation requires the transcription factor XBP-1. *Nature* 412:300–7.

Romero-Ramirez, L., Cao, H., Nelson, D., Hammond, E., Lee, A. H., Yoshida, H., Mori, K., Glimcher, L. H., Denko, N. C., Giaccia, A. J., Le, Q. T., and Koong, A. C. (2004) XBP1 is essential for survival under hypoxic conditions and is required for tumor growth. *Cancer Res* 64:5943–7.

Ron, D., and Habener, J. F. (1992) CHOP, a novel developmentally regulated nuclear protein that dimerizes with transcription factors C/EBP and LAP and functions as a dominant-negative inhibitor of gene transcription. *Genes Dev* 6:439–53.

Roybal, C. N., Yang, S., Sun, C. W., Hurtado, D., Vander Jagt, D. L., Townes, T. M., and Abcouwer, S. F. (2004) Homocysteine increases the expression of vascular endothelial growth factor by a mechanism involving endoplasmic reticulum stress and transcription factor ATF4. *J Biol Chem* 279:14844–52.

Ryu, E. J., Harding, H. P., Angelastro, J. M., Vitolo, O. V., Ron, D., and Greene, L. A. (2002) Endoplasmic reticulum stress and the unfolded protein response in cellular models of Parkinson's disease. *J Neurosci* 22:10690–8.

Sato, N., Urano, F., Yoon, L. J., Kim, S. H., Li, M., Donoviel, D., Bernstein, A., Lee, A. S., Ron, D., Veselits, M. L., Sisodia, S. S., and Thinakaran, G. (2000) Upregulation of BiP and CHOP by the unfolded-protein response is independent of presenilin expression. *Nat Cell Biol* 2:863–70.

Scheuner, D., Mierde, D. V., Song, B., Flamez, D., Creemers, J. W., Tsukamoto, K., Ribick, M., Schuit, F. C., and Kaufman, R. J. (2005) Control of mRNA translation preserves endoplasmic reticulum function in beta cells and maintains glucose homeostasis. *Nat Med* 11:757–64.

Scheuner, D., Song, B., McEwen, E., Liu, C., Laybutt, R., Gillespie, P., Saunders, T., Bonner-Weir, S., and Kaufman, R. J. (2001) Translational control is required for the unfolded protein response and *in vivo* glucose homeostasis. *Mol Cell* 7:1165–76.

Semenza, G. L. (2001) HIF-1, O(2), and the 3 PHDs: How animal cells signal hypoxia to the nucleus. *Cell* 107:1–3.

Semenza, G. L., Agani, F., Feldser, D., Iyer, N., Kotch, L., Laughner, E., and Yu, A. (2000) Hypoxia, HIF-1, and the pathophysiology of common human diseases. *Adv Exp Med Biol* 475:123–30.

Shaffer, A. L., Shapiro-Shelef, M., Iwakoshi, N. N., Lee, A. H., Qian, S. B., Zhao, H., Yu, X., Yang, L., Tan, B. K., Rosenwald, A., Hurt, E. M., Petroulakis, E., Sonenberg, N., Yewdell, J. W., Calame, K., Glimcher, L. H., and Staudt, L. M. (2004) XBP1, downstream of Blimp-1, expands the secretory apparatus and other organelles, and increases protein synthesis in plasma cell differentiation. *Immunity* 21:81–93.

Shen, J., Chen, X., Hendershot, L., and Prywes, R. (2002) ER stress regulation of ATF6 localization by dissociation of BiP/GRP78 binding and unmasking of Golgi localization signals. *Dev Cell* 3:99–111.

Shen, J., Snapp, E. L., Lippincott-Schwartz, J., and Prywes, R. (2005) Stable binding of ATF6 to BiP in the endoplasmic reticulum stress response. *Mol Cell Biol* 25:921–32.

Shen, J. W., Subjeck, J. R., Lock, R. B., and Ross, W. E. (1989) Depletion of topoisomerase II in isolated nuclei during a glucose-regulated stress response. *Mol Cell Biol* 9:3284–91.

Shi, Y., Vattem, K. M., Sood, R., An, J., Liang, J., Stramm, L., and Wek, R. C. (1998) Identification and characterization of pancreatic eukaryotic initiation factor 2 alpha-subunit kinase, PEK, involved in translational control. *Mol Cell Biol* 18:7499–509.

Shiu, R. P., Pouyssegur, J., and Pastan, I. (1977) Glucose depletion accounts for the induction of two transformation-sensitive membrane proteinsin Rous sarcoma virus-transformed chick embryo fibroblasts. *Proc Natl Acad Sci U S A* 74:3840–4.

Shohat, M., Janossy, G., and Dourmashkin, R. R. (1973) Development of rough endoplasmic reticulum in mouse splenic lymphocytes stimulated by mitogens. *Eur J Immunol* 3: 680–7.

Shuda, M., Kondoh, N., Imazeki, N., Tanaka, K., Okada, T., Mori, K., Hada, A., Arai, M., Wakatsuki, T., Matsubara, O., Yamamoto, N., and Yamamoto, M. (2003) Activation of the ATF6, XBP1 and grp78 genes in human hepatocellular carcinoma: A possible involvement of the ER stress pathway in hepatocarcinogenesis. *J Hepatol* 38:605–14.

Sidrauski, C., Cox, J. S., and Walter, P. (1996) tRNA ligase is required for regulated mRNA splicing in the unfolded protein response. *Cell* 87:405–13.

Sidrauski, C., and Walter, P. (1997) The transmembrane kinase Ire1p is a site-specific endonuclease that initiates mRNA splicing in the unfolded protein response. *Cell* 90: 1031–9.

Silva, R. M., Ries, V., Oo, T. F., Yarygina, O., Jackson-Lewis, V., Ryu, E. J., Lu, P. D., Marciniak, S. M., Ron, D., Przedborski, S., Kholodilov, N., Greene, L. A., and Burke, R. E. (2005) CHOP/GADD153 is a mediator of apoptotic death in substantia nigra dopamine neurons in an *in vivo* neurotoxin model of parkinsonism. *J Neurochem* 95:974–86.

Siman, R., Flood, D. G., Thinakaran, G., and Neumar, R. W. (2001) Endoplasmic reticulum stress-induced cysteine protease activation in cortical neurons: Effect of an Alzheimer's disease-linked presenilin-1 knock-in mutation. *J Biol Chem* 276:44736–43.

Siu, F., Bain, P. J., LeBlanc-Chaffin, R., Chen, H., and Kilberg, M. S. (2002) ATF4 is a mediator of the nutrient-sensing response pathway that activates the human asparagine synthetase gene. *J Biol Chem* 277:24120–7.

Song, M. S., Park, Y. K., Lee, J. H., and Park, K. (2001) Induction of glucose-regulated protein 78 by chronic hypoxia in human gastric tumor cells through a protein kinase C-epsilon/ERK/AP-1 signaling cascade. *Cancer Res* 61:8322–30.

Soung, Y. H., Lee, J. W., Kim, S. Y., Park, W. S., Nam, S. W., Lee, J. Y., Yoo, N. J., and Lee, S. H. (2004) Somatic mutations of CASP3 gene in human cancers. *Hum Genet* 115:112–5.

Sriburi, R., Jackowski, S., Mori, K., and Brewer, J. W. (2004) XBP1: A link between the unfolded protein response, lipid biosynthesis, and biogenesis of the endoplasmic reticulum. *J Cell Biol* 167:35–41.

Strom, T. M., Hortnagel, K., Hofmann, S., Gekeler, F., Scharfe, C., Rabl, W., Gerbitz, K. D., and Meitinger, T. (1998) Diabetes insipidus, diabetes mellitus, optic atrophy and deafness (DIDMOAD) caused by mutations in a novel gene (wolframin) coding for a predicted transmembrane protein. *Hum Mol Genet* 7:2021–8.

Sun, M., Rothermel, T. A., Shuman, L., Aligo, J. A., Xu, S., Lin, Y., Lamb, R. A., and He, B. (2004) Conserved cysteine-rich domain of paramyxovirus simian virus 5 V protein plays an important role in blocking apoptosis. *J Virol* 78:5068–78.

Tamatani, M., Matsuyama, T., Yamaguchi, A., Mitsuda, N., Tsukamoto, Y., Taniguchi, M., Che, Y. H., Ozawa, K., Hori, O., Nishimura, H., Yamashita, A., Okabe, M., Yanagi, H., Stern, D. M., Ogawa, S., and Tohyama, M. (2001) ORP150 protects against hypoxia/ischemia-induced neuronal death. *Nat Med* 7:317–23.

Tardif, K. D., Mori, K., Kaufman, R. J., and Siddiqui, A. (2004) Hepatitis C virus suppresses the IRE1-XBP1 pathway of the unfolded protein response. *J Biol Chem* 279:17158–64.

Tessitore, A., Del, P. M., Sano, R., Ma, Y., Mann, L., Ingrassia, A., Laywell, E. D., Steindler, D. A., Hendershot, L. M., and D'Azzo, A. (2004) G(M1)-ganglioside-mediated activation of the unfolded protein response causes neuronal death in a neurodegenerative gangliosidosis. *Mol Cell* 15:753–66.

Tirasophon, W., Welihinda, A. A., and Kaufman, R. J. (1998) A stress response pathway from the endoplasmic reticulum to the nucleus requires a novel bifunctional protein kinase/endoribonuclease (Ire1p) in mammalian cells. *Genes Dev* 12:1812–24.

Travers, K. J., Patil, C. K., Wodicka, L., Lockhart, D. J., Weissman, J. S., and Walter, P. (2000) Functional and genomic analyses reveal an essential coordination between the unfolded protein response and ER-associated degradation. *Cell* 101:249–58.

Ubeda, M., Wang, X. Z., Zinszner, H., Wu, I., Habener, J. F., and Ron, D. (1996) Stress-induced binding of the transcriptional factor CHOP to a novel DNA control element. *Mol Cell Biol* 16:1479–89.

van Anken, E., Romijn, E. P., Maggioni, C., Mezghrani, A., Sitia, R., Braakman, I., and Heck, A. J. (2003) Sequential waves of functionally related proteins are expressed when B cells prepare for antibody secretion. *Immunity* 18:243–53.

Vasudevan, K. M., Gurumurthy, S., and Rangnekar, V. M. (2004) Suppression of PTEN expression by NF-kappa B prevents apoptosis. *Mol Cell Biol* 24:1007–21.

Wang, C. Y., Mayo, M. W., Korneluk, R. G., Goeddel, D. V., and Baldwin, A. S., Jr. (1998a) NF-kappaB antiapoptosis: Induction of TRAF1 and TRAF2 and c-IAP1 and c-IAP2 to suppress caspase-8 activation. *Science* 281:1680–3.

Wang, H. G., Pathan, N., Ethell, I. M., Krajewski, S., Yamaguchi, Y., Shibasaki, F., McKeon, F., Bobo, T., Franke, T. F., and Reed, J. C. (1999) Ca2+-induced apoptosis through calcineurin dephosphorylation of BAD. *Science* 284:339–43.

Wang, X.-Z., Harding, H. P., Zhang, Y., Jolicoeur, E. M., Kuroda, M., and Ron, D. (1998b) Cloning of mammalian Ire1 reveals diversity in the ER stress responses. *EMBO J* 17:5708–17.

Waris, G., Tardif, K. D., and Siddiqui, A. (2002) Endoplasmic reticulum (ER) stress: Hepatitis C virus induces an ER-nucleus signal transduction pathway and activates NF-kappaB and STAT-3. *Biochem Pharmacol* 64:1425–30.

Watowich, S. S., Morimoto, R. I., and Lamb, R. A. (1991) Flux of the paramyxovirus hemagglutinin–neuraminidase glycoprotein through the endoplasmic reticulum activates transcription of the GRP78-BiP gene. *J Virol* 65:3590–7.

Wiest, D. L., Burkhardt, J. K., Hester, S., Hortsch, M., Meyer, D. I., and Argon, Y. (1990) Membrane biogenesis during B cell differentiation: Most endoplasmic reticulum proteins are expressed coordinately. *J Cell Biol* 110:1501–11.

Xu, C., Bailly-Maitre, B., and Reed, J. C. (2005) Endoplasmic reticulum stress: Cell life and death decisions. *J Clin Invest* 115:2656–64.

Yamada, M., Tomida, A., Yun, J., Cai, B., Yoshikawa, H., Taketani, Y., and Tsuruo, T. (1999) Cellular sensitization to cisplatin and carboplatin with decreased removal of platinum-DNA adduct by glucose-regulated stress. *Cancer Chemother Pharmacol* 44:59–64.

Yamamoto, K., Ichijo, H., and Korsmeyer, S. J. (1999) BCL-2 is phosphorylated and inactivated by an ASK1/Jun N-terminal protein kinase pathway normally activated at G(2)/M. *Mol Cell Biol* 19:8469–78.

Yan, W., Frank, C. L., Korth, M. J., Sopher, B. L., Novoa, I., Ron, D., and Katze, M. G. (2002) Control of PERK eIF2alpha kinase activity by the endoplasmic reticulum stress-induced molecular chaperone P58IPK. *Proc Natl Acad Sci U S A* 99:15920–5.

Yang, Y., Turner, R. S., and Gaut, J. R. (1998) The chaperone BiP/GRP78 binds to amyloid precursor protein and decreases Abeta40 and Abeta42 secretion. *J Biol Chem* 273:25552–5.

Ye, J., Rawson, R. B., Komuro, R., Chen, X., Dave, U. P., Prywes, R., Brown, M. S., and Goldstein, J. L. (2000) ER stress induces cleavage of membrane-bound ATF6 by the same proteases that process SREBPs. *Mol Cell* 6:1355–64.

Yoneda, T., Imaizumi, K., Oono, K., Yui, D., Gomi, F., Katayama, T., and Tohyama, M. (2001) Activation of caspase-12, an endoplastic reticulum (ER) resident caspase, through tumor necrosis factor receptor-associated factor 2-dependent mechanism in response to the ER stress. *J Biol Chem* 276:13935–40.

Yoshida, H., Haze, K., Yanagi, H., Yura, T., and Mori, K. (1998) Identification of the cis-acting endoplasmic reticulum stress response element responsible for transcriptional induction of mammalian glucose-regulated proteins. Involvement of basic leucine zipper transcription factors. *J Biol Chem* 273:33741–9.

Yoshida, H., Matsui, T., Yamamoto, A., Okada, T., and Mori, K. (2001) XBP1 mRNA is induced by ATF6 and spliced by IRE1 in response to ER stress to produce a highly active transcription factor. *Cell* 107:881–91.

Yoshida, H., Matsui, T., Hosokawa, N., Kaufman, R. J., Nagata, K., and Mori, K. (2003) A time-dependent phase shift in the mammalian unfolded protein response. *Dev Cell* 4:265–71.

Yoshida, H., Okada, T., Haze, K., Yanagi, H., Yura, T., Negishi, M., and Mori, K. (2000) ATF6 activated by proteolysis binds in the presence of NF-Y (CBF) directly to the cis-acting element responsible for the mammalian unfolded protein response. *Mol Cell Biol* 20:6755–67.

Yoshida, H., Oku, M., Suzuki, M., and Mori, K. (2006) pXBP1(U) encoded in XBP1 pre-mRNA negatively regulates UPR activator pXBP1(S) in mammalian ER stress response. *J Cell Biol* 172:562–575.

Yun, J., Tomida, A., Nagata, K., and Tsuruo, T. (1995) Glucose-regulated stresses confer resistance to VP-16 in human cancer cells through a decreased expression of DNA topoisomerase II. *Oncol Res* 7:583–90.

Zhang, P., McGrath, B., Li, S., Frank, A., Zambito, F., Reinert, J., Gannon, M., Ma, K., McNaughton, K., and Cavener, D. R. (2002) The PERK eukaryotic initiation factor 2 alpha kinase is required for the development of the skeletal system, postnatal growth, and the function and viability of the pancreas. *Mol Cell Biol* 22:3864–74.

Zhang, P. L., Lun, M., Teng, J., Huang, J., Blasick, T. M., Yin, L., Herrera, G. A., and Cheung, J. Y. (2004) Preinduced molecular chaperones in the endoplasmic reticulum protect cardiomyocytes from lethal injury. *Ann Clin Lab Sci* 34:449–57.

Zinszner, H., Kuroda, M., Wang, X., Batchvarova, N., Lightfoot, R. T., Remotti, H., Stevens, J.L., and Ron, D. (1998) CHOP is implicated in programmed cell death in response to impaired function of the endoplasmic reticulum. *Genes Dev* 12:982–95.

II
Molecular Mechanisms of Stress Protein Expression

5
Genetic Models of HSF Function

ANDRÁS OROSZ AND IVOR J. BENJAMIN
University of Utah, Department of Internal Medicine, Division of Cardiology, Salt Lake City, UT 84132-2401

1. Introduction and Historical Overview

Since Ritossa's seminal discovery in 1962 that the puffing pattern changes of *Drosophila* salivary gland polytene chromosomes can be induced by heat shock and chemical treatment (Ritossa, 1962), the heat shock response (HSR) has served as an excellent model and paradigm of inducible gene expression. During the ensuing decades considerable evidence has accumulated, from diverse areas of biology, about the regulation of the stress response during development, homeostatic maintenance of organs and organisms, and pathophysiological conditions. From bacteria to man, environmental stress, cell growth, differentiation, and pathophysiological states are all known to induce the rapid and reversible synthesis of evolutionary conserved set of proteins commonly termed, heat shock proteins (HSPs). HSPs, acting as molecular chaperones, play essential roles in protein folding, trafficking, higher order assembly and degradation of proteins thereby ensuring survival under both stressful and extreme physiological conditions (Lindquist, 1986; Lindquist and Craig, 1988; Morimoto, 1998).

2. Transcriptional Regulation of the Stress Response

While numerous members of the HSP family are constitutively expressed, the capacity for stress-inducible activation of key effectors in response to noxious stimuli significantly enhances the versatility of eukaryotic Hsp gene expression. In higher eukaryotes, the regulation of the heat shock response at the transcriptional level is mediated chiefly by a preexisting transcriptional activator, heat shock factor (HSF), which binds to its regulating heat shock element (HSE) present in all heat shock gene promoters. HSE is characterized as inverse reiterations of the pentanucleotide sequence 5'-nGAAn-3' (Lis and Wu, 1993; Wu, 1995). The structural architecture of HSF molecules is remarkably conserved during evolution and consists of an N-terminal winged helix-turn-helix DNA binding domain (DBD), followed by the oligomerization domain (HR-A/B) of hydrophobic heptad repeats possessing the propensity to form trimers through triple stranded coiled-coil interactions. A third conserved region is located near the carboxy terminus containing another array of conserved hydrophobic heptad repeats (HR-C) that have been implicated in suppression of trimerization. The less conserved transactivator (TA) region has been mapped behind the HR-C hydrophobic heptad repeats (Wisniewski et al., 1996). Under normal conditions, HSF in higher eukaryotes is present in cells as a latent monomer, which binds DNA with low affinity. Upon heat stress or other stressful stimuli, HSF is reversibly converted to a trimeric state and acquires high DNA binding affinity. The maintenance of the monomeric conformation is dependent on the hydrophobic heptad repeats located at the amino and the carboxy-terminal regions, which have been proposed to form an intramolecular coiled-coil structure (Westwood et al., 1991; Baler et al., 1993; Rabindran et al., 1993; Wu, 1995; Zuo et al., 1995; Voellmy, 2004). Negative regulation of DNA-binding and transactivation domains also involves intramolecular interaction with the central "interzipper" domain of HSF and is influenced by phosphorylation of critical serine residues (Knauf et al., 1996; Kline and Morimoto, 1997; Guettouche et al., 2005).

Several investigators have proposed that molecular chaperones, downstream targets of HSF activation, are also key players in keeping HSF in an inactive conformation by shifting the equilibrium toward the monomeric inert state. Specifically, complexes between Hsp70 and the HSF have been detected during the recovery phase of heat shock (Abravaya et al., 1992; Baler et al., 1992; Shi et al., 1998), suggesting an autoregulatory role of chaperones in heat shock response. Other chaperones such as Hsp 90 and Hdj-1/Hsp40 have been also shown to play a role in negative regulation of HSF (Mosser et al., 1993; Rabindran et al., 1994; Ali et al., 1998; Shi et al., 1998; Zou et al., 1998).

Both biochemical separation techniques and indirect immunofluorescent staining of fixed cells addressed the cytoplasmic-nuclear localization as regulatory step of HSF activation during heat shock. Although the importance of cytoplasmic-nuclear relocalization of HSF during stress remains controversial, the emerging consensus emphasizes the importance of the experimental system used: in certain cells and organisms HSF is predominantly nuclear (*Drosophila, Xenopus*) before and after heat shock, while in others (e.g., human or mouse tissue culture cells) cytoplasmic-nuclear relocalization may play considerable role in the regulation of stress response. (Westwood et al., 1991; Zuo et al., 1995; Orosz et al., 1996).

Transcriptionally active HSF purified from yeast, *Drosophila* and mammalian cells is phosphorylated, although the roles of phosphorylation events in HSF activation or deactivation still remain elusive. Maintenance of the inert monomeric conformation of HSF is influenced by constitutive phosphorylation of critical serine residues in higher eukaryotes (Knauf et al., 1996; Kline and Morimoto, 1997). In *S. cerevisiae* phosphorylation facilitates HSF deactivation in the recovery phase of heat stress (Hoj and Jakobsen, 1994), although transcription competence of HSF typically correlates with hyperphosphorylation (Sorger and Pelham, 1988; Cotto et al., 1996). Phosphorylation modifications predominantly earmark serine residues both on the non-stressed and on the heat shocked molecule. In spite of many efforts, the critical residues for constitutive and heat-induced phosphorylation of HSF have not been precisely determined in the context of the whole molecule.

In spite of detailed biochemical analyses, the critical stress signals and the precise molecular mechanisms for *Hsf1* activation have hitherto remained enigmatic until recently. Indeed, there is now compelling evidence, in both intact animals and cultured *Hsf1* null cells, that *Hsf1* activation can be regulated in a redox-dependent manner. Earlier studies by Yan and co-workers (Yan et al., 2002) demonstrated that several members of the HSP family were down-regulated by *Hsf1* deletion in parallel with increased oxidative stress, indicating the requirements of *Hsf1* expression under non-stressed conditions. Thereafter, Ahn and Thiele elegantly showed essential requirements of two critical cysteines located in DNA-binding domain of *Hsf1* to transmit stress-inducible signals (i.e., heat stress or hydrogen peroxide) that mediate DNA-binding competence, stress-inducible nuclear translocation and full cytoprotective effects in vitro (Ahn and Thiele, 2003). At least for mammalian *Hsf1*, such redox-sensing properties are predicted to have important implications for development, normal aging, and pathophysiological states, including ischemic and neurodegenerative conditions.

3. The HSF Family of Transcriptional Regulators

Efforts in the early 1990 to clone HSFs from different species across the phylogenetic spectra lead to the realization that although single cell eukaryotes (*S. cerevisiae, K. lactis, and S. pombe*) and *Drosophila* have only one general HSF, multiple members of HSF families exist in plants, mammalians and other vertebrates. The initial salvo opened with the cloning of three different HSFs from tomato by Southwestern screening (Scharf et al., 1990) and isolation of mammalian *Hsf1* and *Hsf2* from both human and mouse shortly thereafter (Rabindran et al., 1991; Sarge et al., 1991; Schuetz et al., 1991). Subsequently, three different HSFs (*Hsf1, Hsf2* and *Hsf3*) were isolated and analyzed from chicken (Nakai and Morimoto, 1993) with *Hsf4*, the last member of the mammalian HSF family, to be cloned a few years later (Nakai et al., 1997). Existence of several HSF genes in higher eukaryotes immediately raised the question whether one or all members are homologues of the general HSF found in lower eukaryotes. In spite of differences in expression patterns in cell lines or embryonic development, and response to diverse environmental stimuli inducing proteotoxic damage, the unifying theme of HSR induction has emerged that only *Hsf1* responds to heat shock or other physiological and pathophysiological stresses. Therefore, the ubiquitously expressed *Hsf1* is the only true stress-inducible homologue of the general HSF existing in lower eukaryotes.

Conversely, *Hsf2* is active only in certain developmental situations. In human K562 erythroleukemia cell line, hemin induces in vitro differentiation and transient expression of heat shock genes presumably under *Hsf2* control (Sistonen et al., 1992). In mouse developmental pathways, *Hsf2* activity has only been observed during certain stages of embryogenesis and in spermatogenesis in adult animals, but its target genes remained elusive (Sarge et al., 1994; Rallu et al., 1997). *Hsf4* lacks the C-terminal conserved hydrophobic heptad repeat (HR-C) characteristic of other vertebrate HSFs and binds DNA constitutively (Nakai et al., 1997). Alternatively, spliced isoforms of *Hsf4* may exert their function as transcriptional activators or repressors on basal and inducible expression on heat shock proteins and other yet unknown target genes depending on the presence of the DHR domain (downstream of HR-C); DHR is a hydrophobic heptad repeat variant, which functions as part of an transcription activator (Tanabe et al., 1999). A protein alignment of the currently known and predicted mammalian HSFs is shown in Figure 1 compared to vertebrate and other low eukaryotic HSFs. Based on the alignment species-specific relationship of HSFs is depicted in Figure 2 as cladogram.

To understand the specific physiological functions of the HSF family in the context of the whole organism, loss of function (LOF) studies were initiated in several laboratories in the late 1990s. Last year these efforts reached a significant landmark with the completion of knockout mice for all three mammalian HSFs. Important milestones in HSF research are summarized in Table 1. In this chapter, we discuss the emerging picture from the analysis of these animal models containing anticipated, unexpected and, sometimes, contradictory findings. In this regard, a consensus has emerged from a special one day mini-symposium "Making Sense of HSF Mutations in Mice: Does the Field Need Another Knockout Model?" at Salva Regina University, Newport, RI. (Christians and Benjamin, 2005).

4. Targeted Disruption of the Stress-Responsive General HSF: Analysis of *Hsf1* Knockout Mice

The first targeted HSF knockout mouse model was reported from the pioneering work of the Benjamin laboratory (Xiao et al., 1999), and subsequently two other *Hsf1* knockout mice were generated by investigators using similar gene targeting strategies (Zhang et al., 2002; Izu et al., 2004). $Hsf1-/-$ animals can survive to adulthood, indicating that *Hsf1* is dispensable for cell growth and viability in mice. Beside its role in cytoprotection following proteotoxic damage, *Hsf1* has also been implicated in sustaining cell growth and viability in yeast (Sorger and Pelham, 1988), and it is required in *Drosophila,* under normal growth condition, to support oogenesis and early larval development (Jedlicka et al., 1997).

Hsf1-deficient animals show a complete loss of inducible HSP expression and the inability to develop acquired cellular resistance or thermotolerance (McMillan et al., 1998; Xiao et al., 1999), an anticipated finding in line with the evidence from related work in yeast and *Drosophila*, which cannot be compensated by other HSF family members (*Hsf2* or *Hsf4*). Mouse embryonic fibroblast cells (MEFs) derived from mouse embryos also demonstrate noticeable increases in heat-induced apoptosis, a cell death pathway where inducible and constitutive HSPs play vital roles (McMillan et al., 1998). On the other hand, animals lacking *Hsf1* exhibit down regulation of constitutive HSP expression such as Hsp25 in different tissues, indicating for the first time a requirement for *Hsf1* activation in

FIGURE 1. ClustalW alignment of mammalian, vertebratae and lower eukaryotic HSF protein sequences. Abbreviations: m (mouse), h (human), r (rat), d (dog), o (opossum), c (chicken), x (*Xenopu*s), z (zebrafish), f (fugu), dm (*Drosophila*), ce (*C. elegans*), sc (*S. cerevisiae*).

Color codes for amino acids: RED=small+hydrophobic+aromatic-Y, Blue=acidic, MAGENTA+basic, GREEN=hydroxyl+amine+basic-Q.

The conserved DNA binding domain is underlined with red, HR-A/B with green and the HR-C, which is absent in HSF4 is light blue. The multiple alignment was carried out using the ClustalW program from the EMBL European Bioinformatics Institute web site (www.ebi.ac.uk/clustalw/).

```
mHSF2   --------------------------------------------------MKQSS    5
rHSF2   --------------------------------------------------MKQSS    5
hHSF2   --------------------------------------------------MKQSS    5
dHSF2   --------------------------------------------MVTLGRLPISS   11
oHSF2   --------------------------------------------------MKQSS    5
cHSF2   ----------------------------------MKQEPQQQQPAQQPPPAGA     19
xHSF2   --------------------------------------------------MKQNS    5
zHSF2   --------------------------------------------------MKHSS    5
cHSF1   APVLEPPHPYPTAGPHLPAVPPRVASRSPVGIRPRGRRTRARPGLSMREGSALLGAPGAA 180
cHSF3   -------------------------------------------MREGSALLGAPGAA  14
mHSF1   ---------------------------------------------MDLA-VGPG--AAGPS 13
rHSF1   ---------------------------------------------MDLA-VGPG--AAGPS 13
hHSF1   ---------------------------------------------MDLP-VGPG--AAGPS 13
dHSF1   ---------------------------------------------MDLP-VGPG--AAGPS 13
oHSF1   ---------------------------------------------MEFP-GGPG--MAGPS 13
xHSF1   ------------------------------------------------------GGS  3
zHSF1   ------------------------------------------MEVHSVGPGGVVVTGN 16
fHSF1   ------------------------------------------------------S  1
mHSF4   --------------------------------------------MQEAPAALPTEPGPS 15
hHSF4   -------------------------------------------MVQEAPAALPTEPGPS 16
rHSF4   --------------------------------------------MQEAPAALPTEPGPS 15
dHSF4   --------------------------------------------MQEAPAALPTEPGPS 15
oHSF4   -------------------------------------------MQRTGPPSVEVDGYSS 16
dmHSF   ------------------------------MSRERSSAKAVQFKHESEEEEEDEEEQL 28
ceHSF   QQQQQQLIMRVPKQEVSVSGAARRYVQQAPPNRPPRQNHQNGAIGGKKSSVTIQEVPMNA 71
scHSF   QTNLYGHNSRENTNPNSTLLSSKLLAHPPVPYGQNPDLLQHAVYRAQPSSGTTNAQPRQT 162

mHSF2   N---------------VPAFLSKLWTLVEETHTNEFITWSQ----------------- 31
rHSF2   N---------------VPAFLSKLWTLVEETHTNEFITWSQ----------------- 31
hHSF2   N---------------VPAFLSKLWTLVEETHTNEFITWSQ----------------- 31
dHSF2   G---------------LSCVVLKMKGVARVTRDSSSFKYRDSKQNDDDVNSAFAFECIM 55
oHSF2   N---------------VPAFLSKLWTLVEEAHTNEFITWSQ----------------- 31
cHSF2   G---------------VPAFLSKLWALVGEAPSNQLITWSQ----------------- 45
xHSF2   N---------------VPAFLSKLWTLVEDTDTNEFIIWNQ----------------- 31
zHSF2   N---------------VPAFLTKLWTLVEDSDTNEFICWSQ----------------- 31
cHSF1   P---------------VPGFLAKLWALVEDPQSDDVICWSR----------------- 206
cHSF3   P---------------VPGFLAKLWALVEDPQSDDVICWSR----------------- 40
mHSF1   N---------------VPAFLTKLWTLVSDPDTDALICWSP----------------- 39
rHSF1   N---------------VPAFLTKLWTLVSDPDTDALICWSP----------------- 39
hHSF1   N---------------VPAFLTKLWTLVSDPDTDALICWSP----------------- 39
dHSF1   N---------------VPAFLTKLWTLVSDPDTDALICWSP----------------- 39
oHSF1   N---------------VPAFLTKLWTLVGDPDTDPLICWSP----------------- 39
xHSF1   N---------------VPAFLAKLWTLVEDPETDPLICWSP----------------- 29
zHSF1   N---------------VPAFLTKLWTLVEDPDTDPLICWSP----------------- 42
fHSF1   N---------------VPAFLTKLWTLVEDPDTDPLICWSK----------------- 27
mHSF4   P---------------VPAFLGKLWALVGDPGTDHLIRWSP----------------- 41
hHSF4   P---------------VPAFLGKLWALVGDPGTDHLIRWSP----------------- 42
rHSF4   P---------------VPAFLGKLWALVGDPGTDHLIRWSP----------------- 41
dHSF4   P---------------VPAFLGKLWALVGDPGTDHLIRWSP----------------- 41
oHSF4   P---------------VPAFLTKLWTLVGDPETDHLIYWSP----------------- 42
dmHSF   PSRRMHSYGDAAAIGSGVPAFLAKLWRLVDDADTNRLICWTK---------------- 70
ceHSF   YLETLNKSGNNKVDDDKLPVFLIKLWNIVEDPNLQSIVHWDD---------------- 113
scHSF   TRRYQSHKS--------RPAFVNKLWSMLNDDSNTKLIQWAE---------------- 196

mHSF2   -------------------NGQSFLVLDEQRFAKEVLPKYFKHNNMASFVRQLNMYGFRK 72
rHSF2   -------------------NGQSFLVLDEQRFAKEILPKYFKHNNMASFVRQLNMYGFRK 72
hHSF2   -------------------NGQSFLVLDEQRFAKEILPKYFKHNNMASFVRQLNMYGFRK 72
dHSF2   GQRKYKFSEGKDCVCAKDTNGQSFLVLDEQRFAKEILPKYFKHNNMASFVRQLNMYGFRK 115
oHSF2   -------------------NGQSFLVLDEQRFAKEILPKYFKHNNMASFVRQLNMYGFRK 72
cHSF2   -------------------NGQSFLVLDEQRFAKEILPKYFKHNNMASFVRQLNMYGFRK 86
xHSF2   -------------------NGQSFLVLDEQRFAKEILPKFFKHNNMASFVRQLNMYGFRK 72
zHSF2   -------------------EGNSFLVLDEQRFSKEVLPKYFKHNNMASFVRQLNMYGFRK 72
cHSF1   -------------------NGENFCILDEQRFAKELLPKYFKHNNISSFIRQLNMYGFRK 247
cHSF3   -------------------NGENFCILDEQRFAKELLPKYFKHNNISSFIRQLNMYGFRK 81
mHSF1   -------------------SGNSFHVFDQGQFAKEVLPKYFKHNNMASFVRQLNMYGFRK 80
rHSF1   -------------------SGNSFHVFDQGQFAKEVLPKYFKHNNMASFVRQLNMYGFRK 80
hHSF1   -------------------SGNSFHVFDQGQFAKEVLPKYFKHNNMASFVRQLNMYGFRK 80
dHSF1   -------------------SGNSFHVFDQGQFAKEVLPKYFKHNNMASFVRQLNMYGFRK 80
oHSF1   -------------------SGNSFHVFDQGQFAKEVLPKYFKHNNMASFVRQLNMYGFRK 80
xHSF1   -------------------EGNSFHVFDQGQFAKEVLPKYFKHNNMASFVRQLNMYGFRK 70
zHSF1   -------------------NGTSFHVFDQGRFSKEVLPKYFKHNNMASFVRQLNMYGFRK 83
fHSF1   -------------------TGNSFHVFDQGRFSKEILPKFFKHNNMASFIRQLNMYGFRK 68
mHSF4   -------------------SGTSFLVSDQSRFAKEVLPQYFKKKSNMASFVRQLNMYGFRK 82
hHSF4   -------------------SGTSFLVSDQSRFAKEVLPQYFKHSNMASFVRQLNMYGFRK 83
rHSF4   -------------------SGTSFLVSDQSRFAKEVLPQYFKHSNMASFVRQLNMYGFRK 82
dHSF4   -------------------SGTSFLVSDQSRFAKEVLPQYFKHSNMASFVRQLNMYGFRK 82
```

FIGURE 1. (*Continued*)

```
oHSF4    ------------------NGASFHVRDQGRFAKEVLPKYFKHNNMASFVRQLNMYGFRK  83
dmHSF    ------------------DGQSFVIQNQAQFAKELLPLNYKHNNMASFIRQLNMYGFHK 111
ceHSF    ------------------SGASFHISDPYLFGRNVLPHFFKHNNMNSMVRQLNMYGFRK 154
scHSF    ------------------DGKSFIVINREEFVHQILPKYFKHSNFASFVRQLNMYGWHK 237

mHSF2    VVHIESGIIKQERDGP--VEFQHPYFKQGQDDLLENIKRKVSSSKPEENK---------- 120
rHSF2    VVHIESGIVKQERDGP--VEFQHPHFKQGQDDLLENIKRKVSSSKPEENK---------- 120
hHSF2    VVHIDSGIVKQERDGP--VEFQHPYFKQGQDDLLENIKRKVSSSKPEENK---------- 120
dHSF2    VVHIDSGIVKQERDGP--VEFQHPYFKQGQDDLLENIKRKVSSCKPEENK---------- 163
oHSF2    VVHVDSGIVKQERDGP--VEFQHPYFKQGQDDLLENIKRKVSSSKPEETK---------- 120
cHSF2    VVHVDSGIVKLERDGL--VEFQHPYFKQGREDLLENIKRKVSSSRPEENK---------- 134
dHSF2    VVHVDSGIVKQERDGP--VEFQHPFFVKGDELLENIKRKVSSTRPEEGK---------- 120
zHSF2    VMHIDSGIVKQERDGP--VEFQHPYFKHGQDDLLENIKRKVSNARPEESK---------- 120
cHSF1    VVALENGMITAEKNSV--IEFQHPFFKQGNAHLLENIKRKVSAVRTEDLK---------- 295
cHSF3    VVALENGMITAEKNSV--IEFQHPFFKQGNAHLLENIKRKVSAVRTEDLK---------- 129
mHSF1    VVHIEQGGLVKPERDD--TEFQHPCFLRGQEQLLENIKRKVTSVSTLKSEDIK------- 131
rHSF1    VVHIEQGGLVKPERDD--TEFQHPCFLRGQEQLLENIKRKVTSVSTLKSEDIK------- 131
hHSF1    VVHIEQGGLVKPERDD--TEFQHPCFLRGQEQLLENIKRKVTSVSTLKNEDIK------- 131
dHSF1    VVHIEQGGLVKPERDD--TEFQHPCFLRGQEQLLENIKRKVTSVSTLKSEDIK------- 131
oHSF1    VVHIEQGGLVKPERDD--TEFQHPFFIRGQEQLLENIKRKVTSVSSIKHEDIK------- 131
xHSF1    VVHIEQGGLVKPERDD--TEFQHPYFIRGQEQLLENIKRKVNTLSATKSEEVK------- 121
zHSF1    VVHIEQGGLVKPERDD--TEFQHPYFIRGQEQLLENIKRKVITVSNIKHEDYK------- 134
fHSF1    VVHIEQGGLVKPERDD--TEFQHPFFIRGQENLLENIKRKVTNVSAMRQEEVK------- 119
mHSF4    VVSIEQGGLLRPERDH--VEFQHPSFVRGREQLLERVRRKVPALRGDDGR---------- 130
hHSF4    VVSIEQGGLLRPERDH--VEFQHPSFVRGREQLLERVRRKVPALRGDDGR---------- 131
rHSF4    VVSIEQGGLLRPERDH--VEFQHPSFVRGCEQLLERVRRKVPALRGDDTR---------- 130
dHSF4    VVSIEQGGLLRPERDH--VEFQHPSFVRGREQLLERVRRKVPALRGDDGR---------- 130
oHSF4    VVNIEQGGLVKPDLDD--NEFQHQSFLRGHEHLLEQIKRKVSVLRSEENR---------- 131
dmHSF    ITSIDNGGLR-FDRDE--IEFSHPFFKRNSPFFLLDQIKKKISNNKNGDDKGV------- 160
ceHSF    MTPLSQGGLTRTESDQDHLEFSHPSFVQGRPELLSQIKRKQSARTVEDKQVNE------- 207
scHSF    VQDVKSGSIQSSSDDK--WQFENENFIRGREDLLEKIIRQKGSSNNHNSPSGNGNPANGS 295

mHSF2    ---------------------------------------IRQEDLTKIISSAQKVQ 137
rHSF2    ---------------------------------------IRQEDLTKIISSAQKVQ 137
hHSF2    ---------------------------------------IRQEDLTKIISSAQKVQ 137
dHSF2    ---------------------------------------IRQEDLTKIISSAQKVQ 180
oHSF2    ---------------------------------------IRQEDLSKIISSAQKVQ 137
cHSF2    ---------------------------------------ISQEDLSKIISSAQKVE 151
xHSF2    ---------------------------------------VRQEDISKILSNAAKVQ 137
zHSF2    ---------------------------------------IRQDDLSKILTSVQSVH 137
cHSF1    ---------------------------------------VCAEDLHKVLSEVQEMR 312
cHSF3    ---------------------------------------VCAEDLHKVLSEVQEMR 146
mHSF1    ---------------------------------------IRQDSVTRLLTDVQLMK 148
rHSF1    ---------------------------------------IRQDSVTRLLTDVQLMK 148
hHSF1    ---------------------------------------IRQDSVTKLLTDVQLMK 148
dHSF1    ---------------------------------------IRQDSVTKLLTDVQLMK 148
oHSF1    ---------------------------------------VRQDNVTKLLTDVQMMK 148
xHSF1    ---------------------------------------GRQDSVSKLLTDVQSMK 138
zHSF1    ---------------------------------------FSTDDVSKMISDVQHMK 151
fHSF1    ---------------------------------------MSAFEVNKLLSDIHAMK 136
mHSF4    ---------------------------------------WRPEDLSRLLGEVQALR 147
hHSF4    ---------------------------------------WRPEDLGRLLGEVQALR 148
rHSF4    ---------------------------------------WRPEDLGRLLGEVQALR 147
dHSF4    ---------------------------------------WRPEDLGRLLGEVQALR 147
oHSF4    ---------------------------------------LRQEDLSRIICEVQVLR 148
dmHSF    ---------------------------------------LKPEAMSKILTDVKVMR 177
ceHSF    ---------------------------------------QTQQNLEVVMAEMRAMR 224
scHSF    NIPLDNAAGSNNSNNNISSSNSFFNNGHLLQGKTLRLMNEANLGDRNDVTAILGELEQIK 355

mHSF2    IKQETIESRLSELKSENESLWKEVSELRAKHAQQQQVIRKIVQFIVTLV---------- 186
rHSF2    IKQETIESRLSELKSENESLWKEVSELRAKHAQQQQVIRKIVQFIVTLV---------- 186
hHSF2    IKQETIESRLSELKSENESLWKEVSELRAKHAQQQQVIRKIVQFIVTLV---------- 186
dHSF2    IKQETIESRLSELKSENESLWKEVSELRAKHAQQQQVIRKIVQFIVTLV---------- 229
oHSF2    IKQETIESRLTTLKRENESLWREVAELRAKQTQQQQVIRKIVQFIVTLV---------- 186
cHSF2    IKQETIESRLSALKRENESLWREVAELRAKHLRQQQQVIRKIVQFIVTLV---------- 200
xHSF2    VQQETIDSRLFTLKRDNEALWREISDLRNKHAQQQQVIRKIVQFIVTLV---------- 186
zHSF2    EQQENMDARLATLKRENEALWTELSDLRKVHVQQQQVIKELVQFIFTLV---------- 186
cHSF1    EQQNNMDIRLANMKRENKALWKEVAVLRQKHSQQQKLLSKILQFILSLM---------- 361
cHSF3    EQQNNMDIRLANMKRENKALWKEVAVLRQKHSQQQKLLSKILQFILSLM---------- 195
mHSF1    GKQECMDSKLLAMKHENEALWREVASLRQKHAQQQKVVNKLIQFLISLV---------- 197
rHSF1    GKQECMDSKLLAMKHENEALWREVASLRQKHAQQQKVVNKLIQFLISLV---------- 197
hHSF1    GKQECMDSKLLAMKHENEALWREVASLRQKHAQQQKVVNKLIQFLISLV---------- 197
dHSF1    GKQESMDSKLLAMKHENEALWREVASLRQKHAQQQKVVNKLIQFLISLV---------- 197
oHSF1    GKQESMDSKLIAMKHENEALWREVASLRQKHAQQQKVVNKLIQFLISLV---------- 197
xHSF1    GKQETIDKRLLSMKHENEALWREVASLRQKHNQQQKVVNKLIQFLISLV---------- 187
zHSF1    GKQESMDSKISTLKHENEMLWREVATLRQKHSQQQKVVNK-VHFIDANP---------- 199
```

FIGURE 1. (*Continued*)

```
fHSF1   GKQESIDTRIMTMRQENEALWREVASLRQKHAQQQKVVRKLIQFLLSLV-----------  185
mHSF4   GVQESTEARLQELRQQNEILWREVVTLRQSHSQQHRVIGKLIQCLFGPL-----------  196
hHSF4   GVQESTEARLRELRQQNEILWREVVTLRQSHGQQHRVIGKLIQCLFGPL-----------  197
rHSF4   GVQESTEARLQELRQQNEILWREVVTLRQSHSQQHRVIGKLIQCLFGPL-----------  196
dHSF4   GVQEITEARLRELRQQNEILWREVVTLRQSHGQQHRVIGKLIQCLFGPL-----------  196
oHSF4   GQQDSAESQLQDLRQQNEVLWREVMSLRQQHHQQHRVMNKLIHCLFSPI-----------  197
dmHSF   GRQDNLDSRFSAMKQENEVLWREIASLRQKHAKQQQIVNKLIQFLITIV-----------  226
ceHSF   EKAKNMEDKMNKLTKENRDMWTQMGSMRQQHARQQQYFKKLLHFLVSVM-----------  273
scHSF   YNQIAISKDLLRINKDNELLWQENMMARERHRTQQQALEKMFRFLTSIVPHLDPKMIMDG  415

mHSF2   -----------QNNQLVSLRKERP---LLLNTNGAPKKNLYQHIVKEP------TDNHHH  226
rHSF2   -----------QNNQLVSLKRKRP---LLLNTNGAPKKNLFQHIVKEP------TDNHHH  226
hHSF2   -----------QNNQLVSLKRKRP---LLLNTNGAQKKNLFQHIVKEP------TDNHHH  226
dHSF2   -----------QNNQLVSLKRKRP---LLLNTNGAQKKNLFQHIVKEP------ADNHHH  269
oHSF2   -----------QNNQLVSLKRKRP---LLLNTNGSPKTHRFQQIVKDP------VNNHHH  226
cHSF2   -----------QNNQLVSLKRKRP---LLLNTNGPIKSNVFQQIVKEP------ADNNH-  239
xHSF2   -----------QNNRLVSLKRKQP---LLLNTNNSPKSTRLQTIVKET------VEDNHH  226
zHSF2   -----------QNNRLLNLKRKKP---LALNING-KKSKFIKQLFEEP------IDHSK-  224
cHSF1   -----------RGNYIVGVKRKR----SLTDAAGASPSKYSRQYVRIP------VESGQ-  399
cHSF3   -----------RGNYIVGVKRKR----SLTDAAGASPSKYSRQYVRIP------VESGQ-  233
mHSF1   -----------QSNRILGVKR-----KIPLMLSDSNSAHSVPKYGRQY------SLEH--  233
rHSF1   -----------QSNRILGVKR-----KIPLMLSDSSSAHSVPKYGRQY------SLEH--  233
hHSF1   -----------QSNRILGVKR-----KIPLMLNDSGSAHSMPKYSRQF------SLEH--  233
dHSF1   -----------QSNRILGVKR-----KIPLMLNDGSSAHSMPKYGRQY------SLEH--  233
oHSF1   -----------QSNRILGVKR-----KIPLMLNDSSSAHSMPKFSRQF------SLEH--  233
xHSF1   -----------QSNRILGVKR-----KIPLMLNDSTAHPPPKYSRQY------SLEH--  223
zHSF1   -----------HFQ--ITIKW-----HMPLMLNDSSSAHSMPKFSRQY------SLES--  233
fHSF1   -----------QSNGILGLKRKMQVYDIPLMLNDSSTTHSMPKYSRPF------PLES--  226
mHSF4   -----------QTGPSSTGAKR----KLSLMLDEGSASSASAKFNAC-------VSG---  232
hHSF4   -----------QAGPSNAGGER----KLSLMLDEGSSCPTPAKFNTCP------LPG---  233
rHSF4   -----------QTNPSSTGAKR----KLSLMLDEGSASCASAKFNACP------VSG---  232
dHSF4   -----------QTGSSGAGAKR----KLSLMLDEGSSCPTPAKFNTCP------LPG---  232
oHSF4   -----------QAGPSSGASKR----KLPLMLAESTVMHSGVKVSRTS------LPE---  233
dmHSF   -----------QPSRNMSGVKR----HVQLMINNHTPEIDRARTISETE------SESGGG  265
ceHSF   -----------QPSLSKRVAKRGVLEIDFCAANGTAGPNSKRARMNSEESPYKDVCDLLE  322
scHSF   LGDPKVNNEKLNSANNIGLNRDNTGTIDELKSNDSFINDDRNSFTNATTN----ARNNMS  471

mHSF2   KVPHSRTEGLKSRERISDDIIIYDVTDDNVDEE-------NIPVIPETNEDVVVDSS---  276
rHSF2   KVPHSRTEGLKSRERISDDIIIYDVTEDNVDEE-------NIPVIPETNEDVVVDPS---  276
hHSF2   KVPHSRTESLKPRERISDDIIIYDVTDDNADEE-------NIPVIPETNEDVISDPSN--  277
dHSF2   KVPHSRTESLKPRERISDDIIIYDVTDDNADEE-------NIPVIPETNEDVISDPSN--  320
oHSF2   KVPHSRTEGLKPREQISEDIIIYDVTDDNVDEE-------NAPGTPTEMEDATSDTSKHQ  279
cHSF2   -VPLNRTEGLKQREQISDDIIIYDVTEDVADEENTMVDEENAPIPETNEDTTSDSSN--  296
xHSF2   -VPLR-TEGLK-RGDLSEDIVIYDITDNVEFEEEEK-KETATHAVPESSDESTSDIVI-  280
zHSF2   ----TIVNGLKNNSDISDDVIICDITDEDPEVT-------DSISVPDEEDAEIVEITY-  271
cHSF1   -AMAFSEHNSDDEDGNRTGLIIRDIIDTLENATN------GLLAVAHTSGRDRETQTA-  450
cHSF3   -AMAFSEHNSDDEDGNRTGLIIRDIIDTLENATN------GLLAVAHTSGRDRETQTA-  284
mHSF1   --------VHGPGPYSAPSPAYSSSSLYSSDAVTSGPIISDITELAPTSPLASPGRSIDE  286
rHSF1   --------VHGPGPYSAPSPAYSSSSLYSSDAVTSSGPIISDITELAPTSPLASPGRSIDE  286
hHSF1   --------VHGCGPYSAPDPAYGGCELTAFDAVASSGPIISDITELAPASPMASPGGSIDE  286
dHSF1   --------IHGSGPYSAPSPAYSSSSLYSSDAVASSGPIISDITELAPSSPLASPGGSVDE  286
oHSF1   --------VHGSSPYTASSPTYSGSNLYSSDSAANSGPIISDVTELAQSSPSASPSGSVDE  286
xHSF1   --------VHGSTAYPAPVSGFADSSLYSPD--SSAGPIISDVTELAESSPSPSPCLSLEA  274
zHSF1   --------PAPS-------STAFTSTGVFSSESPVKTRGPIISDITELAQSSSPVAT-DEWIED  279
fHSF1   --------IGCL-------LNNPQSALFSPDS-PPSGPIISDITEVHSPVEEEVVTDMTHA  271
mHSF4   --------ALLQDPYFIQSPSP-----GSPSQPR-------WASALTQPE-----GPSS--  267
hHSF4   --------ALLQDPYFIQSPST-----YSLSQRQI------WALALTGPG------APSS--  268
rHSF4   --------ALLQDPYFIQSPLPETTVGLSP-HRARGP-IISDIPEDCPSPEGHRLSPSSA-  282
dHSF4   --------ALLQDPYFIQSPLPETTLGLSSSHRTRGP-IISDIHREDSPSPDGTRLSPSSG-  283
oHSF4   --------LPLGDSYFIQSPSTETESSKRSSSTREVREPIISDISDISPSPE-DIMSPSOVT  285
dmHSF   PVIHELREELLDEVMNPSPAGYTAASHYDSESVSPPAVERPRSNMSISSHNVDYSNQSVE  325
ceHSF   SLQRETQEPFSRRFINNEGPLISEVIDEFGNSPVGRGSAQDLFGDTFGAQSSRYSDGGAT  382
scHSF   PNNDDNSIDTASTNTTNREKNIDENIKNNNDIINDIIFNTNLANNLSNYNSNNNAGSPIR  531

mHSF2   ---NQYPEIVIVEDDNEDEYAPVIQSGEQS-----------EPAREPLRVSSAGSSSPL  321
rHSF2   ---SQYPDIVIVEDDNEDEYAPVIQSGEQS-----------EPA---RVSSAGSSSPL  317
hHSF2   --CSQYPDIVIVEDDNEDEYAPVIQSGEQN-----------EPARESLSSGSDSSS-PL  322
dHSF2   --CSQYPDIVIVEDDNEDEYAPVIQSGDQN-----------EPARESLSSGSDSTS-PL  365
oHSF2   --CSQYPDIVIVEDDNEDDYAAIIQGGDKSSEPAS-------VPACDPISSVSQGAS-PF  329
cHSF2   --CSRSPDIVIVEDDNEEEYAPVIQ-GDKSTESVA------VSANDPLSPVSDSTS-PL  345
xHSF2   --ERLSPEIVIVEDDSDYEDTPVGQ-GNKN------------AETGLRSPAGENTSVPL  324
zHSF2   ---ESSPKTVRQDSAN----GTIVNSTAHQ-----------EAEPKTTPDVSTT---  307
cHSF1   --LDPGLPICQVSQPNELSCAEPIPPVHIN-------------------DVSKPNEM  486
cHSF3   --LDPGLPICQVSQPNELSCAEPIPPVHIN-------------------DVSKPNEM  320
mHSF1   RPLSSSTLVRVKQEPPSPPHSPRVLEASPGRPSS--------MDTPLSPTAFIDSILRESE  339
rHSF1   RPLSSSILVRVKEEPPSPPHSPRVLEASPSRPAS--------MDTPLSPIAFIDSILRESE  339
hHSF1   RPLSSSPLVRVKEEPPSPPQSPRVEEASPNRPSS--------VDTLLSPTALIDSILRESE  339
dHSF1   RPLSSSPVVRVKEEPPSPPFSPRVEEASPGHQSSV-------VEIPLSPTALIDSILRESE  340
oHSF1   R----ISPMIRIKEEPPSPPRSPQPIEETSPGHPPSV------VETPLSPCTFIDSILQESE  337
xHSF1   SP---SPVILIKEEPLTPSHSP---EQSP-APPRV------EDTPVSPSTFIDSILQESE  321
zHSF1   R---TSPLVHIKEEPSSPAHSPEVEEVCPVEVEVGAGSDLPVDTPLSPTIFINSILQESE  336
fHSF1   G---DMQTVNVKEEPSPP-----EADMCPALEVEALP----VDTPLSPTTFINSILQDDI  319
mHSF4   -----LTSQKILHLLKD----TG---------------------FLPP----VVAGAPPP  293
```

FIGURE 1. (*Continued*)

```
hHSF4    -----LTSQKTLHPLRG----PG---------------------FLPP----VMAGAPPP 294
rHSF4    -----GRRVKGLALLKEEPASPGGDGEAG--------------LALAPNECDFCVTAPPP 323
dHSF4    -----GRREKGLALLKEEPASPGGEGEAG--------------LALAPNECDFCVTAPPP 324
oHSF4    ----QRCREKSPMLLKEEPDSPRGKGTTAP-------------SHVSSSDYGFYMTAPPV 328
dmHSF    DLLLQGNGTAGGNILVGGAASPMAQSVSQSPAQHD-------VYTVIEAPDSHVQEVPNS 378
ceHSF    SSREQSPHPIISQPQSNSAGAHGANEQKPDDMYMGSG----PLTHENIHRGISALKRDYQ 438
scHSF    PYKQRYLLKNRANSSTSSENPSLTPFDIESNNDRKISEIPFDDEEEEETDFRPFTSRDPN 591

mHSF2    MSSAVQLNGSSSLTSEDPVTMMDSILNDN------------------------------- 350
rHSF2    MSSAVQLNGSSNLTSEDPVTMMDSILNDN------------------------------- 346
hHSF2    MSSAVQLNGSSSLTSEDPVTMMDSILNDN------------------------------- 351
dHSF2    MSSAVQLNGSSSLTTEDPVTMMDSILNDN------------------------------- 394
oHSF2    MSSAVQLNNSSTLTSDDPVSMMDSILNEN------------------------------- 358
cHSF2    MSSAVQLNNQSTLTAEDPVSMMDSILNENGVISQN------------------------- 380
xHSF2    MSSALQLNNQSSFNQEDPVTMMDSILNESGVISQN------------------------- 359
zHSF2    -NSALQLNKPSCLSLEDPMKLMDSILSENGAISQN------------------------- 341
cHSF1    GNVAVELHTAQANAPEDPVSVIDSILNENNSGNQN------------------------- 521
cHSF3    GNVAVELHTAQANAPEDPVSVIDSILNENNSGNQN------------------------- 355
mHSF1    PTPAASNTAPMDTTG-----AQAPALPTPSTPEK--------CLSVACLD---------- 376
rHSF1    PTPAASNTAPMDTTG-----AQAPAPPAPSTPEK--------CLSVACLD---------- 376
hHSF1    PAPAS-VTALTDARGHTDTEGRPPSPPPTSTPEK--------CLSVACLD---------- 380
dHSF1    PAPAAPTTPLTDAGGRRPS----PSPPPAPAPEK--------CLSVACLDNLARAPQMSG 388
oHSF1    PSTAASVSDTVGGVIS------SPPPGPPQTPEK--------CLSVACLD---------- 373
xHSF1    ASVCPGGNKNEGVSQ---------------SNPPEP--------CLSVACLD---------- 350
zHSF1    PVSALTPPPSEQKCLSVACLDNYPQMSEISRLFS--------GFSTSSLHLR-------- 380
fHSF1    PLTACTSAASPTTASASP----AAVPNSGASPLT--------CQTVACID---------- 357
mHSF4    LPVAVVQAILEGKGSYSPEGPRSVQQPEPRGPR--------------------------- 326
hHSF4    LPVAVVQAILEGKGSFSPEGPRNAQQPEPGDPR--------------------------- 327
rHSF4    LPVAVVQAILEGKGSFSPEGPRSIQQPEPRGPR--------------------------- 356
dHSF4    LSVAVVQAILEGKGNFSPEGPRNAQQPEPRGPR--------------------------- 357
oHSF4    IPVAVVQSVLEGKGSCSPE-QRSAQQPERRDQR--------------------------- 360
dmHSF    PPYYEEQNVLTTPMVREQEQQKRQQLKENNKLRRQAGDVILDAGDILV------------ 426
ceHSF    GASPASGGPSTSSSAPSGAGAGARMAQKRAAPYKN------------------------- 473
scHSF    NQTSENTFDPNRFTMLSDDDLKKDSHTNDNKHNES------------------------- 626

mHSF2    ------------INLLGKVELLDYLDSIDCSLEDFQAMLSGRQFSIDPDLLVD------- 391
rHSF2    ------------INLLGKVELLDYLDSIDCSLEDFQAMLSGRQFSIDPDLLVD------- 387
hHSF2    ------------INLLGKVELLDYLDSIDCSLEDFQAMLSGRQFSIDPDLLVDLFTSSVQ 399
dHSF2    ------------INLLGKVELLDYLDSIDCSLEDFQAMLSGRQFSIDPDLLVDLFTSSVQ 442
oHSF2    ------------INLLGKVELLDYLDSIDCSLEDFQAMLSGRQFSIDPDLLVDLFTSSVQ 406
cHSF2    ------------INLLGKVELLDYLDSIDCSLEDFQAMLSGRQFSIDPDLLFDVI----- 423
xHSF2    ------------LNLLGKVELMDYLDSIDCTLDDFKALLSGRQFSIDRDLLVDVFSSNVQ 407
zHSF2    ------------INLLGKVELMDYLDSIDCSLEDFQAMLYGKQFNIDPDIIEET------ 383
cHSF1    ------------DPLLDREEIQDFLNCIDASLEELQAMLSGKQYSFGSEAFSDVFNPELP 569
cHSF3    ------------DPLLDREEIQDFLNCIDASLEELQAMLSGKQYSFGSEAFSDVFNPELP 403
mHSF1    -----------------KNELSDHLDAMDSNLDNLQTMLTSHGFSVDTSALLD------- 412
rHSF1    -----------------KNELSDHLDAMDSNLDNLQTMLTSHGFSVDTSALLDLFSPSVT 419
hHSF1    -----------------KNELSDHLDAMDSNLDNLQTMLSSHGFSVDTSALLDLFSPSVT 423
dHSF1    VARLFPCPSSLHGRVQPGTELSDHLDAMDSNLDNLQTMLTSKGFSVDTSALLDLFSPSVT 448
oHSF1    -----------------KNELSEHLDTMDSNLDNLQTMLTTHGFSVDTSALLDLFSPSMS 416
xHSF1    -----------------KTELNDHVDSIDFSLDNLQNLLNGQSFSVDTSALLDLFSPSLG 393
zHSF1    --------------PHSGTELHDHLESIDSGLENLQQILNAQSINFDSSPLFDIFSSAAS 426
fHSF1    -----------------RCELSDHVENIESSLENLQNILNTQTFTFDASPLMEVSVSHTC 400
mHSF4    --------------EVPDRGT-LGLDRGNRSPESLLPPMLLRPAPETLEPVAPVDVLGPS 371
hHSF4    --------------EIPDRGP-LGLESGDRSPESLLPPMLLQPPQFSVEPAGPLDVLGPS 372
rHSF4    --------------EVPDRGT-LGLDRGNRSPESLLPPVLLRPPPETVEP---MDALGPS 398
dHSF4    --------------EVPDRGT-LGLDRGARSPENLLPPMLLRAPPESVEPAGPLDVLGPS 402
oHSF4    --------------EPPERELGEPLESEDLSLENLQ--LLLRGQPVTLEPEGGLELSGSG 404
dmHSF    --------------DSSSPKAQRTSIQHSTQPDVMVQPMIIKSEPENSSGLMDLMTPAND 472
ceHSF    ---------------ATRQMAQPQQDYSGGFVNNYSGFMPSDPSMIPYQPSHQYLQPHQK 518
scHSF    --------------DLFWDNVHRNIDEQDARLQNLENMVHILSPGYPNKSFNNKTSSTNT 672
                                     :     :

mHSF2    -----------SENKGLEATKSSVVQHVSEEGRKSKSKPDKQLIQYTAFPLLAFLDGNSA 440
rHSF2    -----------SENKGLEATKNSVVQQVSEEGRKSKSKPDKQLIQYTAFPLLAFLDGNPA 436
hHSF2    MNPTDYINNTKSENKGLETTKNNVVQPVSEEGRKSKSKPDKQLIQYTAFPLLAFLDGNPA 459
dHSF2    MNSTDYINNTKSENKGLETTKNNVVQPVSEEGRKSKPKTDKQLIQYTAFPLLAFLDGNPA 502
oHSF2    MNPTDYISNTKMENKGLETNKNNEVQPTSDDGRNSKPKSDKQLIQYTAFPLLAFLDGNPS 466
cHSF2    ------------------------------------------------------------
xHSF2    MSPSDYLS---DSKVYISKWIQLLCHFPLLQLMDSEEAGYQQLVQYTTFPLLALFDGD-- 462
zHSF2    ----------------ISETKNNNSK-ANKENLDSE-NTDKQLIQYTTCPLMAFLDGC-- 423
cHSF1    -----------ALDMNLMETSPGMENIANMEDSTEDLGASERETAGSKDMQLIQYRANPL 618
cHSF3    -----------ALDMNLMETSPGMENIANMEDSTEDLGASERETAGSK------------ 440
mHSF1    ---------------IQELLSPQEPPRP----IEAENSNPDSGKQLV-----HYTAQPLF 448
rHSF1    MPDMSLPDLDSSLASIQELLSPQEPPRP----IEAENSNADSGKQLV-----HYTAQPLF 470
hHSF1    VPDMSLPDLDSSLASIQELLSPQEPPRP----PEAENSSPDSGKQLV-----HYTAQPLF 474
dHSF1    VPDMSLPDLDSSLASIQELLSPQEPPRP----LEAENSSPDSGKQLV-----HYTAQPLF 499
oHSF1    MTDMSLPDLDSSLASVRSQG-PREVG-------EGQNAYP--GKQLV-----HYTAQPLF 461
xHSF1    IPDLPLPDPDSSLASIQELLSSQDQKAS----ETVESRTPDAGKQIV-----HYTAQPLI 444
zHSF1    DVDLD------SLASIQDLLSPDPVK------ETESGVDTDSGKQLV-----QYTSQPIV 469
fHSF1    KHQQVAPSLPSSFNLTVTVTSCSKNKMVSMDYRTSCFSLIGPGKELV-----HYMATPLQ 455
mHSF4    LHGREWTLMDLDMELSLMQPLAP--ETDEAELTVKELNSSGVGK-------DHTLGTPLM 422
hHSF4    LQGREWTLMDLDMELSLMQPLVP--ERGEPELAVKGLNSPSPGK-------DPTLGAPLL 423
rHSF4    LHGREWTLMDLDMELSLMQPLAP--ERAEAELTVKDLTSSGVGK-------DHTLGTPLM 449
dHSF4    HQGREWTLMDLDMELSLMQPLGP--ERSETELAVKGLNSPGPGK-------DSTLGAPLL 453
```

FIGURE 1. (*Continued*)

```
oHSF4   LSGSEWSLLDMDVGLSLMQQLPKDLEKPETELMSKEPNPPDSGK------DPASPRTFA  457
dmHSF   LYSVNFISEDMPTDIFEDALLPDGVEEAAKLDQQQKFGQSTVSSGKFASNFDVPTNSTLL  532
ceHSF   LMAIEDQHKPTTSTSSTNADPHQNLYSPTLGLSPSFDRQLSQELQEYFTGTDTSLESFRD  578
scHSF   NSNMESAVNVNSPGFNLQDYLTGESNSPNSVHSVPSNGSGSTPLPMPN---DNDTEHAST  729

mHSF2   SAIEQGSTTASSEVVPSVDKPIEVDELLDSSLDPEPTQSKLVRLEPLTEAEASEATLFYL  500
rHSF2   SAVEQSTTTASSEVVPSVDKPIEVDELLDSNLDSEPTQSKLVRLEPLTEAEASEATLFYL  496
hHSF2   SSVEQASTTASSEVLSSVDKPIEVDELLDSSLDPEPTQSKLVRLEPLTEAEASEATLFYL  519
dHSF2   STVEQSS-TASSEVMSSVDKPIEVDELLDSSLDPEPTQSKLVRLEPLTEAEASEATLFYL  561
oHSF2   STVENGN--TTSEAVSSSNKPLGVDELLDASLDPEPTQSKLVRLEPLTEAEASEATLFYL  524
cHSF2   ------------------------------------------------------------
xHSF2   ---------SGNVTTDTPPDKTLEDQLLESNFDSRPPQSKLMRLEPLSEAEASETTLFYL  513
zHSF2   ------------TPLTPSEPQPEDPPDLLDESLEMETPRSSLIRLEPLTEAEANEATLFYL  472
cHSF1   LSLFEELPSGEAGGKTEDPKDLLLAPLEEKPALHPPSGSETVLPLAVPANQAEPPVDALG  678
cHSF3   --------GGQEGTESCDSSVLFQN--------------CVLKWNFSSL----------  467
mHSF1   LLDPDAVDTGSSELPVLFEL--GESSYFSE-GDDYTDDPTISLLTGTEPHKAKDPTVS--  503
rHSF1   LLDPDAVDTGSSELPVLFEL--GESSYFSE-GDDYTDDPTISLLTGTEPHKAKDPTVS--  525
hHSF1   LLDPGSVDTGSNDLPVLFEL--GEGSYFSE-GDSFAEDPTISLLTGSEPPKAKDPTVS--  529
dHSF1   LVDPGSVDMGSSDLPVLFEL--GESSYFSE-GDDYTDDPTISLLTGSEPPKAKDPTVS--  554
oHSF1   LVDSSTLDTTSNDLPIFFEL--GEGPYFSE-GDDYTDDPTISLLSDSEPTTAKDPAIS--  516
xHSF1   LMDSS-----GSDLPILLELE-GDEPYVAEEGDDYSEDPTLSLLCWDPQSKPASSSIS--  496
zHSF1   LFDPLSTDSSSTDLPMLLEL--QDDSYFSS---EPTEDPTIALLNFQPGHRPSRTRIGDP  524
fHSF1   TPELNLLSGGDADMPSLLEL--DEEPFFTT--DLPEDDPTTALLTSSQP-----------  500
mHSF4   LDVQADLEGAALSVPG------ALTLYN-----VTES-NASYLDPGASPSSP-------  462
hHSF4   LDVQAALGGPALGLPG-------ALTIYS-----TPESRTASYLGPEASPSP-------  463
rHSF4   LDVQADLEGAALSVPG-------ALTLYN-----ATES-NASYLDPGASPSSP-------  489
dHSF4   LDVQAALGGPALSLPG-------ALTIYS-----TPES-RANYLGPGANPSP-------  492
oHSF4   MEAHSSFGAPELGVPA-------VQPLYDPSHTLAQNETRTPYLSPGANPDP-------  502
dmHSF   DANQASTSKAAAKAQASEEEG-MAVAKYSGAENGNNRDTNSSQLLRMASVDELHGHLESM  591
ceHSF   LVSNHHNWDDFGNNVPLDDDEEGSEDPLRQLALENAPETSNYDGAEDLLFDNEQQYPENGF  638
scHSF   SVNQGENGSGLTPFLTVDDHTLNDNNTSEGSTRVSPDIKFSATENTKVSDNLPSFNDHSY  789

mHSF2   CELAPAPLDSDMPLLDS-------------------------------------------  517
rHSF2   CELAPAPLDSDMPLLDS-------------------------------------------  513
hHSF2   CELAPAPLDSDMPLLDS-------------------------------------------  536
dHSF2   CELAPAPLDSDMPLLDS-------------------------------------------  578
oHSF2   CELAPAPLDSDMPLLDN-------------------------------------------  541
cHSF2   ------------------------------------------------------------
xHSF2   CELAPAPLDNDMPLLD--------------------------------------------  529
zHSF2   CELNSDLPNADTPPLDI-------------------------------------------  489
cHSF1   MSDPPLLSEDSNGEYKLFPLLLLSPVANPIEEASEIETS---------------------  717
cHSF3   ------------------------------------------------------------
mHSF1   ------------------------------------------------------------
rHSF1   ------------------------------------------------------------
hHSF1   ------------------------------------------------------------
dHSF1   ------------------------------------------------------------
oHSF1   ------------------------------------------------------------
xHSF1   ------------------------------------------------------------
zHSF1   CFKLKKESKR--------------------------------------------------  534
fHSF1   ------------------------------------------------------------
mHSF4   ------------------------------------------------------------
hHSF4   ------------------------------------------------------------
rHSF4   ------------------------------------------------------------
dHSF4   ------------------------------------------------------------
oHSF4   ------------------------------------------------------------
dmHSF   QDELETLKDLLRGDGVAIDQNMLMGLFNDSDLMDNYGLSFPNDSISSERKAPSGSELISY  651
ceHSF   DVFDPNYLPLADEEIFPHSPALRTPSPSDPNLV---------------------------  671
scHSF   STQADTAPENAKKRFVEEIPEPAIVEIQDPTEYNDHRLPKRAKK----------------  833

mHSF2   ------------------------------------------------------------
rHSF2   ------------------------------------------------------------
hHSF2   ------------------------------------------------------------
dHSF2   ------------------------------------------------------------
oHSF2   ------------------------------------------------------------
cHSF2   ------------------------------------------------------------
xHSF2   ------------------------------------------------------------
zHSF2   ------------------------------------------------------------
cHSF1   ------------------------------------------------------------
cHSF3   ------------------------------------------------------------
mHSF1   ------------------------------------------------------------
rHSF1   ------------------------------------------------------------
hHSF1   ------------------------------------------------------------
dHSF1   ------------------------------------------------------------
oHSF1   ------------------------------------------------------------
xHSF1   ------------------------------------------------------------
zHSF1   ------------------------------------------------------------
fHSF1   ------------------------------------------------------------
mHSF4   ------------------------------------------------------------
hHSF4   ------------------------------------------------------------
rHSF4   ------------------------------------------------------------
dHSF4   ------------------------------------------------------------
oHSF4   ------------------------------------------------------------
dmHSF   QPMYDLSDILDTDDQNNDQEASRRQMQTQSSVLNTPRHEL  691
ceHSF   ---------------------------------------
scHSF   ---------------------------------------
```

FIGURE 1. (*Continued*)

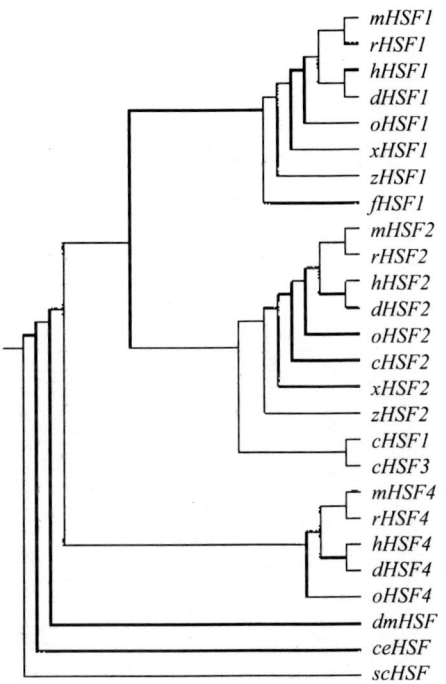

FIGURE 2. Evolutionary relationships among mammalian, vertebratae and lower eukaryotic HSF protein sequences. Abbreviations for the different species are as in Figure 1. Cladogram was generated by a ClustalW multiple alignment program from the web site: http://align.genome.jp/

constitutive HSP expression (Yan et al., 2002; Yan et al., 2005). These results clearly establish *Hsf1* as the major regulator of the HSR as well as requirement for HSP gene expression under non-stressed conditions. Several unexpected findings regarding the role of *Hsf1* in growth and development, male and female fertility, organ maintenance and homeostasis, response to pathophysiological challenges in the context of the whole organism will be discussed in further detail below.

4.1. Hsf1 *Is Required for the Development of Chorioallantoic Placenta*

Heterozygous intercrosses between *Hsf1*+/− germline chimeric males (ES cells, 129SvEv) and females of different genetic backgrounds (129 SvJ, BALB/c and ICR) produced homozygous *Hsf1*−/− progenies that yielded less than the expected Mendelian distribution. The ratio of *Hsf1*−/− progenies at birth varied significantly between 2% and 20% (129SvJX129SvEv and ICRX129SvEv, respectively), suggesting that the penetrance was highly dependent on the genetic background. Prenatal lethality of *Hsf1*−/− animals occurred in the most severely affected (129SvJX129SvEv) background at around mid-gestation (14 d.p.c.) and haploinsufficiency of *Hsf1* affected neither fertilization nor pre-implantation (Xiao et al., 1999). Of interest, Wang and co-workers reported that *Hsf1* knockout animals generated on the C57Bl/6 genetic background produced normal litter sizes

TABLE 1. Milestones in HSF Research

Year	HSF gene isolated	TARGETED HSF knockout generated in whole organism
1988	First HSF gene isolated: *S. cerevisiae* (Sorger and Pelham, 1988; Wiederrecht et al., 1988)	First HSF knockout generated: *S. cerevisiae* (Sorger and Pelham, 1988; Wiederrecht et al., 1988)
1990	First HSF is isolated from a multicellular organism: *Drosophila melanogaster* (Clos et al., 1990) HSF gene family concept arises: cloning 3 tomato HSF genes (Scharf et al., 1990)	
1991	HSF family found in mammals: human HSF1(Rabindran et al., 1991), human HSF2 (Schuetz et al., 1991), mouse HSF1-2 (Sarge et al., 1991)	
1993	HSF family expands and diversifies in vertebrates: cloning chicken 1,2,3 (Nakai and Morimoto, 1993)	First HSF KO from a "true" eukaryotic organism described: *S. pombe* (Gallo et al., 1993)
1997	Third mammalian HSF family member cloned: hHSF4 (Nakai et al., 1997)	First functional KO generated in a multicellular organism: *Drosophila* (Jedlicka et al., 1997)
1999	Genomic era: massive sequencing efforts conducted in different genetic model systems open the way of cloning and predicting functionally homologous HSF genes alongside the evolutionary spectra	First mammalian KO published: mHSF1 (Xiao et al., 1999)
2002		First two independent HSF2 KO published: 1. mHSF2 (McMillan et al., 2002) 2. mHSF2 (Kallio et al., 2002)
2003		Second mHSF1 KO published: mHSF1 (Zhang et al., 2002) Third mHSF1 KO (Inouye et al., 2003) Third mHSF2 KO (Wang et al., 2003)
2004		First two independent mHSF4 KO: 1. mHSF4 (Fujimoto et al., 2004) 2. mHSF4 (Min et al., 2004) First double KO of mHSF1 and mHSF2 generated (Wang et al., 2004)

with no indication of significant embryonic lethality (Wang et al., 2004), suggesting the influence of genetic modifiers and other undetermined factors acting on *Hsf1*−/− phenotypes in different genetic backgrounds.

Consistent with the proposed crucial roles of *Hsf1* requirements in basal HSP expression (Fiorenza et al., 1995), examination of viable embryos with the targeted mutation during the rapid embryonic growth phase, revealed no morphological, histological or microscopic abnormalities in multiple tissues examined at varying developmental stages (Xiao et al., 1999; Zhang et al., 2002). Moreover, no histopathological irregularities, indicative of secondary complications (pulmonary congestion, congestive heart failure or other organ dysfunction) linked to lethality, were discovered in the embryo proper or adult animals. Extensive characterization of extra-embryonic sites in the search of the cause of prenatal lethality associated with *Hsf1* deficiency demonstrated chorioallatoic placental defects in *Hsf1*−/− embryos. Amongst the three morphologically distinct layers of the placenta at mid gestation (decidua, spongiotrophoblast and labyrinthinae trophoblast giant cell layers), only the spongiotrophoblast layer of embryonic origin showed marked thinning with disrupted architecture from 11.5 d.p.c. in *Hsf1*−/− embryos. *In situ* hybridizations with lineage specific markers to both spongiotrophoblast and labyrinthinae trophoblast layers confirmed the findings of the histological studies in both 129-*Hsf1* and C,129-*Hsf1* genetic background. Thus, consistent spongiotrophoblast deficiencies in mid-gestation and further histopathological abnormalities in late gestation period of *Hsf1*-deficient animals on different genetic background demonstrated that *Hsf1* is required to maintain the chorioallatoic placenta and prevent prenatal lethality as a consequence of placental dysfunction. The mechanism(s) for the unanticipated *Hsf1*−/− placental defects remains to be fully elucidated, but the extra-embryonic abnormalities were independent from changes in basal HSP expression levels. Hence, similar to the role of HSF in *Drosophila* oogenesis (Jedlicka et al., 1997), *Hsf1* expression might orchestrate the expression of non HSP target genes in the placenta *in utero*. Genetic rescue experiments of the trophobast defects with ubiquitously expressed or tissue specific *Hsf1* construct(s) would clarify its putative role in extra-embryonic development, while high throughput analysis of gene expression and promoter occupancy changes of *Hsf1* in the placental tissue could identify potential target genes essential for the unforeseen role of *Hsf1* in extra-embryonic development.

4.2. Hsf1 *in Female Reproduction*

While *Hsf1*−/− males are fertile, analysis of all three targeted knockout mice have reported female infertility associated with *Hsf1* loss independent of the genetic background (Xiao et al., 1999; Zhang et al., 2002; Izu et al., 2004). Female infertility was investigated further only by the Benjamin group (Christians et al., 2000), which showed that mouse embryos of *Hsf1*−/− mothers were unable to develop properly beyond the zygotic stage, despite proper ovulation and fertilization of the oocytes. Wild-type spermatozoa do not prevent zygote lethality, designating *Hsf1*−/− as a maternal effect mutation, which controls early post-fertilization development. Mutant embryos of *Hsf1* deficient females can initiate early development, but they perish even after transplantation into the oviduct of an *Hsf1* wild-type female, indicating intrinsic causes of female infertility. Embryos from *Hsf1*−/− female intercrosses with *Hsf1*−/− or *Hsf1*+/+ males die at 1-2 cell stage and never develop to blastocyst stage, further signifying the importance of the maternal genotype. Embryonic death may occur as a result of failure to initiate the zygotic transcriptional program required to complete the switch from maternal to embryonic control of development. This process seems to be governed by some essential, but unknown maternal factors directly under the control of *Hsf1*. Additionally, abnormalities of the ultrastructural nuclear organization of the embryos of *Hsf1* null females at 2 cell stage are clearly present, indicating that both structural deficiencies and altered transcriptional program may contribute to the failure of reproductive success of *Hsf1* deficient females. Transfer of wild-type nucleus to mutant embryo (cloning experiment) provides partial rescue of this defect (Christians et al., 2003). These findings undoubtedly establish *Hsf1* as a candidate gene associated with female infertility in mammals, conceivably in humans, which may be further investigated by searching for genetic polymorphisms in the affected human population.

4.3. Hsf1 *Function in Testis: Quality Control Mechanisms of Male Germ Cells*

It has long been known from experimental observations (e.g., exposure of testes to high temperatures, chryptorchidism) that spermatogenesis is extremely sensitive to elevated temperatures, although the mechanism of injured germ cell elimination remained obscure. Transgenic animals expressing constitutively active *Hsf1* in testes are infertile (Nakai et al., 2000), due to a block in spermatogenesis, suggesting a role of *Hsf1* in male germ cell death induced by high temperature. To investigate *Hsf1* contribution to the quality control of male germ cells, Nakai and colleagues set out to analyze the testes of *Hsf1* null mice exposed to thermal stress (Izu et al., 2004). The key finding of their experiments is that heat shock proteins are not induced in male germ cell under heat stress conditions, in spite of *Hsf1* activation. On the other hand, *Hsf1* appears to promote apoptosis of pachytene spermatocytes undergoing a single exposure to heat shock or in the chryptorchid testes, since *Hsf1*−/− animals had markedly increased survival of these cells compared to wild-type. Surprisingly, *Hsf1* acts as cell-survival factor and saves more immature germ cells, including spermatogonia, from cell death upon exposure to high temperature, indicating a delicate balance between the key cell survival and cell death pathways, most likely in a cell type dependent fashion. This aspect of *Hsf1* expression on male germ cell quality control is reminiscent of the role of p53 during genotoxic stress: depending of the extent of DNA damage, specific cell types may initiate cell cycle arrest to facilitate repair mechanisms, before replication and mitosis or induce apoptosis in case of irreparable DNA injury. Thus, *Hsf1* has two opposing roles in the quality control of male germ cells, both of which are independent of the activation of heat shock genes. Further studies will be needed to identify the unconventional target genes of *Hsf1* in the delicate balance between cell survival and cell death pathways during the complex process of spermatogenesis, which screens out damaged germ cells from transmission to the next generation by sophisticated quality control mechanisms.

4.4. Hsf1 *and Hearing*

Similarly to other organs, heat shock proteins (HSPs) are up-regulated in the rodent cochlea following hyperthermia, ischemia, cisplatin ototoxicity or noise overstimulation and this up-regulation is solely dependent of the presence of the main transcriptional stress regulator, *Hsf1*. Preconditioning stress leading to the activation *Hsf1* and ensuing synthesis of Hsp70 and other HSPs has a protective effect against subsequent damaging noise exposure at the time when Hsp70 expression reaches its plateau (Fairfield et al., 2005). The *Hsf1* deficient mouse model (Xiao et al., 1999) provides an excellent experimental tool to directly address the role of *Hsf1* stress pathways in cochlear protection and recovery following noise stimulation. *Hsf1*−/− mice exhibit normal cochlear morphology and are devoid of any anatomical or microscopic abnormalities, in agreement with the previous report, with no measurable difference in hearing compared with wild-type mice. The authors of this study used moderate-intensity noise exposure to assess the ability of *Hsf1* null mice to recover from a temporary threshold shift (TTS) noise exposure that does not cause permanent damage referred to as permanent threshold shift (PTS) in wild-type animals. *Hsf1*−/− mice displayed significantly decreased recovery after moderate-intensity noise stimulation at all three frequencies tested. The hearing loss persisted 2 weeks after the noise exposure and it was accompanied with outer hair cell loss, mostly in the basal turn of the cochlea that is most sensitive to oxidative damage because of the relative paucity the free radical scavenger glutathione. Interestingly, the threshold shift measured as auditory brainstem response (ABR) also showed excellent correlation with outer hair cell loss, while another study found no such connection (Sugahara et al., 2003). Sugara and co-workers used different experimental protocols to expose *Hsf1* null and wild-type mice to damaging high-intensity noise for one hour and after a six hours recovery period, when Hsp70 synthesis is at its maximum, followed by a second high-intensity noise for 10 hour. Extensive outer hair cell loss (60%) was observed in *Hsf1*−/− mice after 7 days, while wild-type mice showed significant, but decreased damage (20%) at the same time point. The likely disparity in findings between the two studies may be attributed to the different exposure conditions for producing substantial outer hair cell loss in the experimental groups. In conclusion, these findings strongly suggest that stress-inducible *Hsf1*-dependent pathways perform essential functions in cochlear recovery and repair after noise exposure or other physiological stresses through chaperon activities of effector heat shock proteins.

4.5. Hsf1 *Deficiency Alters Cardiac Redox Milieu and Increases Mitochondrial Oxidative Damage*

Recently, an interesting study by Yan and colleagues (Yan et al., 2002) tested the hypothesis whether *Hsf1*-dependent regulation of heat shock proteins synthesis is required to maintain the intracellular redox balance and attenuation of oxidative damage in the absence of stressful and/or pathological conditions in the heart. They found that constitutive level of HSPs is selectively regulated in *Hsf1*-deficient normal adult hearts: while cytoplasmic Hsp90 and mitochondrial Hsp60 protein levels are unaffected, small molecular heat shock proteins Hsp25, αB-crystallin, and cytosolic Hsp70 are significantly down-regulated (72%, 42%, and 34%, respectively). Small heat shock proteins, such as Hsp25 and αB-crystallin have extensively been implicated to influence the cellular redox milieu by raising the intracellular concentration of reduced glutathione (GSH) and maintaining glucose-6-phosphate dehydrogenase (G6PD) activity, the key antioxidative enzyme responsible for reducing glutathione disulphate (GSSG) to GSH via NADPH-dependent mechanism. The GSH:GSSG ratio was significantly decreased (40%) in *Hsf1*-deficient hearts, indicating substantial redox imbalance in the mutant animals. The decrease of GSH:GSSG ratio has been a consequence of reduced G6PD activity, but unchanged protein level, in cardiac mitochondria preparation establishing for the first time a connection between the *Hsf1* regulatory pathway and a key antioxidant enzyme in an intact organ. Decreased GSH:GSSG ratio in *Hsf1*-deficient animals implies the possibility that reactive oxygen species (ROS) may be generated at higher level. Indeed, the rate of superoxide (O_2^-) generation in cardiac sub-mitochondrial particles (SMPs) was 43% higher in *Hsf1* knockout animals compared with wild-type littermate controls. Higher ROS production resulted in increased oxidative damage of several mitochondrial proteins assessed by their carbonyl content. These post-translational modification marks of ROS oxidative attacks lead to the formation of aldehydes and ketones on certain amino acid residues. The identity of two prominently carbonylated *Hsf1*−/− specific mitochondrial proteins were determined by biochemical purification and successive microsequencing, which turned out to be adenine nucleotide translocase 1 (ANT1) and voltage-dependent anion channel 1 (VDAC1); both of them are components of the non-selective mitochondrial permeability transition pore (MPTP) complex. Consequently, oxidative damage of ANT1 due to *Hsf1* deficiency *per se* has indirect functional effects on the catalytic activity of ANT1 without pathological challenges, reducing its activity by 18%. The functional defect of ANT1 in *Hsf1*−/− animals translates to faster MPTP opening, determined by swelling of isolated mitochondria by light scattering, which is independent of the presence of calcium. These results provide evidence that deficiencies in *Hsf1*-dependent regulatory pathways may cause indirect oxidative injury of cardiac ANT1, which in turn leads to mitochondrial functional abnormalities and modifications of cell survival and cell death pathways. Hence, susceptibility of *Hsf1*-deficient mice may be attributed to a partial MPTP malfunction, a key mediator in apoptotic and necrotic cell death. In summary, these results indicate that constitutive expression of certain HSPs requires *Hsf1* expression at normal temperature, and these requirements can be directly and functionally linked to the maintenance of redox homeostasis and antioxidant defense in mouse hearts in the absence of cellular and pathophysiological stress.

4.6. Hsf1 *in Redox Homeostasis*

Inducible HSP synthesis regulated predominantly by *Hsf1* protects cells, organs and organisms from the deleterious effects of oxidative and other stresses under pathophysiological conditions, for instance, ischemia/reperfusion injury. To address the possibility that *Hsf1* exerts such functions under physiological conditions in the kidney, Yan et al. (2005) undertook the examination of the HSP expression profile of wild-type and *Hsf1*−/− mice by Western blot analysis. Their results showed significant down-regulation of Hsp25 and Hsp90 (26% and 50%, respectively), but not in αB-crystallin or Hsp70, which, in turn, was associated with massive decrease in the renal cellular GSH/GSSG ratio (37%) an important indicator of cellular redox status. Superoxide generation through the mitochondrial electron transport chain was increased in the *Hsf1* KO animals by 40% with increased mitochondrial permeability pore opening and mitochondrial membrane potential (48%). This study provides further evidence, that physiological level of heat shock proteins is regulated by *Hsf1* in tissue specific manner. Similar conclusions were drawn in the foregoing experiments regarding constitutive expression of HSPs in wild-type and *Hsf1*−/− hearts (Yan et al., 2002), with the notable exception that in the *Hsf1*-deficient heart Hsp70 and αB-crystallin expression were down-regulated. The common denominator in both organs, however, is the considerably lower level Hsp25 expression. Since Hsp25 is a key player in the maintenance of the normal cellular redox milieu (Preville et al., 1999), it is conceivable that its functional level may contribute significantly to redox homeostasis and oxidative stress protection under physiological conditions both in the heart and kidney.

4.7. *Brain Abnormalities in* Hsf1-*Deficient Mouse*

Earlier work from the Benjamin lab showed that *Hsf1*−/− embryos do not have distinguishable morphological or microscopic alterations in the brain (Xiao et al., 1999), however a detailed characterization of the adult *Hsf1*-null mouse brain is lacking. These considerations prompted investigators to look for anatomical, microscopic, and functional differences in the brains of mice lacking *Hsf1*, the master regulator of the stress response of eukaryotes. To asses for brain abnormalities adult animals, this follow-up study undertook the detailed characterization of brain morphology and cellular expression of selected heat shock proteins in the 1- and 3-month-old adult *Hsf1*-deficient mice under non-stress conditions (Santos and Saraiva, 2004). Both gross morphological and microscopic differences were observed in *Hsf1*-null compared with wild-type mice in both age groups. Specifically, the lateral ventricules were significantly enlarged (ventriculomegaly) and the white matter reduced without the thinning of the cerebral cortex and stenosis of the Sylvius aqueduct, which excludes hydrocephaly. Furthermore, the third ventricule showed minor enlargement, but the fourth ventricule appeared normal. No further morphological abnormalities have been described in brain structures including the hippocampus and cerebellum. Surprisingly, heterozygous mice showed intermediate phenotype between the wild-type and homozygous knockout animals with somewhat enlarged ventricular morphology, suggesting a gene dosage effect. Detailed microscopic analysis revealed spongiosis, amorphous eosinophilic cytoplasm, and pyknotic nuclei in the white matter, specifically in the periventricular region, which also showed signs of astroglyosis and neurodegeneration. Expressions of two known *Hsf1* target genes belonging to the small heat shock protein family were analyzed in details by RT-PCR and semi quantitative imunohystochemistry. RT-PCR has not shown alterations of the level of hsp27, likely due to the mixed cell population used for the preparation of the total RNA. On the other hand, semiquantitative immunohystochemistry revealed higher expression of hsp27 protein in astrocytes localized in the external capsule and striatum, while neurons of the medulla oblongata and cerebellum showed equivalent staining as wild-type brain. These data present the intriguing possibility, that hsp27 may be regulated by different mechanisms in distinct cells of the CNS. αB-crystallin was expressed at a significantly lower level in the cerebellum and in somewhat decreased amounts in medulla oblongata by RT-PCR. This result was corroborated by semi-quantitative immunohystochemistry, which also confirmed considerable decline of αB-crystallin staining in the external capsule.

Further studies using a series of behavioral tests of sensory, motor and cognitive tasks are needed to establish whether these anatomical and microscopic differences in brain morphology translate to functional abnormalities and compromised brain functions in *Hsf1*-deficient animals.

5. Targeted Knockout Models of *Hsf2*, the Enigmatic Member of the Mammalian HSF Family

While the master regulator of eukaryotic stress response genes, *Hsf1*, displays ubiquitous and high-level expression in many tissues, *Hsf2* activity is restricted to developmental pathways during embryogenesis and spermatogenesis in whole animals with unidentified target genes and functions. To investigate the unique and overlapping physiological functions of *Hsf2* in the context of the whole organisms, gene-targeting experiments were performed by three different groups (Kallio et al., 2002; McMillan et al., 2002; Wang et al., 2003) in recent years. In the following paragraphs, we summarize the findings of these experiments with their unanticipated and possibly paradoxical conclusions in the three existing mouse knockout models.

5.1. Gene-Targeting Strategies

To fully understand the phenotypic differences among the three (knockout) KO models it is worthwhile to examine the gene targeting strategies utilized in more details. Generating *Hsf2* null mice by two groups employed reporter gene knock-in approaches in different regions of the genomic sequence (Kallio et al., 2002; Wang et al., 2003), while the third group (McMillan et al., 2002) created a true null allele by the introduction of the neomycin cassette in the *Hsf2* gene. The Morange-Mezger group (Kallio et al., 2002) engineered a β–*geo* gene (a chimera between the lacZ and the G418 resistance genes) into the 5 exon, within the oligomerization domain of *Hsf2*. In this arrangement, homologous recombination leads to the production of a hybrid protein, which retains the DNA binding domain, but disrupts the trimerization domain, effectively eliminating oligomerization and perhaps high affinity DNA binding of *Hsf2*. Nuclear localization of *Hsf2* is also compromised by deletion of one of two nuclear localization signals (NLS), rendering the chimeras predominantly cytoplasmic and incapable of nuclear translocation and other nuclear functions. Expression of *Hsf2*-β–*geo* mimicked *Hsf2* distribution when analyzed for lacZ staining with cytoplasmic localization and band shift experiments lacked DNA binding in E11.5 embryo extract of the chimeras. However, these data do not completely exclude the possibility, that minor fractions of *Hsf2*-β–*geo* below the detection limits of EMSA (band shift) and/or immunohistochemistry would cross the nuclear envelope, bind DNA and perform a task reminiscent of a transcription factor. Additionally, the hybrid protein localized in the cytoplasm may exhibit unanticipated functions alone or in association with other unidentified cellular proteins and influence cellular homeostasis.

Using analogous targeting approach Wang and co-workers (Wang et al., 2003) introduced an EGFP-*neo* (enhanced green fluorescence protein and G418/neomycin) fusion cassette to replace the first exon of *Hsf2* gene resulting in the termination of transcription of *Hsf2*. EGFP-*neo* is expressed by the endogenous *Hsf2* promoter and recapitulates the expression profile of *Hsf2* during embryogenesis and in postnatal animals as judged by EGFP fluorescence. Again, this chimeric protein may arbitrarily participate in cellular processes and protein interactions in the EGFP-*neo* expressing cells leading to unintended consequences.

The third published *Hsf2* knockout mouse (McMillan et al., 2002) was generated by traditional gene targeting strategy effectively disrupting the first exon of *Hsf2* including the translational start codon with a neomycin expression cassette. All experiments involving the three independent *Hsf2* knockout models were performed on F2 mixed C57Bl/6X129Sv genetic background, consequently differences due to genetic polymorphisms likely do not play significant role in the interpretations of the experimental data. The major targeting strategies and HSF phenotypes are summarized in Table 2.

5.2. Hsf2 *Deficiency Influences Embryonic Lethality*

Heterozygous intercrosses of F1 (C57Bl6X129/Sv) mice were analyzed for viability associated with the disruption of *Hsf2* gene. Two groups reported normal litter sizes and the expected Mendelian rations for genotype distribution amongst the F2 progenies, indicating that the *Hsf2* gene is not embryonic lethal and dispensable for normal development. Surprisingly, the third study (Wang et al., 2003) found that the disruption of *Hsf2* gene increases embryonic lethality, resulting in skewed genotype ratios (32.8% +/+, 58.2% +/−, and 8.8% −/−), which is manifested between E7.5 to birth, indicating consequently that haploinsufficiency of *Hsf2*

TABLE 2. Summary of Major Targeting Strategies and HSF Phenotypes in Mice*

Mouse HSFs	HSF1		HSF2		HSF4	
Targeting strategy	Neomycin cassette deleting exons 1–6 (Xiao et al., 1999)	IRES-lacZ-neomycin fused to exon 2 (Zhang et al., 2002)	Neomycin cassette fused to exon 2, deleting exons 2–4 (Inouye et al., 2003)	Neomycin expression cassette, exon1 (McMillan et al., 2002)	In frame fusion of β-geo to exon 5 disrupting the oligomerization domain and creating a fusion protein (Kallio et al., 2002)	EGFP-Neomycin cassette fused to exon 1 (Wang et al., 2003)	Neomycin cassette fused to exon 2, deleting DNA binding and oligomerization domains (Fujimoto et al., 2004)	EGFP-Neomycin cassette fused to exon 1 (Min et al., 2004)
Potential pitfalls of targeting constructs	No	No	No	No	Creates fusion protein with potential DNA binding activity	EGFP-Neo expressiom in Hsf2 dependent fashion	No	EGFP-Neo expressiom in Hsf4 dependent fashion
Genetic background of ES cell	129/Sv KG1	D3	TT2 (C57Bl/6 x CBA F1 origin)	129/Sv, KG1	CK35, derived from 129/SV Pasteur strain	D3	TT2 (C57Bl/6 x CBA F1 origin)	D3
Blastocytes	Balb/c	C57BL/6	ICR	C57BL/6	Not known	C57BL/6	ICR	C57BL/6
Constitutive HSP expression	Tissue-specific reduction	Tissue-specific reduction	ND	NA	ND	NA	Lens specific reduction: Hsp25, γ-crystallin Lens specific increase of some major HSPs	Lens specific reduction: Hsp25, γ-crystallin Lens specific increase of some major HSPs
Stress response	Abolished	Abolished	Abolished	NA	ND	ND	ND	ND

5. Genetic Models of HSF Function   111

	Female deficiency	Female deficiency	Female deficiency		Meiotic deficiency in male and female	Male deficiency		
Gametogenesis	NA	NA	NA	NA	Enlargement of ventricles, small hippocampus	Collapsed ventricular system	NA	NA
Brain formation	NA	NA	NA	NA			NA	NA
Heart formation	NA	NA	NA	NA	ND	NA	NA	NA
Hearing	Decreased recovery after noise overstimulation	ND	Diminished sensory hair cell survival after acoustic overexposure	ND	ND	ND	ND	ND
Extraembryonic development	Reduced spongiotrophoblast	NA	NA	NA	NA	ND	NA	NA
Embryonic lethality	70%	NA	ND	NA	NA	50%	NA	NA
Cataract	No	No	No	No	No	No	Yes	Yes
Disease associated SNPs in human	Not known	Not known	Not known	Not known	Not known	Not known	Yes	Yes

*ND = not determined; NA = not affected.

does not influence fertilization or implantation. Fertility of females, and possibly males, also seems to be affected by the *Hsf2* deficiency, since intercrosses between *Hsf2*−/− females with wild-type or *Hsf2*−/− males significantly decreased litter sizes compared with wild-type females. The mechanisms for different embryonic survival among the *Hsf2* targeted knockout models are currently unknown.

5.3. Regulatory Role of Hsf2 in Brain Development

Based on *Hsf2* expression profile and DNA binding activities in embryonic development and adult animals, it was highly anticipated that *Hsf2* may have essential functions in different phases of mouse brain development and its absence could lead to morphological, histopathological, and functional cerebral abnormalities in postnatal animals. Two research groups (Kallio et al., 2002; Wang et al., 2003), indeed found such abnormalities, although only in a relatively small fraction of the animals (25–30%), indicating low genetic penentrance. In developing embryos, these abnormalities were associated mostly with the ventricular system. One group identified variable defects including collapsed ventricules, hemorrhages in the cerebral region, disruption of the neuroepithelium in the fourth ventricle and incomplete posterior neural tube closure with severe reduction of ventricular space between E9.5 and E13.5 of *Hsf2*−/− embryos (Wang et al., 2003). Finally, these investigators reported that *Hsf2* null mice showed postnatal lethality, since about 50% with extensive cerebral hemorrhages died before reaching 3 weeks of age.

Based on two reports, (Kallio et al., 2002; Wang et al., 2003) adult *Hsf2* null animals showed systematically enlarged lateral and third ventricules; moreover, Kallio and colleagues also described dramatically reduced hippocampus and these morphological abnormalities were interpreted as a consequence of a proliferative defect or increased apoptosis rate, in accord with *Hsf2* expression in the proliferative cells of the ventricular zone. Western blot analysis of representative members of the HSP family at E8.5 to E11.5 showed no difference between wild-type and *Hsf2* null mice, implicating non-HSP target genes in the brain defects associated embryonic lethality (Wang et al., 2003). *Hsf2* null embryos at E8.5 and E10.5 were analyzed by gene expression microarrays, but no overlap has been revealed between the up or down regulated gene sets of E8.5 and E10.5 groups. Several differentially expressed CNS specific genes have been hypothesized to contribute to the *Hsf2*−/− dependent brain abnormalities, but their contributions have not been directly linked to *Hsf2* deficiency.

In contrast to the previous studies, McMillan and colleagues (McMillan et al., 2002) did not observe any histological and morphological abnormalities in the adult *Hsf2* null brain. Since normal brain morphology does not preclude compromised psychomotor function, they have also applied a battery of behavioral test of cognitive, sensory and motor functions. By such analysis wild-type and *Hsf2* knockout mice showed indistinguishable performance, suggesting that *Hsf2* is dispensable to normal brain development. The basis of the significant differences in the three mouse models regarding the role of *Hsf2* in brain development is currently unknown and warrants further investigation. Likely, these differences can be attributed to the different targeting strategies, low genetic penetrance and unidentified environmental factor interactions, which also warrant further analysis.

5.4. Reduced Spermatogenesis Without Fertility Defects in Hsf2−/− Male Mice

Based on expression and localization studies, earlier works (Sarge et al., 1994; Alastalo et al., 1998) have proposed a critical role for *Hsf2* in rodent spermatogenesis; hence, differences in testicular morphology and male fertility have been envisaged in *Hsf2* loss of function mice. Analysis of two targeted knockout models of *Hsf2* (Kallio et al., 2002; Wang et al., 2003), indeed have found significant reduction in testes weights (60% of wild-type), in average diameter of seminiferous tubules, average weight of epididymis and sperm counts (20–40% of wild-type), which were mostly attributed to increased apoptosis in the testes of *Hsf2*−/− mice. Quantitative analysis of apoptotic cells by flow cytometry following annexin V-FITC staining revealed threefold increase in apoptosis of testicular cells in *Hsf2*−/− animals; furthermore, TUNEL assay within the seminiferous tubules also detected clusters of apoptotic cells in both *Hsf2* knockout models. Most of the spermatocytes of *Hsf2*-deficient mice died in a stage specific manner at the late pachytene phase of meiotic prophase or during meiotic division, likely as the result of aberrant synaptonemal complex (SC) formation. Defective germ cell elimination may occur by the mechanism known as "pachytene checkpoint" (Roeder and Bailis, 2000), which, if triggered at late meiotic prophase eliminates defective germ cells via apoptosis. Structural defects in SC formation of *Hsf2*−/− mice may act as a trigger, since several meiosis-associated genes, which may be direct *Hsf2* targets have been significantly down regulated in *Hsf2*-deficient background determined by microarray analysis (Wang et al., 2003). Apoptosis of meiotic M phase spermatocytes may be due to premature separation of the centromeric regions of homologous chromosomes, or, alternatively, *Hsf2* may participate in regulation of meiotic M-phase specific proteins such as protein phosphatase 2A (PP2A), which negatively regulates entry into M phase and has also been postulated to play a role in microtubule dynamics and centrosome function in different model systems (Kallio et al., 2002). PP2A activity is also under the control of *Hsf2* in somatic tissue culture cells (Hong and Sarge, 1999).

Conversely, the *Hsf2* knockout mouse generated by McMillan and colleagues (McMillan et al., 2002) did not show any morphological abnormalities in young adult mice regarding their reproductive development and sexual behavior. Although testicular weights were not notably different those of the wild-type, some degeneration of seminiferous tubules have been observed in animals older than 3 months (unpublished results). Surprisingly, but in agreement with studies from the other two groups *Hsf2* deficiency did not influence fertility: *Hsf2*−/− males produced normal size litters over the observed 8 months period. Expression of testis-specific heat shock proteins (e.g., hsp70.2) were not affected in any of the three *Hsf2* knockout models, further supporting the notions that *Hsf2* defects in spermatogenesis likely to occur through non-conventional target genes whose identity still remain obscure. Of interest, deficiency of both *Hsf1* and *Hsf2* genes causes male infertility, indicating their critical and possibly synergistic roles in mouse spermatogenesis (Wang et al., 2004).

6. *Hsf4*: A Heat Shock Factor with Specialized Functions in the Ocular Lens

The third authentic member of the mammalian HSF family was isolated and analyzed by Nakai and colleagues (1997). Chromosomal localizations, ENSEMBL ID numbers, and gene names of all three known murine HSFs are shown in Figure 3. *Hsf4* expression is tissue restricted and is primarily expressed in the heart, brain, and ocular lens as assessed by RT-PCR (Nakai et al., 1997; Tanabe et al., 1999). Like *Hsf1* and *Hsf2*, *Hsf4* has two splice variants, which have opposing functions: the original *Hsf4*a isoform is transcriptionally inert and likely functions as transcriptional repressor, while *Hsf4*b is a weak activator, which rescues HSF deficiency in yeast (Tanabe et al., 1999). Unlike other mammalian HSFs, *Hsf4* lacks

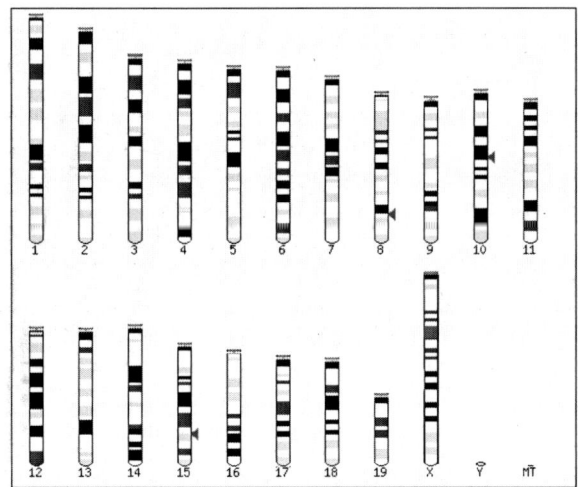

Gene ID	Gene Name	Genome Location	Description(if known)
ENSMUSG00000033249	Hsf4	Chromosome 8: 104.57m	heat shock transcription factor 4 [Source:MarkerSymbol;Acc:MGI:1347058]
ENSMUSG00000019878	Hsf2	Chromosome 10: 57.75m	heat shock factor 2 [Source:MarkerSymbol;Acc:MGI:96239]
ENSMUSG00000022556	Hsf1	Chromosome 15: 76.53m	heat shock factor 1 [Source:MarkerSymbol;Acc:MGI:962A8]

FIGURE 3. Chromosomal localizations of murine *Hsf1*, *Hsf2* and *Hsf4*. Gene IDs, genome locations and gene descriptions are shown. Figure was generated using the http://www.ensembl.org/Mus_musculus/index.html resources.

the carboxy-terminal hydrophobic heptad repeat, which has been proposed to suppress trimerization (Wu, 1995); hence, it forms constitutive trimers and binds DNA with high affinity. Early indications for the potential biological functions of *Hsf4* in the context of the whole organisms came from human genetic studies from Bu and colleagues (2002), who performed whole-genome linkage analysis of affected Chinese individuals with lamellar cataract and mapped the disease gene on the 5.11-cM locus of chromosome 16, the genomic location where *Hsf4* resides. They also established association between missense mutations in the DNA binding domain of *Hsf4* and autosomal dominant lamellar and Marner cataracts in familiar and sporadic cases from Chinese and Danish cohorts suggesting a direct role of *Hsf4* in lens development. Encouraged by the findings of the human studies two groups generated knockout mice with similar targeting strategies to address the molecular underpinnings of *Hsf4* role in lens development and other developmental processes potentially controlled by *Hsf4* (Fujimoto et al., 2004; Min et al., 2004). *Hsf4* mice were born with the expected Mendelian frequencies, suggesting that *Hsf4* is dispensable for embryonic survival. Both groups observed early onset cataracts in mice with disrupted *Hsf4* beginning at early postnatal period of 2 to 5 days after birth; furthermore, lens weights have been significantly reduced by approximately 50%. Pathological abnormalities in histological sections in the fiber cell layers of the lens have also been demonstrated: in 6-week-old animals, wild-type lens fiber cells are flattened by dehydration, while *Hsf4*−/− fiber cells remain engorged with inclusion-like structures in the center. These structures stain heavily with eosin and are likely protein aggregates accumulating the abundant lens proteins αB- and αA-crystallin, although their levels were not significantly increased in the mutant animals. Western blot and semi-quantitative RT-PCR analysis of candidate target genes of *Hsf4* unexpectedly revealed up-regulation of several major Hsps, including Hsp60, Hsp70, Hsp90, and Hsp110 in the fiber cell layer of *Hsf4*−/− animals, while immunohistochemical analysis of the epithelial layer showed increased expression of Hsp70 and Hsp60. These results indicate that *Hsf4* may act as transcriptional repressor for several major Hsps in the ocular lens. Conversely, expression of Hsp25/27, a small Hsp family member and significant player in cell survival/death pathway, is greatly down regulated in the inner layer of the lens, however it is unclear whether its reduced level may lead to protein aggregation. Since crystallins are the main structural components in the lens, their expression levels have been examined in the *Hsf4*−/− mice. αA-crystallins, αB-crystallins, and β-crystallins were unchanged in *Hsf4*−/− animals. Surprisingly, both groups found marked reduction in the γ–crystallin expression, although Min and colleagues (2004) registered changes only in γF expression. The γ-crystallin family consists of six members, γAγBγCγDγEγF, which likely arose by gene duplication with each gene containing multiple HSE consensus sites in their promoter region. Chromatin immunoprecipitation (ChIP) experiments demonstrated *Hsf4* binding to γF-crystallin promoter; moreover, decrease in γ-crystallin expression coincides with the timing of abnormal morphology of fiber cells in the early postnatal stage. In conclusion, these results demonstrate *Hsf4* requirement in the maintenance of γ-crystallin gene expression at the early postnatal stage, which will, in turn, ensure protein solubility and stability in the highly concentrated protein environment of the lens fiber cells. The abnormalities in the lens transparency associated with the lack of *Hsf4* therefore, at least partly, due to the reduced expression of γ-crystallin genes.

In addition to the abnormalities observed in fiber cells, morphological abnormalities have been described in lens epithelial cells of the *Hsf4*−/− animals. Whereas epithelial cells found in the anterior surface of the lens exhibit cuboidal morphology in wild-type mice, *Hsf4*−/− lens epithelial cells are columnar and in 2–6 week-old animals are highly elongated in the bow region, where epithelial cells differentiate into fiber cells. Electron microscopy showed that the fiber cells near the epithelial layer in the *Hsf4* animals are highly rich in organelles such as mitochondria and the denucleation of the fiber cells are also inhibited. Microarray studies by Min et al. (2004) also revealed significant down-regulation of DNase-s and proteases that have likely roles in organelle and DNA degradation during fiber cell differentiation to ensure lens transparency. 5-bromo-2′-deoxyuridine (BrdU) incorporation in the epithelial layer of E18.5 mice was higher in *Hsf4*−/− than wild-type mice, indicating increased proliferation of epithelial cells and premature differentiation of fiber cells. Several humoral factors, including fibroblast growth factors (FGFs), insulin-like growth factor (IGF), and transforming growth factor (TGF-β) regulate growth and differentiation of lens epithelial and fiber cells.

Fujimoto and colleagues (Fujimoto et al., 2004) examined the expression of FGFs by RT-PCR analysis. They found that FGF-1, FGF-4, FGF-7, but not FGF-2, were expressed higher in the *Hsf4* null lens, which was confirmed by in situ hybridization analysis showing increased expression of FGF-1 in *Hsf4* null epithelial and fiber cells. Furthermore, chromatin immunoprecipitation (ChIP) analysis indicated strong binding of *Hsf4* to the upstream region of the FGF-7 gene (-615 to $+10$), where five consensus HSE elements reside. These data provide compelling evidence that *Hsf4* acts as a repressor of FGF genes through physical interaction with the consensus HSE sites present in their promoter.

Hsf1 involvement in FGF gene regulation was assessed in the *Hsf1*−/− background. The authors found normal epithelial cell morphology in *Hsf1*-null lens, but FGF-1 and FGF-7 gene expression was decreased. Since *Hsf1* also binds to the HSE element present in the FGF-7 gene they also elucidated the interplay between *Hsf1* and *Hsf4* in a mouse deficient for both HSF genes. The analysis of the double null animals found that *Hsf1* deficiency at least partially rescues the high level of FGF-1,4,7 expression, the increased epithelial cell number, elongated epithelial cell morphology, and high level of Hsp expression (Hsp70 and Hsp60) of the *Hsf4* null mice, indicating that these transcription factors may compete with one another for the expression of FGFs and Hsps in the ocular lens. Similar data were obtained from other tissues, specifically from lung, which suggests that a complex and finely tuned interplay may exist between the HSF family members in growth factor and Hsp gene expression regulations, which may be a general phenomena not strictly associated with developmental processes, underscoring a possible recruitment for homeostasis of the organisms under normal physiological conditions.

7. Summary and Perspectives

Since the generation of the first targeted deletion of *Hsf1* in the mouse (Xiao et al., 1999) the last several years of HSF research have produced unprecedented progress in the analysis of physiological and pathophysiological functions of the HSF family. The year 2004 marks an important milestone as knockout models have became available for all three known mammalian HSFs in mice. However, the existing knockout models of Hsf 1 exhibit multiple phenotypes including female infertility, placental defects, growth retardation, and increased susceptibility to environmental challenges, precluding unambiguous conclusions about *Hsf1*'s biological functions in vivo. Because the effects of *Hsf1* deficiency on extra-embryonic developments are influenced by the genetic background, sophisticated genetic analyses are undoubtedly needed to unravel the likely possibility of gene modifiers of the *Hsf1* master regulator. In parallel, highly significant and meritorious proposals to generate conditional gene targeting of heat shock transcription factors are urgently needed. Experimental approaches such as Cre recombinase, which are becoming more widespread, have the ability to define both cell- and tissue-specific requirements for HSFs expression during various developmental stages. Without doubt, such innovative approaches will not only accelerate progress on the key role *Hsf1* expression plays during oocyte development but will extend our understanding about the mechanisms of *Hsf1* activation in mitigating neurodegenerative and cardiovascular diseases. The growing list of committed investigators, which now encompasses eminently qualified and internationally recognized leaders, should foster greater collaboration and enrich the environment for attracting outstanding students and trainees into the field. Along with complementary experimental models such as *C.elegans* and possibly zebrafish, the seemingly broad disciplines of heat shock proteins, molecular chaperones, protein folding and genetics seem especially poised to propel new vistas and horizons encompassing drug discovery, development and therapeutic efficacy for common human diseases.

Acknowledgments. Ivor J. Benjamin received support from the Christi T. Smith Foundation and NIH awards RO1HD39404 and RO1HL63834.

References

Abravaya, K., Myers, M. P., Murphy, S. P., and Morimoto, R. I. (1992) The human heat shock protein hsp70 interacts with HSF, the transcription factor that regulates heat shock gene expression. *Genes Dev* 6(7):1153–64.

Ahn, S. G., and Thiele, D. J. (2003) Redox regulation of mammalian heat shock factor 1 is essential for Hsp gene activation and protection from stress. *Genes Dev* 17(4): 516–28.

Alastalo, T. P., Lonnstrom, M., Leppa, S., Kaarniranta, K., Pelto-Huikko, M., Sistonen, L., and Parvinen, M. (1998) Stage-specific expression and cellular localization of the heat shock factor 2 isoforms in the rat seminiferous epithelium. *Exp Cell Res* 240(1): 16–27.

Ali, A., Bharadwaj, S., O'Carroll, R., and Ovsenek, N. (1998). HSP90 interacts with and regulates the activity of heat shock factor 1 in Xenopus oocytes. *Mol Cell Biol* 18:4949–60.

Baler, R., Dahl, G., and Voellmy, R. (1993) Activation of human heat shock genes is accompanied by oligomerization, modification, and rapid translocation of heat shock transcription factor HSF1. *Mol Cell Biol* 13(4):2486–96.

Baler, R., Welch, W. J., and Voellmy, R. (1992) Heat shock gene regulation by nascent polypeptides and denatured proteins: hsp70 as a potential autoregulatory factor. *J Cell Biol* 117(6):1151–9.

Bu, L., Jin, Y., Chu, R., Ban, A., Eiberg, H., Andres, L., Jiang, H., Zheng, G., Qian, M., et al. (2002) Mutant DNA-binding domain of HSF4 is associated with autosomal dominant lamellar and Marner cataract. *Nat Genet* 31(3):276–8.

Christians, E., and Benjamin I. J. (2005) A murine world without HSFs: Meeting report. *Cell Stress Chap* 10(4):265–7.

Christians, E., Davis, A. A., Thomas, S. D., and Benjamin, I. J. (2000) Maternal effect of Hsf1 on reproductive success. *Nature* 407(6805):693–4.

Christians, E. S., Zhou, Q., Renard, J., and Benjamin, I. J. (2003) Heat shock proteins in mammalian development. *Semin Cell Dev Biol* 14(5):283–90.

Clos, J., Westwood, J. T., Becker, P. B., Wilson, S., Lambert, K., and Wu, C. (1990) Molecular cloning and expression of a hexameric *Drosophila* heat shock factor subject to negative regulation. *Cell* 63(5):1085–97.

Cotto, J. J., Kline, M., and Morimoto, R. I. (1996). Activation of heat shock factor 1 DNA binding precedes stress-induced serine phosphorylation. Evidence for a multistep pathway of regulation. *J Biol Chem* 271:3355–8.

Fairfield, D. A., Lomax, M. I., Dootz, G. A., Chen, S., Galecki, A. T., Benjamin, I. J., Dolan, D. F., and Altschuler, R. A. (2005) Heat shock factor 1-deficient mice exhibit decreased recovery of hearing following noise overstimulation. *J Neurosci Res* 81(4):589–96.

Fiorenza, M. T., Farkas, T., Dissing, M., Kolding, D., and Zimarino, V. (1995) Complex expression of murine heat shock transcription factors. *Nucleic Acids Res* 23(3):467–74.

Fujimoto, M., Izu, H., Seki, K., Fukuda, K., Nishida, T., Yamada, S., Kato, K., Yonemura, S., Inouye, S., and Nakai, A. (2004) HSF4 is required for normal cell growth and differentiation during mouse lens development. *EMBO J* 23(21):4297–306.

Gallo, G. J., Prentice, H., and Kingston, R. E. (1993) Heat shock factor is required for growth at normal temperatures in the fission yeast *Schizosaccharomyces pombe*. *Mol Cell Biol* 13(2):749–61.

Guettouche, T., Boellmann, F., Lane, W. S., and Voellmy, R. (2005) Analysis of phosphorylation of human heat shock factor 1 in cells experiencing a stress. *BMC Biochem* 6(1):4.

Hoj, A., and Jakobsen, B. K. (1994). A short element required for turning off heat shock transcription factor: evidence that phosphorylation enhances deactivation. *Embo J* 13:2617–24.

Hong, Y., and Sarge, K. D. (1999) Regulation of protein phosphatase 2A activity by heat shock transcription factor 2. *J Biol Chem* 274(19):12967–70.

Inouye, S., Katsuki, K., Izu, H., Fujimoto, M., Sugahara, K., Yamada, S., Shinkai, Y., Oka, Y., Katoh, Y., and Nakai, A. (2003) Activation of heat shock genes is not necessary for protection by heat shock transcription factor 1 against cell death due to a single exposure to high temperatures. *Mol Cell Biol* 23(16):5882–95.

Izu, H., Inouye, S., Fujimoto, M., Shiraishi, K., Naito, K., and Nakai, A. (2004) Heat shock transcription factor 1 is involved in quality-control mechanisms in male germ cells. *Biol Reprod* 70(1):18–24.

Jedlicka, P., Mortin, M. A., and Wu, C. (1997) Multiple functions of *Drosophila* heat shock transcription factor *in vivo*. *EMBO J* 16(9):2452–62.

Kallio, M., Chang, Y., Manuel, M., Alastalo, T. P., Rallu, M., Gitton, Y., Pirkkala, L., Loones, M. T., Paslaru, L., Larney, S., *et al*. (2002) Brain abnormalities, defective meiotic chromosome synapsis and female subfertility in HSF2 null mice. *EMBO J* 21(11):2591–601.

Kline, M. P., and Morimoto, R. I. (1997) Repression of the heat shock factor 1 transcriptional activation domain is modulated by constitutive phosphorylation. *Mol Cell Biol* 17(4):2107–15.

Knauf, U., Newton, E. M., Kyriakis, J., and Kingston, R. E. (1996) Repression of human heat shock factor 1 activity at control temperature by phosphorylation. *Genes Dev* 10(21):2782–93.

Lindquist, S. (1986) The heat-shock response. *Annu Rev Biochem* 55:1151–91.

Lindquist, S., and Craig, E. A. (1988) The heat-shock proteins. *Annu Rev Genet* 22:631–77.

Lis, J., and Wu, C. (1993) Protein traffic on the heat shock promoter: Parking, stalling, and trucking along. *Cell* 74(1):1–4.

McMillan, D. R., Christians, E., Forster, M., Xiao, X., Connell, P., Plumier, J. C., Zuo, X., Richardson, J., Morgan, S., and Benjamin, I. J. (2002) Heat shock transcription factor 2 is not essential for embryonic development, fertility, or adult cognitive and psychomotor function in mice. *Mol Cell Biol* 22(22):8005–14.

McMillan, D. R., Xiao, X., Shao, L., Graves, K., and Benjamin, I. J. (1998) Targeted disruption of heat shock transcription factor 1 abolishes thermotolerance and protection against heat-inducible apoptosis. *J Biol Chem* 273(13):7523–8.

Min, J. N., Zhang, Y., Moskophidis, D., and Mivechi, N. F. (2004) Unique contribution of heat shock transcription factor 4 in ocular lens development and fiber cell differentiation. *Genesis* 40(4):205–17.

Morimoto, R. I. (1998) Regulation of the heat shock transcriptional response: Cross talk between a family of heat shock factors, molecular chaperones, and negative regulators. *Genes Dev* 12(24):3788–96.

Mosser, D. D., Duchaine, J., and Massie, B. (1993) The DNA-binding activity of the human heat shock transcription factor is regulated in vivo by hsp70. *Mol Cell Biol* 13:5427–38.

Nakai, A., and Morimoto, R. I. (1993) Characterization of a novel chicken heat shock transcription factor, heat shock factor 3, suggests a new regulatory pathway. *Mol Cell Biol* 13(4):1983–97.

Nakai, A., Suzuki, M., and Tanabe, M. (2000) Arrest of spermatogenesis in mice expressing an active heat shock transcription factor 1. *EMBO J* 19(7):1545–54.

Nakai, A., Tanabe, M., Kawazoe, Y., Inazawa, J., Morimoto, R. I., and Nagata, K. (1997) HSF4, a new member of the human heat shock factor family which lacks properties of a transcriptional activator. *Mol Cell Biol* 17(1):469–81.

Orosz, A., Wisniewski, J., and Wu, C. (1996) Regulation of *Drosophila* heat shock factor trimerization: Global sequence requirements and independence of nuclear localization. *Mol Cell Biol* 16(12):7018–30.

Preville, X., Salvemini, F., Giraud, S., Chaufour, S., Paul, C., Stepien, G., Ursini, M. V., and Arrigo, A. P. (1999) Mammalian small stress proteins protect against oxidative stress through their ability to increase glucose-6-phosphate dehydrogenase activity and by maintaining optimal cellular detoxifying machinery. *Exp Cell Res* 247(1):61–78.

Rabindran, S. K., Giorgi, G., Clos, J., and Wu, C. (1991) Molecular cloning and expression of a human heat shock factor, HSF1. *Proc Natl Acad Sci USA* 88(16):6906–10.

Rabindran, S. K., Haroun, R. I., Clos, J., Wisniewski, J., and Wu, C. (1993) Regulation of heat shock factor trimer formation: Role of a conserved leucine zipper. *Science* 259(5092):230–4.

Rabindran, S. K., Wisniewski, J., Li, L., Li, G. C., and Wu, C. (1994). Interaction between heat shock factor and hsp70 is insufficient to suppress induction of DNA-binding activity in vivo. *Mol Cell Biol* 14:6552–60.

Rallu, M., Loones, M., Lallemand, Y., Morimoto, R., Morange, M., and Mezger, V. (1997) Function and regulation of heat shock factor 2 during mouse embryogenesis. *Proc Natl Acad Sci USA* 94(6):2392–7.

Ritossa, F. (1962) A new puffing pattern induced by temperature shock and DNP in *Drosophila*. *Experientia* 18:571–73.

Roeder, G. S., and Bailis, J. M. (2000) The pachytene checkpoint. *Trends Genet* 16(9):395–403.

Santos, S. D., and Saraiva, M. J. (2004) Enlarged ventricles, astrogliosis and neurodegeneration in heat shock factor 1 null mouse brain. *Neuroscience* 126(3):657–63.

Sarge, K. D., Park-Sarge, O. K., Kirby, J. D., Mayo, K. E., and Morimoto, R. I. (1994) Expression of heat shock factor 2 in mouse testis: Potential role as a regulator of heat-shock protein gene expression during spermatogenesis. *Biol Reprod* 50(6): 1334–43.

Sarge, K. D., Zimarino, V., Holm, K., Wu, C., and Morimoto, R. I. (1991) Cloning and characterization of two mouse heat shock factors with distinct inducible and constitutive DNA-binding ability. *Genes Dev* 5(10):1902–11.

Scharf, K. D., Rose, S., Zott, W., Schoffl, F., and Nover, L. (1990) Three tomato genes code for heat stress transcription factors with a region of remarkable homology to the DNA-binding domain of the yeast HSF. *EMBO J* 9(13):4495–501.

Schuetz, T. J., Gallo, G. J., Sheldon, L., Tempst, P., and Kingston, R. E. (1991) Isolation of a cDNA for HSF2: Evidence for two heat shock factor genes in humans. *Proc Natl Acad Sci USA* 88(16):6911–5.

Shi, Y., Mosser, D. D., and Morimoto, R. I. (1998) Molecular chaperones as HSF1-specific transcriptional repressors. *Genes Dev* 12(5):654–66.

Sistonen, L., Sarge, K. D., Phillips, B., Abravaya, K., and Morimoto, R. I. (1992) Activation of heat shock factor 2 during hemin-induced differentiation of human erythroleukemia cells. *Mol Cell Biol* 12(9):4104–11.

Sorger, P. K., and Pelham, H. R. (1988) Yeast heat shock factor is an essential DNA-binding protein that exhibits temperature-dependent phosphorylation. *Cell* 54(6):855–64.

Sugahara, K., Inouye, S., Izu, H., Katoh, Y., Katsuki, K., Takemoto, T., Shimogori, H., Yamashita, H., and Nakai, A. (2003) Heat shock transcription factor HSF1 is required for survival of sensory hair cells against acoustic overexposure. *Hear Res* 182(1–2): 88–96.

Tanabe, M., Sasai, N., Nagata, K., Liu, X. D., Liu, P. C., Thiele, D. J., and Nakai, A. (1999) The mammalian HSF4 gene generates both an activator and a repressor of heat shock genes by alternative splicing. *J Biol Chem* 274(39):27845–56.

Voellmy, R. (2004) On mechanisms that control heat shock transcription factor activity in metazoan cells. *Cell Stress Chap* 9(2):122–33.

Wang, G., Ying, Z., Jin, X., Tu, N., Zhang, Y., Phillips, M., Moskophidis, D., and Mivechi, N. F. (2004) Essential requirement for both hsf1 and hsf2 transcriptional activity in spermatogenesis and male fertility. *Genesis* 38(2):66–80.

Wang, G., Zhang, J., Moskophidis, D., and Mivechi, N. F. (2003) Targeted disruption of the heat shock transcription factor (hsf)-2 gene results in increased embryonic lethality, neuronal defects, and reduced spermatogenesis. *Genesis* 36(1):48–61.

Westwood, J. T., Clos, J., and Wu, C. (1991) Stress-induced oligomerization and chromosomal relocalization of heat-shock factor. *Nature* 353(6347):822–7.

Wiederrecht, G., Seto, D., and Parker, C. S. (1988) Isolation of the gene encoding the *S. cerevisiae* heat shock transcription factor. *Cell* 54(6):841–53.

Wisniewski, J., Orosz, A., Allada, R., and Wu, C. (1996) The C-terminal region of *Drosophila* heat shock factor (HSF) contains a constitutively functional transactivation domain. *Nucleic Acids Res* 24(2):367–74.

Wu, C. (1995) Heat shock transcription factors: Structure and regulation. *Annu Rev Cell Dev Biol* 11:441–69.

Xiao, X., Zuo, X., Davis, A. A., McMillan, D. R., Curry, B. B., Richardson, J. A., and Benjamin, I. J. (1999) HSF1 is required for extra-embryonic development, postnatal growth and protection during inflammatory responses in mice. *EMBO J* 18(21):5943–52.

Yan, L. J., Christians, E. S., Liu, L., Xiao, X., Sohal, R. S., and Benjamin, I. J. (2002) Mouse heat shock transcription factor 1 deficiency alters cardiac redox homeostasis and increases mitochondrial oxidative damage. *EMBO J* 21(19):5164–72.

Yan, L. J., Rajasekaran, N. S., Sathyanarayanan, S., and Benjamin, I. J. (2005) Mouse HSF1 disruption perturbs redox state and increases mitochondrial oxidative stress in kidney. *Antioxid Redox Signal* 7(3–4):465–71.

Zhang, Y., Huang, L., Zhang, J., Moskophidis, D., and Mivechi, N. F. (2002) Targeted disruption of hsf1 leads to lack of thermotolerance and defines tissue-specific regulation for stress-inducible Hsp molecular chaperones. *J Cell Biochem* 86(2):376–93.

Zou, J., Guo, Y., Guettouche, T., Smith, D. F., and Voellmy, R. (1998) Repression of heat shock transcription factor HSF1 activation by HSP90 (HSP90 complex) that forms a stress-sensitive complex with HSF1. *Cell* 94:471–80.

Zuo, J., Rungger, D., and Voellmy, R. (1995) Multiple layers of regulation of human heat shock transcription factor 1. *Mol Cell Biol* 15(8):4319–30.

6
HSF1 and HSP Gene Regulation

RICHARD VOELLMY
HSF Pharmaceuticals S.A., Avenue des Cerisiers 39B, 1009 Pully, Switzerland

1. Introduction: HSP Genes, HSF1, and the Control of HSF1 Activity at Different Levels

Induction of hsp genes by heat and chemicals is strictly dependent on the presence in their non-transcribed regulatory sequences of so-called heat shock elements (abbreviated HSEs) (Mirault et al., 1982; Pelham, 1982). HSEs are arrays of three or more modules of the sequence element NGAAN (or AGAAN) or variations thereof (Amin et al., 1988; Xiao and Lis, 1988). Heat shock factors (HSFs) are defined as proteins that are capable of specifically binding HSE sequences. The first attempts at identifying and/or purifying an HSF were undertaken by Parker and Topol (1984) and Wu (1984). Whereas certain organisms appear to express a single HSF, others, including plants, avian species and mammals express multiple related but distinct factors (reviewed, e.g., in Pirkkala et al., 2001). One of these proteins, HSF1, rapidly took center stage because it was found to be the major heat- and chemically induced HSE DNA-binding protein in mammalian cells and, therefore, likely to be the major transcription factor that activates or enhances transcription of hsp genes in response to these stresses (Baler et al., 1993; Sarge et al., 1993). Gene knockouts confirmed this expected role of HSF1 (McMillan et al., 1998; Zhang et al., 2002; for *Drosophila*, see Jedlicka et al., 1997). As it must be capable of carrying out the same function, the single HSF present in certain organisms will also be referred to as HSF1 herein for the sake of simplicity.

Sequence inspection and structure-function studies revealed that HSF1 has a modular design. It is composed of an HSE DNA-binding domain, two heptad repeat arrays (HR-A and HR-B), a regulatory domain (described in more detail below), a third heptad repeat sequence (HR-C), and transcriptional activation domain(s). HSF1 was found to be predominantly monomeric and essentially incapable of specifically binding DNA in the absence of heat or chemical inducers (with the exception of yeast HSF1 which has a detectable level of uninduced DNA-binding activity). Upon exposure of cells to heat or chemical inducers, HSF1 homo-trimerized and acquired specific DNA-binding and transcription-enhancing activity (Westwood et al., 1992; Baler et al., 1993; Sarge et al., 1993). Mutagensis experiments strongly suggested that gain of HSE DNA-binding activity and factor oligomerization were tightly linked events (see, e.g., Zuo et al., 1994). As will be discussed below, these early experiments failed to discover that a large fraction/all of inactive HSF1 associates in a dynamic fashion with Hsp90-containing heterocomplexes. This aspect will be discussed extensively below.

Different lines of evidence suggested that the activity of HSF1 was controlled at two distinct levels, i.e., oligomerization and transcriptional competence. First, heat treatment of mouse erythroleukemia cells was found to produce HSE DNA-binding but transcriptionally inactive HSF1 (Hensold et al., 1990). Second, a number of compounds were identified that, at least in certain mammalian cell types, were capable of inducing trimerization of the factor, without also enabling it to enhance expression of hsp genes. These compounds included sodium salicylate, menadione, and hydrogen peroxide (Jurivich et al., 1992; Bruce et al., 1993). Third, overexpression of HSF1 in transfected mammalian cells resulted in accumulation of trimeric factor that had only minimal transcriptional activity (Zuo et al., 1995). Fourth, mutagenesis experiments revealed that separate sequence elements within mammalian HSF1 sequences were involved in regulating oligomeric state and transcriptional competence (Zuo et al., 1994, 1995; Green et al., 1995; Shi et al., 1995). Repression of trimerization of human HSF1 in the unstressed cell required the presence of sequences at the beginning of HR-A as well as functional HR-B and HR-C sequences. In contrast, transcription-enhancing activity was largely repressed by the so-called regulatory domain located downstream from HR-B (residues 201–330 of the 529-residue human HSF1 sequence). Finally, transactivation assays with chimeric transcription factors revealed that the regulatory domain of (human) HSF1 was alone capable of conferring stress-sensitive repression of transcriptional competence on a chimeric factor (Newton et al., 1996).

2. Repression of HSF1 Oligomerization

It has long been known that many (but perhaps not all) conditions and chemicals that trigger activation of HSF1 in cells exposed to them are at least marginally damaging to the cells. In large part, this toxicity appears to be related to denaturation of significant amounts of cellular protein and/or to interference with mechanisms of protein homeostasis, including protein synthesis fidelity, protein folding and refolding, maintenance of chaperone-assisted folds, protein trafficking and protein degradation. Heat and many of the chemical inducers of HSF1 activity caused large increases in oxidized glutathione (Zou et al., 1998a). Hence, these inducers appeared to cause denaturation of cellular protein at least in part through the formation of glutathione adducts and intra- and/or inter-protein cross-linking (Freeman et al., 1995; Liu et al., 1996; McDuffee et al., 1997; Senisterra et al., 1997; Zou et al., 1998a). Others like ethanol (which also increased thiol oxidation) and amino acid analogs affected fidelity of protein synthesis, increasing cellular loads of misfolded proteins. Yet other compounds such as the benzoquinone ansamycins (e.g., herbimycin A, geldanamycin) interfered with Hsp90 function, affecting protein folding and refolding, maintenance of chaperone-assisted folds as well as protein trafficking (Murakami et al., 1991; Whitesell et al., 1994; Hedge et al., 1995; Pratt and Toft, 1997). Hence, the net result of exposure to these compounds is an increased accumulation of unfolded protein. However, as will become clear from the discussion below, inactivation of certain Hsps and co-chaperones also has direct effects on HSF1 conformation. Finally, compounds such as lactacystin and MG132 inhibited proteasome-dependent degradation of nonnative proteins, leading to their accumulation (Bush et al., 1997; Ciechanover, 1998; Pirkkala et al., 2000, and references therein). These observations supported the now widely accepted hypothesis that an elevated level of nonnative proteins is the proximal signal for activation of HSF1, which activation begins with factor trimerization. The hypothesis was further strengthened by the demonstration that introduction into cells of nonnative proteins resulted in activation of HSF1 (Ananthan et al., 1986).

A mechanistic explanation for how accumulation of nonnative protein relieves repression of HSF1 oligomerization emerged from studies by Ali et al. (1998), Duina et al. (1998), Zou et al. (1998b), Bharadwaj et al. (1999), and Marchler and Wu (2001). Collectively, these studies demonstrated that inactive HSF1 is bound by an Hsp90-containing multichaperone complex that also includes p23 (in vertebrates) and an immunophilin. Upon activation of HSF1 subsequent to a proteotoxic stress (i.e., exposure to heat or one of the above-described chemical inducers), this interaction all but disappears, leaving most HSF1 molecules in an unbound state in which they can associate with one another to form DNA-binding homotrimers. Because of the dynamic nature of associations of Hsps or Hsp complexes and client proteins, the mechanism of "release" of HSF1 is thought to involve competition for chaperones and co-chaperones participating either in the Hsp90-p23-immunophilin complex or in intermediate chaperone complexes, resulting in a dramatic reduction of chaperone assembly on HSF1 polypeptide. Based on the general hypothesis presented above, it is proposed that this competition is generated by increased levels of stress-denatured proteins which are client proteins for chaperones and co-chaperones. Direct support for this proposition was provided by Zou et al. (1998b) who showed that addition of chemically denatured protein to their in vitro system (derived from HeLa cells) triggered HSF1 oligomerization. Furthermore, Guo et al. (2001) demonstrated that misexpression in mammalian cells of a mutated BSA that accumulated in the cytoplasm in a nonnative form resulted in activation of HSF1. While the above mechanism is capable of explaining stress-induced oligomerization of HSF1, there are reasons for thinking that additional (or even alternative) mechanisms may exist. For example, heat exposure of cells results in activation of HSF1 within minutes. Conceivably, competition for chaperones and co-chaperones may be too slow a mechanism to rapidly rid HSF1 polypeptide of its associated chaperone complex. Other factors may be required that interact with HSF1 or components of the chaperone complex and actively induce the release of chaperone complex from HSF1. While it may normally co-regulate HSF1 in more subtle ways, Chip apparently can function as such a release factor (in the absence of a stress). Upon overexpression in mammalian cells, Chip caused HSF1 to oligomerize and acquire transcriptional competence (Dai et al.,

2003). Chip was also found to associate with HSF1 or an HSF1 complex (Dai et al., 2003; Kim et al., 2005a). It is noted that Chip is known to interact via tetratricopeptide repeats with Hsp70 and Hsp90 and to be capable of down-modulating Hsp70 function and altering the composition of Hsp90-containing chaperone complexes (Hoehfeld et al., 2001; McDonough and Patterson, 2003).

3. Repression, Induction, and Re-Repression of HSF1 Transcriptional Competence

3.1. Repression by Chaperones and Chaperone Complex

Nair et al. (1996) first reported that mature Hsp90-p23-immunophilin complexes as well as intermediate chaperone complexes could be assembled on recombinant human HSF1. Because no active measures were taken to prevent oligomerization, a large fraction of the recombinant HSF1 used in the study could be expected to have been homotrimeric. Therefore, the study suggested the possibility that Hsp90-containing multichaperone complexes also associated with trimeric HSF1. Guo et al. (2001) confirmed that such complexes formed on trimeric HSF1 by repeating the assembly reactions using as template recombinant human HSF1 bound to HSE DNA immobilized on beads. Experiments involving co-immunoprecipitation of HSF1 bound by Hsp90-containing chaperone complex revealed that chaperone complex association did not abrogate the factor's DNA-binding ability. Hence, if the interactions were in fact modulating HSF1 activity, they most likely affected HSF1 transcriptional competence.

In an attempt to identify HSF1 complexes forming in the course of activation in vivo, human HeLa cells were incubated with a cross-linking agent either before or at different times (1–15 min) during a moderately severe heat treatment, and HSF1 complexes were immunoprecipitated using antibodies directed against Hsp90 and co-chaperones (Guo et al., 2001). This experiment revealed that upon heat treatment the HSF1-Hsp90 interaction rapidly declined to a low level. In contrast, an HSF1-FKBP52 interaction formed during but not prior to heat treatment. This interaction was detectable already after 1 minute of heat treatment and increased throughout the period of heat exposure. Taking into consideration that FKBP52 typically interacts with client proteins as part of an Hsp90-p23-FKBP52 complex, these findings were interpreted as evidence that Hsp90-p23-FKBP52 complex also associated with trimeric HSF1 in vivo, albeit only with a fraction of factor under the heat stress conditions employed. The observations also suggested that the Hsp90-p23-immunophilin complex associating with non-(homo)oligomeric HSF1 either did not contain FKBP52 or, if it did, interacted differently with HSF1 polypeptide than with trimeric HSF1 so that FKBP52 could not be cross-linked to HSF1 polypeptide. Irrespective of the reason for the apparent absence of FKBP52 in complexes containing HSF1 polypeptide, the observation provided a means for distinguishing complexes containing non-oligomeric HSF1 and complexes containing trimeric HSF1.

The question whether chaperone complex association affected the transactivation ability of trimeric HSF1 was approached using HeLa cells overexpressing (relative to endogenous HSF1) a LexA-HSF1 chimera (Guo et al., 2001). As discussed before, in cells overexpressing HSF1 from transfected genes, a large fraction of HSF1 is in a DNA binding, trimeric form that essentially lacks transcription-enhancing activity. Cells expressing LexA-HSF1 were subjected to several different maneuvers designed to elevate levels of client proteins of chaperones. These maneuvers reduced the level of interaction between LexA-HSF1 and FKBP52, and increased the transcriptional competence of the chimeric transcription factor. This inverse correlation strongly suggested that assembly of Hsp90-p23-FKBP52 complex on trimeric HSF1 decreases its transcriptional competence. Further support for this notion came from an experiment in which HSF1 deletion mutants were screened for their ability to interact with FKBP52. Results suggested that Hsp90-p23-FKBP52 complex binds to the HSF1 regulatory domain, the domain previously shown to be essential for repression of HSF1 transcriptional activity in the absence of a stress. Based on these data, Guo et al. (2001) proposed that HSF1 activity is repressed by Hsp90-containing multichaperone complexes at the levels of oligomerization as well as acquisition of transcriptional competence. Repression at the latter level may serve to ensure that induced HSF1 activity is proportional to the intensity of the inducing stress.

There appears to exist a second mechanism for chaperone repression of HSF1 transcriptional activity. Shi et al. (1998) used in vitro reconstitution assays to show that Hsp70 and Hsp40 (Hdj1) interacted with the transcriptional activation domain region of mouse HSF1. Both full-length Hsp70 and an isolated substrate-binding domain were capable of associating with the transcriptional activation region. Overexpression of either Hsp70 or Hsp40 (Hdj1) inhibited the transcriptional activity of a co-expressed GAL4-HSF1 activation domain chimera but not of a control GAL4-VP16 fusion protein. HSF1 mutants lacking part of the regulatory domain or the entire domain were active transcription factors in the absence of a stress (Zuo et al., 1995; Xia et al., 1999; Hall and Voellmy, unpublished observations). This finding suggested that repression by an Hsp90-p23-FKBP52 complex that interacts with the HSF1 regulatory domain is more sensitive to chaperone levels than repression by Hsp70 and/or Hsp40 binding to the HSF1 transcriptional activation domain region. Perhaps, this difference in sensitivity reflects the different roles the two mechanisms could be playing. The former mechanism may control levels of HSF1 activation (full repression at levels of chaperones present in the unstressed cell), whereas the latter mechanism may relate to factor inactivation subsequent to a stressful event (full repression at the elevated chaperone levels present in cells recovering from a severe stress).

3.2. Repression, Induction, and Re-Repression Through Phosphorylation

Phosphorylation has turned out to be an important mechanism for controlling HSF1 activity. It was discovered early on that overall phosphorylation of HSF1 increases substantially in cells experiencing a heat stress (Sorger et al., 1987; Sorger and Pelham, 1988). Because such hyper-phosphorylation was also observed for *Saccharomyces cervisiae* HSF1, which factor has HSE DNA-binding activity in the absence of a stress, it was assumed that it served to enhance the transcriptional competence of the factor (but see Giardina and Lis (1995) for a demonstration that HSE DNA-binding activity of yeast HSF1 increases dramatically after heat treatment). Several different types of experiments provided additional indirect support for this notion. For example, as mentioned before, chemicals such as sodium salicylate, menadione or hydrogen peroxide were found to induce oligomerization of a mammalian HSF1 but not its transcriptional activity (Jurivich et al., 1992; Bruce et al., 1993). Subsequent work showed that phosphorylation of human HSF1 was not at all or only weakly induced in cells exposed to the latter chemicals (Jurivich et al., 1995; Cotto et al., 1996; Xia and Voellmy, 1997). Co-exposure of the cells to phosphatase inhibitor calyculin A resulted in an elevated level of phosphorylation of HSF1 and transcriptional competence (Xia and Voellmy, 1997). The fraction of HSF1 accumulating as trimers in human cells overexpressing the factor is DNA binding, but is at most weakly phosphorylated and possesses only marginal transcriptional activity. Exposure of these cells to calyculin A induced both HSF1 phosphorylation and transcriptional activity (Xia and Voellmy, 1997).

Although these types of findings established a correlation between hyper-phosphorylation and transcriptional competence of HSF1, only identification of the specific residues phosphorylated and structure-function studies using HSF1 forms mutated in these residues held the promise of leading to a definitive understanding of how phosphorylation modulates HSF1 activity. A number of such studies were conducted during the last ten years and yielded several unexpected results. While certain phosphorylation sites were previously known, a systematic analysis of HSF1 residues phosphorylated in heat-treated cells was published only recently (Guettouche et al., 2005). The study showed human HSF1 to be phosphorylated on Ser^{121}, Ser^{230}, Ser^{292}, Ser^{303}, Ser^{307}, Ser^{314}, Ser^{319}, Ser^{326}, Ser^{344}, Ser^{363}, Ser^{419}, and Ser^{444}. For seven of these sites there is now evidence that their phosphorylation affects HSF1 activity (discussed below and summarized in Table 1). Ser^{121} appeared to be partially phosphorylated in unstressed cells, and the level of phosphorylation of the residue decreased during heat or salicylate exposure (Wang et al., 2005). Mutation of Ser^{121}-to-Ala increased the residual transcriptional activity of factor expressed from transfected genes. Interestingly, co-expression with wild-type HSF1 of MAPKAP kinase 2 resulted in an elevated level of phosphorylation of HSF1 at Ser^{121} and increased binding of Hsp90 but reduced HSE DNA-binding and transcriptional activity. These observations suggested that, under pro-inflammatory conditions, phosphorylation of Ser^{121} stabilizes non-oligomeric HSF1 complex, tightening repression of factor oligomerization.

Holmberg et al. (2001) found that a Ser^{230}-to-Ala substitution reduced heat-induced activity of human HSF1 by about twofold. HSF1 activity was measured as the ability of wild-type or mutant factor to induce Hsp70 synthesis in mouse $hsf1^{-/-}$ MEF cells. EMSA revealed that HSE DNA binding was not impaired in the mutant factor. A phosphospecific antibody was raised and used to show that Ser^{230} phosphorylation was induced by stress. Ser^{230} is contained within a CaMKII

TABLE 1. Phosphorylation of HSF1

Phosphorylated residue in human HSF1	Likely regulatory role of phosphorylation	Inducibility of phosphorylation
Ser^{121}	Stabilization of non-oligomeric, inactive HSF1	Low-level, constitutive phosphorylation; enhanced under pro-inflammatory conditions
Ser^{230}	Enhancement of HSF1 transcriptional competence	Stress-induced
Ser^{292}	Not known	Not known
Ser^{303}	Re-repression of HSF1 transcriptional competence subsequent to or late during stress exposure; sequestration in the cytoplasm	Stress-induced
Ser^{307}	Re-repression of HSF1 transcriptional competence subsequent to or late during stress exposure; sequestration in the cytoplasm	Stress-induced
Ser^{314}	Not known	Not known
Ser^{319}	Not known	Not known
Ser^{326}	Enhancement of HSF1 transcriptional competence	Stress-enhanced
Ser^{344}	Not known	Not known
Ser^{363}	Re-repression of HSF1 transcriptional competence subsequent to or late during stress exposure	Not known
Ser^{419}	Not known	Not known
Ser^{444}*	Enhancement of HSF1 nuclear translocation	Not known

*Data obtained using mouse HSF1.

consensus site. CaMKII overexpression enhanced heat-induced expression of a reporter gene directed by an hsp70 promoter, and kinase inhibitor KN62 reduced reporter expression. Hence, phosphorylation of Ser^{230} is induced during stressful events and enhances HSF1 transcriptional activity.

The regulatory domain of human HSF1 includes six proline-directed serine motifs. Knauf et al. (1996) examined transactivation by GAL4-HSF1 chimeras lacking some or all of these motifs and found that heat inducibility of transactivation was reduced significantly after elimination of all motifs or selective elimination of Ser^{303} and Ser^{307}. Similar de-repressive effects of Ser^{303} and/or Ser^{307} substitution were reported by Chu et al. (1996), Kline and Morimoto (1997), Xia et al. (1998) and Guettouche et al. (2005). Kline and Morimoto (1997) were the first to demonstrate that Ser^{303} and Ser^{307} were, in fact, phosphorylated in vivo. Phosphorylation at the two residues was induced by stress and was reversed during recovery from stress (Hietakangas et al., 2003). Ser^{307} is a target of MAP kinases (Chu et al., 1996, 1998; Knauf et al., 1996; Kim et al., 1997; He et al., 1998) and Ser^{303} of glycogen synthase kinase 3 (GSK3) (Chu et al., 1996, 1998; He et al., 1998; Xavier et al.,

2000), but other protein kinases may also phosphorylate these sites. Chu et al. (1996; 1998) obtained evidence that Ser^{303} is phosphorylated by GSK3 subsequent to phosphorylation of Ser^{307}. GSK3 activity was found to be heat-induced (He et al., 1998; Xavier et al., 2000). Together, these results strongly suggested that Ser^{303} and Ser^{307} are phosphorylated during stress, likely in a sequential manner by a MAP kinase and GSK3. Phosphorylation of these residues appears to contribute to re-repression of HSF1 transcriptional activity during recovery from stress.

Phosphorylation of Ser^{303} and Ser^{307} may affect HSF1 activity through a second mechanism that does not act at the level of transcriptional competence (Wang et al., 2003; 2004). HSF1 phosphorylated on Ser^{303} and Ser^{307} interacts with a 14-3-3 protein. Binding of 14-3-3 appears to sequester HSF1 in the cytoplasm in the absence of a stress. This mechanism is believed to play a role in re-repression of HSF1 activity subsequent to a stress.

Ser^{326}-to-Ala substitution resulted in a several fold reduction of heat-induced transcriptional activity of human HSF1 when assayed as induced Hsp70 synthesis in hsf1$^{-/-}$ MEF cells (Guettouche et al., 2005). Phosphorylation of Ser^{326} was examined using a phosphospecific antibody and was found to be enhanced in stressed cells. HSE DNA binding and nuclear localization were not affected by the S326A mutation, suggesting that stress-enhanced phosphorylation of Ser^{326} serves to boost the transcriptional activity of HSF1. When S230A and S326A mutants were compared in the same experiment, the transcription-enhancing effect of Ser^{230} phosphorylation was found to be considerably less important than that of Ser^{326} phosphorylation. Ser^{326} does not lie within a consensus phosphorylation site. To date, no information is available on the nature of the protein kinase responsible for the phosphorylation of the residue.

A Ser^{363}-to-Ala mutant was found to disappear from stress granules more slowly than wild-type human HSF1 during recovery from a heat stress (Dai et al., 2000). This observation suggested that phosphorylation of Ser^{363} was an aspect of re-repression of HSF1 activity during recovery from heat stress. It also implied that the residue was inducibly phosphorylated during heat stress. JNK1, PKCα and PKCζ all were capable of phosphorylating the residue and causing a reduction in HSF1 transcriptional activity (Chu et al., 1998; Dai et al., 2000).

Polo-like kinase 1 was shown by Kim et al. (2005b) to be capable of phosphorylating Ser^{418} of mouse HSF1 (corresponding to Ser^{444} of human HSF1). The same kinase was previously known to phosphorylate cyclin B, enabling nuclear translocation of the cyclin. Phosphorylation of Ser^{418} appears to have an analogous facilitating effect on nuclear translocation of HSF1: a Ser^{418}-to-Ala substitution mutant failed to exhibit the typical exclusive nuclear localization of HSF1 in heat-treated cells.

Possible regulatory roles of phosphorylation of (human HSF1 residues) Ser^{292}, Ser^{314}, Ser^{319}, Ser^{344}, and Ser^{419} remain to be elucidated. To summarize what was learned about HSF1 phosphorylation to date, the results of experiments employing a phosphatase inhibitor that led to the discovery of a correlation between hyper-phosphorylation and transcriptional competence of HSF1 were vindicated in part. Two Ser residues of human HSF1, i.e., Ser^{230} and Ser^{326}, are inducibly phosphorylated during stressful events and, upon phosphorylation, enhance HSF1 transcriptional activity. However, other stress-induced phosphorylation, such as that on Ser^{303}, Ser^{307}, and, presumably, also Ser^{363}, may serve to re-repress transcriptional activity of HSF1 during recovery from a stress. Phosphorylation of two other residues only indirectly relates to HSF1 transcriptional activity. Ser^{121} may be phosphorylated constitutively at a low level and may contribute to the stability of the association of monomeric HSF1 and Hsp90-containing chaperone complex. High-level phosphorylation of this residue and, consequently, important repressive effects may occur under pro-inflammatory conditions. Phosphorylation of Ser^{418} of mouse HSF1 (Ser^{444} of human HSF1) appears to facilitate nuclear concentration of HSF1 during stress.

4. Additional Mechanisms: Daxx, Stress Granules

Daxx is a largely nuclear protein that associates predominantly with promyelocytic leukemia oncogenic domains (POD; also referred to as nuclear domains 10) (Ishov et al., 1999; Torii et al., 1999; Zhong et al., 2000). Release of the protein from these subnuclear domains occurs during heat or chemical stress (Maul et al., 1995; Nefkens et al., 2003). Deletion of Daxx, RNA interference depletion or expression of dominant-negative mutants of Daxx substantially reduced stress-induced HSF1 activity in mammalian cells as measured by reporter assays and/or inducible Hsp70 expression in $hsf1^{-/-}$ MEF cells (Boellmann et al., 2004). Although it is not clear how Daxx enhances HSF1 activity, the available data suggest that it functions to relieve repression of transcriptional competence. Daxx interacts with trimeric HSF1. As mentioned before, overexpression from transfected hsf1 genes results in accumulation of HSE DNA-binding but only marginally transcription-competent HSF1 trimers. Co-expression of Daxx strongly enhances transcriptional competence of these HSF1 oligomers. That Daxx co-expression also induces hyper-phosphorylation of the HSF1 molecules suggests the possibility that the protein may function as an adapter of an HSF1 kinase (Boellmann, unpublished results). It is noted that Daxx is known to interact with several protein kinases and, in the case of JNK, acts as an adapter for apoptosis signal-regulating kinase 1 which phosphorylates JNK (Chang et al., 1998; Rochat-Steiner et al., 2000; Ecsedy et al., 2003).

In primate cells exposed to a stress, the bulk of HSF1 appears to concentrate rapidly in a small number of subnuclear domains referred to a stress granules (Cotto et al., 1997). These structures were shown to be associated with a heterochromatic region of (human) chromosome 9 (9q12) (Jolly et al., 2002). As major hsp genes are not located in this particular region (Jolly et al., 1997), this phenomenon is not directly related to hsp gene transcription. However, accumulation of HSF1 in stress granules may well have profound indirect effects on the overall response of cells to stresses. HSF1 in stress granules activates transcription of satellite III repeats (Jolly et al., 2004; Rizzi et al., 2004). The satellite III transcripts that remain associated with the stress granules and HSF1 are both required for the sequestration of specific splicing factors such as hSF2/ASF and hSRp30c (Metz et al., 2004). Furthermore, overexpression of Hsp70 was found to result in a reduced accumulation of HSF1 in stress granules (Jolly et al., 2002). Possibly, this mechanism forms part of a/the pathway of HSF1 inactivation that occurs during stress recovery.

5. Post-Transcriptional Regulation

Expression of hsp genes is also regulated at the levels of transcript stability and translatability (DiDomenico et al., 1982). Hsp mRNAs appear to be translated preferentially in stressed cells. The mechanism underlying this phenomenon is not an enhancement of translation of Hsp mRNAs, but a dramatic inhibition of translation of non-Hsp mRNAs in stressed cells, from which inhibition Hsp mRNAs escape. Hsp mRNAs are translated with similar efficiency in heat-treated and untreated cells (Theodorakis and Morimoto, 1987, and references cited therein). Hsp mRNA stability is actively regulated. Hsp70 mRNA is far more stable in heat- or arsenite-exposed cells than in unstressed cells or in cells recovering from a stress (DiDomenico et al., 1982; Theodorakis and Morimoto, 1987). This stability regulation appears to involve AU-rich sequences in the 3' UTR (Petersen and Lindquist, 1989; Moseley et al., 1993). Interestingly, Henics et al. (1999) more recently reported that both Hsp70 and Hsp110 interacted with such AU-rich sequences. Hence, stability of Hsp mRNAs may conceivably be regulated by the availability of these chaperones for binding to the 3' untranslated regions of the mRNAs. Finally, Xing et al. (2004) provided evidence that HSF1 may also be involved in post-transcriptional regulation through an interaction with Symplekin, a factor known to bind polyadenylation factors (CstF, CPSF) and suspected of being involved in their assembly on pre-mRNA.

6. Concluding Remarks

Figure 1 summarizes the various aspects of regulation of HSF1 activity and hsp gene expression discussed in the preceding sections. Chaperones and chaperone complexes appear to be important regulators of HSF1 activity. They play roles in repression of oligomerization and transcriptional competence of HSF1, and, perhaps, also in post-transcriptional regulation of hsp gene expression. However, this is evidently not all there is to regulation of HSF1 and hsp gene expression. Phosphorylation of HSF1 affects the activity of the factor via several different mechanisms, and Daxx enhances transcriptional competence of HSF1 in mammalian cells. In addition, there are several other factors that were reported to co-regulate mammalian hsp gene expression, including heat shock factor binding protein 1, RAL-binding protein 1, STAT-1, nuclear factor of interleukin 6 and glucocorticoid receptor. Furthermore, mammalian HSF1 is inducibly sumoylated. When primate cells are stressed, a large fraction of HSF1 is sequestered in stress granules. Moreover, there is evidence that other HSFs, i.e., HSF2 and HSF4 in

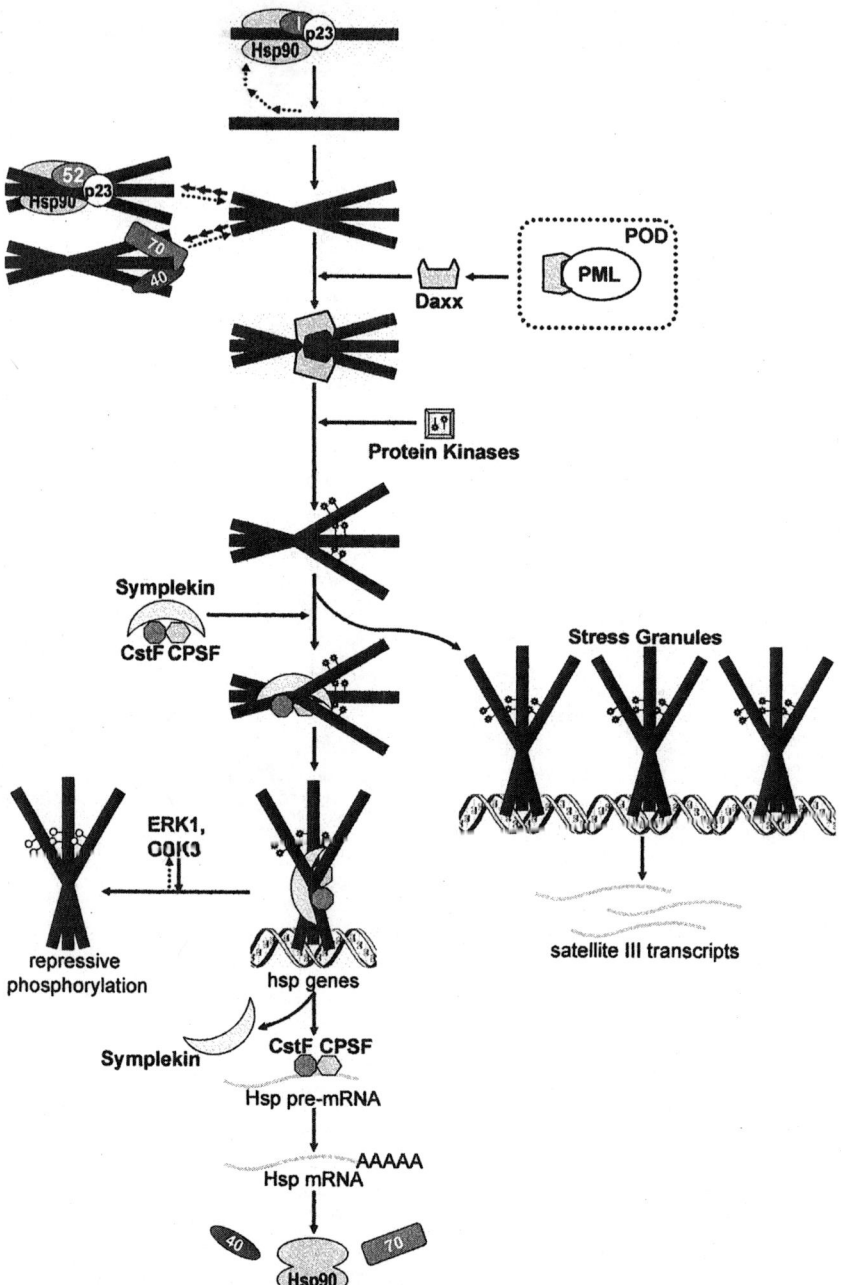

FIGURE 1. Pathway of stress activation of HSF1. Various regulated steps are shown. PML is promyelocytic leukemia protein that organizes the POD.

mammalian cells and HSF3 in avian cells, can modulate HSF1 activity. Finally, several studies concluded that HSF1 directly senses heat and hydrogen peroxide stress in vitro. Although there are good reasons for believing that this is not a key mechanism in vivo, the conformational sensitivity of HSF1 to certain stresses may contribute to the overall regulation of the activity of the factor. As this review attempted to illustrate, HSF1 and hsp genes are subject to a multitude of regulatory inputs as befits a response system of central importance to the cell. If the past is any indication, we can only expect to uncover additional layers of complexity as we continue to probe the regulation of hsp gene expression using improved experimental approaches and tools. To achieve a better understanding of the system will not only be a worthwhile academic goal but could also be of practical importance considering the current thinking that modulation of hsp gene expression could have significant therapeutic benefits. The reader is referred to several excellent earlier reviews for additional information, in particular regarding the topics only mentioned briefly in this section (Wu, 1995; Morimoto, 1998; Morano and Thiele, 1999; Pirkkala et al., 2001; Christians et al., 2002; Holmberg et al., 2002; Voellmy, 2004, 2005; Gaestel, 2006).

Acknowledgments. I thank Alexis Hall for help with Figure 1.

References

Amin, J., Ananthan. J., and Voellmy, R. (1988) Key features of heat shock regulatory elements. *Mol Cell Biol* 8:2761–3769.

Ali, A., Bharadwaj, S., O'Carroll, R., and Ovsenek, N. (1998) Hsp90 interacts with and regulates the activity of heat shock factor 1 in *Xenopus* oocytes. *Mol Cell Biol* 18:4949–60.

Ananthan, J., Goldberg, A. L., and Voellmy, R. (1986) Abnormal proteins serve as eukaryotic stress signals and trigger the activation of heat shock genes. *Science* 232:522–4.

Baler, R., Dahl, G., and Voellmy, R. (1993) Activation of human heat shock genes is accompanied by oligomerization, modification, and rapid translocation of heat shock transcription factor Hsf1. *Mol Cell Biol* 13:2486–96.

Bharadwaj, S., Ali, A., and Ovsenek, N. (1999) Multiple components of the Hsp90 chaperone complex function in regulation of heat shock factor 1 *in vivo*. *Mol Cell Biol* 19:8033–41.

Boellmann, F., Guettouche, T., Guo, Y., Fenna, M., Mnayer, L., and Voellmy, R. (2004) DAXX interacts with heat shock factor 1 during stress activation and enhances its transcriptional activity. *Proc Natl Acad Sci U S A* 101:4100–5.

Bruce, J. L., Price, B. D., Coleman, C. N., and Calderwood, S. K. (1993) Oxidative injury rapidly activates the heat shock transcription factor but fails to increase levels of heat shock proteins. *Cancer Res* 53:12–5.

Bush, K. T., Goldberg, A. L., and Nigam, S. K. (1997) Proteasome inhibition leads to a heat-shock response, induction of endoplasmic reticulum chaperones, and thermotolerance. *J Biol Chem* 272:9086–92.

Chang, H. Y., Nishitoh, H., Yang, X., Ichijo, H., and Baltimore, D. (1998) Activation of apoptosis signal-regulating kinase 1 (ASK 1) by the adapter protein Daxx. *Science* 281:1860–3.

Christians, E. S., Yan, L. J., and Benjamin, I. J. (2002) Heat shock factor 1 and heat shock proteins: Critical partners in protection against acute cell injury. *Crit Care Med* 30:S43–50.

Chu, B., Soncin, F., Price, B. D., Stevenson, M. A., and Calderwood, S. K. (1996) Sequential phosphorylation by mitogen-activated protein kinase and glycogen synthase kinase 3 represses transcriptional activation by heat shock factor-1. *J Biol Chem* 271:30847–57.

Chu, B., Zhong, R., Soncin, F., Stevenson, M. A., and Calderwood, S. K. (1998) Transcriptional activity of heat shock factor 1 at 37 degrees C is repressed through phosphorylation on two distinct serine residues by glycogen synthase kinase 3 and protein kinases Calpha and Czeta. *J Biol Chem* 273:18640–6.

Ciechanover, A. (1998) The ubiquitin–proteasome pathway: On protein death and cell life. *EMBO J* 17:7151–60.

Cotto, J. J., Kline, M., and Morimoto, R. I. (1996) Activation of heat shock factor 1 DNA binding precedes stress-induced serine phosphorylation. Evidence for a multistep pathway of regulation. *J Biol Chem* 271:3355–8.

Cotto, J., Fox, S., and Morimoto, R. (1997) HSF1 granules: A novel stress-induced nuclear component of human cells. *J Cell Sci* 110:2925–34.

Dai, R., Frejtag, W., He, B., Zhang, Y., and Mivechi, N. F. (2000) c-Jun NH2-terminal kinase targeting and phosphorylation of heat shock factor-1 suppress its transcriptional activity. *J Biol Chem* 275:18210–8.

Dai, Q., Zhang, C., Wu, Y., McDonough, H., Whaley, R. A., Godfrey, V., Li, H. H., Madamanchi N., Xu, W., Neckers, L., Cyr, D., and Patterson, C. (2003) Chip activates Hsf1 and confers protection against apoptosis and cellular stress. *EMBO J.* 22:5446–58.

DiDomenico, B. J., Bugaisky, G. E., and Lindquist, S. (1982) The heat shock response is self-regulated at both the transcriptional and post-transcriptional levels. *Cell* 31:593–603.

Duina, A. A., Kalton, H. M., and Gaber, R. F. (1998) Requirement for Hsp90 and a Cyp40-type cyclophilin in negative regulation of the heat shock response. *J Biol Chem* 273:18974–8.

Ecsedy, J. A., Michaelson, J. S., and Leder, P. (2003) Homeodomain-interacting protein kinase 1 modulates Daxx localization, phosphorylation, and transcriptional activity. *Mol Cell Biol* 23:950–60.

Freeman, M. L., Borrelli, M. J., Syed K., Senisterra, G., Stafford, D. M., and Lepock, J. R. (1995) Characterization of a signal generated by oxidation of protein thiols activates the heat shock transcription factor. *J Cell Physiol* 164:356–66.

Gaestel, M. (2006) *Handbook of Experimental Pharmacology*. Vol. 172. Springer, Berlin, Heidelberg, New York.

Giardina, C., and Lis, J. T. (1995) Dynamic protein–DNA architecture of a yeast heat shock promoter. *Mol Cell Biol* 15:2737–44.

Green, M., Schuetz, T. J., Sullivan, E. K., and Kingston, R. E. (1995) A heat-shock-responsive domain of human Hsf1 that regulates transcription activation domain function. *Mol Cell Biol* 15:3354–62.

Guettouche, T., Boellmann., F., Lane, W. S., and Voellmy, R. (2005) Analysis of phosphorylation of human heat shock factor 1 in cells experiencing a stress. *BMC Biochem* 6:4.

Guo, Y., Guettouche, T., Fenna, M., Boellmann, F., Pratt, W. B., Toft, D. O., Smith, D. F., and Voellmy, R. (2001) Evidence for a mechanism of repression of heat shock factor 1 transcriptional activity by a multichaperone complex. *J Biol Chem* 276:45791–9.

He, B., Meng, Y. H., and Mivechi, N. F. (1998) Glycogen synthase kinase 3beta and extracellular signal-regulated kinase inactivate heat shock transcription factor 1 by facilitating the disappearance of transcriptionally active granules after heat shock. *Mol Cell Biol* 18:6624–33.

Hedge, R. S., Zuo, J., Voellmy, R., and Welch, W. J. (1995) Short circuiting stress protein expression via a tyrosine kinase inhibitor, herbimycin A. *J Cell Physiol* 165:186–200.

Henics, T., Nagy, E., Oh, H. J., Csermely, P., von Gabai, A., and Subjeck, J. R. (1999) Mammalian Hsp70 and Hsp110 proteins bind to RNA motifs involved in mRNA stability. *J Biol Chem* 274:17318–24.

Hensold, J. O., Hunt, C. R., Calderwood, S. K., Housman, D. E., and Kingston, R. E. (1990) DNA binding of the heat shock factor to the heat shock element is insufficient for transcriptional activation in murine erythroleukemia cells. *Mol Cell Biol* 10:1600–8.

Hietakangas, V., Ahlskog, J. K., Jakobsson, A. M., Hellesuo, M., Sahlberg, N. M., Holmberg, C. I., Mikhailov, A, Palvimo, J. J., Pirrkala, L., and Sistonen, L. (2003) Phosphorylation of serine 303 is a prerequisite for the stress-inducible SUMO modification of heat shock factor 1. *Mol Cell Biol* 23:2953–68.

Hoehfeld, J., Cyr, D. M., and Patterson, C. (2001) From the cradle to the grave: Molecular chaperones that may choose between folding and degradation. *EMBO Rep* 21:885–90.

Holmberg, C. I., Hietakangas, V., Mikhailov, A., Rantanen, J. O., Kallio, M., Meinander, A., Hellman, J., Morrice, N., MacKintosh, C., Morimoto, R. I., Eriksson, J. E., and Sistonen, L. (2001) Phosphorylation of serine 230 promotes inducible transcriptional activity of heat shock factor 1. *EMBO J* 20:3800–10.

Holmberg, C. I., Tran, S. E. F., Eriksson, J. E., and Sistonen, L. (2002) Multisite phosphorylation provides sophisticated regulation of transcription factors. *Trends Biochem Sci* 27:619–27.

Ishov, A. M., Sotnikov, A. G., Negorev, D., Vladimirova, O. V., Neff, N., Kamitani, T., Yeh, E. T. H., Strauss, J. F. III, and Maul, G. G. (1999) PML is critical for ND10 formation and recruits the PML-interacting protein DAXX to this nuclear structure when modified by SUMO-1. *J Cell Biol* 147:221–33.

Jedlicka, P., Mortin, M. A., and Wu, C. (1997) Multiple functions of *Drosophila* heat shock transcription factor *in vivo*. *EMBO J* 16:2452–62.

Jolly, C., Morimoto, R. I., Robert-Nicoud, M., and Vourc'h, C. (1997) HSF1 transcription factor concentrates in nuclear foci during heat shock: Relationship with transcription sites. *J Cell Sci* 110:2935–41.

Jolly, C., Konecny, L., Grady, D. L., Kurskova, Y. A., Cotto, J. J., Morimoto, R. I., and Vourc'h, C. (2002) *In vivo* binding of active heat shock transcription factor 1 to human chromosome 9 heterochromatin during stress. *J Cell Biol* 156:775–81.

Jolly, C., Metz, A., Govin, J., Vigneron, M., Turner, B. M., Khochbin, S., and Vourc'h, C. (2004) Stress-induced transcription of satellite III repeats. *J Cell Biol* 164:25–33.

Jurivich, D. A., Sistonen, L., Kroes, R. A., and Morimoto, R. I. (1992) Effect of sodium salicylate on the human heat shock response. *Science* 255:1243–5.

Jurivich, D. A., Pachetti, C., Qiu, L., and Welk, J. F. (1995) Salicylate triggers heat shock factor differently than heat. *J Biol Chem* 270:24489–95.

Kim, J., Nueda, A., Meng, Y. H., Dynan, W. S., and Mivechi, N. F. (1997) Analysis of the phosphorylation of human heat shock transcription factor-1 by MAP kinase family members. *J Cell Biochem* 67:43–54.

Kim, S.-A., Yoon, J.-H., Kim, D.-K., Kim, S.-G., and Ahn, S.-G. (2005a) Chip interacts with heat shock factor during heat stress. *FEBS Lett* 579:6559–63.

Kim, S.-A., Yoon, J.-H., Lee, S.-H., and Ahn, S.-G. (2005b) Polo-like kinase 1 phosphorylates HSF1 and mediates its nuclear translocation during heat stress. *J Biol Chem* 280:12653–7.

Kline, M. P., and Morimoto, R. I. (1997) Repression of the heat shock factor 1 transcriptional activation domain is modulated by constitutive phosphorylation. *Mol Cell Biol* 17:2107–15.

Knauf, U., Newton, E. M., Kyriakis, J., and Kingston, R. E. (1996) Repression of human heat shock factor 1 activity at control temperature by phosphorylation. *Genes Dev* 10:2782–93.

Liu, H., Lightfoot, R., and Stevens, J. L. (1996) Activation of heat shock factor by alkylating agents is triggered by glutathione depletion and oxidation of protein thiols. *J Biol Chem* 271:4805–12.

Marchler, G., and Wu, C. (2001) Modulation of *Drosophila* heat shock transcription factor activity by the molecular chaperone DroJ1. *EMBO J.* 20:499–509.

Maul, G. G., Yu, E., Ishov, A. M., and Epstein, A. L. (1995) Nuclear domain 10 (ND10) associated proteins are also present in nuclear bodies and redistribute to hundreds of nuclear sites after stress. *J Cell Biochem* 59:498–513.

McDonough, H., and Patterson, C. (2003) Chip: A link between the chaperone and proteasome systems. *Cell Stress Chap* 8:303–8.

McDuffee, A. T., Senisterra, G., Huntley, S., Lepock, J. R., Sekhar, K. R., Meredith, M. J., Borrelli, M. J., Morrow, J. D., and Freeman, M. L. (1997) Proteins containing non-native disulfide bonds generated by oxidative stress can act as signals for the induction of the heat shock response. *J Cell Physiol* 171:143–51.

McMillan, D. R., Xiao, X., Shao, L., Graves, K., and Benjamin, I. J. (1998) Targeted disruption of heat shock transcription factor 1 abolishes thermotolerance and protection against heat-inducible apoptosis. *J Biol Chem* 273:7523–8.

Metz, A., Soret, J., Vourc'h, C., Tazi, J., and Jolly, C. (2004) A key role for stress-induced satellite III transcripts in the relocalization of splicing factors into nuclear stress granules. *J Cell Sci* 177:4551–8.

Mirault, M.-E., Southgate, R., and Delwart, E. (1982) Regulation of heat shock genes: A DNA sequence upstream of *Drosophila* hsp70 genes is essential for their induction in monkey cells. *EMBO J* 1:1279–85.

Morano, K. A., and Thiele, D. J. (1999) Heat shock factor function and regulation in response to cellular stress, growth, and differentiation signals. *Gene Exp* 7:271–82.

Morimoto, R. I. (1998) Regulation of the heat shock transcriptional response: Cross talk between a family of heat shock factors, molecular chaperones, and negative regulators. *Genes Dev* 12:3788–96.

Moseley, P. L., Wallen, E. S., McCafferty, J. D., Flanagan, S., and Kern, J. A. (1993) Heat stress regulates the human 70-kDa heat-shock gene through the 3'-untranslated region. *Am J Physiol* 264:L533–7.

Murakami, Y., Uehara, Y., Yamamoto, C., Fukazawa, H., and Mizuno, S. (1991) Induction of hsp72/73 by herbimycin A, an inhibitor of transformation by tyrosine kinase oncogenes. *Exp Cell Res* 195:338–44.

Nair, S. C., Toran, E. J., Rimerman R. A., Hjermstad, S., Smithgall, T. E., and Smith, D. F. (1996) A pathway of multi-chaperone interactions common to diverse regulatory proteins: Estrogen receptor, FES tyrosine kinase, heat shock transcription factor 1, and the aryl hydrocarbon receptor. *Cell Stress Chap* 1:237–50.

Nefkens, I., Negorev, D. G., Ishov, A. M., Michaelson, J. S., Yeh, E. T. H., Tanguay, R. M., Mueller, W. E. G., and Maul, G. G. (2003) Heat shock and Cd^{2+} exposure regulate PML and DAXX release from ND10 by independent mechanisms that modify the induction of heat-shock proteins 70 and 25 differently. *J Cell Sci* 116:513–24.

Newton, E. M., Knauf, U., Green, M., and Kingston, R. E. (1996) The regulatory domain of human heat shock factor 1 is sufficient to sense stress. *Mol Cell Biol* 16:839–46.

Parker, C. S., and Topol, J. (1984) A *Drosophila* RNA polymerase II transcription factor binds to he regulatory site of an Hsp70 gene. *Cell* 37:273–83.

Pelham, H. R. B. (1982) A regulatory upstream promoter element in the *Drosophila* hsp70 heat-shock gene. *Cell* 30:517–28.

Petersen, R. B., and Lindquist, S. (1989) Regulation of Hsp70 synthesis by messenger RNA degradation. *Cell Regul* 1:135–49.

Pirkkala, L., Alastalo, T.-P., Zuo, X., Benjamin, I. J., and Sistonen, L. (2000) Disruption of heat shock factor 1 reveals an essential role in the ubiquitin proteolytic pathway. *Mol Cell Biol* 20:2670–5.

Pirkkala, L., Nykanen, P., and Sistonen, L. (2001) Roles of the heat shock transcription factors in regulation of the heat shock response and beyond. *FASEB J* 15:1118–31.

Pratt, W. B., and Toft, D. O. (1997) Steroid receptor interactions with heat shock protein and immunophilin chaperones. *Endocr Rev* 18:306–60.

Rizzi, N., Denegri, M., Chiodi, L., Corioni, M., Valgardsdottir, R., Cobianchi, F., Riva, S., and Biamonti, G. (2004) Transcriptional activation of a constitutive heterochromatic domain of the human genome in response to heat shock. *Mol Biol Cell* 15:543–51.

Rochat-Steiner, V., Becker, K., Micheau, O., Schneider, P., Burns, K., and Tschopp, J. (2000) FIST/HIPK3: A Fas/FADD-interacting serine/threonine kinase that induces FADD phosphorylation and inhibits Fas-mediated Jun NH(2)-terminal kinase activation. *J Exp Med* 192:1165–74.

Sarge, K. D., Murphy, S. P., and Morimoto, R. I. (1993) Activation of heat shock gene transcription by heat shock factor 1 involves oligomerization, acquisition of DNA-binding activity, and nuclear translocation and can occur in the absence of stress. *Mol Cell Biol* 13:1392–407.

Senisterra, G. A., Huntley, S. A., Escaravage, M., Sekhar, K. R., Freeman, M. L., Borrelli. M., and Lepock, J. R. (1997) Destabilization of the Ca^{2+}-ATPase of sarcoplasmic reticulum by thiol-specific, heat shock inducers results in thermal denaturation at 37 degrees C. *Biochemistry* 36:11002–11.

Shi, Y., Kroeger, P. E., and Morimoto, R. I. (1995) The carboxyl-terminal transactivation domain of heat shock factor 1 is negatively regulated and stress responsive. *Mol Cell Biol* 15:4309–18.

Shi, Y., Mosser, D. D., and Morimoto, R. I. (1998) Molecular chaperones as Hsf1-specific transcriptional repressors. *Genes Dev* 12:654–66.

Sorger, P. K., Lewis, M. J., and Pelham, H. R. (1987) Heat shock factor is regulated differently in yeast and HeLa cells. *Nature* 329:81–4.

Sorger P. K., and Pelham, H. R. (1988) Yeast heat shock factor is an essential DNA-binding protein that exhibits temperature-dependent phosphorylation. *Cell* 54:855–64.

Theodorakis, N. G., and Morimoto, R. I. (1987) Posttranscriptional regulation of Hsp70 expression in human cells: Effects of heat shock, inhibition of protein synthesis, and adenovirus infection on translation and mRNA stability. *Mol Cell Biol* 7:4357–68.

Torii, S., Egan, D. A., Evans, R. A., and Reed, J. C. (1999) Human DAXX regulates FAS-induced apoptosis from nuclear PML oncogenic domains (PODs). *EMBO J* 18:6037–49.

Voellmy, R. (1994) On mechanisms that control heat shock transcription factor activity in metazoan cells. *Cell Stress Chap* 9:122–33.

Voellmy, R. (ed.) (2005) *Methods*. Vol. 35. Elsevier Inc., Amsterdam.

Wang, X., Grammatikakis, N., Sikanou, A., and Calderwood, S. K. (2003) Regulation of molecular chaperone gene transcription involves the serine phosphorylation, 14-3-3ε binding, and cytoplasmic sequestration of heat shock factor 1. *Mol Cell Biol* 23:6013–26.

Wang, X., Grammatikakis, N., Sikanou, A., Stevenson, M. A., and Calderwood, S. K. (2004) Interaction between extracellular signal-regulated protein kinase 1, 14-3-3ε, and heat shock factor 1 during stress. *J Biol Chem* 279:49460–9.

Wang, X. Z., Khaleque, M. A., Zhao, M. J., Zhong, R., Gaestel, M., and Calderwood, S. K. (2005) Phosphorylation of HSF1 by MAPKAP kinase 2 on Serine 121 inhibits transcriptional activity and promotes Hsp90 binding. *J Biol Chem* Published on November 8 as Manuscript M505822200.

Westwood, J. T., Clos, J., and Wu, C. (1992) Stress-induced oligomerization and chromosomal relocalization of heat-shock factor. *Nature* 353:822–7.

Whitesell, L., Mimnaugh, E. G., De Costa, B., Myers, C. E., Neckers, L. M. (1994) Inhibition of heat shock protein HSP90-PP60V-SRC heteroprotein complex formation by benzoquinone ansamycins: Essential role for stress proteins in oncogenic transformation. *Proc Natl Acad Sci U S A* 91:8324–8.

Wu, C. (1984) Activating protein factor binds *in vitro* to upstream control sequences in heat shock gene chromatin. *Nature* 311:81–4.

Wu, C. (1995) Heat shock transcription factors: Structure and regulation. *Annu Rev Cell Dev Biol* 11:441–469.

Xia, W., and Voellmy, R. (1997) Hyperphosphorylation of heat shock transcription factor 1 is correlated with transcriptional competence and slow dissociation of active factor trimers. *J Biol Chem* 272:4094–102.

Xia, W., Guo, Y., Vilaboa, N., Zuo, J., and Voellmy, R. (1998) Transcriptional activation of heat shock factor HSF1 probed by phosphopeptide analysis of factor ^{32}P-labeled *in vivo*. *J Biol Chem* 273:8749–55.

Xia, W., Vilaboa, N., Martin, J., Mestril, R., Guo, Y., and Voellmy, R. (1999) Modulation of tolerance by mutant heat shock transcription factors. *Cell Stress Chap* 4:8–18.

Xiao, H., and Lis, J. T. (1988) Germline transformation used to define key features of heat shock response elements. *Science* 239:1139–41.

Xavier, I. J., Mercier, P. A., McLoughlin, C. M., Ali, A., Woodgett, N. Jr., and Ovsenek, N. (2000) Glycogen synthase kinase 3beta negatively regulates both DNA-binding and transcriptional activities of heat shock factor 1. *J Biol Chem* 275:29147–52.

Xing, H., Mayhew, C. N., Cullen, K. E., Park-Sarge, O.-K., and Sarge, K. D. (2004) HSF1 modulation of Hsp70 mRNA polyadenylation via interaction with symplekin. *J Biol Chem* 279:10551–5

Zhang, Y., Huang, L., Zhang, J., Moskophidis, D., and Mivechi. N. F. (2002) Targeted disruption of hsf1 leads to lack of thermotolerance and defines tissue-specific regulation for stress-inducible HSP molecular chaperones. *J Cell Biochem* 86:376–93.

Zhong, S., Salomoni, P., Ronchetti, S., Guo, A., Ruggero, D., and Pandolfi, P. P. (2000) Promyelocytic leukemia protein (PML) and DAXX participate in a novel nuclear pathway for apoptosis. *J Exp Med* 191:631–40.

Zou, J., Salminen, W. F., Roberts, S. M., and Voellmy, R. (1998a) Correlation between glutathione oxidation and trimerization of heat shock factor 1, an early step in stress induction of the Hsp response. *Cell Stress Chap* 3:130–41.

Zou, J., Guo, Y., Guettouche, T., Smith, D. F., and Voellmy, R. (1998b) Repression of heat shock transcription factor Hsf1 activation by Hsp90 (Hsp90 complex) that forms a stress-sensitive complex with HSF1. *Cell* 94:471–80.

Zuo, J., Baler, R., Dahl, G., and Voellmy, R. (1994) Activation of the DNA-binding ability of human heat shock transcription factor 1 may involve the transition from an intramolecular to an intermolecular triple-stranded coiled-coil structure. *Mol Cell Biol* 14:7557–68.

Zuo, J., Rungger, D., and Voellmy, R. (1995) Multiple layers of regulation of human heat shock transcription factor 1. *Mol Cell Biol* 15:4319–30.

III
Cellular Stress Proteins

7
Small Heat Shock Proteins in Physiological and Stress-Related Processes

DIANA OREJUELA, ANNE BERGERON, GENEVIÈVE MORROW, AND ROBERT M. TANGUAY

Laboratoire de génétique cellulaire et développementale, département de médecine et CREFSIP, Pavillon CE Marchand, Université Laval, Québec, Canada G1K 7P4

1. Introduction

The small heat shock proteins (sHsps) family comprises several members found in prokaryotes and eukaryotes with important variations in the number of members between species (e.g., ~10 in mammals, ~20 in plants) (Kappe et al., 2002; Fu et al., 2006). Their common feature is a central α-crystallin domain. The core structure of the α-crystallin domain, rather than its amino acid sequence, is conserved between species. This ~90 amino acid domain presents a well-preserved double β-sheet sandwich structure, which is surrounded by N- and C-terminal domains whose length and sequence can vary extensively between and within species giving rise to proteins ranging from 11 to 42 kilodaltons (kDa). These extensions have been suggested to confer functional specificities to the different sHsps. Another property of sHsps is their propensity to form large oligomers, which are in dynamic equilibrium with smaller subunits (dimers, trimers, or tetramers depending on the sHsp). From X-ray and electron microscopic data, these particles have a diameter of 10–18 nm with a hollow core. The number of subunits can vary from 12 (*Mycobacterium tuberculosis*) to 24 (*Methanocaldo coccus jannaschii* and yeast). The quaternary structure is quite variable with polydisperse complexes in the range of 400 to over 800 kDa (see Sun and MacRae, 2005a; Haslbeck et al., 2005 for recent detailed reviews on the structure of sHsps).

Like the members of the Hsp70 family, several small Hsps have been shown to demonstrate chaperone-like activity in vitro. However in contrast to Hsp70 this activity is not ATP-dependent. As mentioned above, small Hsps form large oligomeric complexes that are in a dynamic equilibrium with smaller structures. Formation of oligomers is governed by domains in both the N- and C-termini of sHsps. These oligomers can either be homo- or hetero-oligomers involving different members of the sHsps family. For example, HspB1 can interact with HspB8 in a similar manner as the αA-αB-crystallin pair (HspB4, HspB5) (Sun et al., 2004). The oligomerization of sHsps, at least in mammals, seems to be regulated by phosphorylation of serine residues (Kato et al., 2002; Gaestel, 2002).

In turn, the chaperone activity is modulated by the dynamic structural organization of sHsps into oligomers, by their post-translational modification, and by factors such as elevated temperature.

As previously mentioned, several sHsps have been shown to possess chaperone-like activities in in vitro folding assays, but there are still few studies proving that this chaperone activity is biologically relevant in vivo. Thus, while sHsps clearly have cytoprotective properties in vivo, whether this is dependent on their chaperone activity or on other properties related to the response to cellular stress, remains largely undocumented.

In spite of the high structural similarities of the conserved α-crystallin domain of sHsps, the various members of the family do not necessarily possess similar functions in vivo. For instance, HspB1 has an anti-apoptotic action both upstream and downstream of the apoptosome while HspB8 shows a proapoptotic activity in a cell-specific manner (Gober et al., 2003).

Since there are recent excellent reviews on the structure of sHsps (Sun and MacRae, 2005a; Haslbeck et al., 2005; Van Montfort et al., 2002; Taylor and Benjamin, 2005), we have focused our attention here on the documented function(s) of these proteins in various cellular processes with a special emphasis on the situation in mammals.

2. sHsps: Oligomerization, Modification, and Catalytic Function as Chaperones

Small Hsps are proteins of a low molecular range that can oligomerize into larger homotypic multimers of up to 800 kDa (Arrigo, 1988; Behlke et al., 1991; Bentley et al., 1992; Groenen et al., 1994; Kato et al., 1994; Lavoie et al., 1995; Leroux et al., 1997). To date, ten sHsps have been described in mammals, and they include Hsp27, MKBP (myotonic dystrophy protein kinase-binding protein), HspB3, αA-crystallin, αB-crystallin, Hsp20, cvHsp (cardiovascular heat shock protein), Hsp22, HspB9, and the recently reported ODF1 (outer dense fiber protein) (Table 1). Each of these proteins harbors a well-conserved α-crystallin domain (see Fig. 1). This domain is believed to hold a dimer interface involved in oligomerization (Wistow, 1993). In contrast, the N- and C-terminal domains of sHsps are highly variable both in sequence and length. The favored role of sHsps is one of chaperone by which they prevent the aggregation of misfolded target proteins by holding them in a folding competent state (Horwitz, 1992; Ehrnsperger et al., 1997; Lee et al., 1997). The chaperone activity ensured by sHsps requires no ATP binding nor ATP hydrolysis (Jakob et al., 1993). However, the renaturation of target substrates into their native state necessitates the action of ATP-dependent chaperones such as Hsp70 (Ehrnsperger et al., 1997).

2.1. Small Hsps Phosphorylation

Although the function of sHsps is one of molecular chaperone at least in vitro, these proteins are also implicated in important cellular processes such as differentiation

TABLE 1. Classification of Vertebrate Small Heat Shock Proteins*

Name	Alternative names	Properties
HSPB1	Hsp25, Hsp27	Ubiquitous expression Stress inducible Involved in multiple cellular processes 3 phosphorylated isoforms
HSPB2	MKBP	Expression in heart and skeletal muscles Not heat-inducible Interacts with HspB3 during myogenic differentiation
HSPB3	HSPL27	Expression in heart and skeletal muscles Not heat-inducible Translocation to the cytoskeleton during stress
HSPB4	αA-crystallin, CRYAA	Expression in eye lenses Not heat-inducible Interacts with CRYAB
HSPB5	αB-crystallin, CRYAB	Ubiquitous expression Stress inducible 3 phosphorylated isoforms
HSPB6	Hsp20, p20	Ubiquitous expression Not heat-inducible Phosphorylated upon stress
HSPB7	cvHsp	Expression in heart and skeletal muscles
HSPB8	Hsp22, H11 kinase	Ubiquitous expression Stress inducible Interacts with itself, HspB1, HspB2, CRYAB, HspB6 and HspB7 in the heart
HSPB9	—	Expression in testis Involved in sperm flagellar integrity
HSPB10	ODF1	Expression in testis Involved in sperm flagellar integrity

*HspB11 to HspB15 are found in birds, amphibians, and fish (Franck et al., 2004).

(Arrigo and Pauli, 1988; Michaud et al., 1997; Davidson et al., 2002), mitogenic stimulation (Moseley, 1998), aging (Morrow and Tanguay, 2003), cell death (Arrigo, 1998; Beere, 2004, 2005), and protection against different types of stress (Parcellier et al., 2005). The various functions of sHsps are regulated by their phosphorylation state. Human Hsp27 may be phosphorylated on residues Ser15, Ser78, and Ser82 (Landry et al., 1992; Stokoe et al., 1992) while murine Hsp27 may be phosphorylated on Ser15 and Ser86 (Gaestel et al., 1991) (see Fig. 1). Phosphorylation of mammalian Hsp27 is mediated by the kinase MAPK-activating protein (MAPKAP) 2/3, a substrate of p38 kinase, and also by protein kinase C (Maizels et al., 1998). Recently, the Akt kinase was also shown to phosphorylate human Hsp27 on Ser82 (Rane et al., 2003). Protein phosphatase 2A is responsible for the dephosphorylation of mammalian Hsp27 in vivo (Cairns et al., 1994). The αB-crystallin protein is also phosphorylated in response to various stresses (Ito et al., 1997) and its phosphorylatable residues include Ser19, Ser45, and Ser59. The p44/42 MAPK kinase was shown to be responsible for Ser45 phosphorylation of

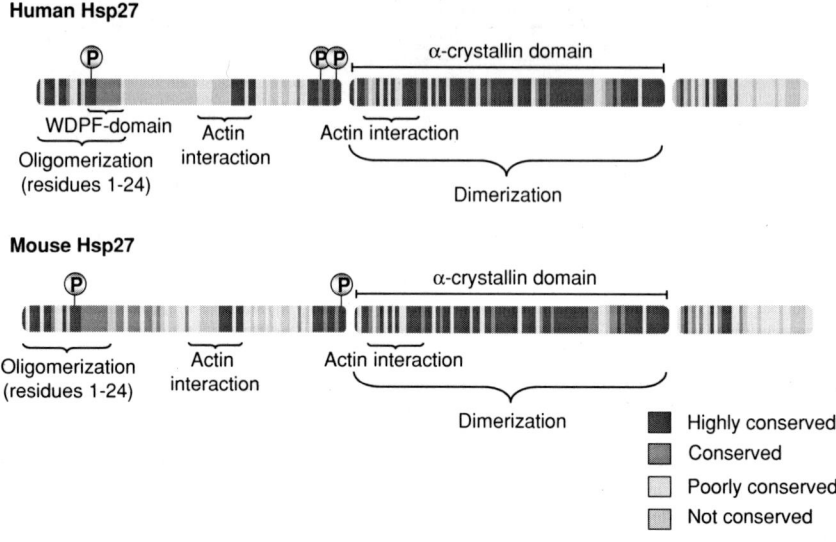

FIGURE 1. Schematic representation of the human and mouse Hsp27 protein. Phosphorylatable residues are indicated by the letter P above the amino acid sequence. [*See* Color Plate II]

αB-crystallin whereas Ser59 is phosphorylated by MAPKAP kinase 2 (Kato et al., 1998; Hoover et al., 2000).

2.2. Cellular Localization of sHsps in Stressed Cells

In addition to phosphorylation, the subcellular localization of sHsps seems to vary depending on physiological and stress conditions but also between cell types. Adhikari et al. (2004) found that in murine C2C12 undifferentiated myoblasts, both αB-crystallin and Hsp27 translocate to nuclear speckles upon heat stress. However, neither αB-crystallin, nor Hsp27 translocate to the nucleus in heat stressed differentiated myotubes. Another study reported that αB-crystallin translocates from the cytoplasm to the nucleus in stressed N1E-115 neuroblastoma cells but the nuclear distribution of the protein was homogeneous in these cells (Wiesmann et al., 1998). In heat-stressed neonatal cardiomyocytes, Hsp27 does not translocate to the nucleus at all (van de Klundert et al., 1998). These results suggest that the cellular localization and the differentiation state may be important for the functions of sHsps.

2.3. Small Hsps as Molecular Chaperones

As previously mentioned, sHsps act as chaperones by binding to non-native proteins in order to promote proper folding in cooperation with high molecular ATP-dependent Hsps (Jakob et al., 1993; Ehrnsperger et al., 1997). It is generally accepted that the oligomerization of sHsps into multimeric complexes is required for the binding of unfolded proteins and for chaperone activity (Leroux et al., 1997). sHsps oligomers form by interaction of unphosphorylated dimers. In hamster, phosphorylation of Ser90 was both sufficient and necessary to induce the dissociation of multimers (Lambert et al., 1999). The two β-sheets structure of the α-crystallin domain is proposed to be involved in the formation of stable dimers (Wistow, 1993; Merck et al., 1993; Berengian et al., 1997; Mchaourab et al., 1997; Koteiche et al., 1998) and the variable N-terminal domain is essential for the oligomerization of sHsps into large multimeric complexes (Leroux et al., 1997; Lambert et al., 1999). Interestingly, the mechanism behind the chaperone activity of yeast Hsp26 is quite different (Haslbeck et al., 1999). The chaperone activity of Hsp26 requires dissociation of multimers into dimers. Hsp26 dimers then bind to denatured proteins to form large globular complexes in which one substrate unit is bound per Hsp26 dimer. A similar mechanism of activation has been described for BiP, an endoplasmic reticulum-resident protein of the Hsp70 family. In the presence of accumulated unfolded proteins, BiP dimers dissociate into monomers (Blond-Elguindi et al., 1993).

2.4. Small Hsps in Cytoskeleton Dynamics

Other important cellular functions also involve phosphorylated sHsps. For example, Hsp27 participates in actin remodeling. Studies showed that purified unphosphorylated mouse and chicken Hsp27 homologs inhibit actin polymerization in vitro (Benndorf et al., 1994; Miron et al., 1991; Wieske et al., 2001). Moreover, Hsp27 phosphorylation was shown to be required for mitogenic stimulation of F-actin formation (Lavoie et al., 1993), stabilization of focal adhesion (Schneider et al., 1998), and cell migration (Rousseau et al., 1997; Hedges et al., 1999). Although phosphorylation of Hsp27 enables actin polymerization, unphosphorylated Hsp27 acts as a CAP-binding protein. Even if Hsp27 inhibits actin polymerization, studies reported that this inhibition activity was highly reduced by phosphorylation (Miron et al., 1991; Benndorf et al., 1994). Based on these observations, a model was proposed to explain the effects of Hsp27 on actin polymerization (Lavoie et al., 1995). Under normal conditions, unphosphorylated Hsp27 binds to the barbed ends of actin microfilaments causing a reduction in the rate of actin polymerization. Therefore, high levels of unphosphorylated Hsp27 would increase the lifetime of Hsp27 at barbed ends and inhibit actin polymerization. As a result, the cellular actin network would become sensitized to stress. Accordingly, increased levels of phosphorylated Hsp27 would favor actin polymerization and would result in actin stabilization and increased survival after stress.

Phosphorylated αB-crystallin was also shown to translocate to the actin cytoskeleton in rat cardiac myoblasts (H9C2 cell line) along with Hsp20, HspB2, and HspB3 (Verschuure et al., 2002). AlphaB-crystallin is involved in intermediate filament assembly as well. Alpha-crystallins inhibit the assembly of both glial fibrillary acidic protein (GFAP) and vimentin and this inhibition is phosphorylation-independent (Nicholl and Quinlan, 1994; Wisniewski and Goldman, 1998). Furthermore, αB-crystallin was shown to associate with intermyofibrillar desmin cytoskeleton in human skeletal muscle (Fisher et al., 2002). Recently, αB-crystallin was suggested to bind to microtubules through interactions with microtubule-associated proteins promoting microtubule stability (Fujita et al., 2004). Hsp27 was also observed to bind to intermediate filaments such as GFAP, vimentin and keratin (Perng et al., 1999a,b). Hirata et al. (2003) showed that Hsp27 is highly expressed in axon outgrowths of injured sciatic nerves of rat suggesting a role of Hsp in cytoskeletal dynamics in the nervous system (Hirata et al., 2003). These associations between sHsps and intermediate filaments are believed to ensure maintenance of intermediate filament integrity (Salvador-Silva et al., 2001) and to protect against intermediate filament aggregations found in some neurodegenerative diseases.

2.5. Small Hsps and Protection Against Oxidative Stress

The observation that sHsps were involved in the protection against oxidative stress highlighted the importance of these proteins in multiple cellular responses. In studies using transfected mammalian cells, Mehlen et al., (1993, 1995) observed that human Hsp27, *Drosophila* Hsp27, and αB-crystallin induced a protection against an oxidative stress caused by hydrogen peroxide or menadione. Later, Rogalla et al. (1999) demonstrated that Hsp27 protected cells against oxidative stress in L929 cells and 13.S.1.24 rat neuroblasts. Using "molecular mimicry" of serine phosphorylation of Hsp27, they further showed that oxidative stress protection was conferred by large oligomers whereas acidic mutants greatly reduced the protection. In another in vitro study, oxidative stress protection by human Hsp27 and its murine homolog in L929 cells was demonstrated to be due to an increase in glucose-6-phosphate dehydrogenase activity which leads to an increase in the intracellular levels of reduced GSH (Préville et al., 1999). These authors suggested that the protection by Hsp27 against oxidative stress-mediated actin disruption, in L929 cells, was indirect and attributable to the maintenance of reduced GSH levels, on the basis of the observation that GSH was able to protect the cellular morphology of stressed cells. Thus, there is an obvious contradiction between the model proposed by Préville and colleagues concerning the protection against oxidative stress-induced actin disruption and the proposed model for actin microfilament protection in which the phosphorylated form of Hsp27 seems to intervene. Although there is as yet no explanation for this apparent contradiction, one hypothesis could be that the type of protection against oxidative stress is mediated by the levels of cellular Hsp27.

3. sHsps in Protein Quality Control

3.1. Small Hsps in the Endoplasmic Reticulum Stress Response

Very recently, Hsp27 was reported to be induced as a result of endoplasmic reticulum (ER) stress (Ito et al., 2005). Conditions that lead to perturbations in ER homeostasis, such as protein misfolding, changes in calcium intracellular levels, glucose deprivation, mutations in secreted proteins genes, viral infections, heme deficiency, or oxidative stress, elicit an ER stress response in order to cope with the protein load in the organelle (reviewed in Zhang and Kaufman, 2004). The ER stress response causes a general attenuation of translation, a transcriptional activation of chaperones and folding catalyst genes, and the ER associated degradation (ERAD) of misfolded proteins by the proteasome. In a study using tunicamycin (an inhibitor of protein N-glycosylation) and thapsigargin (an inhibitor of the ER calcium ATPase pump) as inducers of ER stress, Ito et al. (2005) showed that Hsp27 was phosphorylated in different mammalian cell types. The phosphorylation of Hsp27 was blocked by the p38 MAP kinase inhibitor SB203580.

Hsp27 was also described to be involved in proteasome-mediated protein degradation in vitro and in vivo (Parcellier et al., 2003). Hsp27 has the ability to bind to polyubiquitinated proteins and to the 26S proteasome. In fact, the degradation of phosphorylated I-κBα was shown to depend on the interaction of Hsp27 with the 26S proteasome in etoposide and TNFα treated cells. Of interest is the observation that phosphorylated Hsp27 and αB-crystallin are recruited to protein aggresomes in muscle atrophy (Kato et al., 2002). This finding is suggested to be attributable to Hsp27's ability to bind to ubiquitin (Parcellier et al., 2005).

In addition to providing a protection against protein aggregation, heat shock, microfilament disruption, oxidative stress and ER stress, sHsps have also been reported to protect against apoptosis. Interestingly, apoptosis inhibition may be achieved by large unphosphorylated oligomers (Bruey et al., 2000) as well as by phosphorylated Hsp27 (Charette and Landry, 2000, Charette et al., 2000). The sHsp-mediated protection against apoptosis will be discussed in further details in the next section. However, in light of all the data gathered from the study of sHsps, it is clear that the oligomeric/phosphorylation state is crucial in the regulation of the diverse functions of the sHsps.

4. sHsps in Cell Life and Death Processes

Damaging stimuli have two paradoxical effects on cells: on one hand, they elicit a stress response to facilitate recovery and maintain survival, and on the other hand, they initiate cell death processes to remove damaged cells and avoid inflammation. Heat shock proteins comprise several different families of proteins that have dual roles as regulators of protein conformation and stress sensors. Under normal conditions, Hsps are able to change protein conformation, promote multiprotein complex assembly or disassembly, regulate protein degradation via the proteasome pathway, facilitate protein translocation between organelles, and ensure proper folding of polypeptides during translation (Garrido et al., 2001; Takayama et al., 2003). Their expression can be induced in response to environmental, physical and chemical stresses to limit the cellular damages and allow recovery. Interestingly, the stress response has evolved as a mechanism to prevent inappropriate activation of cell death signaling pathways (Xanthoudakis and Nicholson, 2000; Garrido et al., 2001; Mosser and Morimoto, 2004; Beere, 2005). It is in fact highly likely that molecules such as chaperones, which influence assembly, transport and folding of other proteins, may directly affect the coordinated, multistep signaling of programmed cell death. Apoptosis or programmed cell death is a controlled, energy-dependent mechanism that removes supernumerary cells during development and adult homeostasis. Although the classification of Hsps in different major families is based on their size or structure, it does not define their function (Lindquist and Craig, 1988); thus the anti-apoptotic Hsps do not belong to one single family. In addition to Hsp70 and Hsp90, members of the sHsps family like Hsp27 and αB-crystallin have emerged as anti-apoptotic effectors.

4.1. Hsp27 and αB-crystallin: Molecular Chaperones in Apoptosis

While some Hsps are constitutively expressed and increase their expression in response to stress, others are only expressed following exposure to external stresses. The cellular response to stress depends on the severity of injury. However, when the molecular damage becomes too extensive and/or is sustained for too long to allow repair mechanisms to restore homeostasis balance, the cell undergoes apoptosis. Hsp27 and αB-crystallin proteins are closely related members (sharing ~40% of amino acid identity) that protect cells from stress-induced apoptosis at multiple signaling levels (Clark and Muchowski, 2000). αB-crystallin is mainly expressed in the lens but is also constitutively expressed in many other tissues such as heart and skeletal muscle; its expression is also induced by diverse cellular stresses (Klemenz et al., 1993; Clark and Muchowski, 2000). Hsp27 is an inducible protein whose levels within unstressed cells are generally low and whose normal conformation is a large oligomeric unit (Concannon et al., 2003). However, the large unphosphorylated oligomers of Hsp27 can intervene in stressed cells as well and can exert anti-apoptotic activity by different mechanisms. First, Hsp27 has the capacity to increase intracellular levels of glutathione (Mehlen et al., 1996a), and thus, to enhance the antioxidant defence of cells. This attribute is important as reactive oxygen species (ROS), which rapidly increase with mitochondrial dysfunction, play a central role in cell death induced by many stimuli (Concannon et al., 2003). Oxidative stress induced by the TNF-α cytokine, H_2O_2 or menadione can also be blocked by αB-crystallin, even though Hsp27 seems to be twice as efficient in the same conditions (Mehlen et al., 1995). More precisely, the protection conferred against hydrogen peroxide by overexpression of Hsp27/25 has been suggested to be mediated by regulation of glucose-6-phosphate dehydrogenase (Préville et al., 1999), which underlies an anti-apoptotic effect linked to the chaperone properties of Hsps.

In addition to oxidative-stress-induced cell death, Hsp27 has been shown to block other kinds of stress-induced apoptosis, like apoptosis induced by heat, by Fas ligand and by several anticancer drugs (Richards et al., 1996). As discussed above, this cytoprotective effect of Hsp27 also depends on its capacity to modulate the actin network, and to prevent the disruption of the cytoskeleton (Guay et al., 1997), or on its capacity to modulate the redox state of the cell (Préville et al., 1998).

Another mechanism involved in the anti-apoptotic activity of sHsps is the neutralization of damaged or misfolded toxic proteins by their chaperone activity, for example, by limiting levels of misfolded proteins or preventing their irreversible aggregation (Concannon et al., 2003). Indeed, in many cases, damaged proteins can be the apoptotic signal triggering cell death signaling. However, regardless of their chaperone function, it has been shown that sHsps may also exert regulation of apoptosis through direct protein–protein interaction.

In contrast to necrosis which is a passive, pathological form of cell death, apoptosis is an energy-dependent mechanism, involving signaling transduction in both initiation and execution steps. In mammalian cells, apoptosis is mediated by two main pathways depending on death receptor (extrinsic) or mitochondrion (intrinsic) signaling. These two pathways converge on the proteolytic activation of caspase-3, one of the major members of the caspase family. Caspases are cysteine proteases produced as inactive zymogens that become catalytically active after cleavage. They are classified as "initiator" caspases (caspase-2, -8, and -9) or "executioner" caspases (caspase-3, -6, and -7) and their function is to ensure the apoptotic program in a multistep mechanism (Xanthoudakis and Nicholson, 2000; Beere, 2004).

4.2. Direct Inhibition of Mitochondria-Mediated Apoptosis (Intrinsic)

The intrinsic pathway of apoptosis is based on the permeabilization of the mitochondrial outer membrane (MOMP) and release in the cytosol of several molecules such as cytochrome C, flavoprotein AIF (apoptosis inducing factor), Smac (second-mitochondria-derived activator of caspases)/DIABLO protein, EndoG, and HtrA2/Omi (Adams, 2003; Bohm and Schild, 2003). These molecules allow a sequential activation of initiating and executioning caspases which finally cleave different substrates like iCAD (inhibitor of caspase-activated DNase), PARP (poly(ADP-ribose)polymerase), fodrin, actin and focal adhesion kinase (FAK) (Beere, 2005). Release of cytochrome C is the key hallmark event of the apoptotic cascade, and it is tightly regulated by several members of the Bcl-2 family like Bax, Bak, Bik, Bad and Bid (pro-apoptotic) or Bcl-2 and Bcl-Xl (anti-apoptotic) (Green and Reed, 1998; Gross et al., 1999). The apoptotic phenotype is characterized by cell blebbing, condensation and fragmentation of chromatin and formation of the so-called apoptotic bodies, which harbor intracellular lipids in the outer plasma membrane and are eliminated by phagocytosis.

Hsp27 can exert its inhibitory effect on intrinsic apoptosis at two levels of mitochondria: through direct interaction with cytochrome C (Garrido et al., 1999) or by preventing the activation of procaspase-3 (Pandey et al., 2000). The protein-protein interaction between Hsp27 and cytochrome C sequesters cytochrome C from Apaf-1 (apoptosis activating factor-1) thereby inhibiting the activation of procaspase-9, the third component of the apoptosome complex (Zou et al., 1999; Bruey et al., 2000). Interestingly, recent reports explain the Hsp27-mediated prevention of cytochrome C release by the ability of Hsp27 to maintain the integrity of the actin network, which would block translocation of pro-apoptotic factors to the mitochondrion (Paul et al., 2002). Whatever the exact mechanism, Hsp27 seems to interfere with cytosolic cytochrome C, and as a consequence inhibits assembly and function of the apoptosome. A second mechanism used by Hsp27 to inhibit intrinsic apoptosis is, as previously mentioned, by direct binding to procaspase-3 (Pandey et al., 2000). This interaction would affect the proteolytic activation on procaspase-3 by active caspase-9.

As for αB-crystallin, this protein can also bind to procaspase-3 and partially processed caspase-3 to inhibit their final cleavage to activated caspase-3, as it has been shown for H_2O_2-induced apoptosis (Mao et al., 2001). Nevertheless, this mode of inhibition by αB-crystallin differs from that of Hsp27, as it concerns autoproteolytic maturation of the p24 partially processed caspase-3 intermediate (Kamradt et al., 2001). Autocatalytic maturation of caspase-3 is required for the intrinsic as well as for the extrinsic apoptosis, and thus, αB-crystallin can account as a key regulator of both pathways of programmed cell death. Nonetheless, αB-crystallin has not only one but several roles in the inhibition of intrinsic apoptosis as it can interact with Bax and BCl-Xs to prevent their translocation to mitochondria and inhibit staurosporine-induced apoptosis (Mao et al., 2004). As a result of this binding, release of cytochrome C is blocked, activation of caspase-3 is repressed and degradation of PARP is not achieved.

To the same extent as Bax and BCl-Xs, the regulation of cytochrome C release depends on Bid. This proapoptotic member of the BCl-2 family is cleaved by caspase-8 and integrates the death receptor and the mitochondria-mediated pathways (Adams, 2003). It has been reported that phosphorylated Hsp27 can prevent translocation of Bid to the mitochondria, most probably by stabilization of the cytoskeleton (Paul et al., 2002; Guay et al., 1997). This latter result suggests that Hsp27 can inhibit mitochondria-mediated apoptosis by direct interaction with three different mediators of the intrinsic pathway: cytochrome C, p24 caspase-3 intermediate and pro-apoptotic Bid.

4.3. Direct Inhibition of Receptor-Mediated Apoptosis (Extrinsic)

Death receptors belong to the large family of plasma membrane TNF-receptor-related members composed of TNFR1, CD95/APO-1/Fas receptor, TRAIL-R1, TRAIL-R2, DR3, and DR6, among others (Nagata, 1997). Typical extrinsic apoptosis is triggered by engagement of death receptors, followed by trimerization and association of their death domain (DD) to the DD of the cytosolic adaptor proteins TRADD (TNF receptor death domain protein) or FADD (Fas associated death domain protein). The protein complex formed by these interactions is the so-called death-inducing signaling complex (DISC). TRADD and FADD adaptor proteins are then able to interact with procaspase-8 (and perhaps procaspase-10) to elicit their proteolytic autoactivation, and thereby, allow downstream caspase-dependent signaling. One of these downstream cascades involves activating cleavage of proapoptotic Bid, therefore linking extrinsic and mitochondria-mediated apoptosis. Several studies have reported an important regulation of Fas-, TNF- and TRAIL-mediated apoptosis by Hsp27 and αB-crystallin (Mehlen et al., 1996b). However, the mechanisms concerned seem to be very complex and Hsps can trigger apoptotic cascades, as well as signaling pathways which rather inhibit cell death. As an example of this dual action, the adaptor molecule TRADD which activates caspase-8 to induce apoptosis, can also recruit a TRAF/RIP complex and relieve the repression on the survival protein NFκB (Hsu et al., 1996).

In the case of Fas receptor-induced apoptosis, a supplementary caspase-independent, Daxx-dependent process, can be regulated by Hsp27 (Chang et al., 1998; Charette et al., 2000). Indeed, binding of the Fas ligand stimulates recruitment of the Daxx adapter protein to the trimerized receptor. Phosphorylated dimers of Hsp27 can interact with Daxx to prevent its translocation from the nucleus to the cytosol (Charette et al., 2000). To elicit apoptosis, Daxx activates ASK-1 (apoptosis signal-regulating kinase 1), which leads to a downstream induction of the two main apoptotic MAPK, p38 and JNK, via activation of MKK3/6 and MKK4/7 respectively (Ichijo et al., 1997). Activation of the JNK pathway can mediate both pro- and anti-apoptotic effects depending on the signal and the cellular type involved. Nevertheless, the apoptotic effects of JNK are most significant as this kinase promotes mitochondria-mediated cell death in multiple ways: by positive regulation of p53 and c-myc transcription, by direct inhibition of anti-apoptotic BCl-2 and BCl-Xl, by activation of pro-apoptotic Bad, Bim and Bmf, and by enabling Bax translocation via phosphorylation of 14-3-3 proteins. Thus, repression of Daxx-mediated extrinsic apoptosis by Hsp27 affects JNK signaling which can also be a key and decisive event for the intrinsic pathway.

Compared to the well-known role of Hsp27 in the inhibition of extrinsic cell death, the apoptotic capacity of αB-crystallin in death receptor-mediated apoptosis is still poorly documented. However, it has recently been reported that expression of αB-crystallin correlates with TRAIL resistance in a large number of human cancers. This observation led to the finding that αB-crystallin inhibits TRAIL-induced apoptosis even though, the exact mechanism for this negative regulation is yet unknown (Kamradt et al., 2005).

4.4. Regulation of Other Survival Signaling Pathways

Besides the direct action of Hsp27 and αB-crystallin in the regulation of apoptotic pathways, the cytoprotective effect of these two sHsps can also be mediated by the modulation of several survival signaling cascades, including those mediated by Akt, NFκB, and Erk1/2. The Akt kinase is considered as one of the most important regulators of cell survival and proliferation (Datta et al., 1999; Song et al., 2005). It has been demonstrated that Akt can be stabilized and activated through direct interaction with unphosphorylated Hsp27 in response to diverse cellular stresses like heat and hydrogen peroxide (Konishi et al., 1997). This activation depends on the presence of a signaling complex composed by Akt, Hsp27, p38 MAPK, and MAPK-activated protein kinase-2 (MAPKAPK-2). Within this complex, Akt is able to phosphorylate Hsp27 on Ser82, and this phenomenon leads to disruption of the Akt-Hsp27 interaction and loss of Akt activation (Rane et al., 2003). Thus, Hsp27 tightly regulates Akt activation, and by the same means a great number of Akt substrates, like Bad, caspase-9, GSK3β and FKHR that are all involved in apoptosis. Moreover, activation of Akt, which is normally induced by cytokines including IGF-1 (insulin-like growth factor 1), NGF (nerve growth factor), and PDGF (platelet-derived growth factor), can be maintained by Hsps in the absence of cytokine signaling (Rane et al., 2003).

Several extracellular stimuli, including inflammatory cytokines, growth factors and chemokines, induce phosphorylation, ubiquitylation and subsequent proteasomal degradation of IκB, the specific inhibitor for the NFκB transcription factor. NFκB is then free to translocate to the nucleus, where it activates a variety of target genes encoding proteins involved in the immune or inflammation responses and in cell growth and death control. Phosphorylation of IκB is ensured by the IKK complex, composed of a regulatory subunit IKKγ/NEMO and two catalytic subunits, IKKα and IKKβ, that both interact with Hsp27. The Hsp27-IKKβ interaction, enhanced by phosphorylation of Hsp27 by MAPKAPK-2, is thought to inhibit IKK kinase activity and the consequent repression of NFκB (Park et al., 2003a,b). Moreover, this regulation of the NFκB pathway represents the best explanation for the Hsp27-induced, TNF α-mediated cell survival (Park et al., 2003a,b). However, the effect of Hsp27 on NFκB signaling is controversial, as another published work demonstrates that the survival-promoting regulation endorsed by Hsp27 in response to TNF-α stimulus, depends directly on the induction of the proteasomal degradation of IκB (Parcellier et al., 2003). Finally, the concomitant induction of Hsp27 and activation of the NFκB pathway has been observed under exhaustive GSH depletion by BSO treatment (Filomeni et al., 2005). This result provides evidence for Hsp27-regulated NFκB signaling in cell adaptation and survival against oxidative injury, linking the chaperone and anti-apoptotic functions of Hsp27 in its cytoprotective effect.

The last survival cascade that has been shown to be directly regulated by sHsps is the Raf/Mek/Erk pathway, activated downstream of receptor tyrosine kinases after stimulation by growth factors. The Erk1/2-mediated pathway plays an essential role in promoting cell-cycle progression and is involved in proliferative signaling and cell survival. The stress response depends on oncogenic Ras, which is a GTP-binding protein that requires a guanine nucleotide exchange to be activated and allow the activation of the Raf MAPKKK. This causes the sequential activation of Mek MAPKK and Erk1/2 MAPK which translocates to the nucleus to regulate differentiation or proliferation genes (Marshall, 1995). In lens epithelial cells, it has been shown that overexpression of αB-crystallin blocks the activation of Ras and consequently inhibits Erk1/2 activation to attenuate calcimycin-induced apoptosis (Li et al., 2005) or UVA-induced apoptosis (Liu et al., 2004). Regulation of MAPK signaling by αB-crystallin is a complex phenomenon since its overexpression can also activate Raf/Mek/Erk, at least in human mammary epithelial cells (Moyano et al., 2006). However, no matter how the regulation of MAPK cascades is achieved, all these results show repression of stress-induced apoptosis by αB-crystallin.

Thus, the anti-apoptotic effects of Hsp27 and αB-crystallin are insured not only by their chaperone activity and their direct roles in intrinsic and extrinsic apoptosis pathways, but also by an indirect action through different survival kinases signaling cascades.

4.5. Oncogenic Potential of Small Hsps and Cancer Therapy

Apoptosis is the negative counterpart of proliferation, as well as the predominant form of cell death triggered by cytotoxic drugs in tumor cells; therefore, defects in apoptosis are associated with survival of the transformed cells and cancer (Evan and Littlewood, 1998). The ability of Hsps to protect cells from stressful stimuli and apoptosis, combined to the observation of high levels of Hsp27 or αB-crystallin in a wide range of cancers implies an important role of Hsps in tumorigenicity (see Ciocca and Calderwood, 2005, for review). For this reason considerable efforts have been focussed on targeting heat shock protein expression or regulating their function as an effective anticancer therapy. Increased levels of Hsp27 have been found in breast (Love and King, 1994; Oesterreich et al., 1993), prostate (Cornford et al., 2000), gastric (Ehrenfried et al., 1995), and ovarian (Langdon et al., 1995) cancers, as well as in leukemia (Fuller et al., 1994). However, Hsp27 cannot be defined as a universal marker of poor prognosis as some controversial results have been found in breast tumors treated with estrogens. Experiments on tumor cells from nude mice point to a correlation between Hsp27 expression and two tumorigenesis-related features: increase of metastatic potential and resistance to therapy (Blackburn et al., 1997). Nevertheless, establishing a link between Hsp27 levels and clinical prognosis is not possible at this time. Regarding αB-crystallin, its expression has been reported in human brain tumors (Aoyama et al., 1993), in renal carcinomas (Pinder et al., 1994), and, recently, in breast cancer (Moyano et al., 2006). Overexpression of αB-crystallin does not correlate with that of Hsp27, and its clinical significance is still poorly understood.

At this time, it is unclear why Hsps levels are high in tumor cells and whether they only support malignancy or have an essential role in development of the transformed phenotype. Resistance to doxorubicin, cisplatin and several commonly used anticancer agents has been associated with induction of different Hsps (Ciocca et al., 1992; Richards et al., 1996). Indeed, it seems that administrating these drugs at levels not high enough to achieve apoptosis may actually enhance the expression of Hsps and enable cells to resist conventional chemotherapy. Targeting of heat-shock proteins is currently being tested in clinical trials, but, unfortunately, no specific inhibitor of either Hsp27 or αB-crystallin has been identified so far. Inducing the death of cancer cells by pharmacological targeting of Hsps is not the only approach tested. Increased levels of Hsps may also lead to tumor cell sensitization against immune attacks: tumor cells could express Hsps on their surface and enhance their recognition by the natural killer cells of the immune system (Multhoff and Hightower, 1996). Specific antitumor immunity could then be developed by Hsps-based vaccines against tumors using Hsps as antigens (Multhoff and Hightower, 1996; Ferrarini et al., 1992). Hsps are

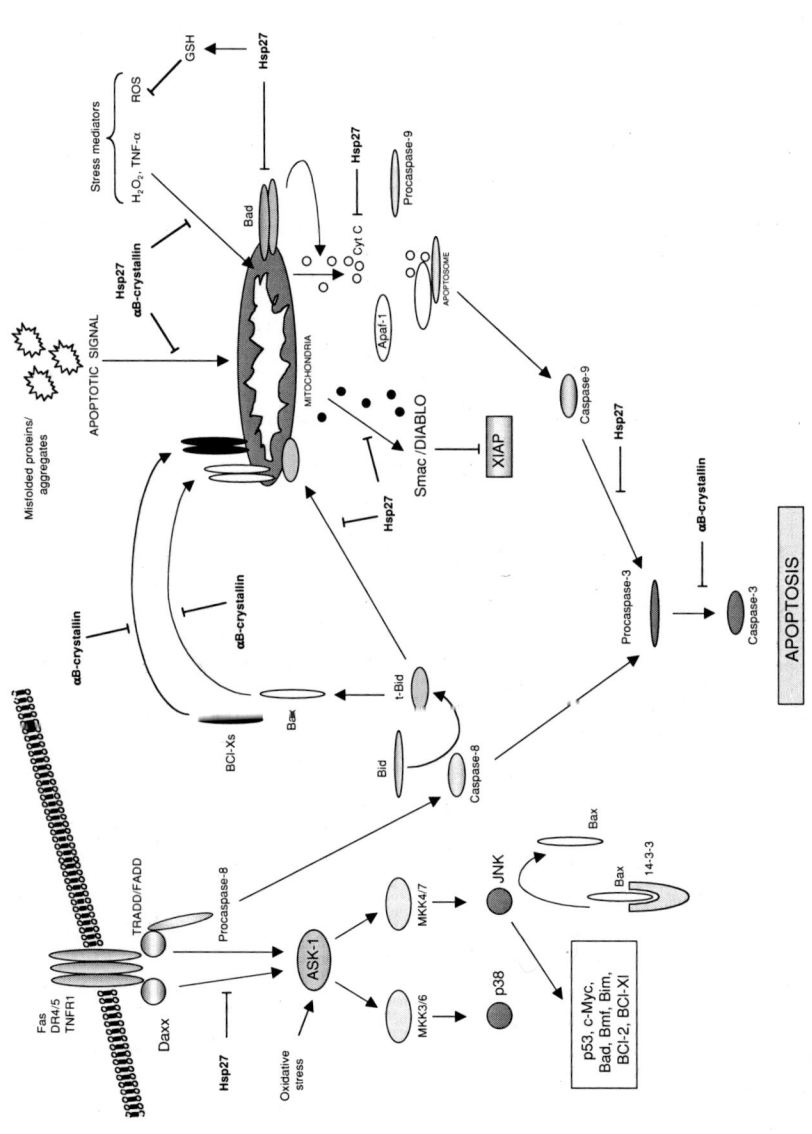

mainly cytosolic or nuclear-localized proteins but some Hsps can be directed to membranes; hence the use of Hsp-peptide complexes as a source for tumor-specific antigens remains an interesting alternative for cancer therapy (Jolly and Morimoto, 2000).

In summary, protection against cell stress and cell death are very closely related events, such that molecular chaperones induced in response to stress can act as key regulators of the apoptotic pathways. One of the anti-apoptotic mechanisms of sHsps depends on their chaperone function, which can modulate protein folding and aggregation, ubiquitin-dependent degradation, or intracellular translocation of proteins (Takayama et al., 2003). It is now clear that Hsp27 is a highly adapted and dynamic chaperone with several anti-apoptotic sites of action. These are summarized in Figure 2. Hsp27 can interact with cytochrome C to prevent activation of procaspase-9, bind to caspase-3 and modulate its activity, regulate the intracellular distribution of Bid, act as a critical redox modulator in oxidative stress-induced apoptosis, and protect integrity of the actin network. When overexpressed in tumor cells, Hsp27 exerts a cytoprotective effect against apoptosis induced by a wide range of stimuli from hyperthermia, oxidative stress (Mehlen et al., 1995; Préville et al., 1999), staurosporine, and Fas ligand to chimiotherapeutic drugs (Garrido et al., 1997).

The second anti-apoptotic, small heat shock protein, αB-crystallin acts to block cell death induced by oxidative injury but can also inhibit intrinsic apoptosis by direct control of the autoproteolytic maturation of p24 caspase-3 intermediate and translocation of Bax and BCl-Xs to mitochondria. The action of αB-crystallin in the extrinsic pathway of cell death is much less important and only inhibition of TRAIL-induced apoptosis has been reported to date. On the contrary, essential roles in positive and negative regulation of Raf/Mek/Erk survival pathway lead to repression of several stress-induced types of apoptosis.

Even if both Hsp27 and αB-crystallin are able to act at multiple levels of the apoptotic pathways (Fig. 2), they inhibit stress-induced apoptosis by largely

←

FIGURE 2. Suppression of stress-induced apoptosis by small heat shock proteins (sHsps). Hsp27 and αB-crystallin act at multiple levels of both, the intrinsic (mitochondria-mediated), and the extrinsic (death receptor-mediated) apoptotic pathways. Hsp27 acts in a direct fashion on several anti-apoptotic targets: it interacts with cytochrome C to prevent the activation of procaspase-9, it binds to caspase-3 and regulates its activity, it regulates the intracellular distribution of the proapoptotic Bid protein, it inhibits the release of the mitochondrial Smac/DIABLO factor, and it inhibits the interaction of Daxx and ASK-1. Hsp27 can also suppress programmed cell death by an indirect regulation of cellular GSH contents, by preventing aggregation of misfolded proteins, by modulation of the redox state of the cell and/or by protecting the integrity of the actin network. AlphaB-crystallin triggers its anti-apoptotic function by inhibiting the translocation to mitochondria of proapoptotic proteins, either Bax or BCl-Xs and by suppressing the last step of the final activation of caspase-3: autocleavage of the caspase-3 intermediate. An Hsp27-like indirect action on apoptosis is also observed for αB-crystallin: αB-crystallin is able to inhibit protein aggregation and to modulate the redox state of the cell in response to oxidative stress.

distinct mechanisms (Kamradt et al., 2001). This somehow unexpected result is supported by differentiation-induced caspase-3 activation, which can be repressed by αB-crystallin, but not by Hsp27 (Kamradt et al., 2002). Impairment of this specific regulation is affected in Desmin-like myopathy, providing evidence that Hsps are implicated in other diseases than cancer. Nevertheless, the implication of Hsps in cancer studies has become essential by their capacity of stabilizing potentially oncogenic mediators, as well as regulating transmission of many apoptotic signals, a powerful property which can easily become a danger for the cell. Thus, Hsps emerge as cellular lifeguards and homeostasis keepers. Obviously, many additional studies are needed to get a clearer picture of the optimal levels of Hsp27 and/or αB-crystallin needed to ensure a proper cytoprotective effect on stressed cells without disrupting the balance between survival and cell death. Moreover, in contrast to Hsp27 and αB-crystallin, two other Hsps, Hsp60 and Hsp10, have been shown to promote, rather than inhibit, apoptosis by regulating caspase-3 activation (Samali et al., 1999), increasing the complexity of the Hsp network.

5. sHsps and Aging

Aging is a complex process involving hereditary, environmental, and life-style factors (Sun and Tower, 1999; Vieira et al., 2000). Genome-wide studies in lower eukaryotes and in mammals have revealed that the aging mechanisms are, at least in part, regulated and conserved (Rogina and Helfand, 1995; Pletcher et al., 2002; Seroude, 2002; Hekimi and Guarente, 2003; Helfand and Rogina, 2003; Hozenberg et al., 2003; Tatar et al., 2003). Reactive oxygen species (ROS), which are mainly produced by mitochondria, are thought to be at the center of lifespan determination (Hekimi and Guarente, 2003; Stadtman, 2004). ROS are highly reactive molecules that react with DNA, lipids and proteins to inhibit and/or disrupt their functions (Sun et al., 1999; Raha and Robinson, 2000; 2001; Hrelia et al., 2002; Chomyn and Attardi, 2003; Djojosubroto et al., 2003; Liu et al., 2003; Park et al., 2003a,b; Sastre et al., 2003; Wright, 1991; Stadtman, 2004). An imbalance between ROS production and removal would lead to the accumulation of cellular damages and subsequent shortening of lifespan (Butov et al., 2001; Sohal, 2002; Stadtman, 2004). Therefore, as stated in the free radical theory of aging, the ability of organisms to cope with random somatic damages induced by ROS would determine lifespan (Sohal, 2002). Interestingly, long-lived mutants, obtained in *Caenorhabditis elegans* and *Drosophila melanogaster*, display increased thermotolerance and resistance to stress (Lin et al., 1998; Rogina et al., 2000; Ekengren et al., 2001).

5.1. Small Hsps Expression During Aging

Hsps expression during aging is tightly regulated and the heat shock response following exposure to stress is also modified. While old *Drosophila* display an increased production of heat shock proteins following stresses (Niedzwiecki et al., 1991; Fleming et al., 1988), aged rats and humans fail to activate the heat shock transcription factor 1, which is responsible for the stress-induced Hsp expression, in multiple tissues (Heydari et al., 1993; Jurivich et al., 1997). Moreover, a reduced ability of old individuals to induce Hsps expression in response to heat has been observed in peripheral lymphocytes (Rao et al., 2003) and in skin cells (Muramatsu et al., 1996).

The regulation of sHsps expression during aging appears to be tissue- and cell-type specific. In *D. melanogaster* dmhsp22 mRNA is up-regulated by 60-fold in the head of flies between day 6 and 35, while it is only up-regulated by 16-fold in the thorax (King and Tower, 1999). Dmhsp23 mRNA is the only other sHsp to be up-regulated in aged drosophila and displays a 5% up-regulation in thoraces extracts. Interestingly, tissues displaying the preferential up regulation of shsp genes either have an increased sensitivity to stress (like neurons, (Boulianne, 2001) (or are submitted to intensive stress (like flying muscles cells, (Sohal and Brunk, 1992)). In mammals, HspB1, HspB5, and HspB6 are ubiquitously expressed and are up-regulated with aging in skeletal muscles (Piec et al., 2005; Chung and Ng, 2005) and in the brain (Schultz et al., 2001). Thus HspB1 displays a preferential up regulation (105.5%) over Hsp60 (61.3%) and Hsp70 (29.5%) in rat skeletal muscles from 29-month-old rats compared to 16-month-old rats (Chung and Ng, 2005). A decrease of HspB1 in cultured dermal fibroblasts from humans of different ages (Boraldi et al., 2003) and a decreased phosphorylation of this protein in gastrointestinal smooth muscles of old rats (Bitar and Patil, 2004) have been reported. This apparent decrease in HspB1 phosphorylation would be associated with a reduced association of this small Hsp with tropomyosin and actin and could be responsible, at least in part, for the reduced contractile response of gastrointestinal smooth muscles with age (Bitar and Patil, 2004).

5.2. *Post-Translational Modifications of sHsps During Aging*

The α-crystallins HspB4 and HspB5 are two major proteins of the lens that have been extensively studied in aging due to their role in cataract formation. With age, HspB4 undergoes multiple post-translational modifications, such as deamidation, acetylation, oxidation, phosphorylation, and truncation (Takemoto and Boyle, 1994; Takemoto and Gopalakrishnan, 1994; Kamei et al., 1997; Hanson et al., 2000; Colvis and Garland, 2002; Ueda et al., 2002; Kapphahn et al., 2003). Deamidation and oxidation of HspB4 and HspB5 both alter their structural integrity and function (Miesbauer et al., 1994; Takemoto, 1998; Hanson et al., 2000), but the exact effect of phosphorylation on their function is still controversial (Wang et al., 1995; Carver et al., 1996; van Boekel et al., 1996). The most important truncation of HspB4 is a C-terminal deletion of 22 amino acids, which results in large sHsps aggregates with no chaperone function (Takemoto et al., 1993; Takemoto and Gopalakrishnan, 1994; Andley et al., 1996; Hanson et al., 2000; Colvis and Garland, 2002). N-terminal truncations of HspB4 and HspB5 are also found in retina from old rats and humans and result in decreased chaperone activity (Smith et al., 1996; Kamei et al., 1997; Kumar and Rao, 2000).

In older lenses, a high-molecular-weight protein complex (HMW) formed by α-crystallins appears. This HMW protein complex contains partly unfolded HspB4 and HspB5 (Liang et al., 1985; Messmer and Chakrabarti, 1988; Chiou and Azari, 1989; Liang and Rossi, 1989) and displays a decreased chaperone-like activity (Takemoto and Boyle, 1994; Derham and Harding, 1997a;b). Native HspB4 and HspB5 (without post-translational modifications or involvement in HMW protein) maintain their chaperone-like activity during aging (Carver et al., 1996). The decrease in chaperone activity of HspB4 and HspB5 with age is specific to the intracellular compartment, α-crystallins from the nucleus of old human and rabbit lenses displaying decreased chaperone activity while α-crystallins from the cytoplasm maintain their activity (Derham and Harding, 1997a;b).

5.3. sHsps Have Beneficial Effects on the Aging Process

The first evidence that Hsps could have a beneficial effect on aging was suggested from mild stress experiments, which activate the stress response without causing cellular damages (Minois, 2000). Organisms submitted to repeated mild heat stresses display increased Hsps expression, which results in increased lifespan and cell survival to otherwise lethal conditions (Khazaeli et al., 1997; Rattan, 1998; Minois, 2000; Le Bourg et al., 2001; Fonager et al., 2002; Hercus et al., 2003; Rattan, 2004).

While the overexpression of DmHsp70 failed to increase lifespan substantially (Tatar et al., 1997; Minois et al., 2001), overexpressing the sHsps in *D. melanogaster* has been found more advantageous. Indeed, overexpressing DmHsp22, DmHsp26, or DmHsp27 in all tissues by means of the GAL4/UAS system (Brand and Perrimon, 1993) resulted in a mean 30% increase in lifespan (Morrow et al., 2004; Wang et al., 2004). Moreover, targeting the overexpression of DmHsp22 to motor neurons was sufficient to increase *Drosophila* lifespan and resistance to oxidative stress by 30% pointing at the importance of these cells in the aging process (Morrow et al., 2004). In mammals, overexpression of Hsp70 and Hsp27 in neurons have both been shown to be beneficial but by different mechanisms. While Hsp70 would prevent apoptosis, Hsp27 would inhibit necrosis (Alsbury et al., 2004). A beneficial role of sHsps in the aging process has also been reported in the worm *C. elegans*. Indeed, it was shown that introducing extra copies of the gene encoding Hsp16 was sufficient to confer thermotolerance and increase worms longevity (Walker and Lithgow, 2003) while treatment of worms with Hsp16 RNAi resulted in the shortening of lifespan (Hsu et al., 2003). Altogether, these results argue in favor of an increase in protein damages during aging. Since sHsps are molecular chaperones, they could act on the aging process by preventing protein aggregation and maintaining misfolded protein in a refoldable state (Haslbeck, 2002).

6. sHsps and Neurodegenerative Diseases

In the late 90s, mutations in αB-crystallin genes were reported to be involved in a desmin-related myopathy (Vicart et al., 1998). Similar mutations in sHsps were soon reported in other myofibrillar myopathies (Selcen and Engel, 2003).

Mutations in the αA- and αB-crystallins were also reported to be associated with congenital cataracts in humans (reviewed in Benndorf and Welsh, 2004; Sun and MacRae, 2005b). This led to further investigations on association between mutations in sHsps, and a number of inherited neuropathies. HspB1 and HspB8 were reported to harbor mutations within their crystallin domain that caused Charcot-Marie-Tooth disease and distal hereditary motor neuropathy (Evgrafov et al., 2004; Irobi et al., 2004). All these mutations are located in residues that are expected to disturb the secondary and/or tertiary structure of the corresponding sHsps and likely lead to defective chaperone activity. A list of the mutations described so far and their association with specific diseases can be found in Table 2. However whether these mutations actually disturb chaperone activity has not been tested in all cases.

An interesting feature of these diseases is that they are caused either by mutations in small Hsps or mutations in intermediate filament proteins, e.g., desmin or neurofilaments, supporting evidence from in vitro studies in which interactions were described between both classes of molecules. However whether the disease phenotype is due to defective filament assembly caused by mutated chaperones or whether misfolded chaperones might themselves form aggregates that would induce misfolding or aggregation of other proteins is not clear. Chavez Zobel et al. (2003) showed that expression of an R120G mutated αB-crystallin in transfected mammalian cells led to the formation of aggregates even if the cells used had no desmin. However as pointed out by the authors, this does not exclude the possibility that desmin might be unstable or disorganized in the presence of the mutated sHsp. Another possibility is based on the observation that different members of the sHsp family interact with each other by forming hetero-oligomers and that mutations in one member may change its interaction with its sHsp partners thereby causing interference with their function in vivo. An increased interaction between the K141N mutant of HspB8 and HspB1 was previously shown (Irobi et al., 2004). This is consistent with the finding that mutations in HspB1 and HspB8 can be associated with the same disease (see Dierick et al., 2005, for review).

As neuromuscular diseases such as desmin-related myopathy are characterized by the formation of protein aggregates, a link was rapidly inferred between these structures and the possible implications of the small Hsps in neurodegenerative diseases. Most neurodegenerative diseases are characterized by the presence of insoluble toxic protein aggregates (Bucciantini et al., 2002). The recognized role of sHsps as reservoirs of misfolded proteins in a folding-competent state, and the observation that sHsps tend to colocalize with cellular inclusion bodies has triggered further investigation into the possible implications of small Hsps chaperones in these diseases. As this topic has recently been reviewed in the case of Alzheimer's disease, and in amyotrophic lateral sclerosis (Sun and MacRae, 2005b), we will simply summarize here recent observations mostly on the polyglutamine proteins (poly Q) involved in Huntington and other neurodegenerative diseases (ataxias). HspB1 (Hsp27) was shown to prevent poly Q toxicity in transfected cell assays (Wyttenbach et al., 2002) but without preventing poly Q aggregation. The authors suggested that HspB1 protects against reactive oxygen species generated by the mutant poly Q protein aggregates rather than by its chaperone function. Thus the

TABLE 2. sHsps in Neurodegenerative Diseases*

Hsp	Mutation and localization in the sequence			Associated disease	Affected tissue	Ref.
	N-termini	Crystallin domain	C-termini			
HspB1		R127W		DHMN	Motor neuron	Evgrafov et al., 2004
HspB1		R127W		CMT2	Motor and sensory neurons	Tang et al., 2005b
HspB1		S135F		DHMN	Motor neuron	Evgrafov et al., 2004
HspB1		S135F		CMT2	Motor and sensory neurons	Evgrafov et al., 2004
HspB1		R136W		CMT2	Motor and sensory neurons	Evgrafov et al., 2004
HspB1		T151I		DHMN	Motor neuron	Evgrafov et al., 2004
HspB1			P 82L	DHMN	Motor neuron	Evgrafov et al., 2004
HspB1			P 82S	DHMN	Motor neuron	Kijima et al., 2005
HspB4	M1 oxidation			RHC	Eyes	Fujii et al., 2004
HspB4	W9X			ARC	Eyes	Pras et al., 2000
HspB4	R49C			ADCC	Eyes	Mackay et al., 2003
HspB4	D58 isom. racem.			Cataract	Eyes	Fujii et al., 2001
HspB4		R116C		ADCC	Eyes	Litt et al., 1998
HspB4		V124E		ADCC	Eyes	Graw et al., 2001
HspB4		C131 oxidation		Cataract	Eyes	Takemoto, 1996
HspB4		C142 oxidation		Cataract	Eyes	Takemoto, 1996
HspB4		D151 isom. racem.		RHC	Eyes	Fujii et al., 2004
HspB4			Δ	Cataract	Eyes	Takemoto, 1997
HspB4			Δs.	RHC	Eyes	Takeuchi et al., 2004
HspB5		R120G		DRC	Heart cardiac muscle	Wang et al., 2001

HspB5	R120G		Cataract	Eyes	Perng et al., 1999a,b
HspB5	R120G		DRM	Muscles	Vicart et al., 1998
HspB5	D140N		ADCC	Eyes	Liu et al., 2006
HspB5	N146 deamidation		Cataract	Eyes	Srivastava and Srivastava, 2003
HspB5	R157H		DCM	Heart muscles	Inagaki et al., 2006
HspB5		Δ25	DRM	Muscles	Selcen and Engel, 2003
HspB5		Δ13	DRM	Muscles	Selcen and Engel, 2003
HspB5		Δs	RHC	Eyes	Takeuchi et al., 2004
HspB8	K141N		DHMN	Motor neuron	Irobi et al., 2004
HspB8	K141E		DHMN	Motor neuron	Irobi et al., 2004
HspB8	K141N		CMT2		Tang et al., 2005a

*DHMN = Distal hereditary motor neuropathy, CMT2 = Charcot-Marie-Tooth disease type 2, RHC = rats hereditary cataract, ARC = autosomal recessive cataract, ADCC = autosomal dominant congenital cataract, isom. racem. = isomerization and racemization, DRC = desmin-related cardio-myopathy, DRM = desmin-related myopathy, DCM = dilated cardiomyopathy importance of the chaperone property of sHsps in these diseases remains unclear. Recently, it was shown that one member of the family, HspB8 (but not HspB1 nor HspB5), could indeed act as a chaperone in vivo and prevent the formation of poly Q-containing aggresomes (Carra et al., 2005). Another interesting observation is that HspB8 interacts only with the mutant form of αB-crystallin (R120G) but not with wild type crystallin (Chavez Zobel et al., 2003). This again points at the complexity and specificity of the different members of the small Hsp family, a phenomenon well documented for the sHsps of *Drosophila* melanogaster that are localized in different cell compartments, and show distinct chaperone-like activities in in vitro assays (Michaud et al., 2002; Morrow et al., 2006).

In summary, members of the small Hsp family in different kingdoms show divergence in sequence rather than in structure. sHsps all share an α-crystallin-like core domain with a well conserved β-sheet structure bordered by variable length non-conserved N- and C-terminal extensions. In addition to their role in oligomerization, these extensions have been suggested to provide specificity of functions to the various sHsps. This is certainly consistent with the multiplicity of sites of action of sHsps in physiological and stress-related processes and with the wide range of cellular functions of proteins found associated with sHsps (Basha et al., 2004). The recent reports that mutations in sHsps are associated with neuromuscular disorders, and that they seem to play a role in proteins aggregates in many neurodegenerative diseases, have triggered a new and well-deserved interest on this family of stress proteins. More structure-function studies will be needed to understand the mechanisms of action of these proteins in the process of apoptosis, and aging for example but it is also important that such studies are conducted for many organisms as there is biological evidence for their involvement in many other cell-specific processes.

Acknowledgments. The authors would like to thank the Canadian Institutes of Health Research for their continued support through grants (R.M.T) and studentships (A.B., G.M.).

References

Adams, J. M. (2003) Ways of dying: Multiple pathways to apoptosis. *Genes Dev* 17(20):2481–95.

Adhikari, A. S., Rao, K. S., Rangaraj, N., Parnaik, V. K., and Rao, C. M. (2004) Heat stress-induced localization of small heat shock proteins in mouse myoblasts: Intranuclear lamin A/C speckles as target for aB-crystallin and Hsp25. *Exp Cell Res* 299:393–403.

Alsbury, S., Papageorgiou, K., and Latchman, D. S. (2004) Heat shock proteins can protect aged human and rodent cells from different stressful stimuli. *Mech Ageing Dev* 125:201–9.

Andley, U. P., Mathur, S., Griest, T. A., and Petrash, J. M. (1996) Cloning, expression, and chaperone-like activity of human alphaA-crystallin. *J Biol Chem* 271:31973–80.

Aoyama, A., Steiger, R. H., Frohli, E., Schafer, R., von Deimling, A., Wiestler, O. D., and Klemenz, R. (1993) Expression of alpha B-crystallin in human brain tumors. *Int J Cancer* 55(5):760–4.

Arrigo, A. P. (1998) Small stress proteins: Chaperones that act as regulators of intracellular redox state and programmed cell death. *Biol Chem* 379:19–26.

Arrigo, A. P., and Pauli, D. (1988) Characterization of HSP27 and three immunologically related polypeptides during *Drosophila* development. *Exp Cell Res* 175:169–83.

Basha, E., Lee, G. J., Breci, L. A., Hausrath, A. C., Buan, N. R., Giese, K. C., and Vierling, E. (2004) The identity of proteins associated with a small heat shock protein during heat stress *in vivo* indicates that these chaperones protect a wide range of cellular functions. *J Biol Chem* 279:7566–75.

Beere, H. M. (2004) "The stress of dying": The role of heat shock proteins in the regulation of apoptosis. *J Cell Sci* 117:2641–51.

Beere, H. M. (2005) Death versus survival: Functional interaction between the apoptotic and stress-inducible heat shock protein pathways. *J Clin Invest* 115:2633–9.

Behlke, J., Lutsch, G., Gaestel, M., and Bielka, H. (1991) Supramolecular structure of the recombinant murine small heat shock protein hsp25. *FEBS Lett* 288:119–22.

Benndorf, R., and Welsh, M. J. (2004) Shocking degeneration. *Nat Genet* 36:547–8.

Benndorf, R., Hayess, K., Ryazantsev, S., Wieske, M., Behlke, J., and Lutsch, G. (1994) Phosphorylation and supramolecular organization of murine small heat shock protein HSP25 abolish its actin polymerization-inhibiting activity. *J Biol Chem* 269:29780–4.

Bentley, N. J., Fitch, I. T., and Tuite, M. F. (1992) The small heat-shock protein Hsp26 of *Saccharomyces cerevisiae* assembles into a high molecular weight aggregate. *Yeast* 8:95–106.

Berengian, A. R., Bova, M. P., and Mchaourab, H. S. (1997) Structure and function of the conserved domain in alphaA-crystallin. Site-directed spin labeling identifies a beta-strand located near a subunit interface. *Biochemistry* 36:9951–7.

Bitar, K. N., and Patil, S. B. (2004) Aging and gastrointestinal smooth muscle. *Mech Ageing Dev* 125:907–10.

Blackburn, R. V., Galoforo, S. S., Berns, C. M., Armour, E. P., McEachern, D., Corry, P. M., and Lee, Y. J. (1997) Comparison of tumor growth between hsp25- and hsp27-transfected murine L929 cells in nude mice. *Int J Cancer* 72:871–7.

Blond-Elguindi, S., Fourie, A. M., Sambrook, J. F., and Gething, M. J. (1993) Peptide-dependent stimulation of the ATPase activity of the molecular chaperone BiP is the result of conversion of oligomers to active monomers. *J Biol Chem* 268:12735–9.

Bohm, I., and Schild, H. (2003) Apoptosis: The complex scenario for a silent cell death. *Mol Imaging Biol* 5:2–14.

Boraldi, F., Bini, L., Liberatori, S., Armini, A., Pallini, V., Tiozzo, R., Pasquali-Ronchetti, I., and Quaglino, D. (2003) Proteome analysis of dermal fibroblasts cultured *in vitro* from human healthy subjects of different ages. *Proteomics* 3:917–29.

Boulianne, G. L. (2001) Neuronal regulation of lifespan: Clues from flies and worms. *Mech Ageing Dev* 122:883–94.

Brand, A. H., and Perrimon, N. (1993) Targeted gene expression as a means of altering cell fates and generating dominant phenotypes. *Development* 118:401–15.

Bruey, J. M., Ducasse, C., Bonniaud, P., Ravagnan, L., Susin, S. A., Diaz-Latoud, C., Gurbuxani, S., Arrigo, A. P., Kroener G., Solary, E., Garrido, C. (2000) Hsp27 negatively regulates cell death by interacting with cytochrome c. *Nat Cell Biol* 2:645–52.

Bucciantini, M., Giannoni, E., Chiti, F., Baroni, F., Formigli, L., Zurdo, J., Taddei, N, Ramponi, G., Dobson, C. M., and Stefani, M. (2002) Inherent toxicity of aggregates implies a common mechanism for protein misfolding diseases. *Nature* 416:507–11.

Butov, A., Johnson, T., Cypser, J., Sannikov, I., Volkov, M., Sehl, M., and Yashin, A. (2001) Hormesis and debilitation effects in stress experiments using the nematode worm

Caenorhabditis elegans: The model of balance between cell damage and HSP levels. *Exp Gerontol* 37:57–66.

Cairns, J., Qin, S., Philp, R., Tan, Y. H., and Guy, G. R. (1994) Dephosphorylation of the small heat shock protein Hsp27 *in vivo* by protein phosphatase 2A. *J Biol Chem* 269:9176–83.

Carra, S., Sivilotti, M., Chavez Zobel, A. T., Lambert, H., and Landry, J. (2005) HspB8, a small heat shock protein mutated in human neuromuscular disorders, has *in vivo* chaperone activity in cultured cells. *Hum Mol Genet* 14:1659–69.

Carver, J. A., Nicholls, K. A., Aquilina, J. A., and Truscott, R. J. (1996) Age-related changes in bovine alpha-crystallin and high-molecular-weight protein. *Exp Eye Res* 63:639–47.

Chang, H. Y., Nishitoh, H., Yang, X., Ichijo, H., and Baltimore, D. (1998) Activation of apoptosis signal-regulating kinase 1 (ASK1) by the adapter protein Daxx. *Science* 281:1860–3.

Charette, S. J., and Landry, J. (2000) The interaction of HSP27 with Daxx identifies a potential regulatory role of HSP27 in Fas-induced apoptosis. *Ann NY Acad Sci* 926:126–31.

Charette, S. J., Lavoie, J. N., Lambert, H., and Landry, J. (2000) Inhibition of Daxx-mediated apoptosis by heat shock protein 27. *Mol Cell Biol* 20:7602–12.

Chavez Zobel, A. T., Loranger, A., Marceau, N., Theriault, J. R., Lambert, H., and Landry, J. (2003) Distinct chaperone mechanisms can delay the formation of aggresomes by the myopathy-causing R120G alphaB-crystallin mutant. *Hum Mol Genet* 12:1609–20.

Chiou, S. H., and Azari, P. (1989) Physicochemical characterization of alpha-crystallins from bovine lenses: Hydrodynamic and conformational properties. *J Protein Chem* 8:1–17.

Chomyn, A., and Attardi, G. (2003) MtDNA mutations in aging and apoptosis. *Biochem Biophys Res Commun* 304:519–29.

Chung, L., and Ng, Y. C. (2005) Age-related alterations in expression of apoptosis regulatory proteins and heat shock proteins in rat skeletal muscle. *Biochem Biophys Acta* 1762:103–9.

Ciocca, D., and Calderwood, S. K. (2005) Heat shock proteins in cancer: Diagnostic, prognostic, predictive and treatment implications. *Cell Stress and Chap* 10:86–103.

Ciocca, D. R., Fuqua, S. A., Lock-Lim, S., Toft, D. O., Welch, W. J., and McGuire, W. L. (1992) Response of human breast cancer cells to heat shock and chemotherapeutic drugs. *Cancer Res* 52:3648–54.

Clark, J. I., and Muchowski, P. J. (2000) Small heat-shock proteins and their potential role in human disease. *Curr Opin Struct Biol* 10(1):52–9.

Colvis, C., and Garland, D. (2002) Posttranslational modification of human alphaA-crystallin: Correlation with electrophoretic migration. *Arch Biochem Biophys* 397:319–23.

Concannon, C. G., Gorman, A. M., and Samali, A. (2003) On the role of Hsp27 in regulating apoptosis. *Apoptosis* 8:61–70.

Cornford, P. A., Dodson, A. R., Parsons, K. F., Desmond, A. D., Woolfenden, A., Fordham, M., Neoptolemos, J. P., Ke, Y., and Foster, C. S. (2000) Heat shock protein expression independently predicts clinical outcome in prostate cancer. *Cancer Res* 60:7099–105.

Datta, S. R., Brunet, A., and Greenberg, M. E. (1999) Cellular survival: A play in three *Akts Genes Dev* 13:2905–27.

Davidson, S.M, Loones, M. T., Duverger, O., and Morange, M. (2002) The developmental expression of small HSP. *Prog Mol Subcell Biol* 28:103–28.

Derham, B. K., and Harding, J. J. (1997a) Effect of aging on the chaperone-like function of human alpha-crystallin assessed by three methods. *Biochem J* 328:763–8.

Derham, B. K., and Harding, J. J. (1997b) The effects of ageing on the chaperone-like function of rabbit alpha-crystallin, comparing three methods of assay. *Biochim Biophys Acta* 1336:187–94.

Dierick, I., Irobi, J., De Jonghe, P., and Timmerman, V. (2005) Small heat shock proteins in inherited peripheral neuropathies. *Ann Med* 37:413–22.

Djojosubroto, M. W., Choi, Y. S., Lee, H. W., and Rudolph, K. L. (2003) Telomeres and telomerase in aging, regeneration and cancer. *Mol Cell* 15:164–75.

Ehrenfried, J. A., Herron, B. E., Townsend, C. M. Jr., and Evers, B. M. (1995) Heat shock proteins are differentially expressed in human gastrointestinal cancers. *Surg Oncol* 4:197–203.

Ehrnsperger, M., Graber, S., Gaestel, M., and Buchner, J. (1997) Binding of non-native protein to Hsp25 during heat shock creates a reservoir of folding intermediates for reactivation. *EMBO J* 16:221–9.

Ekengren, S., Tryselius, Y., Dushay, M. S., Liu, G., Steiner, H., and Hultmark, D. (2001) A humoral stress response in *Drosophila*. *Curr Biol* 11:714–8.

Evan, G., and Littlewood, T. (1998) A matter of life and cell death. *Science* 281:1317–22.

Evgrafov, O. V., Mersiyanova, I., Irobi, J., Van Den Bosch, L., Dierick, I., Leung, C. L., Schagina, O., Verpoorten, N., Van Impe, K., Fedotov, V., Dadali, E., Auer-Grumbach, M., Windpassinger, C., Wagner, K., Mitrovic, Z., Hilton-Jones, D., Talbot, K., Martin, J. J., Vasserman, N., Tverskaya, S., Polyakov, A., Liem, R. K. H., Gettemans, J., Robberecht, W., De Jonghe, P., and Timmerman, V. (2004) Mutant small heat-shock protein 27 causes axonal Charcot-Marie-Tooth disease and distal hereditary neuropathy. *Nat Genet* 36:602–6.

Ferrarini, M., Heltai, S., Zocchi, M. R., and Rugarli, C. (1992) Unusual expression and localization of heat-shock proteins in human tumor cells. *Int J Cancer* 51:613–9.

Filomeni, G., Aquilano, K., Rotilio, G., and Ciriolo, M. R. (2005) Antiapoptotic response to induced GSH depletion: Involvement of heat shock proteins and NF-kappaB activation. *Antioxid Redox Signal* 7:446–55.

Fisher, D., Matten, J., Reimann, J., Bonnemann, C., and Schroder, R. (2002) Expression, localization and functional divergence of alphaB-crystallin and heat shock protein 27 in core myopathies and neurogenic atrophy. *Acta Neuropathol (Berl)* 104:297–304.

Fleming, J. E., Walton, J. K., Dubitski, R., and Bensch, K. G. (1988) Aging results in an unusual expression of *Drosophila* heat shock proteins. *Proc Natl Acad Sci USA* 85:4099–103.

Fonager, J., Beedholm, R., Clark, B. F., and Rattan, S. I. (2002) Mild stress-induced stimulation of heat-shock protein synthesis and improved functional ability of human fibroblasts undergoing aging *in vitro*. *Exp Gerontol* 37:1223–8.

Franck, E., Madsen, O., van Rheede, T., Ricard, G., Huynen, M. A., de Jong, W. W. (2004) Evolutionary diversity of vertebrate small heat shock proteins. *J Mol Evol* 59:792–805.

Fu, X., Jiao, W., and Chang, Z. (2006) Phylogenetic and biochemical studies reveal a potential evolutionary origin of small heat shock proteins of animals from bacterial class A. *J Mol Evol* 62:257–66.

Fujii, N., Matsumoto, S., Hiroki, K., and Takemoto, L. (2001) Inversion and isomerization of Asp-58 residue in human alphaA-crystallin from normal aged lenses and cataractous lenses. *Biochim Biophys Acta* 1549:179–87.

Fujii, N., Takeuchi, N., Fujii, N., Tezuka, T., Kuge, K., Takata, T., Kamei, A., and Saito, T. (2004) Comparison of post-translational modifications of alpha A-crystallin from normal and hereditary cataract rats. *Amino Acids* 26:147–52.

Fujita, Y., Ohto, E., Katayama, E., and Atomi, Y. (2004) alphaB-crystallin-coated MAP microtubule resists nocodazole and calcium-induced disassembly. *J Cell Sci* 117:1719–26.

Fuller, K. J., Issels, R. D., Slosman, D. O., Guillet, J. G., Soussi, T., and Polla, B. S. (1994) Cancer and the heat shock response. *Eur J Cancer* 30A:1884–91.

Gaestel, M. (2002) sHsp-phosphorylation: Enzymes, signaling pathways and functional implications. *Prog Mol Subcell Biol* 28:151–69.

Gaestel, M., Schroder, W., Benndorf, R., Lippmann, C., Buchner, K., Hucho, F., Erdmann, V. A., and Bielka, H. (1991) Identification of the phosphorylation sites of the murine small heat shock protein hsp25. *J Biol Chem* 266:14721–4.

Garrido, C., Ottavi, P., Fromentin, A., Hammann, A., Arrigo, A. P., Chauffert, B., and Mehlen, P. (1997) HSP27 as a mediator of confluence-dependent resistance to cell death induced by anticancer drugs. *Cancer Res* 57:2661–7.

Garrido, C., Bruey, J. M., Fromentin, A., Hammann, A., Arrigo, A. P., and Solary, E. (1999) HSP27 inhibits cytochrome c-dependent activation of procaspase-9. *FASEB J* 13:2061–70.

Garrido, C., Gurbuxani, S., Ravagnan, L., and Kroemer, G. (2001) Heat shock proteins: Endogenous modulators of apoptotic cell death. *Biochem Biophys Res Commun* 28:433–42.

Gober, M. D., Smith, C. C., Ueda, K., Toretsky, J. A., and Aurelian, L. (2003) Forced expression of the H11 heat shock protein can be regulated by DNA methylation and trigger apoptosis in human cells. *J Biol Chem* 278:37600–9.

Graw, J., Loster. J., Soewarto, D., Fuchs, H., Meyer, B., Reis, A., Wolf, E., Balling, R., and Hrabe de Angelis, M. (2001) Characterization of a new, dominant V124E mutation in the mouse alphaA-crystallin-encoding gene. Invest. *Ophthalmol Vis Sci* 42:2909–15.

Green, D. R., and Reed, J. C. (1998) Mitochondria and apoptosis. *Science* 281:1309–12.

Groenen, P. J., Merck, K. B., de Jong, W. W., and Bloemendal, H. (1994) Structure and modifications of the junior chaperone alpha-crystallin. From lens transparency to molecular pathology. *Eur J Biochem* 225:1–19.

Gross, A., McDonnell, J. M., and Korsmeyer, S. J. (1999) BCL 2 family members and the mitochondria in apoptosis. *Genes Dev* 13:1899–911.

Guay, J., Lambert, H., Gingras-Breton, G., Lavoie, J. N., Huot, J., and Landry, J. (1997) Regulation of actin filament dynamics by p38 map kinase-mediated phosphorylation of heat shock protein 27. *J Cell Sci* 110:357–68.

Hanson, S. R., Hasan, A., Smith, D. L., and Smith, J. B. (2000) The major *in vivo* modifications of the human water-insoluble lens crystallins are disulfide bonds, deamidation, methionin oxidation and backbone cleavage. *Exp Eye Res* 71:195–207.

Haslbeck, M. (2002) sHsps and their role in the chaperone network. *Cell Mol Life Sci* 59:1649–57.

Haslbeck, M., Walke, S., Stromer, T., Ehrnsperger, M., White, H. E., Chen, S., Saibil, H. R., and Buchner, J. (1999) Hsp26: A temperature-regulated chaperone. *EMBO J* 18:6744–51.

Haslbeck, M., Franzman, T., Weinfurtner D, Buchner J. (2005) Some like it hot: The structure and function of small heat-shock proteins. *Nat Struct Mol Biol* 12:842–6.

Hedges, J. C., Dechert, M. A., Yamboliev, I. A., Martin, J. L., Hickey, E., Weber, L. A., and Gerthoffer, W. T. (1999) A role for p38 (MAPK)/HSP27 pathway in smooth muscle cell migration. *J Biol Chem* 274:24211–9.

Hekimi, S., and Guarente, L. (2003) Genetics and the specificity of the aging process. *Science* 299:1351–4.

Helfand, S. L., and Rogina, B. (2003) Molecular genetics of aging in the fly: Is this the end of the beginning. *BioEssays* 25:134–41.

Hercus, M. J., Loeschcke, V., and Rattan, S. I. (2003) Lifespan extension of *Drosophila melanogaster* through hormesis by repeated mild heat stress. *Biogerontology* 4:149–56.

Heydari, A. R., Wu, B., Takahashi, R., Strong, R., and Richardson, A. (1993) Expression of heat shock protein 70 is altered by age and diet at the level of transcription. *Mol Cell Biol* 13:2909–18.

Hirata, K., He, J., Hirakawa, Y., Liu, W., Wang, S., and Kawabuchi, M. (2003) HSP27 is markedly induced in Schwann cell columns and associated regenerating axons. *Glia* 42:1–11.

Hoover, H. E., Thuerauf, D. J., Martindale, J. J., and Glembotski, C. C. (2000) AlphaB-crystallin gene induction and phosphorylation by MKK6-activated p38. A potential role for alphaB-crystallin as a target of the p38 branch of the cardiac stress response. *J Biol Chem* 275:23825–33.

Horwitz, J. (1992) Alpha-crystallin can function as a molecular chaperone. *Proc Natl Acad Sci U S A* 89:10449–53.

Hozenberg, M., Dupont, J., Ducos, B., Leneuve, P., Géloën, A., Even, P. C., Cervera, P., and Le Bouc, Y. (2003) IGF-1 receptor regulates lifespan and resistance to oxidative stress in mice. *Nature* 421:182–7.

Hrelia, S., Fiorentini, D., Maraldi, T., Angeloni, C., Bordoni, A., Biagi, P. L., and Hakim, G. (2002) Doxorubicin induces early lipid peroxidation associated with changes in glucose transport in cultured cardiomyocytes. *Biochim Biophys Acta* 1567:150–6.

Hsu, A. L., Murphy, C. T., and Kenyon, C. (2003) Regulation of aging and age-related disease by DAF-16 and heat-shock factor. *Science* 300:1142–5.

Hsu, H., Huang, J., Shu, H. B., Baichwal, V., and Goeddel, D. V. (1996) TNF-dependent recruitment of the protein kinase RIP to the TNF receptor-1 signaling complex. *Immunity* 4:387–96.

Ichijo, H., Nishida, E., Irie, K., ten Dijke, P., Saitoh, M., Moriguchi, T., Takagi, M., Matsumoto, K., Miyazono, K., and Gotoh, Y. (1997) Induction of apoptosis by ASK1, a mammalian MAPKKK that activates SAPK/JNK and p38 signaling pathways. *Science* 275:90–4.

Irobi, J., Van Impe, K., Seeman, P., Jordanova, A., Dierick, I., Verpoorten, N., Michalik, A., De Vriendt, F., Jacobs, A., VanGerwen, V., Vennekens, K., Mazanecx, R., Tournev, I., Hilton-Jones, D., Talbot, K., Kremensky, I., Van Den Bosch, I., Robberecht, W., Vandekerckhove J., Van Broeckhoven, C., Gettemans, J., De Jonghe, P, and Timmerman, V. (2004) Hot-spot residue in small heat-shock protein 22 causes distal motor neuropathy. *Nat Genet* 36:597–601.

Inagaki, N., Hayashi, T., Arimura, T., Koga, Y., Takahashi, M., Shibata, H., Teraoka, K., Chikamori, T., Yamashina, A., and Kimura, A. (2006) AlphaB-crystallin mutation in dilated cardiomyopathy. *Biochem Biophys Res Commun* 342:379–86.

Ito, H., Okamoto, K., Nakayama, H., Isobe, T., and Kato, K. (1997) Phosphorylation of alphaB-crystallin in response to various types of stress. *J Biol Chem* 272:29934–41.

Ito, H., Iwamoto, I., Inaguma, Y., Takizawa, T., Nagata, K., Asano, T, and Kato, K. (2005) Endoplasmic reticulum stress induces the phosphorylation of small heat shock protein, Hsp27. *J Cell Biochem* 95:932–41.

Jakob, U., Gaestel, M., Engel, K., and Buchner, J. (1993) Small heat shock proteins are molecular chaperones. *J Biol Chem* 268:1517–20.

Jolly, C., and Morimoto R. I. (2000) Role of the heat shock response and molecular chaperones in oncogenesis and cell death. *J Natl Cancer Inst* 92:1564–72.

Jurivich, D. A., Qiu, L., and Welk, J. F. (1997) Attenuated stress responses in young and old human lymphocytes. *Mech Ageing Dev* 94:233–49.

Kamei, A., Iwase, H., and Masuda, K. (1997) Cleavage of amino acid residue(s) from the N-terminal region of alpha A- and alpha B-crystallins in human crystalline lens during aging. *Biochem Biophys Res Commun* 231:373–8.

Kamradt, M. C., Chen, F., and Cryns, V. L. (2001) The small heat shock protein alpha B-crystallin negatively regulates cytochrome c- and caspase-8-dependent activation of caspase-3 by inhibiting its autoproteolytic maturation. *J Biol Chem* 276: 16059–63.

Kamradt, M. C., Chen, F., Sam, S., and Cryns, V. L. (2002) The small heat shock protein alpha B-crystallin negatively regulates apoptosis during myogenic differentiation by inhibiting caspase-3 activation. *J Biol Chem* 277:38731–6.

Kamradt, M. C., Lu, M., Werner, M. E., Kwan, T., Chen, F., Strohecker, A., Oshita, S., Wilkinson, J. C., Yu, C., Oliver, P. G., Duckett, C. S., Buchsbaum, D. J., LoBouglio, A. F., Jordan, V. C., and Cryns, V. L. (2005) The small heat shock protein alpha B-crystallin is a novel inhibitor of TRAIL-induced apoptosis that suppresses the activation of caspase-3. *J Biol Chem* 280:11059–66.

Kappe, G., Leunissen, J. A. M., and de Jong, W. W. (2002) Evolution and diversity of prokaryotic small heat shock proteins. *Prog Mol Subcell Biol* 28:1–17.

Kapphahn, R. J., Ethen, C. M., Peters, E. A., Higgins, L., and Ferrington, D. A. (2003) Modified alphaA-crystallin in the retina: Altered expression and truncation with aging. *Biochemistry* 42:15310–25.

Kato, K., Hasegawa, K., Goto, S., and Inaguma, Y. (1994) Dissociation as a result of phosphorylation of an aggregated form of the small stress protein, hsp27. *J Biol Chem* 269:11274–8.

Kato, K., Ito, H., Kamei, K., Inaguma, Y., Iwamoto, I., and Saga, S. (1998) Phosphorylation of alphaB-crystallin in mitotic cells and identification of enzymatic activities responsible for phosphorylation. *J Biol Chem* 273:28346–54.

Kato, K., Ito, H., and Inaguma, Y. (2002) Expression and phosphorylation of mammalian small heat shock proteins. *Prog Mol Subcell Biol* 28:129–50.

Kato, K., Ito, H., Kamei, K., Iwamoto, I., and Inaguma, Y. (2002) Innervation-dependent phosphorylation and accumulation of alphaB-crystallin and Hsp27 as insoluble complexes in disused muscle. *Faze J* 16:1432–4.

Khazaeli, A. A., Tatar, M., Pletcher, S. D., and Curtsinger, J. W. (1997) Heat-induced longevity extension in *Drosophila*. I. Heat treatment, mortality, and thermotolerance. *J Gerontol A Biol Sci Med Sci* 52:B48–52.

Kijima, K., Numakura, C., Goto, T., Takahashi, T., Otagiri, T., Umetsu, K., and Hayasaka, K. (2005) Small heat shock protein 27 mutation in a Japanese patient with distal hereditary motor neuropathy. *J Hum Genet* 50:473–6.

King, V., and Tower, J. (1999) Aging-specific expression of *Drosophila* Hsp22. *Dev Biol* 207:107–18.

Klemenz, R., Andres, A. C., Frohli, E., Schafer, R., and Aoyama, A. (1993) Expression of the murine small heat shock proteins hsp25 and alpha B crystallin in the absence of stress. *J Cell Biol* 120:639–45.

Konishi, H., Matsuzaki, H., Tanaka, M., Takemura, Y., Kuroda, S., Ono, Y., and Kikkawa, U. (1997) Activation of protein kinase B (Akt/RAC-protein kinase) by cellular stress and its association with heat shock protein Hsp27. *FEBS Lett* 410:493–8.

Koteiche, H. A., Berengian, A. R., and Mchaourab, H. S. (1998) Identification of protein folding patterns using site-directed spin labeling. Structural characterization of beta-sheet and putative substrate binding regions in the conserved domain of alpha A-crystallin. *Biochemistry* 37:12681–8.

Kumar, L. V., and Rao, C. M. (2000) Domain swapping in human alpha A and alpha B crystallins affects oligomerization and enhances chaperone-like activity. *J Biol Chem* 275:22009–13.

Lambert, H., Charette, S. J., Bernier, A. F., Guimond A., and Landry, J. (1999) Hsp27 multimerization mediated by phosphorylation-sensitive intermolecular interactions at the amino terminus. *J Biol Chem* 274:9378–85.

Landry, J., Lambert, H., Zhou, M., Lavoie, J. N., Hickey, E., Weber, L. A., Anderson, C. W. (1992) Human Hsp27 is phosphorylated at serines 78 and 82 by heat shock and mitogen-activated kinases that recognize the same amino acid motif as S6 kinase II. *J Biol Chem* 267:794–803.

Langdon, S. P., Rabiasz, G. J., Hirst, G. L., King, R. J., Hawkins, R. A., Smyth, J. F., and Miller, W. R. (1995) Expression of the heat shock protein HSP27 in human ovarian cancer. *Clin Cancer Res* 1:1603–9.

Lavoie, J. N., Hickey, E., Weber, L. A., and Landry, J. (1993) Modulation of actin microfilament dynamics and fluid phase pinocytosis by phosphorylation of heat shock protein 27. *J Biol Chem* 268:24210–4.

Lavoie, J. N., Lambert, H., Hickey, E., Weber, L. A., and Landry, J. (1995) Modulation of cellular thermoresistance and actin filament stability accompanies phosphorylation-induced changes in the oligomeric structure of heat shock protein 27. *Mol Cell Biol* 15:505–16.

Le Bourg, E., Valenti, P., Lucchetta, P., and Payre, F. (2001) Effects of mild heat shocks at young age on aging and longevity in *Drosophila* melanogaster. *Biogerontology* 2:155–64.

Lee, G. J., Roseman, A. M., Saibil, H. R., and Vierling, E. (1997) A small heat shock protein stably binds heat-denatured model substrates and can maintain a substrate in a folding-competent state. *EMBO J* 16:659–71.

Leroux, M. R., Melki, R., Gordon, B., Batelier, G., and Candido, E. P. (1997) Structure-function studies on small heat shock protein oligomeric assembly and interaction with unfolded polypeptides. *J Biol Chem* 272:24646–56.

Li, D. W., Liu, J. P., Mao, Y. W., Xiang, H., Wang, J., Ma, W. Y., Dong, Z., Pike, H. M., Brown, R. E., and Reed, J. C. (2005) Calcium-activated RAF/MEK/ERK signaling pathway mediates p53-dependent apoptosis and is abrogated by alpha B-crystallin through inhibition of RAS activation. *Mol Biol Cell* 16:4437–53.

Liang, J. N., Bose, S. K., and Chakrabarti, B. (1985) Age-related changes in protein conformation in bovine lens crystallins. *Exp Eye Res* 40:461–9.

Liang, J., and Rossi, M. (1989) Near-ultraviolet circular dichroism of bovine high molecular weight alpha-crystallin. *Invest Ophthalmol Vis Sci* 30:2065–8.

Lin, Y. J., Seroude, L., and Benzer, S. (1998) Extended life-span and stress resistance in the *Drosophila* mutant methuselah. *Science* 282:943–6.

Lindquist, S., and Craig, E. A. (1988) The heat-shock proteins. *Annu Rev Genet* 22:631–77.

Litt, M., Kramer, P., LaMorticella, D. M., Murphey, W., Lovrien, E. W., Weleber, R. G. (1998) Autosomal dominant congenital cataract associated with a missense mutation in the human alpha crystallin gene CRYAA. *Hum Mol Genet* 7:471–4.

Liu, J. P., Schlosser, R., Ma, W. Y., Dong, Z., Feng, H., Li, L., Huang, X. Q., Liu, Y., and Li D. W. (2004) Human alphaA- and alphaB-crystallins prevent UVA-induced apoptosis through regulation of PKCalpha, RAF/MEK/ERK and AKT signaling pathways. *Exp Eye Res* 79:393–403.

Liu, L., Trimarchi, J. R., Navarro, P., Blasco, M. A., and Keefe, D. L. (2003) Oxidative stress contributes to arsenic-induced telomere attrition, chromosome instability and apoptosis. *J Biol Chem* 278:31998–2004.

Liu, Y., Zhang, X., Luo, L., Wu, M., Zeng, R., Cheng, G., Hu, B., Liu, B., Liang, J. J., and Shang, F. (2006) A Novel {alpha}B-Crystallin mutation associated with autosomal dominant congenital lamellar cataract. *Invest Ophthalmol Vis Sci* 47: 1069–75.

Love, S., and King, R. J. (1994) A 27 kDa heat shock protein that has anomalous prognostic powers in early and advanced breast cancer. *Br J Cancer* 69:743–8.

Mackay, D. S., Andley, U. P., and Shiels, A. (2003) Cell death triggered by a novel mutation in the alphaA-crystallin gene underlies autosomal dominant cataract linked to chromosome 21q. *Eur J Hum Genet* 11:784–93.

Maizels, E. T., Peters, C. A., Kline, M., Cutler, R. E., Shanmugam, M., and Hunzicker-Dunn, M. (1998) Heat-shock protein-25/27 phosphorylation by the d isoform of protein kinase C. *Biochem J* 332:703–12.

Mao, Y. W., Liu, J. P., Xiang, H., and Li, D. W. (2004) Human alphaA- and alphaB-crystallins bind to Bax and Bcl-X(S) to sequester their translocation during staurosporine-induced apoptosis. *Cell Death Differ* 11:512–26.

Mao, Y. W., Xiang, H., Wang, W., Korsmeyer, S. J., Reddan, J., and Li, D. W. (2001) Human bcl-2 gene attenuates the ability of rabbit lens epithelial cells against H2O2-induced apoptosis through down-regulation of the alpha-B-crystallin gene. *J Biol Chem* 278:43435–45.

Marshall, C. J. (1995) Specificity of receptor tyrosine kinase signaling: Transient versus sustained extracellular signal-regulated kinase activation. *Cell* 80:179–85.

Mchaourab, H. S., Berengian, A. R., and Koteiche, H. A. (1997) Site-directed spin-labeling study of the structure and subunit interactions along a conserved sequence in the alpha-crystallin domain of heat-shock protein 27. Evidence of a conserved subunit interface. *Biochemistry* 36:14627–34.

Mehlen, P., Briolay, J., Smith, L., Diaz-latoud, C., Fabre, N., Pauli, D., and Arrigo, A. P. (1993) Analysis of the resistance to heat and hydrogen peroxide stresses in COS cells transiently expressing wild type or deletion mutants of the *Drosophila* 27-kDa heat-shock protein. *Eur J Biochem* 215:277–84.

Mehlen, P., Kretz-Remy, C., Preville, X., and Arrigo, A. P. (1996a) Human hsp27, *Drosophila* hsp27 and human alphaB-crystallin expression-mediated increase in glutathione is essential for the protective activity of these proteins against TNFalpha-induced cell death. *EMBO J* 15:2695–706.

Mehlen, P., Preville, X., Chareyron, P., Briolay, J., Klemenz, R., and Arrigo, A. P. (1995) Constitutive expression of human hsp27, *Drosophila* hsp27, or human alpha B-crystallin confers resistance to TNF- and oxidative stress-induced cytotoxicity in stably transfected murine L929 fibroblasts. *J Immunol* 154:363–74.

Mehlen, P., Schulze-Osthoff, and Arrigo, A. P. (1996b) Small stress proteins as novel regulators of apoptosis. Heat shock protein 27 blocks Fas/APO-1 and staurosporine-induced cell death. *J Biol Chem* 271:16510–4.

Merck, K. B., Horwitz, J., Kersten, M., Overkamp, P., Gaestel, M., Bloemendal, H., and de Jong, W. W. (1993) Comparison of the homologous carboxy-terminal domain and tail of alpha-crystallin and small heat shock protein. *Mol Biol Rep* 18:209–15.

Messmer, M., and Chakrabarti, B. (1988) High-molecular-weight protein aggregates of calf and cow lens: Spectroscopic evaluation. *Exp Eye Res* 47:173–83.

Michaud, S., Marin, R, and Tanguay, R. M. (1997) Regulation of heat shock gene induction and expression during *Drosophila* development. *Cell Mol Life Sci* 53:104–13.

Michaud, S., Morrow, G., Marchand, J., and Tanguay R. M. (2002) *Drosophila* small heat shock proteins: Cell and organelle-specific chaperones. *Prog Mol Subcell Biol* 28: 79–101.

Miesbauer, L. R., Zhou, X., Yang, Z., Yang, Z., Sun, Y., Smith, D. L., and Smith, J. B. (1994) Post-translational modifications of water-soluble human lens crystallins from young adults. *J Biol Chem* 269:12494–502.

Minois, N. (2000) Longevity and aging: Beneficial effects of exposure to mild stress. *Biogerontology* 1:15–29.

Minois, N., Khazaeli, A. A., and Curtsinger, J. W. (2001) Locomotor activity as a function of age and lifespan in *Drosophila* melanogaster overexpressing hsp70. *Exp Gerontol* 36:1137–53.

Miron, T., Vancompernolle, K., Vandekerckhove, J., Wilchek, M., and Geiger, B. (1991) A 25-kD inhibitor of actin polymerization is a low molecular mass heat shock protein. *J Cell Biol* 114:255–61.

Morrow, G., and Tanguay, R. M. (2003) Heat shock proteins and aging in *Drosophila* melanogaster. *Sem Cell Dev Biol* 14:291–9.

Morrow, G., Heikkila, J. J., and Tanguay, R. M. (2006) Differences in the chaperone-like activities of the four main small heat shock proteins of *Drosophila* melanogaster. *Cell Stress Chap* 11:51–60.

Morrow, G., Samson, M., Michaud, S., and Tanguay, R. M. (2004) Overexpression of the small mitochondrial Hsp22 extends *Drosophila* life span and increases resistance to oxidative stress. *FASEB J* 18:598–9.

Moseley, P. L. (1998) Heat shock proteins and the inflammatory response. *Ann NY Acad Sci* 856:206–13.

Mosser, D. D., and Morimoto, R. I. (2004) Molecular chaperones and the stress of oncogenesis. *Oncogene* 23:2907–18.

Moyano, J. V., Chen, F., Lu, M., Werner, M. E., Yehiely, F., Diaz, L. K., Turbin, D., Karaca, G. Wiley, E., Nielsen, T. O., Perou, C. M., and Cryns, V. L. (2006) AlphaB-crystallin is a novel oncoprotein that predicts poor clinical outcome in breast cancer. *J Clin Invest* 116:261–70.

Multhoff, G., and Hightower, L. E. (1996) Cell surface expression of heat shock proteins and the immune response. *Cell Stress Chap* 1:167–76.

Muramatsu, T., Hatoko, M., Tada, H., Shirai, T., and Ohnishi, T. (1996) Age-related decrease in the inductability of heat shock protein 72 in normal human skin. *Br J Dermatol* 134:1035–108.

Nagata, S. (1997) Apoptosis by death factor. *Cell* 88:355–65.

Nicholl, I. D., and Quinlan, R. A. (1994) Chaperone activity of alpha-crystallins modulates intermediate filament assembly. *EMBO J* 13:945–53.

Niedzwiecki, A., Kongpachith, A. M., and Fleming, J. E. (1991) Aging affects expression of 70-kDa heat shock proteins in *Drosophila*. *J Biol Chem* 266:9332–8.

Oesterreich, S., Weng, C. N., Qiu, M., Hilsenbeck, S. G., Osborne, C. K., and Fuqua, S. A. (1993) The small heat shock protein hsp27 is correlated with growth and drug resistance in human breast cancer cell lines. *Cancer Res* 53:4443–8.

Pandey, P., Farber, R., Nakazawa, A., Kumar, S., Bharti, A., Nalin, C., Weichselbam, R., Kufe, D., and Kharbanda, S. (2000) Hsp27 functions as a negative regulator of cytochrome c-dependent activation of procaspase-3. *Oncogene* 19:1975–81.

Parcellier, A., Schmitt, E., Gurbuxani, S., Seigneurin-Berny, D., Pance, A., Chantome, A., Plenchette, S., Khochbin, S., Solary, E., and Garrido, C. (2003) HSP27 is an ubiquitin-binding protein involved in I-kappaBalpha proteasomal degradation. *Mol Cell Biol* 23:5790–802.

Parcellier, A., Schmitt, E., Brunet, M., Hammann, A., Solary, E., and Garrido, C. (2005) Small heat shock proteins Hsp27 and aB-crystallin: Cytoprotective and oncogenic functions. *Antioxid Redox Signal* 7:404–13.

Park, J. E., Yang, J. H., Yoon, S. J., Lee, J. H., Yang, E. S., and Park, J. W. (2003a) Lipid peroxidation-mediated cytotoxicity and DNA damage in U937 cells. *Biochimie* 84:1198–204.

Park, K. J., Gaynor, R. B., and Kwak, Y. T. (2003b). Heat shock protein 27 association with the I- kappa-B kinase complex regulates tumor necrosis factor alpha-induced NF-kappa B activation. *J Biol Chem* 278:35272–8.

Paul, C., Manero, F., Gonin, S., Kretz-Remy, C., Virot, S., and Arrigo, A. P. (2002) Hsp27 as a negative regulator of cytochrome C release. *Mol Cell Biol* 22:816–34.

Perng, M. D., Cairns, L., van den, I. J., Prescott, A., Hutcheson, A. M., and Quinlan, R. A. (1999a) Intermediate filament interactions can be altered by HSP27 and alphaB-crystallin. *J Cell Sci* 112:2099–112.

Perng, M. D., Muchowski, P. J., van Den Ijssel, P., Wu, G. J., Hutcheson, A. M., Clark, J. I., and Quinlan, R. A. (1999b) The cardiomyopathy and lens cataract mutation in alphaB-crystallin alters its protein structure, chaperone activity, and interaction with intermediate filaments *in vitro*. *J Biol Chem* 274:33235–43.

Piec, I., Listrat, A., Alliot, J., Chambon, C., Taylor, R. G., and Bechet, D. (2005) Differential proteome analysis of aging in rat skeletal muscle. *FASEB J* 19:1143–5.

Pinder, S. E., Balsitis, M., Ellis, I. O., Landon, M., Mayer, R. J., and Lowe, J. (1994) The expression of alpha B-crystallin in epithelial tumors: A useful tumor marker? *J Pathol* 174:209–25.

Pletcher, S. D., Macdonald, S. J., Marguerie, R., Certa, U., Stearns, S. C., Goldstein, D. B., and Partridge, L. (2002) Genome-wide transcripts profiles in aging and calorically restricted *Drosophila* melanogaster. *Curr Biol* 12:712–23.

Pras, E., Frydman, M., Levy-Nissenbaum, E., Bakhan, T., Raz, J., Assia, E. I., Goldman, B., and Pras, E. (2000) A nonsense mutation (W9X) in CRYAA causes autosomal recessive cataract in an inbred Jewish Persian family. *Invest Ophthalmol Vis Sci* 41:3511–5.

Preville, X., Gaestel, M., and Arrigo, A. P. (1998) Phosphorylation is not essential for protection of L929 cells by Hsp25 against H2O2-mediated disruption actin cytoskeleton, a protection which appears related to the redox change mediated by Hsp25. *Cell Stress Chap* 3:177–87.

Preville, X., Salvemini, F., Giraud, S., Chaufour, S., Paul, C., Stepien, G., Ursini, M. V., and Arrigo, A. P. (1999) Mammalian small stress proteins protect against oxidative stress through their ability to increase glucose-6-phosphate dehydrogenase activity and by maintaining optimal cellular detoxifying machinery. *Exp Cell Res* 247:61–78.

Raha, S., and Robinson, B. H. (2000) Mitochondria, oxygen free radicals, disease and ageing. *Trends Biochem Sci* 25:502–8.

Raha, S., and Robinson, B. H. (2001) Mitochondria, oxygen free radicals, and apoptosis. *Am J Med Genet* 106:62–70.

Rane, M. J., Pan, Y., Singh, S., Powell, D. W., Wu, R., Cummins, T., Chen, Q., McLeish, K. R., and Klein, J. B. (2003) Heat shock protein 27 controls apoptosis by regulating Akt activation. *J Biol Chem* 278:27828–35.

Rattan, S. I. (1998) Repeated mild heat shock delays ageing in cultured human skin fibroblasts. *Biochem Mol Biol Int* 45:753–9.

Rattan, S. I. (2004) Mechanisms of hormesis through mild heat stress on human cells. *Ann NY Acad Sci* 1019:554–8.

Richards, E. H., Hickey, E., Weber, L., and Master, J. R. (1996) Effect of overexpression of the small heat shock protein HSP27 on the heat and drug sensitivities of human testis tumor cells. *Cancer Res* 56:2446–51.

Rogalla, T., Ehrnsperger, M., Preville, X., Kotlyarov, A., Lutsch, G., Ducasse, C., Paul, C., Wieske, M., Arrigo, A. P., Buchner, J., and Gaestel, M. (1999) Regulation of Hsp27 oligomerization, chaperone function, and protective activity against oxidative stress/tumor necrosis factor a by phosphorylation. *J Biol Chem* 274:18947–56.

Rogina, B., and Helfand, S. L. (1995) Regulation of gene expression is linked to life span in adult *Drosophila*. *Genetics* 141:1043–8.

Rogina, B., Reenan, R. A., Nilsen, S. P., Helfand, S. L. (2000) Extended life-span conferred by cotransporter gene mutations in *Drosophila*. *Science* 290:2137–40.

Rousseau, S., Houle, F., Landry, J., and Huot, J. (1997) p38 MAP kinase activation by vascular endothelial growth factor mediates actin reorganization and cell migration in human endothelial cells. *Oncogene* 15:2169–77.

Salvador-Silva, M., Ricard, C. S., Agapova, O. A., Yang, P., and Hernandez, M. R. (2001) Expression of small heat shock proteins and intermediate filaments in the human optic nerve head astrocytes exposed to elevated hydrostatic pressure *in vitro*. *J Neurosci Res* 66:59–73.

Samali, A., Zhivotovsky, B., Jones, D. P., and Orrenius, S. (1999) Presence of a pre-apoptotic complex of pro-caspase-3, Hsp60 and Hsp10 in the mitochondrial fraction of jurkat cells. *EMBO J* 18:2040–8.

Sastre, J., Pallardó, F. V., and Viña, J. (2003) The role of mitochondrial oxidative stress in aging. *Free Rad Biol Med* 35:1–8.

Schneider, G. B., Hamano, H., and Cooper, L. F. (1998) *In vivo* evaluation of hsp27 as an inhibitor of actin polymerization: hsp27 limits actin stress fiber and focal adhesion formation after heat shock. *J Cell Physiol* 177:575–84.

Schultz, C., Dick, E. J., Cox, A. B., Hubbard, G. B., Braak, E., and Braak, H. (2001) Expression of stress proteins alpha B-crystallin, ubiquitin, and hsp27 in pallido-nigral spheroids of aged rhesus monkeys. *Neurobiol Aging* 22:677–82.

Selcen, D., and Engel, A. G. (2003) Myofibrillar myopathy caused by novel dominant negative alphaB-crystallin mutations. *Ann Neurol* 54:804–10.

Seroude, L. (2002) Differential gene expression and aging. *Scientific World Journal* 2:618–31.

Smith, J. B., Liu, Y., and Smith, D. L. (1996) Identification of possible regions of chaperone activity in lens alpha-crystallin. *Exp Eye Res* 63:125–8.

Sohal, R. S. (2002) Role of oxidative stress and protein oxidation in the aging process. *Free Rad Biol Med* 33:37–44.

Sohal, R. S., and Brunk, U. T. (1992) Mitochondrial production of pro-oxidants and cellular senescence. *Mutat Res* 275:295–304.

Song, G., Ouyang, G., and Bao, S. (2005) The activation of Akt/PKB signaling pathway and cell survival. *J Cell Mol Med* 9:59–71.

Srivastava, O. P., and Srivastava, K. (2003) Existence of deamidated alphaB-crystallin fragments in normal and cataractous human lenses. *Mol Vis* 9:110–8.

Stadtman, E. R. (2004) Role of oxidant species in aging. *Curr Med Chem* 11:1105–12.

Stokoe, D., Engel, K., Campbell, D. G., Cohen, P., and Gaestel, M. (1992) Identification of MAPKAP kinase 2 as a major enzyme responsible for the phosphorylation of the small mammalian heat shock proteins. *FEBS Lett* 313:307–13.

Sun, H., Gao, J., Ferrington, D. A., Biesiada, H., Williams, T. D., and Squier, T. C. (1999) Repair of oxidized calmodulin by methionine sulfoxide reductase restores ability to activate the plasma membrane Ca-ATPase. *Biochemistry* 38:105–12.

Sun, Y, and MacRae, T. H. (2005a) Small heat shock proteins: Molecular structure and chaperone function. *Cell Mol Life Sci* 62:2460–76.

Sun, Y, and MacRae, T. H. (2005b) The small heat shock proteins and their role in human disease. *FEBS J* 272:2613–27.

Sun, J., and Tower, J. (1999) FLP recombinase-mediated induction of Cu/Zn-superoxide dismutase transgene expression can extend the lifespan of adult *Drosophila* melanogaster flies. *Mol Cell Biol* 19:216–28.

Sun, X., Fontaine, J. M., Rest, J. S., Shelden, E. A., Welsh, M. J., and Benndorf, R. (2004) Interaction of human HSP22 (HSPB8) with other small heat shock proteins. *J Biol Chem* 279:2394–402.

Takayama, S., Reed, J. C., and Homma, S. (2003) Heat-shock proteins as regulators of apoptosis. *Oncogene* 22:9041–7

Takemoto, L. J. (1996) Oxidation of cysteine residues from alpha-A crystallin during cataractogenesis of the human lens. *Biochem Biophys Res Commun* 223:216–20.

Takemoto, L. J. (1997) Changes in the C-terminal region of alpha-A crystallin during human cataractogenesis. *Int J Biochem Cell Biol* 29:311–5.

Takemoto, L. J. (1998) Quantitation of asparagine-101 deamidation from alpha-A crystallin during aging of the human lens. *Curr Eye Res* 17:247–50.

Takemoto, L., and Boyle, D. (1994) Molecular chaperone properties of the high molecular weight aggregate from aged lens. *Curr Eye Res* 13:35–44.

Takemoto, L., and Gopalakrishnan, S. (1994) Alpha-A crystallin: Quantitation of C-terminal modification during lens aging. *Curr Eye Res* 13:879–83.

Takemoto, L., Emmons, T., and Horwitz, J. (1993) The C-terminal region of alpha-crystallin: Involvement in protection against heat-induced denaturation. *Biochem J* 294:435–8.

Takeuchi, N., Ouchida, A., and Kamei, A. (2004) C-terminal truncation of alpha-crystallin in hereditary cataractous rat lens. *Biol Pharm Bull* 27:308–14.

Tang, B. S., Zhao, G. H., Luo, W., Xia, K., Cai, F., Pan, Q., Zhang, R. X., Zhang, F. F., Liu, X. M., Chen, B., Zhang, C., Shen, L., Jiang, H., Long, Z. G., and Dai, H. P. (2005a) Small heat-shock protein 22 mutated in autosomal dominant Charcot-Marie-Tooth disease type 2L. *Hum Genet* 116:222–4.

Tang, B., Liu, X., Zhao, G., Luo, W., Xia, K., Pan, Q., Cai, F., Hu, Z., Zhang, C., Chen, B., Zhang, F., Shen, L., Zhang, R., and Jiang, H. (2005b) Mutation analysis of the small heat shock protein 27 gene in chinese patients with Charcot Marie-Tooth disease. *Arch Neurol* 62.1201–7.

Tatar, M., Khazaeli, A. A., and Curtsinger, J. W. (1997) Chaperoning extended life. *Nature* 390:30.

Tatar, M., Bartke, A., and Antebi, A. (2003) The endocrine regulation of aging by insulin-like signals. *Science* 299:1346–51.

Taylor, R. P., and Benjamin, I. J. (2005) Small heat shock proteins: A new classification scheme in mammals. *J Mol Cell Cardiol* 38:433–44.

Ueda, Y., Duncan, M. K., and David, L. L. (2002) Lens proteomics: The accumulation of crystallin modifications in the mouse lens with age. *Invest Ophthalmol Vis Sci* 43: 205–15.

van Boekel, M. A., Hoogakker, S. E., Harding, J. J., and de Jong, W. W. (1996) The influence of some post-translational modifications on the chaperone-like activity of alpha-crystallin. *Ophthalmic Res* 28:32–8.

van de Klundert, F. A., Gijsen, M. L., van den IJssel, P. R., Snoeckx, L. H., and de Jong, W. W. (1998) Alpha B-crystallin and hsp25 in neonatal cardiac cells- differences in cellular localization under stress conditions. *Eur J Cell Biol* 75:38–45.

Van Montfort, R., Slingsby, C., and Vierling, E. (2002) Structure and function of the small heat shock protein/α-crystallin family of molecular chaperones. In Horwich, A. (ed.): *Advances in Protein Chemistry. Protein folding in the Cell.* Vol. 59. Academic Press, San Diego, pp. 105–56.

Verschuure, P., Croes, Y., van den IJssel, P. R., Quinlan, R. A., de Jong, W. W., and Boelens, W. C. (2002) Translocation of small heat shock proteins to the actin cytoskeleton upon proteasomal inhibition. *Mol Cell Cardiol* 34:117–28.

Vicart, P., Caron, A., Guicheney, P., Prévost, M. C., Faure, A., Chateau, D., Chapon, F., Tomé, F., Dupret, J. M., Paulin, D., and Fardeau, M. (1998) A missense mutation in the αB-crystallin chaperone gene causes a desmin-related myopathy. *Nat Genet* 20:92–5.

Vieira, C., Pasyukova, E. G., Zeng, Z. B., Hackett, J. B., Lyman, R. F., and Mackay, T. F. (2000) Genotype-environment interaction for quantitative trait loci affecting lifespan in *Drosophila* melanogaster. *Genetics* 154:213–27.

Visala Rao, D., Boyle, G. M., Parsons, P. G., Watson, K., Jones, G. L. (2003) Influence of ageing, heat shock treatment and *in vivo* total antioxidant status on gene-expression profile and protein synthesis in human peripheral lymphocytes. *Mech Ageing Dev* 124:55–69.

Walker, G. A., and Lithgow, G. J. (2003) Lifespan extension in *C. elegans* by a molecular chaperone dependent upon insulin-like signal. *Aging Cell* 2:131–9.

Wang, H. D., Kazemi-Esfarjani, P., and Benzer, S. (2004) Multiple-stress analysis for isolation of *Drosophila* longevity genes. *Proc Natl Acad Sci U S A* 101:12610–5.

Wang, K., Ma, W., and Spector, A. (1995) Phosphorylation of alpha-crystallin in rat lenses is stimulated by H2O2 but phosphorylation has no effect on chaperone activity. *Exp Eye Res* 61:115–24.

Wang, X., Osinska, H., Klevitsky, R., Gerdes, A. M., Nieman, M., Lorenz, J., Hewett, T., and Robbins, J. (2001) Expression of R120G-alphaB-crystallin causes aberrant desmin and alphaB-crystallin aggregation and cardiomyopathy in mice. *Circ Res* 89:84–91.

Wieske, M., Benndorf, R., Behlke, J., Dolling, R., Grelle, G., Bielka, H., and Lutsch, G. (2001) Defined sequence segments of the small heat shock proteins HSP25 and alphaB-crystallin inhibit actin polymerization. *Eur J Biochem* 268:2083–9.

Wiesmann, K. E., Coop, A., Goode, D., Hepburne-Scott, H. W., Crabbe, M. J. (1998) Effect of mutations of murine lens alphaB crystalline on transfected neural cell viability and cellular translocation in response to stress. *FEBS Lett* 438:25–31.

Wisniewski, T., and Goldman, J. E. (1998) Alpha B-crystallin is associated with intermediate filaments in astrocytoma cells. *Neurochem Res* 23:385–92.

Wistow, G. (1993) Lens crystallins: Gene recruitment and evolutionary dynamism. *Trends Biochem Sci* 18:301–6.

Wright, H. T. (1991) Nonenzymatic deamination of asparaginyl and glutaminyl residues in proteins. *Crit Rev Biochem Mol Bio* 26:1–52.

Wyttenbach, A., Sauvageot, O., Carmichael, J., Diaz-Latoud, C., Arrigo, A. P., and Rubinsztein, D. C. (2002) Heat shock protein 27 prevents cellular polyglutamine toxicity and suppresses the increase of reactive oxygen species caused by huntingtin. *Hum Mol Genet* 11:1137–51.

Xanthoudakis, S., and Nicholson D. W. (2000) Heat-shock proteins as death determinants. *Nat Cell Biol* 2:E163–5.

Zhang, K., and Kaufman, R. J. (2004) Signaling the unfolded protein response from the endoplasmic reticulum. *J Biol Chem* 279:25935–8.

Zou, H., Li, Y., Liu, X., and Wang, X. (1999) An APAF-1.cytochrome c multimeric complex is a functional apoptosome that activates procaspase-9. *J Biol Chem* 274:11549–56.

8
Large Mammalian hsp70 Family Proteins, hsp110 and grp170, and Their Roles in Biology and Cancer Therapy

XIANG-YANG WANG,[1,2] DOUGLAS P. EASTON,[3] AND JOHN R. SUBJECK[1]

[1]*Department of Cell Stress Biology, Roswell Park Cancer Institute, Buffalo, New York 14263*
[2]*Urologic Oncology, Roswell Park Cancer Institute, Buffalo, New York 14263*
[3]*Department of Biology, State University of New York College at Buffalo, NY 14222*

1. Introduction

All living organisms respond to conditions such as mild heat shock, oxidative stress, reperfusion injury, or other stressful situations by increasing the expression of specific sets of protective proteins that have been commonly referred to for more than 30 years as heat shock proteins (hsps). Most, if not all, of these proteins are also expressed in the absence of stress. Many of these highly conserved proteins function as molecular chaperones to guide changes in conformational states that are critical to the synthesis, folding, translocation, assembly, and degradation of other proteins (Hartl, 1996). Additionally, they can act to inhibit the irreversible aggregation of denatured proteins caused by protein-damaging stresses and, in some instances, assist in the refolding of denatured proteins (Craig et al., 1993; Gething and Sambrook, 1992). The principal hsps of mammalian cells can be classified into several sequence-related families that are characterized by molecular size, i.e., the hsp25/hsp27 (small heat shock protein), the hsp40 (J-domain proteins), the hsp60, the hsp70, the hsp90, and the hsp110/Sse family. The regulation of hsps is coordinated by heat shock transcription factors (HSFs) that interact with heat shock elements (HSEs) in the promoters of the hsp genes (Morimoto et al., 1997). The hsps are principally found in the cytoplasm, nucleus, and mitochondria.

There is a second set of stress proteins referred to as glucose-regulated proteins (grps), which are localized in the endoplasmic reticulum (ER) and the associated lumen of the nuclear envelope. The principal proteins in this group are grp78 (also referred to as the immunoglobulin binding protein or BiP), grp94/gp96 and grp170. The "grp" nomenclature derives from the fact that these stress proteins were first identified as being induced by glucose deprivation (Pouyssegur et al., 1977; Shiu et al., 1977). Since then additional inducers of grps were identified such as chronic anoxia/hypoxia, calcium ionophores, inhibitors of glycosylation (Lin et al., 1993; Sciandra and Subjeck, 1983; Sciandra et al., 1984). A secondary nomenclature that refers to this group of stress proteins as oxygen regulated protein (orp) has been used in some instances (Ikeda et al., 1997; Kuwabara et al., 1996). We will continue to adhere to the original "grp" nomenclature. While grps are functionally and structurally related to the hsps, they are induced by stresses that disrupt the ER function which generally leads to the accumulation of misfolded proteins in the ER (Kaufman, 1999). Interestingly, this does not include heat shock. Induction of grps has been used extensively as a marker for the unfolded protein response (UPR), an adaptive process conserved from yeast to human cells that acts in response to the disruption of ER homeostasis. The existence of an independent stress response for the ER speaks to the significance of this cellular compartment in the evolution of eukaryotic cells.

While the hsp70 family has been perhaps the most intensely studied group of stress proteins, the much larger and distant relatives of this "super-family" have been largely overlooked. Despite their high relative abundance in cells and frequent appearance in numerous studies over many years, the large stress proteins, called hsp110 and grp170, were almost entirely ignored until the early 1990's. By cloning and sequencing cDNA coding for hsp110 and grp170 of mammals, we determined that these large molecular weight stress proteins were related to the hsp70 family (Chen et al., 1996; Lee-Yoon et al., 1995). A number of additional sequences coding for these large stress proteins were subsequently cloned. Sequence analysis of grp170 and hsp110 homologues from a large and phylogenetically diverse collection of organisms indicates that hsp110 and grp170 are highly diverged and distant relatives of the hsp70 family (Easton et al., 2000). Taken together the hsp70, hsp110, and grp170 proteins comprise a group described as the hsp70 superfamily (Easton et al., 2000). In this article we review aspects of the structure and function of these large stress proteins. The role of the hsp110 and grp170 as molecular chaperones in the biology of cell stress, and prospects for their use in clinical cancer immunotherapy will be discussed.

2. The Large Mammalian hsp70 Family Proteins: Distribution, Induction, and Regulation

2.1. The Mammalian Members of the hsp110 Family

Hsp110 has been characterized best in mammals (mouse, hamster, and humans) (Kaneko et al., 1997a,c; Kojima et al., 1996; Lee-Yoon et al., 1995; Storozhenko et al., 1996; Yagita et al., 1999; Nonoguchi et al., 1999; Yasuda et al., 1995). A given

TABLE 1. Mammalian Members of hsp110/Sse and grp170/Lhs1 Stress Protein Families

Family	Alternative names	Phylogeneic distribution
hsp110/Sse family	hsp105, osp94, irp94	S. pombe, S. cerevisiae, Neurospora, Drosophila, Arabidopsis, Caenorhabditis, rodents, humans
hsp110		
apg-1		
apg-2		
family grp170	orp150	S. pombe, S. cerevisiae, Drosophila, Caenorhabditis, rodents, humans

species has more than one hsp110 family cognate (Table 1) (Easton et al., 2000). These proteins have been found in both the nucleus and cytoplasm. They are co-regulated with the other major hsps by a specific set of stress conditions including hyperthermia, ethanol, oxidative reagents, recovery from anoxia (i.e. reperfusion injury), and inflammation (Black and Subjeck, 1991; Lindquist and Craig, 1988). Hsp110 must provide critical functions independent of hsp70 in eukaryotes, since hsp110 gene knockout in both yeast and *Drosophila* are lethal (Trott et al., 2005).

The hsp110 protein, also referred to as hsp105, has been noted in a vast number of studies using many different cell types (Hightower, 1980; Landry et al., 1982; Levinson et al., 1980; Subjeck et al., 1982a,b; Tomasovic et al., 1983). As one of the most abundant hsps in most mammalian cell lines and tissues, hsp110 is easily detected, although its expression can vary significantly (Lee-Yoon et al., 1995; Subjeck et al., 1982a). The constitutive expression of hsp110 is lowest in heart and skeletal muscle, whereas it is strongly expressed in liver and brain (Lee-Yoon et al., 1995; Yasuda et al., 1995). Specifically, mammalian cerebellum expresses little hsp110, whereas expression in other brain regions is highly abundant (Hylander et al., 2000). Curiously, the cerebellum is specifically sensitive to heat stroke and alcohol-associated toxic effects (Albukrek et al., 1997; Manto, 1996). Fujita and coworkers (Kaneko et al., 1997b,c; Nonoguchi et al., 1999) have cloned hsp110 cognate cDNA and have designated them apg-1 and apg-2 because of their concentration in testis, particularly in germ cells. Apg-1 is developmentally expressed in human testicular germ cells and sperm, suggesting its role in spermatogenesis and fertilization (Kaneko et al., 1997b,c; Nonoguchi et al., 2001).

The molecular mechanism for hsp gene induction in eukaryotes is relatively well understood. The binding of an activated HSF to a heat shock element in the upstream region of hsps initiates transcription of the downstream hsp gene (Morimoto et al., 1997). Analysis of mouse hsp110 genomic sequence indicates that the promoter region contains two consensus HSE sequences, which can be activated by HSFs in response to heat shock and are also sufficient for constitutive expression of the gene (Yasuda et al., 1999). Studies of HSF knockout mice indicate that HSF1 is required for induction of hsp110, as well as hsp70 and hsp25/27, by thermal stress (Zhang et al., 2002). Interestingly, two alternative forms of hsp110, 105α and 105β (43 fewer amino acids than 105α), are observed in mouse FM3A cells (Yasuda et al., 1995). Alternative splicing presumably accounts for this. However, the differential roles (if any) played by these two versions of hsp110 remain unknown. It was recently found that hsp110/105 can be phosphorylated by casein kinase II. Peptide mapping analysis and use of various deletion and substitution mutants revealed that hsp110/105 is phosphorylated at Ser (509) in the β-sheet domain (Ishihara et al., 2003a). Phosphorylated hsp110/105 is especially prominent in the brain compared to other tissues of mice and rats, suggesting that the phosphorylation of hsp110/105 may be physiologically significant (Ishihara et al., 2000).

2.2. Mammalian Members of the grp170 Family

As with hsp110, several grp170 cDNA have been cloned and analyzed (Chen et al., 1996; Easton et al., 2000). Grp170 has also been referred to as orp150 due to its induction by oxygen deprivation (Ikeda et al., 1997; Kuwabara et al., 1996). We will unify terms here and use the original and broadly recognized grp nomenclature (Table 1). Like other ER resident grps (e.g., grp78, grp94), the members of the grp170 family are characterized by C-terminal ER retention sequences and are ER localized in yeasts and mammals. Grp170 is inducible by glucose starvation, anoxia, calcium ionophores, low pH, chronic anoxia/hypoxia, a variety of reducing conditions, and other stresses that disrupt the function of the ER (Lin et al., 1993; Sciandra et al., 1984), indicating that it may participate in quality control of protein folding in the ER. Sequence analysis of grp170 has suggested that it, like hsp110, is a highly diverged relative of the hsp70 family. However, although they have a somewhat greater degree of sequence similarity to the hsp110/Sse family sequences than to members of the hsp70 family, the grp170s of mammals are essentially as diverged from the hsp110 family of proteins as they are from the hsp70s (Chen et al., 1996; Craven et al., 1997; Easton et al., 2000). As an ER-resident protein, grp170 is recognized to be co-regulated with other major grps (i.e., grp78 and grp94) by the UPR (Kozutsumi et al., 1988). These UPR-inducible genes are downstream from mammalian ER stress-response elements (ERSE) that are necessary for ER-stress-induced gene transcription mediated by key transcription factors, including the X-box DNA-binding protein 1 (XBP-1) and ATF6 (Feldman et al., 2005; Nozaki et al., 2004; Yoshida et al., 1998; Yoshida et al., 2001).

3. Cellular Functions of the Large Mammalian hsp70 Family Proteins

3.1. Role of hsp110: Chaperoning and Cytoprotection

Information concerning hsp110 and its intracellular functions is limited and a description of earlier studies can be found in the review by Easton et al (2000). Overexpression of hsp110 alone in Rat-1 cells provided 2-logs greater cell survival at 45°C for 60 minutes than that of control cells, whereas the fully induced heat shock response yields a 3-log greater survival (Oh et al., 1997). Although the precise mechanisms of this protection are not well understood at present, the ability of hsp110 to confer thermal tolerance in vivo is presumably mediated through its protein chaperoning activity. In this context, thermal stress is thought to result in destabilization of protein tertiary or quaternary structure, leading to the exposure of interactive surfaces. Hsp110 as a chaperone is able to bind the exposed surfaces and protect the damaged proteins from aggregation. Using luciferase as a reporter protein, we found that hsp110 can prevent aggregation of denatured proteins in vitro. Yamagishi et al. reported recently that hsp110 suppressed the aggregation of heat-denatured protein in the presence of ADP rather than ATP and suggested that hsp110 functions under severe stress, in which the cellular ATP levels decrease markedly (Yamagishi et al., 2003). Although hsp110 cannot by itself refold luciferase, luciferase bound to hsp110 can refold in the presence of a rabbit reticulocyte lysate (RRL) (Oh et al., 1997). Similar results have been observed in yeast hsp110 (Sse1). In this case, yeast cytosol containing an ATP-generating system and an active hsp70 are required for refolding (Brodsky et al., 1999), suggesting that, although hsp110 holds luciferase in a folding competent state, other chaperones (specifically hsp70) are required for refolding.

Of significant interest is the relationship between hsp110 function and cellular reproduction and programmed cell death. Hsp110 synthesis can be induced by the human papilloma virus oncoprotein, E7, which is a viral transcription factor (Morozov et al., 1995). This induction requires the presence of the E7 conserved region 2, which is essential for the binding of E7 to retinoblastoma family proteins, suggesting that hsp110 induction may be coordinated with the initiation of the cell cycle and involved in viral cell transformation. Another recent study found that hsp105, hsp90, and hsp27 were up-regulated in K562 cells by ectopic expression of the oncogene c-Myc, whose overexpression is associated with the resistance of cell to apoptosis. This observation further strengthens the association of hsp110 with a protective role in cell survival (Ceballos et al., 2005). Apg-2, another hsp110 family member, has also been shown to inhibit apoptosis in tumor cells (Gotoh et al., 2004). More studies are needed to determine whether the effect hsp110 has on apoptosis is physiologically relevant.

The expression of polyglutamine-expanded mutant proteins in Huntington's disease and other neurodegenerative disorders is associated with the formation of intra-neuronal inclusions described as oligomers, protein aggregates, fibrils, or plaques (Kakizuka, 1998). Spinal and bulbar muscular atrophy (SBMA) is a neurodegenerative disorder caused by the expansion of a polyglutamine tract in the androgen receptor (AR). When hsp110 was overexpressed with truncated ARs in COS-7 cells, the aggregation and cell toxicity caused by expansion of the polyglutamine tract were markedly reduced. Thus, the elevated levels of hsp110 may be able to reduce or dampen aggregate formation and cellular degeneration. The enhanced expression of hsp110 in brain may provide an effective therapeutic approach for this neurodegenerative disease (Ishihara et al., 2003b).

In addition to acting as a molecular chaperone in protein folding (and assembly), hsp110 may also function as a RNA chaperone. Recent studies have shown that in addition to their peptide-binding properties, hsp110 and hsp70 preferentially bind AU-rich regions of RNA in vitro, suggesting that hsp110 and hsp70 may have in vivo RNA-chaperoning properties, regulating mRNA degradation and/or translation of lymphokine and other short-lived messages (Henics et al., 1999). This is an important new area of hsp function that has not been well studied.

3.2. Role of grp170: Chaperoning, Cytoprotection, and Polypeptide Transportation

Investigations of the conditions that induce grp170 under physiological stress such as hypoxia and low pH have provided some insight into potential protective functions (Sciandra et al., 1984; Whelan and Hightower, 1985). Cai et al. demonstrated that grps (including grp170) were induced in radiation induced fibrosarcoma (RIF) murine tumors during tumor growth in a manner that correlates with the development of hypoxia and ischemia as tumors increased in size (Cai et al., 1993). These observations suggested a protective role for grp170 in cells under anoxic and ischemic stress based on the survival of cancer cells detected in the visibly necrotic material of the large tumors expressing grp170. Another area where hypoxia and ischemia are of great importance is in the response of brain tissue to stroke and heart muscle in myocardial infarction. Grp170/orp150 has been observed to be induced in cultured rat astrocytes exposed to hypoxia, astrocytes in ischemic mouse brain (Kuwabara et al., 1996), and mouse neurons following cerebral artery occlusion (Matsushita et al., 1998). To study the role of grp170 directly in cytoprotection, Ozawa et al. established human embryonic kidney (HEK) cells stably transfected with grp170 antisense RNA. Although hypoxia-mediated enhancement of grp78 and grp94 was maintained, these cells with decreased grp170 expression displayed reduced viability when subjected to hypoxia (Ozawa et al., 1999). Another study by Tamatani et al. demonstrated that cultured neurons overexpressing grp170 showed suppressed caspase-3-like activity and were resistant to hypoxic stress, whereas astrocytes with inhibited grp170 expression were more vulnerable (Tamatani et al., 2001). Strikingly, mice with targeted neuronal overexpression of grp170 had smaller strokes compared with wild-type controls (Tamatani et al., 2001). Most observations concerning physiological responses and protective phenomenon are believed to reflect the coordinated functions of the entire set of ER chaperones (Easton et al., 2000). However, these specific studies strongly argue for a major role played by grp170 in cytoprotective pathways of ischemic neural tissue.

Like hsp110, the chaperoning activity of grp170 could be a major cytoprotective factor against the various stress conditions which induce its synthesis. The ability of grp170 to bind to and chaperone protein substrates in vitro has recently been characterized (Park et al., 2003). Grp170 is capable of preventing the aggregation of heat-denatured luciferase and maintains it in a folding competent state in vitro. The *S. cerevisiae* grp170 (i.e., Lhs1p) has also been shown to be required for solubilization and refolding of a heat-denatured maker enzyme in vivo, further supporting its role as a protein chaperone (Saris et al., 1997). Molecular chaperones such as hsp70 use ATP binding and hydrolysis to prevent aggregation and the efficient folding of newly translated and stress-denatured polypeptides. Grp170 binds ATP readily, while hsp110 has a very weak ATP binding ability (Chen et al., 1996; Oh et al., 1999). Despite this major difference, neither hsp110 nor grp170 require ATP for binding unfolded protein in vitro and both chaperones require RRL to promote refolding of the bound protein suggesting the necessity of a co-chaperone like Hdj-1 (Oh et al., 1997; Park et al., 2003).

Co-immunoprecipitation studies indicated that grp170 associates with the other major grps in the ER (Lin et al., 1993). Additionally, grp170 was found to interact in vivo with immunoglobulin chains in B-cell hybridomas (Lin et al., 1993; Melnick et al., 1994) or with thyroglobulin, a major protein secreted by thyroid epithelial cells (Kuznetsov et al., 1994), suggesting that grp170 may play a role in folding/assembly of secretory proteins in concert with other ER chaperones. In addition, a possible function for grp170 in polypeptide translocation has also been described. In these studies grp170 was found to be the most efficient ATP-binding protein in a microsomal extract (Dierks et al., 1996). The addition of purified grp78 (putative ATPase for protein import into microsomes) as the only ATP-binding component in the in vitro reconstituted microsomes, did not restore import function to proteoliposomes depleted of ATP-binding proteins. It was, therefore, suggested that grp170 might be the ATPase responsible for efficient import of proteins from the cytosol into the ER. Studies of yeast grp170 (referred to Lhs1; Lumenal hsp seventy) also indicated its involvement in signal recognition particle-independent post-translational translocation of proteins into the ER (Craven et al., 1996; Craven et al., 1997; Tyson and Stirling, 2000). In addition, grp170 was identified as an ER chaperone involved with peptide transport into the ER via the transporter associated with antigen processing (TAP) (Spee et al., 1999), which suggests that grp170 may be involved in the antigen presentation pathway. Most recently, a cytoplasmic form of grp170, which lacks the ER translocation signal peptide, was found to be required for nuclear localization sequence-dependent nuclear protein import (Yu et al., 2002). More studies are required to firmly establish such a putative function for grp170.

Several lines of studies recently described the potential involvement of grp170 in angiogenic response due to its chaperoning property (Miyagi et al., 2001; 2002; Ozawa et al., 2001a,b). Ozawa et al first showed that grp170 co-localized with macrophages in neovasculature (Ozawa et al., 2001a). Inhibition of grp170 expression in cultured human macrophages, caused retention of VEGF antigen within the ER, while overexpression of grp170 promoted the secretion of VEGF into hypoxic culture supernatants. It suggests that grp170 may contribute to revascularization during wound healing. Studies from the same group further demonstrated a link between angiogenesis and grp170 in human glioblastoma (Ozawa et al., 2001b). In these studies, grp170 antisense-transfected C6 glioma cells showed a significant growth delay in nude mice, in association with reduced angiogenesis as indicated by decreased density of platelet/endothelial cell adhesion molecule 1-positive structures within the tumor bed. In agreement with their earlier studies, inhibition of grp170 expression was found to inhibit VEGF release into culture supernatants in vitro. These findings demonstrate a critical role for the grp170 in tumor-mediated angiogenesis via processing of VEGF for secretion.

Emerging evidence also points to a role for grp170 in insulin release (Kobayashi and Ohta, 2005) or insulin resistance in diabetes (Nakatani et al., 2005; Ozawa et al., 2005). It was observed that grp170 overexpression in the liver of obese diabetic mice significantly improved insulin resistance and markedly ameliorated glucose tolerance in diabetic animals. Conversely, expression of antisense grp170 in the liver of normal mice decreased insulin sensitivity (Nakatani et al., 2005). Although the molecular mechanisms remain to be further studied, these observations imply that the grp170 could be a potential therapeutic target for diabetes treatment.

Most recently, Kitao et al. showed that grp170 is highly expressed in both the human brain after a seizure and in the mouse hippocampus after kainate administration (Kitao et al., 2001). A possible role for the ER stress in glutamate toxicity was subsequently examined using grp170 deficient mice. Hippocampal neurons from these animals showed increased vulnerability to glutamate-induced cell death in vitro and decreased survival to kainate in vivo. In contrast, targeted neuronal overexpression of grp170 enhanced neuronal and animal survival in parallel with diminished seizure intensity. Further studies using cultured hippocampal neurons showed that grp170 regulates activation of proteolytic pathways causing cell death as well as the levels of free cytosolic calcium, in agreement with the previous finding that grp170 is a major calcium-binding protein in the ER (Naved et al., 1995). These studies underscore the significant contribution of grp170 to the stress response in neuronal cells.

4. Modeling and Secondary Structure Prediction for hsp110 and grp170

The 70-kDa stress proteins have been well studied and include representatives in eukaryotes and bacteria (DnaK) (Zhu et al., 1996). X-ray diffraction studies of crystallized DnaK have established its three-dimensional structure (Zhu et al., 1996). The sequence similarity between hsp110 and DnaK has allowed us to align the predicted secondary structure of hsp110 to the crystallographic structure of DnaK in order to establish a three-dimensional structural model of hsp110 (Oh et al., 1999). The N-terminal ATP-binding domain (residues 1–394) of hsp110 shows 34% identity in amino acid sequence to the same region of DnaK. The predicted peptide-binding domain (residues 394–509) consists of 7 major β-strands arranged as a β-sandwich and structurally aligns with the corresponding region of the DnaK. The following loop domain (residues 510–608) are composed of a number of negatively charged residues that computer analysis fails to predict as having any obvious secondary structure. Finally, distal to the loop domain the C-terminal residues of hsp110 are predicted to form a series of α-helices (residues 608–858 or domain H). Interestingly, the hsp110 family members exhibit a high degree of sequence homology among themselves in this region (Lee-Yoon et al., 1995). Lastly, all hsp110-like proteins possess a C-terminal domain (DLD, DVD, etc.) that is highly similar to the EEVD motif, which is responsible for hsp70's interaction with proteins containing tetratricopeptide repeat (TPR) domains.

Secondary structure predictions of grp170 suggest that the overall organization of grp170 is similar to that of hsp110 (Park et al., 2003): it contains an N-terminal, 40-kDa globular, ATP-binding domain which has 30% identity with the same region of DnaK. This is followed by a β-sheet domain consisting of approximately seven β-strands, followed by a long acidic loop domain which is unique to grp170 (Chen et al., 1996; Craven et al., 1997). The expansion of grp170 relative to hsp110 occurs mainly in the central loop domain; the sequences of these loop domains are divergent in members of both the hsp110 and grp170 families. The H-domain is predicted to form approximately 5 α-helixes joined by random coils and ends in KNDEL, an ER retention sequence. The C-terminal H domain (residues 700–900) shares 19% identity with the carboxyl terminal 200 amino acids of bovine hsc70 and 18% identity with the corresponding segment of grp78 but exhibits 40% identity with hsp110 (residues 600–800).

It should be noted that hsp110 and grp170, as well as hsp70, have two unique conserved consensus motifs in the helix B and helix C of the H domain. The first motif was designated as "Magic" (LEKERNDAKNAVEECVY) and the second motif as "TedWlyee" (TEDWLYEEGEDQAKQAY) (Easton et al., 2000). Since many of the proteins containing these motifs specifically bind to other proteins, it is tempting to suggest that they may be involved in substrate or co-chaperone recognition and binding. It has been established that the β-sheet domain in DnaK serves as the peptide-binding site, whereas the helix-turn-helix structure forms a lid above the β-sheet domain, regulating entry and/or exit of the substrate (Zhu et al., 1996). The highly expanded version of this lid in both the hsp110 and grp170 family may represent elaboration on the functional properties of the homologous DnaK structure.

5. Studies of Chaperoning Functions of hsp110 and grp170 Using Structural Deletion Mutants

Based on the similarities between the predicted structural elements of hsp110 and the actual structure of DnaK, we constructed a set of deletion mutants which provided the first view of structure-function relations for hsp110 (Oh et al., 1999). Two assays were used to identify the functional domains of hsp110 required for chaperoning activity: (1) the ability of the mutants to prevent luciferase aggregation induced by heat treatment (i.e., holding) and (2) the ability of mutants to maintain the denatured luciferase in a folding competent state for subsequent activity recovering (i.e., refolding). As discussed above, hsp110 (and grp170) can be divided into four principal structural domains. These are the ATP-binding domain (A), the peptide-binding/β-sheet domain (B), the loop domain (L), and the helical domain (H). As shown in Figure 1, mutant BLH containing only the β-sheet peptide-binding domain (B), loop (L) and α-helical domain (H) can prevent the aggregation of heat-denatured luciferase as efficiently as does wild-type hsp110, indicating that the ATP-binding domain is dispensable for this function. However, mutant LH does not prevent luciferase aggregation, which strongly argues that β-sheet domain is the putative substrate-binding site. Mutants AB and ABL are not functional, indicating that in addition to the B domain, more C-terminal domains of hsp110 are required for its holding ability. The deletion of L from hsp110 renders it more comparable in its holding efficiency to hsc70 and makes it structurally more similar to hsc70. Thus, the loop domain (L), although not essential, significantly influences the efficiency with which hsp110 stabilizes luciferase in a folding competent form. Finally, a truncated form of hsp110 containing the first

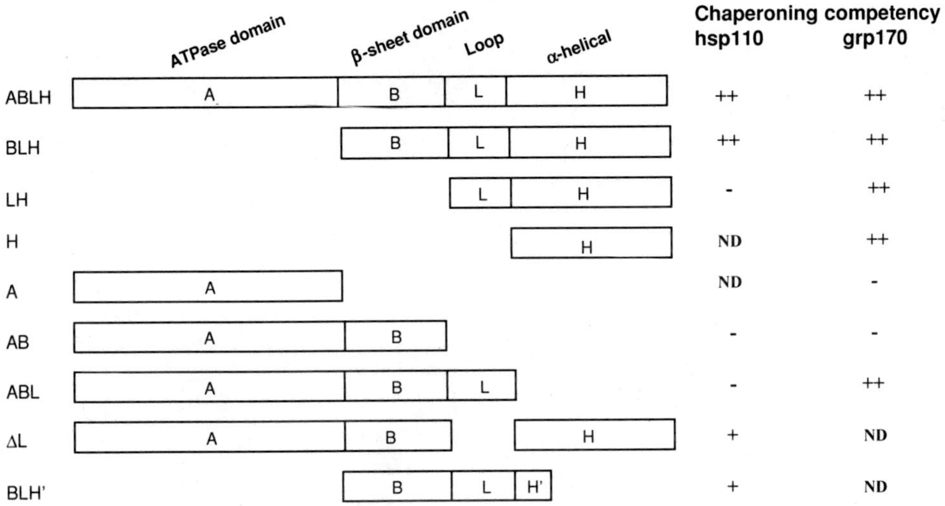

	ATPase domain	β-sheet domain	Loop	α-helical	Chaperoning competency hsp110	grp170
ABLH	A	B	L	H	++	++
BLH		B	L	H	++	++
LH			L	H	–	++
H				H	ND	++
A	A				ND	–
AB	A	B			–	–
ABL	A	B	L		–	++
ΔL	A	B		H	+	ND
BLH'		B	L	H'	+	ND

FIGURE 1. Schematic diagram of hsp110 and grp170 deletion mutants. Models for hsp110 and grp170 were based on the structure of DnaK (modified from Oh et al., 1999; Park et al., 2003). The predicted domains are denoted as the following: A, ATP binding domain; B, β-sheet domain; L, acidic loop; H, α-helices; H', first two α-helices. ND, not determined.

part of the lid (residues 614–655) shows 50% activity in holding heated luciferase compared with the wild-type protein, strongly suggesting that the complete lid domain is needed for complete functionality. The ability of these mutants to refold the heat-denatured luciferase in the presence of RRL directly correlates with their holding ability. Our studies indicate that the ATP-binding domain of hsp110 is not required for its holding functions in vitro, despite its possessing a conserved ATP-binding domain. Thus, the regulatory mechanism for substrate binding by hsp110 may differ from the present models for hsp70, in which the binding and hydrolysis of ATP are thought to regulate access to the peptide-binding site (Zhu et al., 1996). While the wild-type hsp110 does not bind ATP in vitro, the ATP-binding domain mutant (A) does (Oh et al., 1999), which argues strongly that hsp110 may exhibit ATP binding activity in vivo. Thus, it appears that hsp110 binds to substrate in a way similar to that of DnaK, but the details of this chaperone's substrate specificity and regulation are probably different from those of hsp70. This is consistent with the observations that in yeast all Sse1 mutants predicted to abolish ATP hydrolysis (D8N, K69Q, D174N, D203N) complemented the temperature sensitivity of Sse1 deficient cells and lethality of both Sse1- and Sse2-deficient cells (note that budding yeast has 2 hsp110 members called Sse1 and Sse2) (Shaner et al., 2004). However, these investigators found that residues responsible for binding ATP were still required for Sse1 function in vivo, indicating that ATP binding as distinct from ATP hydrolysis plays a crtitical role in Sse1 function.

Efforts have also been made to determine the structural domains of grp170 involved in its chaperoning activity in vitro (Park et al., 2003). Studies using various deletion mutants showed that neither the ATP binding domain (A) nor the mutant AB containing the ATP-binding domain and the β-sheet domain prevented the heat-induced aggregation of luciferase. However, the mutant lacking the ATP-binding domain (BLH) exhibits the full ability to stabilize heat-denatured luciferase. The mutants LH and α-helical domain only mutant H can still function as a chaperone, even though they lack the β-sheet domain that is critical to the chaperoning function of hsp110 (Oh et al., 1999), hsp70 (Freeman et al., 1995), and DnaK (Zhu et al., 1996). In addition, mutant ABL, lacking the H domain, is also capable of holding heat-denatured luciferase. That LH and H mutants possess equivalent chaperoning activity suggests that H is the active element of the LH mutant. Moreover, the ABL mutant is active while AB is not, suggesting that the L region is necessary for the β-sheet (B) region to exhibit activity. This is similar to our earlier studies with hsp110 when some C-terminal sequence to the B domain was required for function (Oh et al., 1999). These studies demonstrate that grp170 has two domains that can bind to and prevent the heat-induced aggregation of luciferase, and suggests that one is the B (-L) region while the second is the H domain. However, the question still remains as to why grp170 in the ER has two protein binding domains while cytosolic hsp110 does not. The specific basis for the chaperoning ability of grp170-H and the precise peptide binding motifs in each of these domains requires a more detailed analysis. The presence of two strong peptide-binding regions in such a large protein may facilitate the interactions and assembly of two substrate proteins. Further study will be required to clarify the role played by grp170 in vivo and to correlate this information with the structural and substrate binding characteristics of this large molecular weight ER resident molecular chaperone.

Comparison of the results from studies of domain deletion mutants led to several interesting findings. Both hsp110 and grp170 contain the predicted β-sheet domains which are homologous to the peptide binding domain of DnaK (Zhu et al., 1996). However, the ABL mutant of hsp110 is inactive in binding luciferase, while the comparable ABL form of grp170 has chaperoning activity. Notably, the L domain of grp170 is approximately twice as large as it is in hsp110. Inclusion of the acidic loop domain in grp170 has significance for substrate binding. Importantly, unlike the C-terminal α-helical region of hsp70 and hsp110, H mutant of grp170 can bind denatured polypeptides and holds them in a folding competent state. It may be due to the presence of heptad repeats in the helical domain which might promote formation of a coiled-coil structure (Park et al., 2003), which has been shown to mediate the interaction between a eukaryotic chaperone protein prefoldin (PFD) and its substrates (Siegert et al., 2000). Lastly and importantly, both hsp110 and grp170 are significantly more efficient than hsp70/hsc70 in stabilizing heat-denatured protein. This observation is not restricted to the reporter protein luciferase (Oh et al., 1997; Park et al., 2003; Wang et al., 2003a). The large size of these two chaperones due to the expansion of the C-terminal half of the protein (L and H domains) may be an important factor in this.

6. Interaction of hsp110 and grp170 with Other Chaperones

Intracellular stress proteins have been shown to form a chaperone network or machinery that interacts with newly translated polypeptides during their biogenesis. However, interplay between these different chaperone proteins remains ill-defined. Recent efforts to examine the native interactions of large stress proteins in situ have indicated that they do not act alone. An earlier study by Ishihara et al. (1999) showed that hsp110 exists in large multi-protein complexes that contain multiple copies of hsp110 and hsc70 in FM3A cells. Our studies similarly showed that anti-hsp110 antisera can pull-down hsc70 in an immunoprecipitation assay as was also described earlier (Oh et al., 1997; Wang et al., 2000). This has been observed in several cell types and murine tissues. Although less obvious, hsp25 interaction with the hsp110-hsc70 complex can also be detected. The fact that these chaperones interact may represent a general phenomenon. Plesofsky-Vig recently showed that the small hsp of *Neurospora crassa*, hsp30, binds to two cellular proteins, hsp70 and hsp88 (Plesofsky-Vig and Brambl, 1998). Cloning and analysis of hsp88 have shown that it represents the hsp110 of *N. crassa*, suggesting that the interactions described above are phylogenetically conserved.

Purification of native hsp110 using ion-exchange and size-exclusion chromatography indicates that this hsp110 complex exists as a 400–700-kDa native composite (Wang et al., 2000). In fact, hsp110's ability to function cooperatively in the folding of luciferase with Hdj-1 (a DnaJ homolog) in the presence of hsc70, indicates that an interaction between hsp110 and Hsc70/Hdj-1 does occur in vitro (Oh et al., 1997). Moreover, this complex forms in vitro in the absence of an added substrate, but substrate (luciferase) can be induced to migrate into the complex following heat shock, arguing for an active chaperoning activity of the complex in vivo. One suggestion is that hsc70 may specifically bind to hsp110, which is the more efficient holding chaperone, in a manner that allows transfer of substrate from hsp110 to hsc70, with subsequent folding in conjunction with DnaJ homologues and other chaperones. It is also possible that hsp110 may itself interact with DnaJ proteins in folding with the addition of a yet to be identified chaperone that acts to open the hsp110 ATP-binding domain as suggested above. Recent studies by Yamagishi et al suggest that hsp110 regulates the substrate binding cycle of hsp70/hsc70 by inhibiting the ATPase activity of hsp70/hsc70, thereby functioning as a negative regulator of the hsp70/hsc70 chaperone system in mammalian cells (Yamagishi et al., 2000, 2004). Interestingly, it was found that hsp110 is phosphorylated by protein kinase CK2 at Ser (509) and that the inhibition of hsp70-mediated refolding by hsp110 was phosphorylation-dependent. Phosphorylated hsp110 had no effect on the refolding of heat-denatured luciferase, whereas a non-phosphorylatable mutant of hsp110 suppressed the hsp70-mediated luciferase refolding (Ishihara et al., 2003a). Regulation of hsp110 by phosphorylation, therefore, may play an important role in the regulation of its interactions with other chaperones.

In yeast, hsp110 Sse1 forms heterodimeric complexes with the abundant cytosolic hsp70s Ssa and Ssb in vivo (Yam et al., 2005). The absence of Sse1 enhances polypeptide binding to both Ssa and Ssb and impairs cell growth, implicating that hsp110 is an important regulator of hsp70-substrate interactions. The ATPase domain of Sse1 was found to be critical for interaction as inactivating point mutations severely reduced ATP interactions with Ssa and Ssb (Shaner et al., 2005). Sse1 stimulated Ssa1 ATPase activity synergistically with the co-chaperone Ydj1, and stimulation required complex formation. These data suggest that the hsp110 chaperone operates in concert with hsp70 in yeast, and that this collaboration is required for cellular hsp70 functions. It appears that hsp110 suppresses ATPase activity of hsp70 in mammalian system, whereas hsp110 (Sse 1) stimulates ATPase activity of hsp70 (Ssa1) in yeast system. Further studies need to be carried out to examine the regulatory effects mediated by hsp110 on the hsp70 chaperoning system. It addition, Sse1 has also been shown to be important for the function of exogenous glucocorticoid receptor and physically associates with the hsp90 in yeast. Deletion of the Sse1 gene rendered cells susceptible to the hsp90 inhibitors, suggesting functional interaction between Sse1 and hsp90 (Liu et al., 1999). However, interaction of hsp110 with hsp90 in mammalian cells was not detectable by a coimmunoprecipitation assay (Wang et al., 2001). Immunoprecipitation as a method for identification of interaction partners by protein gels has significant limitations and a more careful study looking for hsp110 in the glucocorticoid receptor is necessary.

7. Large Stress Proteins in Tumor Immunity and Their Use for Vaccine Design

While studies of biological or biochemical functions of stress protein still remain a mainstream in the field, emerging evidence supporting the roles of stress proteins in host immunity has opened a door to explore different aspects of these ubiquitous and conserved proteins. Search for the immunogenic components within tumor cells that could confer immunity led to the identification of stress proteins as tumor rejection antigens (Srivastava et al., 1986; Udono and Srivastava, 1993; Ullrich et al., 1986). The immunogenicity of these tumor-derived chaperones was subsequently attributed to the individually distinct array of antigenic peptides associated with these chaperone proteins (Udono and Srivastava, 1993). Despite many questions concerning how peptides interact with chaperones in vivo, structural studies (Arnold et al., 1995; Breloer et al., 1998; Grossmann et al., 2004; Ishii et al., 1999; Meng et al., 2002; Nieland et al., 1996) and considerable immunological evidence (Castelli et al., 2001; Noessner et al., 2002; Rivoltini et al., 2003; Srivastava, 2002) suggest that peptides associated with these chaperone proteins contribute significantly to the immunogenicity of stress proteins. Our studies have shown that vaccination of mice with hsp110 or grp170 purified from methylcholanthrene-induced fibrosarcoma resulted in a complete regression of the tumor (Wang et al., 2001). Administration of hsp110 or grp170 derived from Colon 26 tumors can significantly extend the life span of Colon 26 tumor-bearing mice. It is of interest that the vaccine efficacy of hsp110 and hsp70, but not grp170, is enhanced, when these chaperones are prepared from mild, fever-like hyperthermia-treated tumor, suggesting that peptide profile carried by tumor-derived chaperones may be altered upon thermal stress (Wang et al., 2001). Using a more aggressive and less immunogenic B16 melanoma model, we further demonstrated that immunization with grp170 preparations from autologous tumor significantly delayed progression of the primary cancer and reduced established pulmonary metastases (Wang et al., 2003b).

Studies using colon-26 tumor cells stably transfected with autologous hsp110 showed that hsp110 overexpression significantly enhanced the immunogenicity of the tumor in vivo. Immunization of mice with irradiated CT26-hsp110 cells elicited an increased frequency of tumor-specific T cells, which is associated with the growth inhibition of wild-type CT26 tumor in challenge studies (Wang et al., 2002a). Regardless of the immunological mechanisms involved, these findings suggest that the induction or manipulation of certain chaperones may help break host tolerance to tumor antigens that otherwise remain immunologically silent in the progressively growing tumor. Interestingly, Miyazaki et al recently showed that vaccination with hsp110 cDNA leads to growth inhibition of both colorectal CT26 and melanoma B16 tumors through activation of both hsp110-specific CD4 and CD8 T cells (Miyazaki et al., 2005). Significantly, evidence of autoimmune reactions was not present in surviving mice. In agreement with its finding that hsp110 itself may be immunogenic, Minohara et al. demonstrated the immune responses to hsp110 in multiple sclerosis (MS) and experimental autoimmune encephalomyelitis (EAE). These results suggest that hsp110 itself may play a regulatory role in inflammation (Minohara, 2003).

Although the intracellular trafficking and processing pathway involved in the presentation of chaperone-associated antigens is not clear, evidence accumulated during the last decade indicates that chaperone proteins are capable of shuttling antigens into the endogenous antigen presentation pathway of professional APCs. When bone-marrow-derived dendritic cells (DCs) pulsed with hsp110 or grp170 preparations from tumor are injected into animals, they can generate a tumor-specific immune response, suggesting that DCs may mediate the vaccine effects of these two large stress proteins (Wang et al., 2001). Recent studies have documented that specific receptor-mediated endocytosis of chaperone complexes by APCs is critical for the cross-presentation of stress protein-chaperoned peptides (Arnold-Schild et al., 1999; Castellino et al., 2000; Singh-Jasuja et al., 2000b; Wassenberg et al., 1999). Indeed, several receptors, e.g., CD91, LOX1, scavenger receptor class-A (SR-A), and SREC have been identified as, or suggested to be, involved in the cross-priming event (Basu et al., 2001; Berwin et al., 2003; 2004; Binder et al., 2000; Delneste et al., 2002).

In addition to generation of antigen-specific immune responses, APC-chaperone interaction can promote phenotypic and functional maturation of APCs such as DCs or monocytes (Asea et al., 2000; Kuppner et al., 2001; Singh-Jasuja et al., 2000a; Wang et al., 2002b). Low endotoxin hsp110 and grp170 have been shown to induce DCs to up-regulate the expression of MHC class II, CD40 and CD86 molecules, and to secrete pro-inflammatory cytokines IL-6, IL-12, and TNF-α

(Manjili et al., 2005, 2006). Interestingly, hsp110 also induced mouse mammary carcinoma (MMC) to secrete IL-12 and elevate secretion of IL-6 and expression of CD40 molecule. The involvement of the large stress proteins in the induction of innate immunity suggests a role for these molecules as endogenous danger signals.

Hsp110 was recently found to be a major functional component of the lumenal contents of the normal human intestine, where it contributes to CD1d surface regulation in intestinal epithelial cells (IECs). CD1d is expressed on the surface of APCs, including IECs, and is involved in the presentation of glycolipid-based antigens. These data support the idea for the presence of a novel autocrine pathway of CD1d regulation by hsp110 (Colgan et al., 2003). However, it is not clear whether hsp110 plays a role in CD1 presentation of glycolipid antigens.

Hsp70 has been shown to be a cell-surface immune mediator acting as a target of NK cells, resulting in proliferation of and cytotoxic activity of NK cells (Gross et al., 2003; Multhoff, 2002; Multhoff et al., 1999). Although an in vivo antibody depletion assay indicates that NK cells are required for hsp110 and grp170-mediated anti-tumor activities (Wang et al., 2003a;b), direct evidence showing the direct interaction between the large stress proteins and NK cells is still lacking.

The ability of stress proteins to activate both innate and adaptive arm of immune system provides a novel approach for cancer immunotherapy (Tamura et al., 1997). Purification of chaperones from a tumor is believed to co-purify an antigenic peptide "fingerprint" of the cell of origin. Thus, chaperone-peptide complexes derived from patient tumor represent a patient-specific polyvalent vaccine. While potentially powerful, the use of this approach clinically has several limitations. It is very likely that only a small percentage of the chaperone-associated peptides would be capable of eliciting tumor-specific CTLs. A major limitation of this approach is that vaccine preparation is time-consuming and requires a patient specimen (Oki and Younes, 2004; Gordon, 2004 #1043). As a result, some patients are unable to participate in this autologous vaccine trial (Belli et al., 2002). The lack of information on targeted antigens also limits the ability to monitor immune responses generated by this autologous vaccine approach.

It has been shown that tumor-derived hsp110 or grp170 elicited a more potent anti-tumor response on a molar basis than hsp70 (Wang et al., 2001). The enhanced immunogenicity may be attributed to the more efficient chaperoning capability of hsp110 and grp170, because several reports suggest that the affinity with which the chaperone binds antigen significantly contributes to its ability to generate a CTL response (MacAry et al., 2004; Moroi et al., 2000; Tobian et al., 2004). Indeed, molecular chaperone hsp110 or grp170 binds to and stabilizes full-length heat-denatured proteins fourfold more efficiently than hsp70 (Oh et al., 1997; Park et al., 2003; Wang et al., 2003a). Since these large chaperones are able to complex with large proteins at a 1:1 molar ratio, a highly concentrated recombinant vaccine could be prepared when tumor antigen forms a complex with hsp110 or grp170 under heat shock conditions (i.e., conditions under which the antigen would begin to denature and thus form a tight natural complex with hsp110 or grp170). While heat is used as the denaturant of the mature protein in these studies, such chaperone complexes reflect folding and transport intermediates with nascent proteins characteristic of the natural functions of some molecular chaperones (Oh et al., 1997). This recombinant chaperone vaccine approach has several unique and significant advantages: (1) hsp110 and grp170 efficiently bind proteins at 1:1 molar ratio. Therefore, a highly concentrated vaccine would be presented to the immune system. (2) The whole protein antigen employed in this approach contains a large reservoir of potential peptides that allow the individual's own MHC alleles to select the appropriate epitope for presentation. Such a chaperone complex vaccine not only increases the chance of polyepitope-directed T- and B-cell responses, but also circumvents HLA restriction and allows most patients to be eligible for this form of vaccine treatment. (3) Since a tumor specimen is not required for vaccine production, patients with no measurable disease or inaccessible tumor can still be treated using this approach. This makes it an ideal adjuvant therapy for patients with completely resected disease. Furthermore, this approach is theoretically applicable as a preventive therapy for patients at high risk for cancer or for recurrence of cancer. (4) Production of the recombinant chaperone complex is less time-consuming and far less expensive than the production of allogeneic vaccines, autologous vaccines, or tumor-derived chaperone vaccines. And vaccine can be generated in considerable quantities with significant uniformity from batch to batch. (5) Using well-defined antigens as opposed to whole tumor cells or lysate allows better characterization of the resultant immune responses. (6) This synthetic approach of combining antigen with chaperone can be used to develop vaccines against different protein antigen targets, either alone or in combination. That is, hsp110 binds to melanoma associated antigens gp100 and Trp2. Thus, an hsp110-gp100 vaccine and an hsp110-Trp2 vaccine can be prepared separately and combined to create a vaccine that now targets two major protein antigens. The concept can then be extended to additional melanoma antigens such as the MAGE antigens to generate a powerful, multi-targeted melanoma vaccine.

Studies by Manjili et al. (2002) demonstrated that a "natural chaperone complex" between hsp110 and the intracellular domain (ICD) of human epidermal growth factor receptor 2 protein (HER-2)/neu elicited both CD8+ and CD4+ T-cell responses against ICD. The hsp110-ICD complex also significantly enhanced ICD-specific antibody responses relative to that seen with ICD alone. However, no CD8+ T cell or antibody response was detected against hsp110 (Manjili et al., 2002). Subsequent studies showed that the hsp110-ICD complex was capable of breaking tolerance against the rat neu protein and inhibiting spontaneous mammary tumor development in FVB-neu (FVBN202) transgenic mice (Manjili et al., 2003).

In a parallel study, melanoma-associated antigen gp100 complexed with hsp110 was found to inhibit growth of established B16 melanoma tumors. Moreover, it was shown that only chaperone complexes are immunologicaly active in protecting mice against a tumor challenge, whereas mixing antigen and hsp110 is not effective against this B16 model tumor. The significantly higher efficiency of hsp110 in holding/binding tumor antigens in comparison to hsp70 indicates that these large chaperones are the best candidates for use in this vaccine approach. Most strikingly, mouse hsp110 (in mouse) is more effective as an adjuvant than is Complete Freud's Adjuvant (CFA), suggesting potentially significant clinical applications in cancer therapy as well as in treatment of infectious diseases (Wang et al., 2003a).

8. Molecular Chaperoning is Essential for Generation of Antigen-Specific Tumor Immunity by hsp70 Superfamily Members

Based on the studies that characterized the relationship between structure and molecular chaperoning function for grp170 (Park et al., 2003) and the finding that the chaperoning ability of the large stress proteins can be successfully used for cancer vaccine development (Wang et al., 2003a), efforts have been made to identify the functional domains of grp170 that are required for the activity of the chaperone vaccine and to address whether the immunoadjuvant function of grp170 is related to its chaperoning feature (Park et al., 2006).

First, various domain deletion mutants of grp170 were examined for their chaperoning abilities employing the melanoma-associate antigen gp100. Full-length grp170 (described as ABLH as noted above) and mutants BLH, LH, H, and ABL were found to form efficient "chaperone complexes" with gp100 following heat shock and protect it from heat-induced aggregation. However, the AB mutant did not inhibit gp100 aggregation, despite the fact that it contains, based on homology modeling with hsp70, the putative "peptide-binding domain" (B). These results are consistent with the chaperoning studies using luciferase as a reporter protein (Park et al., 2003). Secondly, the immunogenicity of gp100, heat shocked in the presence of grp170 or its deletion mutants were determined in vitro by ELISPOT assay for cytokine production. An antigen-specific immune response was observed in animals immunized with gp100 complexed with full length grp170, as well as all chaperoning functional mutants (i.e., BLH, LH, H, and ABL). However, AB-gp100 immunized mice produced significantly lower IFN-γ than those immunized with grp170 or its functional mutants. Subsequent tumor challenge studies demonstrated that gp100-grp170 complexes with chaperoning functional mutants elicited a potent anti-tumor response, whereas gp100 mixed with AB mutant only exhibited a marginal activity. The ability of grp170 and its mutants to bind to RAW264.7 macrophages was also determined, since the interaction between a stress protein and APCs logically plays a critical role in presentation of the stress protein-chaperoned antigen and initiation of anti-tumor immunity. It was found that grp170 and all of its chaperoning functional mutants bound to RAW264.7 macrophages at 4°C in a concentration dependent and saturable manner. The chaperone null mutant, AB, did not. The binding can be reduced by competition with excess unlabeled grp170, characteristic of a receptor-ligand interaction. These results demonstrate a direct correlation between the abilities of grp170 or its mutants to chaperone gp100, bind to APCs, elicit an antigen specific immune response, and inhibit growth of antigen-expressing tumor. Remarkably, the two active grp170 deletion mutants studied, i.e., ABL and H, differ totally in sequence. Yet each exhibits a similar ability to chaperone gp100, bind to APCs, and elicit a specific immune response. Further studies using ABL and H mutants showed that these two entirely different proteins were able to cross-compete with one another for APC surface binding. In view of the structural similarity between grp170 and other hsp70 superfamily members, i.e., hsp110 and hsp70 (Easton et al., 2000), mutants containing the C-terminal α-helical domain (H) derived from chaperones hsp110 and hsp70 were examined for their ability to bind to APCs. As expected, both full-length hsp110 and hsp70 bind to RAW264.7 cells. However, their H domain mutants show little or no binding ability. Both hsp110 and hsp70 can inhibit heat induced protein aggregation (Oh et al., 1997). However, while grp170-H can bind to and protect heat-denatured protein (e.g., gp100), hsp110-H and hsp70-H cannot (i.e., they do no possess notable molecular-chaperoning ability (Freeman et al., 1995; Park et al., 2003). These data again demonstrate a relationship between ability to act as a chaperone and the ability to bind to macrophages. It is possible that the substrate binding site itself may be able to bind to more than a single substrate protein, thus accommodating both the protein antigen and the receptor simultaneously at the same site. Therefore, these studies suggest that molecular chaperoning is involved in stress protein interactions with APCs, antigen binding, and in eliciting anti-tumor immunity (Table 2). Earlier studies have shown that tumor derived stress proteins including grp94/gp96, hsc70, calreticulin, hsp110, and grp170 can each trigger anti-tumor immunity (Nair et al., 1999; Tamura et al., 1997; Udono and Srivastava, 1994; Wang et al., 2001). That these chaperones have no common sequence elements and the means by which entirely different autologous chaperones mediate antigen presentation has raises important questions as to how APC binding and antigen presentation occurs. The studies outlined here on grp170 suggest that the ancient property of molecular chaperoning appears to be the common denominator underlying all of these observations.

TABLE 2. Ability of grp170, hsp110, hsp70, and their mutants to function as molecular chaperone correlates with competitive binding to antigen-presenting cells (APCs) and vaccine activity

Structural deletion mutants	Ability to chaperone antigen	Binding to macrophages	Vaccine activity
grp170	+	+	+
grp170-BLH	+	+	+
grp170-LH	+	+	+
grp170-H	+	+	+
grp170-ABL	+	+	+
grp170-AB	−	−	−
hsp110	+	+	+
hsp110-H	−	−	ND*
hsp70	+	+	ND
hsp70-H	−	−	ND

*ND: Not determined.

9. Conclusions

Both the grp170 and hsp110 families represent relatively conserved and distinct sets of stress proteins, within a more diverse category that also includes the hsp70s. The large hsp70-like proteins exhibit structural similarities as well as differences from the hsp70s. The structural differences detailed in this review result in functional differences between these large members of the hsp70 superfamily (grp170 and hsp110) and the hsp70s themselves, the most distinctive being the increased ability of these large proteins to bind (hold) denatured polypeptides, perhaps a consequence of the enlarged C-terminal helical domain that is characteristic of these large hsp70 chaperones. In addition to the familiar general role of these large stress proteins as molecular chaperones in assisting protein folding/assembly or transportation, accumulating evidence suggests that hsp110 and grp170 may exhibit other functions in eukaryotes not effectively performed by hsp70s (including grp78). The large chaperones may cooperate with or regulate members of the hsp70 family in their functional activities. Future studies addressing the physiological roles played by these conserved large molecular chaperones in vivo will further enhance our understanding of the importance of these proteins in human development and in disease.

Studies in the last few years have established the roles of hsp110 and grp170 as immunoadjuvants in the host immune response. These large stress proteins, upon interaction with APCs, not only provide "danger signals" but also shuttle chaperoned antigens into antigen-processing and presentation pathways. Based on the highly efficient protein holding property of these large chaperones, a novel vaccine approach using tumor antigen complexed with hsp110 or grp170 by heat shock has been developed. Preclinical studies in animal models demonstrate the therapeutic efficacy of these large chaperone formulated vaccines, indicating the potential translation of this immunotherapeutic approach into the clinic. Studies of structure and immunological function of grp170 suggest that it is not a specific common sequence or specific protein domain that is integral to a stress protein's ability to bind to APCs and elicit an immune response, but only the ability to function as a molecular chaperone. The molecular mechanisms by which the hsp110 and grp170 trigger the activation of a potent anti-tumor immune response is a major area of interest, especially since these two chaperone families are so effective in the formulation of anti-cancer vaccines using large protein antigen substrates. Furthermore, this new technology is not restricted to cancer therapy, but may be equally as applicable to the treatment of infectious diseases such as tuberculosis and even influenza.

NOTE ADDED IN PROOF: Since this review was prepared, the following studies indicating that grp170 and hsp110 can be serve as nucleotide exchange factors for other chaperones have appeared.

1. Dragovic, Z., Broadley, S. A., Shomura, Y., Bracher, A., Hartl, F. U. (Jun 7, 2006) Molecular chaperones of the Hsp110 family act as nucleotide exchange factors of Hsp70s. *EMBO J.* 25(11):2519–28.

2. Raviol, H., Sadlish, H., Rodriguiz, F., Mayer, M. P., Bukau, B. (Jun 7, 2006) Chaperone network in the yeast cytosol: Hsp110 is revealed as an Hsp70 nucleotide exchange factor. *EMBO J.* 25(11):2510–8.

3. Weitzmann, A., Volkmer, J., Zimmermann, R. (Oct 2, 2006) The nucleotide exchange factor activity of Grp170 may explain the non-lethal phenotype of loss of Sil1 function in man and mouse. *FEBS Lett.* 580(22):5237–40.

References

Albukrek, D., Bakon, M., Moran, D. S., Faibel, M., and Epstein, Y. (1997) Heat-stroke-induced cerebellar atrophy: Clinical course, CT and MRI findings. *Neuroradiology* 39:195–7.

Arnold, D., Faath, S., Rammensee, H., and Schild, H. (1995) Cross-priming of minor histocompatibility antigen-specific cytotoxic T cells upon immunization with the heat shock protein gp96. *J Exp Med* 182:885–9.

Arnold-Schild, D., Hanau, D., Spehner, D., Schmid, C., Rammensee, H. G., de la Salle, H., and Schild, H. (1999) Cutting edge: Receptor-mediated endocytosis of heat shock proteins by professional antigen-presenting cells. *J Immunol* 162:3757–60.

Asea, A., Kraeft, S. K., Kurt-Jones, E. A., Stevenson, M. A., Chen, L. B., Finberg, R. W., Koo, G. C., and Calderwood, S. K. (2000) HSP70 stimulates cytokine production through a CD14-dependant pathway, demonstrating its dual role as a chaperone and cytokine. *Nat Med* 6:435–42.

Basu, S., Binder, R. J., Ramalingam, T., and Srivastava, P. K. (2001) CD91 is a common receptor for heat shock proteins gp96, hsp90, hsp70, and calreticulin. *Immunity* 14:303–13.

Belli, F., Testori, A., Rivoltini, L., Maio, M., Andreola, G., Sertoli, M. R., Gallino, G., Piris, A., Cattelan, A., Lazzari, I., et al. (2002) Vaccination of metastatic melanoma patients with autologous tumor-derived heat shock protein gp96-peptide complexes: Clinical and immunologic findings. *J Clin Oncol* 20:4169–80.

Berwin, B., Delneste, Y., Lovingood, R. V., Post, S. R., and Pizzo, S. V. (2004) SREC-1, a type F scavenger receptor, is an endocytic receptor for calreticulin. *J Biol Chem* 279:51250–7.

Berwin, B., Hart, J. P., Rice, S., Gass, C., Pizzo, S. V., Post, S. R., and Nicchitta, C. V. (2003) Scavenger receptor-A mediates gp96/GRP94 and calreticulin internalization by antigen-presenting cells. *EMBO J* 22:6127–36.

Binder, R. J., Han, D. K., and Srivastava, P. K. (2000) CD91: A receptor for heat shock protein gp96. *Nat Immunol* 1:151–5.

Black, A. R., and Subjeck, J. R. (1991) The biology and physiology of the heat shock and glucose-regulated stress protein systems. *Methods Achiev Exp Pathol* 15:126–66.

Breloer, M., Marti, T., Fleischer, B., and von Bonin, A. (1998) Isolation of processed, H-2Kb-binding ovalbumin-derived peptides associated with the stress proteins HSP70 and gp96. *Eur J Immunol* 28:1016–21.

Brodsky, J. L., Werner, E. D., Dubas, M. E., Goeckeler, J. L., Kruse, K. B., and McCracken, A. A. (1999) The requirement for molecular chaperones during endoplasmic reticulum-associated protein degradation demonstrates that protein export and import are mechanistically distinct. *J Biol Chem* 274:3453–60.

Cai, J. W., Henderson, B. W., Shen, J. W., and Subjeck, J. R. (1993) Induction of glucose regulated proteins during growth of a murine tumor. *J Cell Physiol* 154:229–37.

Castelli, C., Ciupitu, A. M., Rini, F., Rivoltini, L., Mazzocchi, A., Kiessling, R., and Parmiani, G. (2001) Human heat shock protein 70 peptide complexes specifically activate antimelanoma T cells. *Cancer Res* 61:222–7.

Castellino, F., Boucher, P. E., Eichelberg, K., Mayhew, M., Rothman, J. E., Houghton, A. N., and Germain, R. N. (2000) Receptor-mediated uptake of antigen/heat shock protein complexes results in major histocompatibility complex class I antigen presentation via two distinct processing pathways. *J Exp Med* 191:1957–64.

Ceballos, E., Munoz-Alonso, M. J., Berwanger, B., Acosta, J. C., Hernandez, R., Krause, M., Hartmann, O., Eilers, M., and Leon, J. (2005) Inhibitory effect of c-Myc on p53-induced apoptosis in leukemia cells. Microarray analysis reveals defective induction of p53 target genes and upregulation of chaperone genes. *Oncogene* 24: 4559–71.

Chen, X., Easton, D., Oh, H. J., Lee-Yoon, D. S., Liu, X., and Subjeck, J. (1996) The 170 kDa glucose regulated stress protein is a large HSP70-, HSP110-like protein of the endoplasmic reticulum. *FEBS Lett* 380:68–72.

Colgan, S. P., Pitman, R. S., Nagaishi, T., Mizoguchi, A., Mizoguchi, E., Mayer, L. F., Shao, L., Sartor, R. B., Subjeck, J. R., and Blumberg, R. S. (2003) Intestinal heat shock protein 110 regulates expression of CD1d on intestinal epithelial cells. *J Clin Invest* 112:745–54.

Craig, E. A., Gambill, B. D., and Nelson, R. J. (1993) Heat shock proteins: Molecular chaperones of protein biogenesis. *Microbiol Rev* 57:402–14.

Craven, R. A., Egerton, M., and Stirling, C. J. (1996) A novel Hsp70 of the yeast ER lumen is required for the efficient translocation of a number of protein precursors. *EMBO J* 15:2640–50.

Craven, R. A., Tyson, J. R., and Colin J., Stirling, C. J. (1997) A novel subfamily of Hsp70s in the endoplasmic reticulum. *Trends Cell Biol* 7:277–82.

Delneste, Y., Magistrelli, G., Gauchat, J., Haeuw, J., Aubry, J., Nakamura, K., Kawakami-Honda, N., Goetsch, L., Sawamura, T., Bonnefoy, J., and Jeannin, P. (2002) Involvement of LOX-1 in dendritic cell-mediated antigen cross-presentation. *Immunity* 17:353–62.

Dierks, T., Volkmer, J., Schlenstedt, G., Jung, C., Sandholzer, U., Zachmann, K., Schlotterhose, P., Neifer, K., Schmidt, B., and Zimmermann, R. (1996) A microsomal ATP-binding protein involved in efficient protein transport into the mammalian endoplasmic reticulum. *EMBO J* 15:6931–42.

Easton, D. P., Kaneko, Y., and Subjeck, J. R. (2000) The hsp110 and Grp170 stress proteins: Newly recognized relatives of the Hsp70s. *Cell Stress Chap* 5:276–90.

Feldman, D. E., Chauhan, V., and Koong, A. C. (2005) The unfolded protein response: A novel component of the hypoxic stress response in tumors. *Mol Cancer Res* 3:597–605.

Freeman, B. C., Myers, M. P., Schumacher, R., and Morimoto, R. I. (1995) Identification of a regulatory motif in Hsp70 that affects ATPase activity, substrate binding and interaction with HDJ-1. *EMBO J* 14:2281–92.

Gething, M. J., and Sambrook, J. (1992) Protein folding in the cell. *Nature* 355:33–45.

Gotoh, K., Nonoguchi, K., Higashitsuji, H., Kaneko, Y., Sakurai, T., Sumitomo, Y., Itoh, K., Subjeck, J. R., and Fujita, J. (2004) Apg-2 has a chaperone-like activity similar to Hsp110 and is overexpressed in hepatocellular carcinomas. *FEBS Lett* 560:19–24.

Gross, C., Hansch, D., Gastpar, R., and Multhoff, G. (2003) Interaction of heat shock protein 70 peptide with NK cells involves the NK receptor CD94. *Biol Chem* 384:267–79.

Grossmann, M. E., Madden, B. J., Gao, F., Pang, Y. P., Carpenter, J. E., McCormick, D., and Young, C. Y. (2004) Proteomics shows Hsp70 does not bind peptide sequences indiscriminately *in vivo*. *Exp Cell Res* 297:108–17.

Hartl, F. U. (1996) Molecular chaperones in cellular protein folding. *Nature* 381:571–9.

Henics, T., Nagy, E., Oh, H. J., Csermely, P., von Gabain, A., and Subjeck, J. R. (1999) Mammalian Hsp70 and Hsp110 proteins bind to RNA motifs involved in mRNA stability. *J Biol Chem* 274:17318–24.

Hightower, L. E. (1980) Cultured animal cells exposed to amino acid analogues or puromycin rapidly synthesize several polypeptides. *J Cell Physiol* 102:407–27.

Hylander, B. L., Chen, X., Graf, P. C., and Subjeck, J. R. (2000) The distribution and localization of hsp110 in brain. *Brain Res* 869:49–55.

Ikeda, J., Kaneda, S., Kuwabara, K., Ogawa, S., Kobayashi, T., Matsumoto, M., Yura, T., and Yanagi, H. (1997) Cloning and expression of cDNA encoding the human 150 kDa oxygen-regulated protein, ORP150. *Biochem Biophys Res Commun* 230:94–9.

Ishihara, K., Yamagishi, N., and Hatayama, T. (2003a) Protein kinase CK2 phosphorylates Hsp105 alpha at Ser509 and modulates its function. *Biochem J* 371:917–25.

Ishihara, K., Yamagishi, N., Saito, Y., Adachi, H., Kobayashi, Y., Sobue, G., Ohtsuka, K., and Hatayama, T. (2003b) Hsp105alpha suppresses the aggregation of truncated androgen receptor with expanded CAG repeats and cell toxicity. *J Biol Chem* 278:25143–50.

Ishihara, K., Yasuda, K., and Hatayama, T. (1999) Molecular cloning, expression and localization of human 105 kDa heat shock protein, hsp105. *Biochim Biophys Acta* 1444:138–42.

Ishihara, K., Yasuda, K., and Hatayama, T. (2000) Phosphorylation of the 105-kDa heat shock proteins, HSP105alpha and HSP105beta, by casein kinase II. *Biochem Biophys Res Commun* 270:927–31.

Ishii, T., Udono, H., Yamano, T., Ohta, H., Uenaka, A., Ono, T., Hizuta, A., Tanaka, N., Srivastava, P. K., and Nakayama, E. (1999) Isolation of MHC class I-restricted tumor antigen peptide and its precursors associated with heat shock proteins hsp70, hsp90, and gp96. *J Immunol* 162:1303–9.

Kakizuka, A. (1998) Protein precipitation: A common etiology in neurodegenerative disorders? *Trends Genet* 14:396–402.

Kaneko, Y., Kimura, T., Kishishita, M., Noda, Y., and Fujita, J. (1997a) Cloning of apg-2 encoding a novel member of heat shock protein 110 family. *Gene* 189:19–24.

Kaneko, Y., Kimura, T., Nishiyama, H., Noda, Y., and Fujita, J. (1997b) Developmentally regulated expression of APG-1, a member of heat shock protein 110 family in murine male germ cells. *Biochem Biophys Res Commun* 233:113–6.

Kaneko, Y., Nishiyama, H., Nonoguchi, K., Higashitsuji, H., Kishishita, M., and Fujita, J. (1997c) A novel hsp110-related gene, apg-1, that is abundantly expressed in the testis responds to a low temperature heat shock rather than the traditional elevated temperatures. *J Biol Chem* 272:2640–5.

Kaufman, R. J. (1999) Stress signaling from the lumen of the endoplasmic reticulum: Coordination of gene transcriptional and translational controls. *Genes Dev* 13:1211–33.

Kitao, Y., Ozawa, K., Miyazaki, M., Tamatani, M., Kobayashi, T., Yanagi, H., Okabe, M., Ikawa, M., Yamashima, T., Stern, D. M., et al. (2001) Expression of the endoplasmic reticulum molecular chaperone (ORP150) rescues hippocampal neurons from glutamate toxicity. *J Clin Invest* 108:1439–50.

Kobayashi, T., and Ohta, Y. (2005) 150-kD oxygen-regulated protein is an essential factor for insulin release. *Pancreas* 30:299–306.

Kojima, R., Randall, J., Brenner, B. M., and Gullans, S. R. (1996) Osmotic stress protein 94 (Osp94). A new member of the Hsp110/SSE gene subfamily. *J Biol Chem* 271:12327–32.

Kozutsumi, Y., Segal, M., Normington, K., Gething, M. J., and Sambrook, J. (1988) The presence of malfolded proteins in the endoplasmic reticulum signals the induction of glucose-regulated proteins. *Nature* 332:462–4.

Kuppner, M. C., Gastpar, R., Gelwer, S., Nossner, E., Ochmann, O., Scharner, A., and Issels, R. D. (2001) The role of heat shock protein (hsp70) in dendritic cell maturation: hsp70 induces the maturation of immature dendritic cells but reduces DC differentiation from monocyte precursors. *Eur J Immunol* 31:1602–9.

Kuwabara, K., Matsumoto, M., Ikeda, J., Hori, O., Ogawa, S., Maeda, Y., Kitagawa, K., Imuta, N., Kinoshita, T., Stern, D. M., et al. (1996) Purification and characterization of a novel stress protein, the 150-kDa oxygen-regulated protein (ORP150), from cultured rat astrocytes and its expression in ischemic mouse brain. *J Biol Chem* 271:5025–32.

Kuznetsov, G., Chen, L. B., and Nigam, S. K. (1994) Several endoplasmic reticulum stress proteins, including ERp72, interact with thyroglobulin during its maturation. *J Biol Chem* 269:22990–5.

Landry, J., Bernier, D., Chretien, P., Nicole, L. M., Tanguay, R. M., and Marceau, N. (1982) Synthesis and degradation of heat shock proteins during development and decay of thermotolerance. *Cancer Res* 42:2457–61.

Lee-Yoon, D., Easton, D., Murawski, M., Burd, R., and Subjeck, J. R. (1995) Identification of a major subfamily of large hsp70-like proteins through the cloning of the mammalian 110-kDa heat shock protein. *J Biol Chem* 270:15725–33.

Levinson, W., Oppermann, H., and Jackson, J. (1980) Transition series metals and sulfhydryl reagents induce the synthesis of four proteins in eukaryotic cells. *Biochim Biophys Acta* 606:170–80.

Lin, H. Y., Masso-Welch, P., Di, Y. P., Cai, J. W., Shen, J. W., and Subjeck, J. R. (1993) The 170-kDa glucose-regulated stress protein is an endoplasmic reticulum protein that binds immunoglobulin. *Mol Biol Cell* 4:1109–19.

Lindquist, S., and Craig, E. A. (1988) The heat-shock proteins. *Annu Rev Genet* 22:631–77.

Liu, X. D., Morano, K. A., and Thiele, D. J. (1999) The yeast Hsp110 family member, Sse1, is an Hsp90 cochaperone. *J Biol Chem* 274:26654–60.

MacAry, P. A., Javid, B., Floto, R. A., Smith, K. G., Oehlmann, W., Singh, M., and Lehner, P. J. (2004) HSP70 peptide binding mutants separate antigen delivery from dendritic cell stimulation. *Immunity* 20:95–106.

Manjili, M. H., Henderson, R., Wang, X. Y., Chen, X., Li, Y., Repasky, E., Kazim, L., and Subjeck, J. R. (2002) Development of a recombinant HSP110-HER-2/neu vaccine using the chaperoning properties of HSP110. *Cancer Res* 62:1737–42.

Manjili, M. H., Park, J., Facciponte, J. G., and Subjeck, J. R. (2005) HSP110 induces "danger signals" upon interaction with antigen presenting cells and mouse mammary carcinoma. *Immunobiology* 210:295–303.

Manjili, M. H., Park, J., Facciponte, J. G., Wang, X.-Y., and Subjeck, J. R. (2006) Immunoadjuvant chaperone, GRP170, induces "danger signals" upon interaction with dendritic cells. *Immmunol Cell Biol* in press.

Manjili, M. H., Wang, X. Y., Chen, X., Martin, T., Repasky, E. A., Henderson, R., and Subjeck, J. R. (2003) HSP110-HER2/neu chaperone complex vaccine induces protective immunity against spontaneous mammary tumors in HER-2/neu transgenic mice. *J Immunol* 171:4054–61.

Manto, M. U. (1996) Isolated cerebellar dysarthria associated with a heat stroke. *Clin Neurol Neurosurg* 98:55–6.

Matsushita, K., Matsuyama, T., Nishimura, H., Takaoka, T., Kuwabara, K., Tsukamoto, Y., Sugita, M., and Ogawa, S. (1998) Marked, sustained expression of a novel 150-kDa

oxygen-regulated stress protein, in severely ischemic mouse neurons. *Brain Res Mol Brain Res* 60:98–106.
Melnick, J., Dul, J. L., and Argon, Y. (1994) Sequential interaction of the chaperones BiP and GRP94 with immunoglobulin chains in the endoplasmic reticulum. *Nature* 370:373–5.
Meng, S. D., Song, J., Rao, Z., Tien, P., and Gao, G. F. (2002) Three-step purification of gp96 from human liver tumor tissues suitable for isolation of gp96-bound peptides. *J Immunol Methods* 264:29–35.
Minohara, M. (2003) Heat shock protein 105 in multiple sclerosis. *Nippon Rinsho* 61:1317–22.
Miyagi, T., Hori, O., Egawa, M., Kato, H., Kitagawa, Y., Konaka, H., Ozawa, K., Koshida, K., Uchibayashi, T., Ogawa, S., and Namiki, M. (2001) Antitumor effect of reduction of 150-kDa oxygen-regulated protein expression in human prostate cancer cells. *Mol Urol* 5:79–80.
Miyagi, T., Hori, O., Koshida, K., Egawa, M., Kato, H., Kitagawa, Y., Ozawa, K., Ogawa, S., and Namiki, M. (2002) Antitumor effect of reduction of 150-kDa oxygen-regulated protein expression on human prostate cancer cells. *Int J Urol* 9:577–85.
Miyazaki, M., Nakatsura, T., Yokomine, K., Senju, S., Monji, M., Hosaka, S., Komori, H., Yoshitake, Y., Motomura, Y., Minohara, M., et al. (2005) DNA vaccination of HSP105 leads to tumor rejection of colorectal cancer and melanoma in mice through activation of both CD4 T cells and CD8 T cells. *Cancer Sci* 96:695–705.
Morimoto, R. I., Kline, M. P., Bimston, D. N., and Cotto, J. J. (1997) The heat-shock response: Regulation and function of heat-shock proteins and molecular chaperones. *Essays Biochem* 32:17–29.
Moroi, Y., Mayhew, M., Trcka, J., Hoe, M. H., Takechi, Y., Hartl, F. U., Rothman, J. E., and Houghton, A. N. (2000) Induction of cellular immunity by immunization with novel hybrid peptides complexed to heat shock protein 70. *Proc Natl Acad Sci U S A* 97:3485–90.
Morozov, A., Subjeck, J., and Raychaudhuri, P. (1995) HPV16 E7 oncoprotein induces expression of a 110 kDa heat shock protein. *FEBS Lett* 371:214–8.
Multhoff, G. (2002) Activation of natural killer cells by heat shock protein 70. *Int J Hyperthermia* 18:576–85.
Multhoff, G., Mizzen, L., Winchester, C. C., Milner, C. M., Wenk, S., Eissner, G., Kampinga, H. H., Laumbacher, B., and Johnson, J. (1999) Heat shock protein 70 (Hsp70) stimulates proliferation and cytolytic activity of natural killer cells. *Exp Hematol* 27:1627–36.
Nair, S., Wearsch, P. A., Mitchell, D. A., Wassenberg, J. J., Gilboa, E., and Nicchitta, C. V. (1999) Calreticulin displays *in vivo* peptide-binding activity and can elicit CTL responses against bound peptides. *J Immunol* 162:6426–32.
Nakatani, Y., Kaneto, H., Kawamori, D., Yoshiuchi, K., Hatazaki, M., Matsuoka, T. A., Ozawa, K., Ogawa, S., Hori, M., Yamasaki, Y., and Matsuhisa, M. (2005) Involvement of endoplasmic reticulum stress in insulin resistance and diabetes. *J Biol Chem* 280:847–51.
Naved, A. F., Ozawa, M., Yu, S., Miyauchi, T., Muramatsu, H., and Muramatsu, T. (1995) CBP-140, a novel endoplasmic reticulum resident Ca(2+)-binding protein with a carboxy-terminal NDEL sequence showed partial homology with 70-kDa heat shock protein (hsp70). *Cell Struct Funct* 20:133–41.
Nieland, T. J., Tan, M. C., Monne-van Muijen, M., Koning, F., Kruisbeek, A. M., and van Bleek, G. M. (1996) Isolation of an immunodominant viral peptide that is endogenously bound to the stress protein GP96/GRP94. *Proc Natl Acad Sci U S A* 93:6135–9.
Noessner, E., Gastpar, R., Milani, V., Brandl, A., Hutzler, P. J., Kuppner, M. C., Roos, M., Kremmer, E., Asea, A., Calderwood, S. K., and Issels, R. D. (2002) Tumor-derived

heat shock protein 70 peptide complexes are cross-presented by human dendritic cells. *J Immunol* 169:5424–32.

Nonoguchi, K., Itoh, K., Xue, J. H., Tokuchi, H., Nishiyama, H., Kaneko, Y., Tatsumi, K., Okuno, H., Tomiwa, K., and Fujita, J. (1999) Cloning of human cDNAs for Apg-1 and Apg-2, members of the Hsp110 family, and chromosomal assignment of their genes. *Gene* 237:21–8.

Nonoguchi, K., Tokuchi, H., Okuno, H., Watanabe, H., Egawa, H., Saito, K., Ogawa, O., and Fujita, J. (2001) Expression of Apg-1, a member of the Hsp110 family, in the human testis and sperm. *Int J Urol* 8:308–14.

Nozaki, J., Kubota, H., Yoshida, H., Naitoh, M., Goji, J., Yoshinaga, T., Mori, K., Koizumi, A., and Nagata, K. (2004) The endoplasmic reticulum stress response is stimulated through the continuous activation of transcription factors ATF6 and XBP1 in Ins2+/Akita pancreatic beta cells. *Gene Cell* 9:261–70.

Oh, H. J., Chen, X., and Subjeck, J. R. (1997) Hsp110 protects heat-denatured proteins and confers cellular thermoresistance. *J Biol Chem* 272:31636–40.

Oh, H. J., Easton, D., Murawski, M., Kaneko, Y., and Subjeck, J. R. (1999) The chaperoning activity of hsp110. Identification of functional domains by use of targeted deletions. *J Biol Chem* 274:15712–8.

Oki, Y., and Younes, A. (2004) Heat shock protein-based cancer vaccines. *Expert Rev Vaccines* 3:403–11.

Ozawa, K., Kondo, T., Hori, O., Kitao, Y., Stern, D. M., Eisenmenger, W., Ogawa, S., and Ohshima, T. (2001a) Expression of the oxygen-regulated protein ORP150 accelerates wound healing by modulating intracellular VEGF transport. *J Clin Invest* 108:41–50.

Ozawa, K., Kuwabara, K., Tamatani, M., Takatsuji, K., Tsukamoto, Y., Kaneda, S., Yanagi, H., Stern, D. M., Eguchi, Y., Tsujimoto, Y., et al. (1999) 150-kDa oxygen-regulated protein (ORP150) suppresses hypoxia-induced apoptotic cell death. *J Biol Chem* 274:6397–404.

Ozawa, K., Miyazaki, M., Matsuhisa, M., Takano, K., Nakatani, Y., Hatazaki, M., Tamatani, T., Yamagata, K., Miyagawa, J., Kitao, Y., et al. (2005) The endoplasmic reticulum chaperone improves insulin resistance in type 2 diabetes. *Diabetes* 54:657–63.

Ozawa, K., Tsukamoto, Y., Hori, O., Kitao, Y., Yanagi, H., Stern, D. M., and Ogawa, S. (2001b) Regulation of tumor angiogenesis by oxygen-regulated protein 150, an inducible endoplasmic reticulum chaperone. *Cancer Res* 61:4206–13.

Park, J., Easton, D. P., Chen, X., MacDonald, I. J., Wang, X. Y., and Subjeck, J. R. (2003) The chaperoning properties of mouse grp170, a member of the third family of hsp70 related proteins. *Biochemistry* 42:14893–902.

Park, J., Facciponte, J. G., Chen, X., MacDonald, I. J., Repasky, E., Manjili, M. H., Wang, X. Y., and Subjeck, J. (2006) Chaperoning function of stress protein grp170, a member of the hsp70 superfamily, is responsible for its immunoadjuvant activity. *Cancer Res.* in press.

Plesofsky-Vig, N., and Brambl, R. (1998) Characterization of an 88-kDa heat shock protein of Neurospora crassa that interacts with Hsp30. *J Biol Chem* 273:11335–41.

Pouyssegur, J., Shiu, R. P., and Pastan, I. (1977) Induction of two transformation-sensitive membrane polypeptides in normal fibroblasts by a block in glycoprotein synthesis or glucose deprivation. *Cell* 11:941–7.

Rivoltini, L., Castelli, C., Carrabba, M., Mazzaferro, V., Pilla, L., Huber, V., Coppa, J., Gallino, G., Scheibenbogen, C., Squarcina, P., et al. (2003) Human tumor-derived heat shock protein 96 mediates *in vitro* activation and *in vivo* expansion of melanoma- and colon carcinoma-specific T cells. *J Immunol* 171:3467–74.

Saris, N., Holkeri, H., Craven, R. A., Stirling, C. J., and Makarow, M. (1997) The Hsp70 homologue Lhs1p is involved in a novel function of the yeast endoplasmic reticulum, refolding and stabilization of heat-denatured protein aggregates. *J Cell Biol* 137: 813–24.

Sciandra, J. J., and Subjeck, J. R. (1983) The effects of glucose on protein synthesis and thermosensitivity in Chinese hamster ovary cells. *J Biol Chem* 258:12091–3.

Sciandra, J. J., Subjeck, J. R., and Hughes, C. S. (1984) Induction of glucose-regulated proteins during anaerobic exposure and of heat-shock proteins after reoxygenation. *Proc Natl Acad Sci U S A* 81:4843–7.

Shaner, L., Trott, A., Goeckeler, J. L., Brodsky, J. L., and Morano, K. A. (2004) The function of the yeast molecular chaperone Sse1 is mechanistically distinct from the closely related hsp70 family. *J Biol Chem* 279:21992–2001.

Shaner, L., Wegele, H., Buchner, J., and Morano, K. A. (2005) The yeast HSP110 SSE1 functionally interacts with the HSP70 chaperones SSA and SSB. *J Biol Chem* 280:41262–9.

Shiu, R. P., Pouyssegur, J., and Pastan, I. (1977) Glucose depletion accounts for the induction of two transformation-sensitive membrane proteinsin *Rous sarcoma* virus-transformed chick embryo fibroblasts. *Proc Natl Acad Sci U S A* 74:3840–4.

Siegert, R., Leroux, M. R., Scheufler, C., Hartl, F. U., and Moarefi, I. (2000) Structure of the molecular chaperone prefoldin: Unique interaction of multiple coiled coil tentacles with unfolded proteins. *Cell* 103:621–32.

Singh-Jasuja, H., Scherer, H. U., Hilf, N., Arnold-Schild, D., Rammensee, H. G., Toes, R. E., and Schild, H. (2000a) The heat shock protein gp96 induces maturation of dendritic cells and down-regulation of its receptor. *Eur J Immunol* 30:2211–5.

Singh-Jasuja, H., Toes, R. E., Spee, P., Munz, C., Hilf, N., Schoenberger, S. P., Ricciardi-Castagnoli, P., Neefjes, J., Rammensee, H. G., Arnold-Schild, D., and Schild, H. (2000b) Cross-presentation of glycoprotein 96-associated antigens on major histocompatibility complex class I molecules requires receptor-mediated endocytosis. *J Exp Med* 191:1965–74.

Spee, P., Subjeck, J., and Neefjes, J. (1999) Identification of novel peptide binding proteins in the endoplasmic reticulum: ERp72, calnexin, and grp170. *Biochemistry* 38:10559–66.

Srivastava, P. (2002) Interaction of heat shock proteins with peptides and antigen presenting cells: Chaperoning of the innate and adaptive immune responses. *Annu Rev Immunol* 20:395–425.

Srivastava, P. K., DeLeo, A. B., and Old, L. J. (1986) Tumor rejection antigens of chemically induced sarcomas of inbred mice. *Proc Natl Acad Sci U S A* 83:3407–11.

Storozhenko, S., De Pauw, P., Kushnir, S., Van Montagu, M., and Inze, D. (1996) Identification of an *Arabidopsis thaliana* cDNA encoding a HSP70-related protein belonging to the HSP110/SSE1 subfamily. *FEBS Lett* 390:113–8.

Subjeck, J. R., Sciandra, J. J., Chao, C. F., and Johnson, R. J. (1982a) Heat shock proteins and biological response to hyperthermia. *Br J Cancer Suppl* 45:127–31.

Subjeck, J. R., Sciandra, J. J., and Johnson, R. J. (1982b) Heat shock proteins and thermotolerance; A comparison of induction kinetics. *Br J Radiol* 55:579–84.

Tamatani, M., Matsuyama, T., Yamaguchi, A., Mitsuda, N., Tsukamoto, Y., Taniguchi, M., Che, Y. H., Ozawa, K., Hori, O., Nishimura, H., et al. (2001) ORP150 protects against hypoxia/ischemia-induced neuronal death. *Nat Med* 7:317–23.

Tamura, Y., Peng, P., Liu, K., Daou, M., and Srivastava, P. K. (1997) Immunotherapy of tumors with autologous tumor-derived heat shock protein preparations. *Science* 278:117–20.

Tobian, A. A., Canaday, D. H., and Harding, C. V. (2004) Bacterial heat shock proteins enhance class II MHC antigen processing and presentation of chaperoned peptides to CD4+ T cells. *J Immunol* 173:5130–7.

Tomasovic, S. P., Steck, P. A., and Heitzman, D. (1983) Heat-stress proteins and thermal resistance in rat mammary tumor cells. *Radiat Res* 95:399–413.

Trott, A., Shaner, L., and Morano, K. A. (2005) The molecular chaperone Sse1 and the growth control protein kinase Sch9 collaborate to regulate protein kinase A activity in *Saccharomyces cerevisiae*. *Genetics* 170:1009–21.

Tyson, J. R., and Stirling, C. J. (2000) LHS1 and SIL1 provide a lumenal function that is essential for protein translocation into the endoplasmic reticulum. *EMBO J* 19:6440–52.

Udono, H., and Srivastava, P. K. (1993) Heat shock protein 70-associated peptides elicit specific cancer immunity. *J Exp Med* 178:1391–6.

Udono, H., and Srivastava, P. K. (1994) Comparison of tumor-specific immunogenicities of stress-induced proteins gp96, hsp90, and hsp70. *J Immunol* 152:5398–403.

Ullrich, S. J., Robinson, E. A., Law, L. W., Willingham, M., and Appella, E. (1986) A mouse tumor-specific transplantation antigen is a heat shock-related protein. *Proc Natl Acad Sci U S A* 83:3121–5.

Wang, X. Y., Chen, X., Manjili, M. H., Repasky, E., Henderson, R., and Subjeck, J. R. (2003a) Targeted immunotherapy using reconstituted chaperone complexes of heat shock protein 110 and melanoma-associated antigen gp100. *Cancer Res* 63:2553–60.

Wang, X. Y., Chen, X., Oh, H. J., Repasky, E., Kazim, L., and Subjeck, J. (2000) Characterization of native interaction of hsp110 with hsp25 and hsc70. *FEBS Lett* 465:98–102.

Wang, X. Y., Kazim, L., Repasky, E. A., and Subjeck, J. R. (2001) Characterization of heat shock protein 110 and glucose-regulated protein 170 as cancer vaccines and the effect of fever-range hyperthermia on vaccine activity. *J Immunol* 166:490–7.

Wang, X. Y., Kazim, L., Repasky, E. A., and Subjeck, J. R. (2003b) Immunization with tumor-derived ER chaperone grp170 elicits tumor-specific CD8+ T-cell responses and reduces pulmonary metastatic disease. *Int J Cancer* 105:226–31.

Wang, X. Y., Li, Y., Manjili, M. H., Repasky, E. A., Pardoll, D. M., and Subjeck, J. R. (2002a) Hsp110 over-expression increases the immunogenicity of the murine CT26 colon tumor. *Cancer Immunol Immunother* 51:311–9.

Wang, Y., Kelly, C. G., Singh, M., McGowan, E. G., Carrara, A. S., Bergmeier, L. A., and Lehner, T. (2002b) Stimulation of Th1-polarizing cytokines, C–C chemokines, maturation of dendritic cells, and adjuvant function by the peptide binding fragment of heat shock protein 70. *J Immunol* 169:2422–9.

Wassenberg, J. J., Dezfulian, C., and Nicchitta, C. V. (1999) Receptor mediated and fluid phase pathways for internalization of the ER Hsp90 chaperone GRP94 in murine macrophages. *J Cell Sci* 112:2167–75.

Whelan, S. A., and Hightower, L. E. (1985) Differential induction of glucose-regulated and heat shock proteins: Effects of pH and sulfhydryl-reducing agents on chicken embryo cells. *J Cell Physiol* 125:251–8.

Yagita, Y., Kitagawa, K., Taguchi, A., Ohtsuki, T., Kuwabara, K., Mabuchi, T., Matsumoto, M., Yanagihara, T., and Hori, M. (1999) Molecular cloning of a novel member of the HSP110 family of genes, ischemia-responsive protein 94 kDa (irp94), expressed in rat brain after transient forebrain ischemia. *J Neurochem* 72:1544–51.

Yam, A. Y., Albanese, V., Lin, H. T., and Frydman, J. (2005) HSP110 cooperates with different cytosolic HSP70 systems in a pathway for de novo folding. *J Biol Chem* 280:41252–61.

Yamagishi, N., Ishihara, K., and Hatayama, T. (2004) Hsp105alpha suppresses Hsc70 chaperone activity by inhibiting Hsc70 ATPase activity. *J Biol Chem* 279:41727–33.

Yamagishi, N., Ishihara, K., Saito, Y., and Hatayama, T. (2003) Hsp105 but not Hsp70 family proteins suppress the aggregation of heat-denatured protein in the presence of ADP. *FEBS Lett* 555:390–6.

Yamagishi, N., Nishihori, H., Ishihara, K., Ohtsuka, K., and Hatayama, T. (2000) Modulation of the chaperone activities of Hsc70/Hsp40 by Hsp105alpha and Hsp105beta. *Biochem Biophys Res Commun* 272:850–5.

Yasuda, K., Ishihara, K., Nakashima, K., and Hatayama, T. (1999) Genomic cloning and promoter analysis of the mouse 105-kDa heat shock protein (HSP105) gene. *Biochem Biophys Res Commun* 256:75–80.

Yasuda, K., Nakai, A., Hatayama, T., and Nagata, K. (1995) Cloning and expression of murine high molecular mass heat shock proteins, HSP105. *J Biol Chem* 270:29718–23.

Yoshida, H., Haze, K., Yanagi, H., Yura, T., and Mori, K. (1998) Identification of the cis-acting endoplasmic reticulum stress response element responsible for transcriptional induction of mammalian glucose-regulated proteins. Involvement of basic leucine zipper transcription factors. *J Biol Chem* 273:33741–9.

Yoshida, H., Okada, T., Haze, K., Yanagi, H., Yura, T., Negishi, M., and Mori, K. (2001) Endoplasmic reticulum stress-induced formation of transcription factor complex ERSF including NF-Y (CBF) and activating transcription factors 6alpha and 6beta that activates the mammalian unfolded protein response. *Mol Cell Biol* 21:1239–48.

Yu, L. G., Andrews, N., Weldon, M., Gerasimenko, O. V., Campbell, B. J., Singh, R., Grierson, I., Petersen, O. H., and Rhodes, J. M. (2002) An N-terminal truncated form of Orp150 is a cytoplasmic ligand for the anti-proliferative mushroom *Agaricus bisporus* lectin and is required for nuclear localization sequence-dependent nuclear protein import. *J Biol Chem* 277:24538–45.

Zhang, Y., Huang, L., Zhang, J., Moskophidis, D., and Mivechi, N. F. (2002) Targeted disruption of hsf1 leads to lack of thermotolerance and defines tissue-specific regulation for stress-inducible Hsp molecular chaperones. *J Cell Biochem* 86:376–93.

Zhu, X., Zhao, X., Burkholder, W. F., Gragerov, A., Ogata, C. M., Gottesman, M. E., and Hendrickson, W. A. (1996) Structural analysis of substrate binding by the molecular chaperone DnaK. *Science* 272:1606–14.

IV
Molecular Chaperones and Protein Folding

Color Plate I

Stress Conditioning Protects
Rodent Kidneys Against 48
Hours of Cold Storage

Sham-Heat Shock   Heat Shock

Fifteen minutes following reperfusion, the sham-HS kidney (*right organ in left photo) is dark and cyanotic compared to the HS kidney (*left organ in right photo), which is pink and well perfused. Reproduced with permission (Perdrizet, 1997). [*See* page 22]

Color Plate II

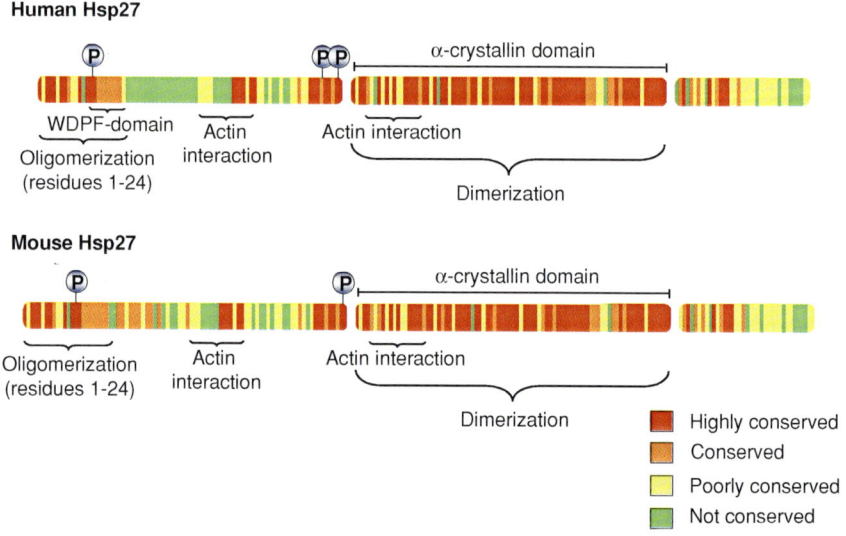

Schematic representation of the human and mouse Hsp27 protein. Phosphorylatable residues are indicated by the letter P above the amino acid sequence. [*See* page 146]

9
Regulation of Hsp70 Function: Hsp40 Co-Chaperones and Nucleotide Exchange Factors

ROBERT T. YOUKER[1,2] AND JEFFREY L. BRODSKY[1]
[1]Department of Biological Sciences, University of Pittsburgh, Pittsburgh, PA;
[2]The Vollum Institute, Oregon Health & Science University, Portland, OR

1. Introduction

Anfinsen discovered over 30 years ago that the information necessary to dictate the tertiary structure of a protein was contained in the primary amino acid sequence (Anfinsen, 1973). However, Afinsen's experiments were performed in vitro with dilute solutions of a small globular protein (ribonuclease A), and these conditions are distinct from the highly crowded environment inside a cell where protein conformations vary and protein concentrations may be as high as 300 mg/mL (Zimmerman and Trach, 1991). In fact, most newly synthesized proteins inside a cell would fail to fold efficiently without the assistance of additional "machines." Important components of these machines are molecular chaperones, and three of the most abundant classes of molecular chaperones are the Hsp70s, Hsp40s, and Hsp90s. In part, the Hsp70 and Hsp40 chaperones prevent protein aggregation and catalyze polypeptide folding because they bind to hydrophobic patches on unfolded or misfolded proteins. Hsp70 function can be regulated by specific Hsp40 partners and by nucleotide exchange factors (NEFs). Hsp90 chaperones are regulated by a distinct group of proteins, and although they also associate with polypeptides, Hsp90s do not bind preferentially to exposed hydrophobic amino acid patches (Joachimiak, 1997).

In this chapter, the function and regulation of the Hsp70 class of molecular chaperones will be discussed. Specifically, we will describe the general features of Hsp70s and the Hsp40 and NEFs that impact the Hsp70 catalytic cycle. Next, data on the interaction between Hsp70 and some of these co-chaperones with a mediator of protein degradation in the cell will be presented. And finally, as a paradigm for Hsp70-Hsp40-NEF function, current knowledge on how these chaperones and co-chaperones catalyze the ER-associated protein degradation (ERAD) of aberrant proteins in the early secretory pathway will be discussed.

TABLE 1. Select Hsp70/Hsc70 Molecular Chaperones and Their Co-Chaperones.

Yeast	Mammals	Class	Location	Function
BiP	BiP	Hsp70	ER lumen	Translocation/retro-translocation Protein folding
Ssa1p	Hsc/Hsp70	Hsp70	Cytosol	Protein folding/degradation
Ydj1p	Hdj2	Hsp40/J-protein	ER membrane/cytosol	Protein folding/degradation
Hlj1p	—	Hsp40/J-protein	ER membrane/cytosol	Protein folding/degradation
—	CSP1	Hsp40/J-protein	Cytosol	Protein folding/Exocytosis
Fes1p	HspBP1	Nucleotide exchange factor (NEF)	Cytosol	Stimulates ADP release from Hsp70
Snl1p	BAG-1	Nucleotide exchange factor (NEF)	ER membrane/cytosol	Stimulates ADP release from Hsp70
—	CHIP	E3 Ligase	cytosol	Protein degradation
Scj1p	ERdj3	Hsp40/J-protein	ER lumen	Protein folding/degradation
Jem1p	—	Hsp40/J-protein	ER lumen	Protein folding/degradation
Sec63p	Sec63	Hsp40/J-protein	ER lumen	Translocation
Sls1p/Sil1p	BAP	Nucleotide exchange factor (NEF)	ER lumen	ER quality control?

1.1. Hsp70

The Hsp70 class of molecular chaperones reside in nearly every compartment of the cell and participate in a wide variety of processes, including the folding of newly synthesized proteins, the prevention of protein aggregation, the refolding of misfolded proteins, the translocation of proteins across organellar membranes, and the association and dissociation of protein complexes (see Table 1).

Hsp70s are composed of a ~45-kDa NH_2-terminal ATPase domain, a ~15-kDa peptide binding domain, and a ~10-kDa COOH-terminal lid domain. Like many other chaperone classes, there is a stress-inducible form of Hsp70 and a constitutively expressed form, known as Hsc70. Their functions are largely interchangeable and the only differences between Hsp70 and Hsc70 seem to be that their levels of expression are differentially regulated and that Hsp70 resides primarily in the nucleus (Pelham, 1984). Hsp70 is expressed at low levels under normal physiological conditions but its expression is induced to high levels during times of stress. Conversely, Hsc70 expression is not stress inducible and its levels are constant. Both proteins bind to short hydrophobic stretches of amino acids normally buried in the native conformations of substrate proteins, and Hsp70 homologues catalyze their diverse functions through cycles of substrate binding and release, which in turn is regulated by ATP hydrolysis (Mayer and Bukau, 2005). In the ATP-bound state, Hsp70 binds with low affinity to its substrates due to a fast peptide off-rate. In the ADP-bound state, Hsp70 binds with high affinity to its substrates due to a

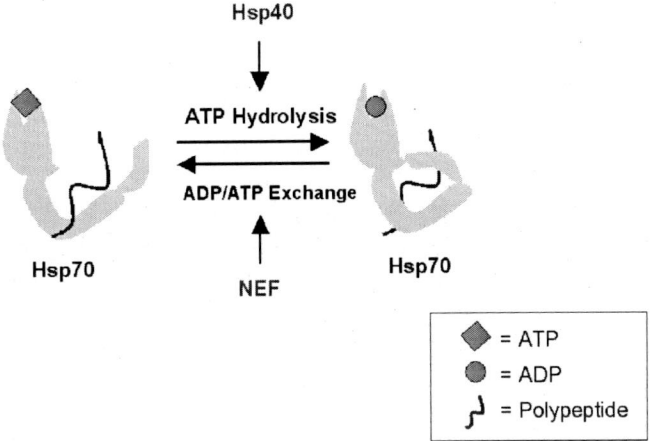

FIGURE 1. Hsp70 ATPase cycle. In the ATP-bound state, Hsp70 binds peptide weakly and the COOH-terminal lid domain is open. Upon stimulation by Hsp40, ATP is hydrolyzed and the COOH-terminal lid closes, locking the peptide onto Hsp70 to favor tight binding. Nucleotide exchange factors (NEFs) release ADP to allow ATP to bind to Hsp70. Upon ATP binding the COOH-terminal lid on Hsp70 opens and the peptide is released.

slow peptide off-rate. The ATPase activity of Hsp70 is inherently weak (3×10^{-4} to 1.6×10^{-2} s^{-1}) (Zylicz et al., 1983; McCarty et al., 1995; Bukau, 1999), and this rate is too slow to promote substrate binding and drive productive folding. However, substrate binding is catalyzed by Hsp40 co-chaperones, which stimulate Hsp70 ATP hydrolysis (see Section 1.2.1 and Figure 1). The first Hsp40 identified was the DnaJ protein in bacteria, and addition of DnaJ to a bacterial Hsp70, DnaK, enhances DnaK's ATPase activity (Liberek et al., 1991). In addition, whereas all Hsp70 family members require ATP hydrolysis for chaperone activity, there are clear differences in the rates of ATP hydrolysis and ADP dissociation rates amongst Hsp70/Hsc70 homologues. These differences are at least partly explained by subtle alterations in an exposed loop and the absence or presence of salt bridges in the clefts of their ATPase domains (Mayer and Bukau, 2005).

The X-ray crystal structures of several Hsp70 ATPase domains have been solved, including bovine Hsp70 (Flaherty et al., 1990), and the structures indicate that the ATPase domain is composed of two globular subdomains with a deep cleft formed at the interface. A nucleotide is bound at the bottom of the cleft—in complex with one Mg^{2+} and two K^+ ions—by four loops: Two loops bind the β phosphate and two loops bind the γ phosphate. The adenosine ring sits in a hydrophobic pocket. A crystal structure of the peptide binding domain of a bacterial Hsp70 (DnaK) complexed with peptide was also solved (Zhu et al., 1996). This structure revealed that the peptide-binding domain is composed of β-sheets, which form a cleft that can accommodate an ∼7 amino acid peptide. The COOH-terminal domain is α-helical and forms a lid over the peptide-binding domain.

Due to the high concentrations of cytoplasmic ATP, ADP release becomes rate-limiting in the bacterial Hsp70 (DnaK) hydrolytic cycle. In effect, the cleft in the ATPase domain of DnaK must open in order to release ADP and permit re-binding of another ATP. This event is catalyzed by Hsp70 co-chaperones, known as nucleotide exchange factors (NEFs, also see Section 1.2.2). The first NEF for Hsp70 identified was the bacterial protein GrpE, which as predicted opens the nucleotide cleft to facilitate ADP release (Harrison et al., 1997). It is important to note that the NEFs do not actively "choose" to bind ATP after ADP is released, but ATP is preferentially bound to the opened site because the cellular concentration of ATP is much higher than ADP. Although the addition of a NEF modestly enhances the overall rate of Hsp70-mediated ATP hydrolysis, GrpE and DnaJ working together can stimulate the ATPase activity of DnaK by up to 5,000-fold (Karzai and McMacken, 1996).

The inability to identify a GrpE homologue and the significantly high spontaneous rate of ADP release from mammalian Hsp70 led to the assumption that the eukaryotic cytosol lacked NEFs. However, several groups simultaneously identified the Bcl-2 athanogene 1 (BAG-1) as a NEF for mammalian Hsc70, and it was shown that purified BAG-1 accelerates Hsp70 ATPase activity by enhancing ADP release (Hohfeld and Jentsch, 1997; Packham et al., 1997; Takayama et al., 1997; Zeiner et al., 1997; Takayama et al., 1998). At about the same time, a negative regulator of Hsc70 was also discovered and called Hsc70 interacting protein (HIP). In contrast to the NEFs, HIP stabilizes the ADP bound form of Hsc70—thus slowing the ATPase cycle—and plays an important role during the cooperation of Hsc70 with other chaperone systems, such as the Hsp90 chaperone machine (Ziegelhoffer et al., 1996; Frydman and Hohfeld, 1997).

1.2. Regulators of Hsp70 Function

1.2.1. Hsp40 Chaperones

As mentioned above, the founding member of this chaperone class is the *E. coli* DnaJ protein, which stimulates the ATPase activity of DnaK. This canonical Hsp40 contains the ~70 residue NH_2-terminal J-domain, an adjacent glycine/phenylalanine-rich domain (G/F), a cysteine-rich zinc finger domain, and a COOH-terminal domain. As a result, Hsp40s are sometimes referred to as "DnaJ proteins" or simply "J proteins." It is the J-domain that contacts and stimulates the ATPase domain of Hsp70, and it is composed of four α-helices (I, II, III, and IV) that form a finger-like projection (Szyperski et al., 1994; Hill et al., 1995; Qian et al., 1996). Helices II and III are anti-parallel amphipatic α-helices that are tightly packed (Szyperski et al., 1994; Qian et al., 1996). A loop connecting helices II and III contains the invariant tripeptide, HPD, which is required to stimulate Hsp70 ATPase activity (Greene et al., 1998). The G/F domain is a flexible linker that connects the J-domain to the rest of the protein. A second function of some Hsp40 chaperones is to deliver peptide substrates to Hsp70. The zinc finger domain, which is part of the larger central peptide-binding domain, can bind to polypeptides and is

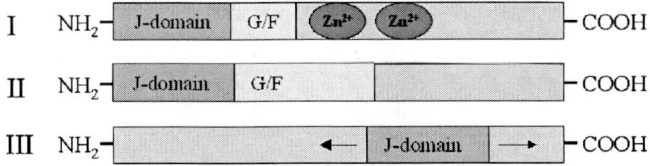

FIGURE 2. Three classes of J-domain-containing proteins. Class I proteins, whose domain organization is most like bacterial DnaJ, are composed of a J-domain, a G/F domain, and two zinc finger (Zn_{2+}) domains. Class II proteins are composed of a J-domain and G/F domain. Class III proteins are composed only of a J-domain that can be anywhere in the protein but generally is near the NH_2 terminus (left arrowhead is larger to indicate this fact). This nomenclature was first proposed by Cheetham and Caplan (1998).

required to present substrates to Hsp70 (Banecki et al., 1996; Szabo et al., 1996). In fact, the zinc finger domain of DnaJ can prevent the in vitro aggregation of denatured rhodanese, but the entire full-length DnaJ protein is required for protein refolding (Banecki et al., 1996). The function of the COOH-terminal domain is less well characterized but is also thought to function in substrate binding (Szabo et al., 1994; Banecki et al., 1996; Fan et al., 2005).

A large number of J-domain proteins have been identified in multiple organisms, and Cheetham and Caplan (1998) therefore developed a nomenclature to organize these proteins based on their similarities to DnaJ (see Figure 2). Class I proteins contain the NH_2-terminal J-domain, G/F domain, and a zinc finger domain, like DnaJ. Class II Hsp40s contain an NH_2-terminal J-domain and an enlarged G/F domain, but they lack a zinc finger domain. Class III proteins only contain a J-domain, and the domain may be located anywhere in the protein. As a result, Class III Hsp40s perform functions that are quite distinct from the Class I and Class II chaperones: It is known that the G/F domain is required for the formation of a Hsp70-peptide-DnaJ ternary complex but the J-domain alone can stimulate the ATPase activity of a preformed Hsp70-peptide complex (Wall et al., 1995). This result suggests that the G/F domain interacts with DnaK and that peptide binding mimics this interaction. Since class III J-proteins lack the G/F domain, they may be unable to present substrates to Hsp70, and in some cases they tether Hsp70s to sites where unfolded polypeptides congregate, such as in the lumen of the endoplasmic reticulum (ER, Brodsky, 1996). However, this may not be true for all class III J-proteins because cysteine-string protein (CSP) can stimulate Hsp70 in a substrate-independent manner, similar to type I J-proteins (Braun et al., 1996; Chamberlain and Burgoyne, 1997). In any event, a more detailed analysis of Class I-III proteins will be required to clearly define their substrate specificity and interaction with Hsp70s.

A given J-protein is not able to stimulate the ATPase activity of every Hsp70, and there appears instead to be partner specificity amongst the universe of potential Hsp70-Hsp40 complexes. For example, in yeast the cytosolic Hsp40, Ydj1p, stimulates the steady-state ATPase activity of the cytosolic yeast Hsp70, Ssa1p, up to ten-fold in vitro, but weakly stimulates (≤ two-fold) the activity of the ER lumenal Hsp70, Kar2p (McClellan et al., 1998). Conversely, the ER lumenal Hsp40, Sec63p, weakly stimulates Ssa1p's ATPase activity in vitro (McClellan et al., 1998). Moreover, Hsp70s and their potential Hsp40 partners in the cytosol may be targeted to distinct locations and cellular machines. Notably, the yeast Hsp70 chaperone, Ssz1p, interacts with Zuo1p (an Hsp40) on translating ribosomes (Walsh et al., 2004). In contrast, Ssa1p associates with diverse Hsp40 and Hsp70 homologues on translating ribosomes, in the cytoplasm, on clathrin coats, and at the ER membrane to engineer a variety of cellular phenomena (Horton et al., 2001; Lemmon, 2001; Youker et al., 2004; Shaner et al., 2005; Yam et al., 2005). It is not completely understood how the specificity between Hsp70/Hsp40 pairs are mediated within each compartment (e.g., the cytosol, mitochondria, and ER), although it has been suggested that this is conferred by residues on the surface of the J-domain (Schlenstedt et al., 1995; Pellecchia et al., 1996). In fact, Kelley and colleagues identified mutations in DnaJ that abolish J-domain function and that map to a small solvent-exposed region in helix II and III (Genevaux et al., 2002). The same mutations in a chimeric J-protein also abolished function, suggesting that this surface may represent an evolutionarily conserved motif that cements the Hsp70-J-domain complex (Genevaux et al., 2002).

It is thought that the J-domain contacts the bottom side of the ATP-binding domain in Hsp70, and that this interaction is essential to stimulate Hsp70's ATPase activity; mutations in residues that define a "groove" in this region in DnaK abolish J-domain stimulation of ATP hydrolysis (Gassler et al., 1998; Suh et al., 1998). However, the stimulation of an Hsp70's ATPase activity by Hsp40 is also abolished if the four COOH-terminal, conserved amino acids (EEVD) of Hsp70 are deleted (Freeman et al., 1995). These residues are also essential to mediate the interaction between Hsp70 and tetratricopeptide repeat (TPR) domain-containing proteins, which are important for coupling Hsp70s to the Hsp90 chaperone machinery (Smith, 2004; also see Section 1.3, below). In any event, this result indicates that Hsp40 interacts with at least two regions in Hsp70. Overall, the protein surfaces that dictate specificity between Hsp40 and Hsp70 remain to be elucidated, but whichever surface(s) mediate Hsp70-Hsp40 specificity their interaction is likely to require complex multivalent binding sites.

1.2.2. Nucleotide Exchange Factors

The BAG (Bcl-2–associated athanogene) family of proteins is conserved from yeast to humans and in mammals they play an important role in diverse cellular processes, including cell differentiation, migration, division and apoptosis. The founding family member, BAG-1 (also known as RAP46/HAP46), was identified in a screen for Bcl-2 binding proteins (Takayama et al., 1995). Four human BAG-1 isoforms are generated by alternate translation initiation (BAG-1S, BAG-1, BAG-1M, BAG-1L) (Zeiner and Gehring, 1995; Packham et al., 1997; Takayama et al., 1998; Yang et al., 1998). Even though the BAG-1 isoforms possess different NH_2-terminal regions they all contain a ubiquitin-like domain (UBL). Additional human

BAG family members (BAG-2, BAG-3, BAG-4, BAG-5) have also been discovered (Takayama et al., 1999), and there are at least seven BAG proteins in humans (not counting the isoforms), one in *S. cerevisiae*, two in *S. pombe*, two in *C. elegans*, and one each in *Drosophila, Xenopus*, and *A. thaliana* (Takayama and Reed, 2001). All family members possess a BAG domain at the COOH-terminus that binds to the ATPase domain of Hsc70 and stimulates nucleotide release, but other domains are present in distinct members: Notably, the UBL domain mentioned above that resides in BAG-1 and in BAG-6, a nuclear localization domain in BAG-1L, and a WW protein interaction domain in BAG-3. These domains facilitate target protein binding and/or deliver BAG proteins to different locations in the cell. For example, BAG-1 targets Hsc70 to the proteasome—a multi-catalytic, cytosolic protease—through its UBL domain, where BAG-1 is thought to facilitate the release and subsequent destruction of Hsc70-bound, unfolded polypeptides (Luders et al., 2000). Because the binding sites are non-overlapping, Hsp70 is able to interact with both Hsp40 and BAG-1 (Demand et al., 1998). Hence, BAG-1 may convert Hsc/Hsp70 from a protein-folding machine to a protein-degrading machine (Luders et al., 2000; Demand et al., 2001; Hohfeld et al., 2001).

Consistent with its binding to Bcl-2, cells over-expressing BAG-1 are more resistant to apoptotic-inducing stimuli (Takayama et al., 1995). It has been suggested that BAG-1, in conjunction with Hsp70, induces a conformational change in Bcl-2 that is required to regulate apoptotic pathways (Takayama et al., 1997). In addition, the serine/threonine kinase, Raf-1, competes with Hsp70 for BAG-1 binding. During times of cellular stress Hsp70 levels increase and there is a shift from BAG-1/Raf-1 to BAG-1/Hsp70 complex formation. The reduction in the amount of the BAG-1/Raf-1 complex may lead to depressed Raf-1 signaling and cell growth inhibition. Morimoto et al. hypothesize that this mode of competition represents a molecular switch to control cell growth (Song et al., 2001).

The mechanisms by which BAG-1 and GrpE—the founding, bacterial Hsp70 NEF—stimulate ADP release from Hsp70 are distinct. First, in the presence of inorganic phosphate BAG-1 can stimulate the release of ADP from Hsc/Hsp70 only up to 100-fold, but in the absence of inorganic phosphate the protein enhances nucleotide release by ~600-fold (Gassler et al., 2001). In contrast, GrpE activity is unaffected by inorganic phosphate. Second, BAG-1 exclusively stimulates the release of ATP from Hsc/Hsp70, whereas GrpE enhances the release of both ADP and ATP from Hsp70. And third, structural studies of NEF-Hsp70 complexes suggest that BAG-1, GrpE, and HspBP1—another cytosolic NEF for mammalian Hsp70 (see below)—effect nucleotide release through different conformational perturbations of the nucleotide binding pocket in Hsp70 (Shomura et al., 2005).

More recently, additional NEFs have been identified in mammalian cells and yeast. The yeast proteins Sls1p and Lhs1p are exchange factors for the ER lumenal Hsp70, BiP. Deletion of both exchange factors results in synthetic lethality, highlighting the importance of nucleotide exchange during the BiP ATPase cycle and for BiP function (Kabani et al., 2000; Tyson and Stirling, 2000). Furthermore, it was shown that Lhs1p—which is a distant Hsp70 homologue—possesses ATPase activity, and that BiP activates Lhs1p-mediated ATP hydrolysis (Steel et al., 2004).

These data suggest that the Lhs1p-BiP pair coordinately regulate one another's functions, perhaps to facilitate substrate transfer. It is currently unknown if other Lhs1p homologues act identically. In addition, Hendershot and colleagues identified an Sls1p homologue, the BiP-associated-protein (BAP), in a yeast two-hybrid screen and demonstrated that it functions as a NEF for mammalian BiP (Chung et al., 2002). And finally, a yeast cytosolic exchange factor Fes1p and a mammalian Fes1p homologue, Hsp70-binding-protein 1 (HspBP1), are also homologous to Sls1p/BAP and stimulate nucleotide release from cytosolic Hsp70 but not from the lumenal Hsp70, BiP (Kabani et al., 2002a). Interestingly, both HspBP1 and Fes1p inhibit Hsp40-dependent stimulation of Hsp70, suggesting that they may be negative regulators of Hsp70s (Kabani et al., 2002b). Indeed, BAG family members have been shown to act as either positive or negative regulators of Hsp70-catalyzed protein refolding in vitro, depending on the protein and phosphate concentrations and the Hsp40 co-chaperone used in the assay (Zeiner et al., 1997; Takayama et al., 1999; Thress et al., 2001; Gassler et al., 2001). Overall, it is likely that many undiscovered Hsp70 NEFs remain to be identified and characterized, and it will be exciting to uncover their distinct modes of action in the cell.

1.3. Hsp70 Chaperone and Co-Chaperone Interactions with CHIP, a Mediator of Protein Degradation

CHIP (COOH-terminal Hsc70 interacting protein) was identified in a screen for human TPR domain-containing proteins and was shown to associate with the COOH-terminus of Hsc70 via this domain both in vitro and through the use of a yeast two-hybrid system (Ballinger et al., 1999). As introduced above, the TPR domain is a protein-protein interaction motif and is composed of incompletely conserved 34 amino acid repeats. CHIP contains three TPR domains at its NH_2-terminus and also harbors a COOH-terminus U-box that binds ubiquitin, a protein "tag" that can be used to deliver proteins to the proteasome for degradation (Hatakeyama et al., 2001).

The combination of TPR and U-box domains in one protein strongly suggested that CHIP couples the chaperone and ubiquitin-proteasome machineries; indeed, CHIP-induced degradation requires both a functional U-box and TPR domain, and mutation of the U-box leads to a dominant-negative effect in vivo (Connell et al., 2001). These data are consistent with a role for CHIP in targeting Hsp70-bound substrates to the proteasome, and it has been shown that CHIP functions as either an E3 ubiquitin ligase or as a ubiquitin chain assembly factor, which is sometimes referred to as an "E4" (Demand et al., 2001; Jiang et al., 2001; Murata et al., 2001; Pringa et al., 2001; Hatakeyama et al., 2001). To facilitate the ubiquitination of proteasome-targeted substrates, CHIP also prevents premature release of Hsp70 substrates by blocking the chaperone's ATPase cycle (Ballinger et al., 1999; Connell et al., 2001). Not surprisingly, over-expression of CHIP in mammalian cells leads to the enhanced degradation of a number of chaperone substrates, including the glucocorticoid receptor, the cystic fibrosis transmembrane conductance regulator (CFTR), and p53 (Meacham et al., 2001; Galigniana et al., 2004; Esser et al., 2005).

CHIP-mediated ubiquitination of heat-denatured firefly luciferase in vitro requires either Hsp70 or Hsp90 (Murata et al., 2001), and the E3 ubiquitin ligase activity exhibited by CHIP is specific because native luciferase is not ubiquitinated. However, CHIP ubiquitinates some substrates in the absence of Hsp70/Hsp90 (Demand et al., 2001), suggesting direct interactions with substrates and perhaps chaperone-like activity (He et al., 2004). Since both CHIP and BAG-1 are involved in targeting chaperone-substrate complexes to the proteasome, it is plausible that the proteins interact with one other. In support of this hypothesis, BAG-1 and CHIP interact in vitro and BAG-1/Hsp70/CHIP complexes have been isolated from mammalian cells (Demand et al., 2001). Hsp70 might serve as a bridge for these proteins because BAG-1 binds to the NH_2-terminus and CHIP binds to the COOH-terminus of Hsp70 (Ballinger et al., 1999; Sondermann et al., 2001). Although CHIP-induced degradation of the glucocorticoid receptor can be stimulated by BAG-1—providing further evidence for the cooperation of these two proteins in the sorting of chaperone substrates to the proteasome (Demand et al., 2001)—BAG-1 is dispensable for CHIP-induced degradation of other proteins in vitro (Xu et al., 2002). Therefore, it remains to be seen whether BAG-1 is an essential cellular component for CHIP's ubiquitin ligase activity, and/or is a co-chaperone for CHIP's chaperone-like activity.

Another regulator of CHIP is the cytoplasmic NEF, HspBP1, which binds to the ATPase domain of Hsp70 and enhances CHIP association with Hsp70. The outcome of ternary complex formation is that CHIP's E3 ligase activity is inhibited, and thus HspBP1 over-expression prevents the CHIP-induced ubiquitination and degradation of CFTR; in turn, this promotes CFTR maturation in mammalian cells (Alberti et al., 2004). Other groups have reported that CHIP over-expression aids in the maturation or activation of the androgen receptor (Cardozo et al., 2003), endothelial nitric oxide synthase (Jiang et al., 2003) and the heat shock transcription factor (Dai et al., 2003), suggesting further that CHIP might be a bona fide chaperone and that HspBP1 may regulate this activity.

CHIP not only ubiquitinates substrates bound by chaperones but directly ubiquitinates chaperones and chaperone co-factors, notably Hsc/Hsp70 and BAG-1 (Jiang et al., 2001; Alberti et al., 2002). CHIP-catalyzed ubiquitination of BAG-1 targets BAG-1 to the proteasome, but curiously proteasome-associated and ubiquitinated Hsc/Hsp70/BAG-1 are not necessarily degraded. One explanation for this apparent paradox is that the linkages between the ubiquitin moeities in the polyubiquitin chain formed by CHIP are at the non-canonical Lys27 instead of at Lys48, which serves as a signal for proteolysis. Instead, Lys27 ubiquitination might help anchor these factors to the proteasome (Esser et al., 2004), perhaps via their interactions with any one of a number of potential ubiquitin receptors in the 19S "cap" of the proteasome (Deveraux et al., 1994; Young et al., 1998; Hiyama et al., 1999; Lam et al., 2002).

Together, the emerging picture is that Hsp70 and other chaperone complexes function as either protein folding or protein degrading machines, depending on the co-chaperones with which they associate. For example, Hsp70 co-factors HiP (Hsc70 interacting protein; see Section 1.1) and Hsp70-Hsp90 organizing protein (HOP) promote protein folding (Pratt and Welsh, 1994; Smith et al., 1995; Pratt and Toft, 2003). In contrast, BAG-1 and CHIP promote degradation, converting Hsp70 from a protein folding to a protein degrading machine (Esser et al., 2004). Interestingly, the pro-folding and pro-degrading co-factors appear to compete for the same binding sites on Hsp70. CHIP and HOP bind to the COOH-terminal EEVD motif on Hsp70 through their TPR motifs (Connell et al., 2001), whereas BAG-1 and HIP compete for binding at the NH_2-terminal ATPase domain of Hsp70 (Hohfeld and Jentsch, 1997; Takayama et al., 1999). Thus, the levels of the co-factor may set the balance between protein folding and degradation in the cell. Because the cellular levels of BAG-1 and CHIP are low compared to HIP or HOP, under normal conditions the balance favors protein folding. However, it is not difficult to imagine that conditions that stress or harm cells might tip the balance in the opposition direction—toward protein degradation—simply by altering the ratios of BAG-1/CHIP to HIP or HOP. Therefore, an elucidation of the mechanisms that regulate the expression levels of these co-factors will provide insight into how the cell "decides" to fold or degrade proteins.

1.4. The Function of Hsp70 and Hsp70 Regulators in ER Quality Control

Nearly all secreted and membrane proteins are targeted to the endoplasmic reticulum (ER) concomitant with or soon after synthesis, and are imported or "translocated" into the lumen of the ER or into the ER membrane. As a result, a high concentration of chaperones and co-chaperones are housed within the ER or reside at the ER membrane. Although the primary role for these factors is to catalyze secreted and membrane protein folding, they also ensure that aberrant, misfolded ER substrates are destroyed before they may harm the cell. The process by which these misfolded proteins are targeted and destroyed by the cytosolic proteasome has been termed ER Associated protein Degradation (ERAD), and Hsp70 and regulators of Hsp70 play an important role in catalyzing ERAD (Fewell et al., 2001).

Studies conducted in yeast suggested that the export, or "retro-translocation" of ERAD substrates requires the Sec61p channel, which is also employed during translocation (Pilon et al., 1997) (Plemper et al., 1997; Zhou and Schekman, 1999); however, genetic data suggest that the mechanisms for retro-translocation and translocation are mechanistically distinct (Zhou and Schekman, 1999; Brodsky et al., 1999). Nevertheless, in both cases the substrate protein must be retained in a soluble, unfolded conformation to allow transport through the confines of the channel. During translocation, the Ssa1p cytoplasmic Hsp70 chaperone retains polypeptides in solution (Chirico et al., 1988; Deshaies et al., 1988) and functions with a J-domain protein, Ydj1p (Caplan et al., 1992a). The ER lumenal Hsp70, BiP, prevents the aggregation of retro-translocating polypeptides during ERAD. BiP also helps drive the import of proteins into the ER during translocation,

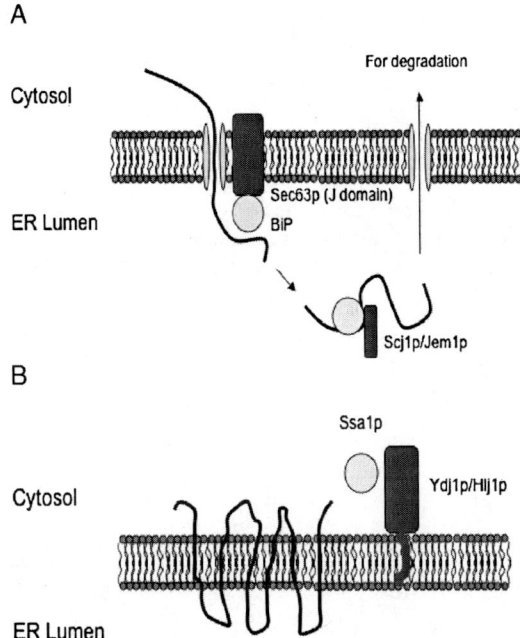

FIGURE 3. Interactions between ER lumenal and cytoplasmic Hsp70-Hsp40 chaperones during ER-associated degradation (ERAD). A: Soluble protein translocation into the yeast ER is driven by the BiP (Hsp70)-Sec63p (Class III Hsp40) complex. If the protein cannot fold or folds slowly, it is retained in a soluble state prior to retro-translocation by BiP-Scj1p/Jem1p. B: Integral membrane proteins that are defective for folding in many cases require the Ssa1p (Hsp70)-Ydj1p/Hlj1p chaperone pair for degradation; however, BiP activity is dispensable for the degradation of membrane proteins in every case thus far examined. Hsp70 chaperones are depicted as circles and Hsp40 chaperones are shown as rectangles. (A) Proposed model for the ERAD of soluble proteins. (B) Proposed model for the ERAD of polytopic integral membrane proteins.

acting either as a molecular ratchet or as a force-generating engine (Brodsky, 1996; Nishikawa et al., 2001). Although BiP participates both in translocation and retro-translocation, different co-chaperones are employed depending upon whether a polypeptide is entering or leaving the ER (see Figure 3A). Specifically, the ER membrane-anchored J-domain protein, Sec63p, is required for protein translocation but is largely dispensable for ERAD (Pilon et al., 1997; Plemper et al., 1997; Nishikawa et al., 2001). In contrast, the ER lumenal J-domain proteins Scj1p and Jem1p facilitate the ERAD of soluble proteins but are not required for translocation (Nishikawa et al., 2001). Together, it appears that the distinct functions performed by BiP are dictated—at least in part—by its location and co-chaperone interactions within the ER.

Are cytosolic Hsp70 chaperones also involved in the ERAD of soluble proteins? Most likely not, since disabling the activity of the major cyotosolic Hsp70 in yeast, Ssa1p, has no effect on the ERAD of three proteins examined (Brodsky et al., 1999; Huyer et al., 2004). In contrast, Ssa1p facilitates the degradation of integral membrane proteins in yeast that are targeted for ERAD. These include CFTR, a mutant form of the yeast ABC transporter, Ste6p (Ste6p*), a mutant form of the yeast multi-drug transporter Pdr5p (Pdr5p*), and an orphaned subunit of the vacuolar ATPase Vph1p (Plemper et al., 1998; Hill and Cooper, 2000; Zhang et al., 2001; Huyer et al., 2004). The ERAD of two of these proteins also requires the participation of two J-domain proteins, Hlj1p and Ydj1p (Youker et al., 2004; Huyer et al., 2004; see Table 1 and Fig. 3B). Hlj1p is a COOH terminally anchored membrane protein, and Ydj1p—which is also required for translocation (see above)—is farnesylated, a modification that tethers it to the ER membrane (Caplan et al., 1992b). The requirement for this cytosolic Hsp70-Hsp40 complex during the ERAD of polytopic membrane proteins might derive from the need to keep the proteins soluble prior to proteasome-delivery, or from the need to deliver these proteins to the ubiquitination machinery that also resides at the ER membrane.

Are other Hsp70 co-chaperones involved in ERAD or translocation? Yeast contain a BAG-1 homologue, Snl1p, that localizes to the ER/nuclear membrane (Sondermann et al., 2002), but the deletion of the gene encoding this protein—either alone or when combined with a *fes1* deletion mutant (see Table 1)—has no effect on ERAD or protein import into the ER (J. Bennett, J. Young, and J. L. B., unpublished observations). In contrast, the NEFs in the yeast ER, Sls1p and Lhs1p (see Section 1.2.2), appear to facilitate protein translocation. Most notably, yeast deleted both for *LHS1* and for *SLS1* exhibit strong translocation defects (Tyson and Stirling, 2000), and another yeast species expressing an Sls1 mutant that is defective for BiP interaction displays a translocation defect (Boisrame et al., 1998). It is possible that these NEFs augment BiP's catalytic cycle, which may be important for maximal translocation.

Another important question in this field regards the potential interaction between the chaperone machinery and the translocation/retro-translocation channel. It has been suggested that BiP is important for helping target soluble ERAD substrates back to the Sec61p-containing channel, although it remains unclear how it plays this role and whether J-domain proteins participate in this process (Gillece et al., 1999; Schmitz et al., 1995; Brodsky et al., 1999; Skowronek et al., 1998). However, it should be noted that the Sec61p translocon is not required for the ERAD of Ste6p* (Huyer et al., 2004), suggesting that some membrane proteins might be "shaved" by the proteasome and do not have to threaded back out through the channel. But yeast contains at least two other putative translocation/retro-translocation channels, so it cannot be completely ruled-out that retro-translocation is dispensable for Ste6p* degradation (Finke et al., 1996; Walter et al., 2001; Hitt and Wolf, 2004). Clearly, much more work is needed better to comprehend the interplay between the chaperone, translocation, and proteasome machineries during ERAD.

2. Conclusions

Diverse cellular processes from DNA replication to the degradation of aberrant proteins rely on Hsp70. The biological function of Hsp70 is tightly coupled to its

ATPase activity, which is modulated by a growing family of regulators. Hsp70 regulators can enhance or block the ATPase activity of Hsp70. In addition, regulators can link Hsp70 to different folding, or degrading machinery in the cell. One current model suggests that Hsp70 can be switched from a pro-folding to pro-degrading machine depending on the cellular levels of positive and negative regulators, and it will be interesting to see if this model is upheld. The identification of additional Hsp70 regulators in conjunction with biochemical and genetic experiments will aid in elucidating the complex regulation of Hsp70-dependent processes in the cell.

References

Alberti, S., Bohse, K., Arndt, V., Schmitz, A., and Hohfeld, J. (2004) The cochaperone HspBP1 inhibits the CHIP ubiquitin ligase and stimulates the maturation of the cystic fibrosis transmembrane conductance regulator. *Mol Biol Cell* 15:4003–10.

Alberti, S., Demand, J., Esser, C., Emmerich, N., Schild, H., and Hohfeld, J. (2002) Ubiquitylation of BAG-1 suggests a novel regulatory mechanism during the sorting of chaperone substrates to the proteasome. *J Biol Chem* 277:45920–7.

Anfinsen, C. B. (1973) Principles that govern the folding of protein chains. *Science* 181:223–30.

Ballinger, C. A., Connell, P., Wu, Y., Hu, Z., Thompson, L. J., Yin, L. Y., and Patterson, C. (1999) Identification of CHIP, a novel tetratricopeptide repeat-containing protein that interacts with heat shock proteins and negatively regulates chaperone functions. *Mol Cell Biol* 19:4535–45.

Banecki, B., Liberek, K., Wall, D., Wawrzynow, A., Georgopoulos, C., Bertoli, E., Tanfani, F., and Zylicz, M. (1996) Structure-function analysis of the zinc finger region of the DNAJ molecular chaperone. *J Biol Chem* 271:14840–8.

Boisrame, A., Kabani, M., Beckerich, J. M., Hartmann, E., and Gaillardin, C. (1998) Interaction of Kar2p and Sls1p is required for efficient co-translational translocation of secreted proteins in the yeast *Yarrowia lipolytica*. *J Biol Chem* 273:30903–8.

Braun, J. E., Wilbanks, S. M., and Scheller, R. H. (1996) The cysteine string secretory vesicle protein activates Hsc70 ATPase. *J Biol Chem* 271:25989–93.

Brodsky, J. L. (1996) Post-translational protein translocation: Not all Hsc70s are created equal. *Trends Biochem Sci* 21:122–6.

Brodsky, J. L., Werner, E. D., Dubas, M. E., Goeckeler, J. L., Kruse, K. B., and McCracken, A. A. (1999) The requirement for molecular chaperones during endoplasmic reticulum-associated protein degradation demonstrates that protein export and import are mechanistically distinct. *J Biol Chem* 274:3453–60.

Bukau, B. (1999) *Molecular Chaperones and Folding Catalysts: Regulation, Cellular Function, and Mechanisms.* Harwood Academic Publishers, Amsterdam.

Caplan, A. J., Cyr, D. M., and Douglas, M. G. (1992a) YDJ1 facilitates polypeptide translocation across different intracellular membranes by a conserved mechanism. *Cell* 71:1143–55.

Caplan, A. J., Tsai, J., Casey, P. J., and Douglas, M. G. (1992b) Farnesylation of YDJ1p is required for function at elevated growth temperatures in *Saccharomyces cerevisiae*. *J Biol Chem* 267:18890–5.

Cardozo, C. P., Michaud, C., Ost, M. C., Fliss, A. E., Yang, E., Patterson, C., Hall, S. J., and Caplan, A. J. (2003) C-terminal Hsp-interacting protein slows androgen receptor synthesis and reduces its rate of degradation. *Arch Biochem Biophys* 410:134–40.

Chamberlain, L. H., and Burgoyne, R. D. (1997) Activation of the ATPase activity of heat-shock proteins Hsc70/Hsp70 by cysteine-string protein. *Biochem J* 322(Pt 3):853–8.

Cheetham, M. E., and Caplan, A. J. (1998) Structure, function, and evolution of DnaJ: Conservation and adaptation of chaperone function. *Cell Stress Chap* 3:28–36.

Chirico, W. J., Waters, M. G., and Blobel, G. (1988) 70K heat shock related proteins stimulate protein translocation into microsomes. *Nature (Lond.)* 332:805–10.

Chung, K. T., Shen, Y., and Hendershot, L. M. (2002) BAP, a mammalian BiP-associated protein, is a nucleotide exchange factor that regulates the ATPase activity of BiP. *J Biol Chem* 277:47557–63.

Connell, P., Ballinger, C. A., Jiang, J., Wu, Y., Thompson, L. J., Hohfeld, J., and Patterson, C. (2001) The co-chaperone CHIP regulates protein triage decisions mediated by heat-shock proteins. *Nat Cell Biol* 3:93–6.

Dai, Q., Zhang, C., Wu, Y., McDonough, H., Whaley, R. A., Godfrey, V., Li, H. H., Madamanchi, N., Xu, W., Neckers, L., Cyr, D., and Patterson, C. (2003) CHIP activates HSF1 and confers protection against apoptosis and cellular stress. *EMBO J* 22:5446–58.

de Virgilio, M., Weninger, H., and Ivessa, N. E. (1998) Ubiquitination is required for the retro-translocation of a short-lived luminal endoplasmic reticulum glycoprotein to the cytosol for degradation by the proteasome. *J Biol Chem* 273:9734–43.

Demand, J., Alberti, S., Patterson, C., and Hohfeld, J. (2001) Cooperation of a ubiquitin domain protein and an E3 ubiquitin ligase during chaperone/proteasome coupling. *Curr Biol* 11:1569–77.

Demand, J., Luders, J., and Hohfeld, J. (1998) The carboxy-terminal domain of Hsc70 provides binding sites for a distinct set of chaperone cofactors. *Mol Cell Biol* 18:2023–8.

Deshaies, R. J., Koch, B. D., Werner-Washburne, M., Craig, E. A., and Schekman, R. (1988) A subfamily of stress proteins facilitates translocation of secretory and mitochondrial precursor proteins. *Nature (Lond.)* 332:800–5.

Deveraux, Q., Ustrell, V., Pickart, C., and Rechsteiner, M. (1994) A 26 S protease subunit that binds ubiquitin conjugates. *J Biol Chem* 269:7059–61.

Esser, C., Alberti, S., and Hohfeld, J. (2004) Cooperation of molecular chaperones with the ubiquitin/proteasome system. *Biochim Biophys Acta* 1695:171–88.

Esser, C., Scheffner, M., and Hohfeld, J. (2005) The chaperone-associated ubiquitin ligase CHIP is able to target p53 for proteasomal degradation. *J Biol Chem* 280:27443–8.

Fan, C. Y., Ren, H. Y., Lee, P., Caplan, A. J., and Cyr, D. M. (2005) The type I Hsp40 zinc finger-like region is required for Hsp70 to capture non-native polypeptides from Ydj1. *J Biol Chem* 280:695–702.

Fewell, S. W., Travers, K. J., Weissman, J. S., and Brodsky, J. L. (2001) The action of molecular chaperones in the early secretory pathway. *Annu Rev Genetics* 35:149–91.

Finke, K., Plath, K., Panzner, S., Prehn, S., Rapoport, T. A., Hartmann, E., and Sommer, T. (1996) A second trimeric complex containing homologs of the Sec61p complex functions in protein transport across the ER membrane of *S. cerevisiae*. *EMBO J* 15:1482–94.

Flaherty, K. M., DeLuca-Flaherty, C., and McKay, D. B. (1990) Three-dimensional structure of the ATPase fragment of a 70K heat-shock cognate protein. *Nature* 346:623–8.

Freeman, B. C., Myers, M. P., Schumacher, R., and Morimoto, R. I. (1995) Identification of a regulatory motif in Hsp70 that affects ATPase activity, substrate binding and interaction with HDJ-1. *EMBO J* 14:2281–92.

Frydman, J., and Hohfeld, J. (1997) Chaperones get in touch: The Hip-Hop connection. *Trends Biochem Sci* 22:87–92.

Galigniana, M. D., Harrell, J. M., Housley, P. R., Patterson, C., Fisher, S. K., and Pratt, W. B. (2004) Retrograde transport of the glucocorticoid receptor in neurites requires dynamic assembly of complexes with the protein chaperone hsp90 and is linked to the CHIP component of the machinery for proteasomal degradation. *Brain Res Mol Brain Res* 123:27–36.

Gassler, C. S., Wiederkehr, T., Brehmer, D., Bukau, B., and Mayer, M. P. (2001) Bag-1M accelerates nucleotide release for human Hsc70 and Hsp70 and can act concentration-dependent as positive and negative cofactor. *J Biol Chem* 276:32538–44.

Gassler, C. S., Buchberger, A., Lauffen, T., Mayer, M. P., Schroder, H., Valencia, A., and Bukau, B. (1998) Mutations in the DnaK chaperone affecting interaction with the DnaJ cochaperone. *Proc Natl Acad Sci U S A* 95:15229–34.

Genevaux, P., Schwager, F., Georgopoulos, C., and Kelley, W. L. (2002) Scanning mutagenesis identifies amino acid residues essential for the *in vivo* activity of the *Escherichia coli* DnaJ (Hsp40) J-domain. *Genetics* 162:1045–53.

Greene, M. K., Maskos, K., and Landry, S. J. (1998) Role of the J-domain in the cooperation of Hsp40 with Hsp70. *Proc Natl Acad Sci U S A* 95:6108–13.

Harrison, C. J., Hayer-Hartl, M., Di Liberto, M., Hartl, F., and Kuriyan, J. (1997) Crystal structure of the nucleotide exchange factor GrpE bound to the ATPase domain of the molecular chaperone DnaK. *Science* 276:431–5.

Hatakeyama, S., Yada, M., Matsumoto, M., Ishida, N., and Nakayama, K. I. (2001) U box proteins as a new family of ubiquitin–protein ligases. *J Biol Chem* 276:33111–20.

He, B., Bai, S., Hnat, A. T., Kalman, R. I., Minges, J. T., Patterson, C., and Wilson, E. M. (2004) An androgen receptor NH2-terminal conserved motif interacts with the COOH terminus of the Hsp70-interacting protein (CHIP). *J Biol Chem* 279:30643–53.

Hill, K., and Cooper, A. A. (2000) Degradation of unassembled Vph1p reveals novel aspects of the yeast ER quality control system. *EMBO J* 19:550–61.

Hill, R. B., Flanagan, J. M., and Prestegard, J. H. (1995) 1H and 15N magnetic resonance assignments, secondary structure, and tertiary fold of *Escherichia coli* DnaJ (1–78). *Biochemistry* 34:5587–96.

Hitt, R., and Wolf, D. H. (2004) Der1p, a protein required for degradation of malfolded soluble proteins of the endoplasmic reticulum: Topology and Der1-like proteins. *FEMS Yeast Res* 4:721–9.

Hiyama, H., Yokoi, M., Masutani, C., Sugasawa, K., Maekawa, T., Tanaka, K., Hoeijmakers, J. H., and Hanaoka, F. (1999) Interaction of hHR23 with S5a. The ubiquitin-like domain of hHR23 mediates interaction with S5a subunit of 26 S proteasome. *J Biol Chem* 274:28019–25.

Hohfeld, J., Cyr, D. M., and Patterson, C. (2001) From the cradle to the grave: Molecular chaperones that may choose between folding and degradation. *EMBO Rep* 2:885–90.

Hohfeld, J., and Jentsch, S. (1997) GrpE-like regulation of the hsc70 chaperone by the anti-apoptotic protein BAG-1. *EMBO J* 16:6209–16.

Horton, L. E., James, P., Craig, E. A., and Hensold, J. O. (2001) The yeast hsp70 homologue Ssa is required for translation and interacts with Sis1 and Pab1 on translating ribosomes. *J Biol Chem* 276:14426–33.

Huyer, G., Piluek, W. F., Fansler, Z., Kreft, S. G., Hochstrasser, M., Brodsky, J. L., and Michaelis, S. (2004) Distinct machinery is required in *Saccharomyces cerevisiae* for the endoplasmic reticulum-associated degradation of a multispanning membrane protein and a soluble lumenal protein. *J Biol Chem.* 279(37):38369–78.

Jiang, J., Ballinger, C. A., Wu, Y., Dai, Q., Cyr, D. M., Hohfeld, J., and Patterson, C. (2001) CHIP is a U-box-dependent E3 ubiquitin ligase: Identification of Hsc70 as a target for ubiquitylation. *J Biol Chem* 276:42938–44.

Jiang, J., Cyr, D., Babbitt, R. W., Sessa, W. C., and Patterson, C. (2003) Chaperone-dependent regulation of endothelial nitric-oxide synthase intracellular trafficking by the co-chaperone/ubiquitin ligase CHIP. *J Biol Chem* 278:49332–41.

Joachimiak, A. (1997) Capturing the misfolds: Chaperone-peptide-binding motifs. *Nat Struct Biol* 4:430–4.

Kabani, M., Beckerich, J. M., and Brodsky, J. L. (2002a) Nucleotide exchange factor for the yeast Hsp70 molecular chaperone Ssa1p. *Mol Cell Biol* 22:4677–89.

Kabani, M., Beckerich, J. M., and Gaillardin, C. (2000) Sls1p stimulates Sec63p-mediated activation of Kar2p in a conformation-dependent manner in the yeast endoplasmic reticulum. *Mol Cell Biol* 20:6923–34.

Kabani, M., McLellan, C., Raynes, D. A., Guerriero, V., and Brodsky, J. L. (2002b) HspBP1, a homologue of the yeast Fes1 and Sls1 proteins, is an Hsc70 nucleotide exchange factor. *FEBS Lett* 531:339–42.

Karzai, A. W., and McMacken, R. (1996) A bipartite signaling mechanism involved in DnaJ-mediated activation of the *Escherichia coli* DnaK protein. *J Biol Chem* 271:11236–46.

Lam, Y. A., Lawson, T. G., Velayutham, M., Zweier, J. L., and Pickart, C. M. (2002) A proteasomal ATPase subunit recognizes the polyubiquitin degradation signal. *Nature* 416:763–7.

Lemmon, S. K. (2001) Clathrin uncoating: Auxilin comes to life. *Curr Biol* 11:R49–52.

Liberek, K., Marszalek, J., And, D., Georgopoulos, C., and Zylicz, M. (1991) *Escherichia coli* DnaJ and GrpE heat shock proteins jointly stimulate ATPase activity of DnaK. *Proc Natl Acad Sci U S A* 88:2874–8.

Luders, J., Demand, J., and Hohfeld, J. (2000) The ubiquitin-related BAG-1 provides a link between the molecular chaperones Hsc70/Hsp70 and the proteasome. *J Biol Chem* 275:4613–7.

Mayer, M. P., and Bukau, B. (2005) Hsp70 chaperones: Cellular functions and molecular mechanism. *Cell Mol Life Sci* 62:670–84.

McCarty, J. S., Buchberger, A., Reinstein, J., and Bukau, B. (1995) The role of ATP in the functional cycle of the DnaK chaperone system. *J Mol Biol* 249:126–37.

McClellan, A. J., Endres, J. B., Vogel, J. P., Palazzi, D., Rose, M. D., and Brodsky, J. L. (1998) Specific molecular chaperone interactions and an ATP-dependent conformational change are required during posttranslational protein translocation into the yeast ER. *Mol Biol Cell* 9:3533–45.

Meacham, G. C., Patterson, C., Zhang, W., Younger, J. M., and Cyr, D. M. (2001) The Hsc70 co-chaperone CHIP targets immature CFTR for proteasomal degradation. *Nat Cell Biol* 3:100–5.

Murata, S., Minami, Y., Minami, M., Chiba, T., and Tanaka, K. (2001) CHIP is a chaperone-dependent E3 ligase that ubiquitylates unfolded protein. *EMBO Rep* 2:1133–8.

Nishikawa, S. I., Fewell, S. W., Kato, Y., Brodsky, J. L., and Endo, T. (2001) Molecular chaperones in the yeast endoplasmic reticulum maintain the solubility of proteins for retrotranslocation and degradation. *J Cell Biol* 153:1061–70.

Packham, G., Brimmell, M., and Cleveland, J. L. (1997) Mammalian cells express two differently localized Bag-1 isoforms generated by alternative translation initiation. *Biochem J* 328(Pt 3):807–13.

Pelham, H. R. (1984) Hsp70 accelerates the recovery of nucleolar morphology after heat shock. *EMBO J* 3:3095–100.

Pellecchia, M., Szyperski, T., Wall, D., Georgopoulos, C., and Wuthrich, K. (1996) NMR structure of the J-domain and the Gly/Phe-rich region of the *Escherichia coli* DnaJ chaperone. *J Mol Biol* 260:236–50.

Pilon, M., Schekman, R., and Romisch, K. (1997) Sec61p mediates export of a misfolded secretory protein from the endoplasmic reticulum to the cytosol for degradation. *EMBO J* 16:4540–8.

Plemper, R. K., Bohmler, S., Bordallo, J., Sommer, T., and Wolf, D. H. (1997) Mutant analysis links the translocon and BiP to retrograde protein transport for ER degradation. *Nature* 388:891–5.

Plemper, R. K., Egner, R., Kuchler, K., and Wolf, D. H. (1998) Endoplasmic reticulum degradation of a mutated ATP-binding cassette transporter Pdr5 proceeds in a concerted action of Sec61 and the proteasome. *J Biol Chem* 273:32848–56.

Pratt, W. B., and Toft, D. O. (2003) Regulation of signaling protein function and trafficking by the hsp90/hsp70-based chaperone machinery. *Exp Biol Med* (*Maywood*) 228: 111–33.

Pratt, W. B., and Welsh, M. J. (1994) Chaperone functions of the heat shock proteins associated with steroid receptors. *Semin Cell Biol* 5:83–93.

Pringa, E., Martinez-Noel, G., Muller, U., and Harbers, K. (2001) Interaction of the ring finger-related U-box motif of a nuclear dot protein with ubiquitin-conjugating enzymes. *J Biol Chem* 276:19617–23.

Qian, Y. Q., Patel, D., Hartl, F. U., and McColl, D. J. (1996) Nuclear magnetic resonance solution structure of the human Hsp40 (HDJ-1) J-domain. *J Mol Biol* 260:224–35.

Schlenstedt, G., Harris, S., Risse, B., Lill, R., and Silver, P. A. (1995) A yeast DnaJ homologue, Scj1p, can function in the endoplasmic reticulum with BiP/Kar2p via a conserved domain that specifies interactions with Hsp70s. *J Cell Biol* 129:979–88.

Schmitz, A., Maintz, M., Kehle, T., and Herzog, V. (1995) *In vivo* iodination of a misfolded proinsulin reveals co-localized signals for Bip binding and for degradation in the ER. *EMBO J* 14:1091–8.

Shaner, L., Wegele, H., Buchner, J., and Morano, K. A. (2005) The yeast Hsp110 Sse1 functionally interacts with the Hsp70 chaperones Ssa and Ssb. *J Biol Chem* 280(50):41262–9.

Shomura, Y., Dragovic, Z., Chang, H. C., Tzvetkov, N., Young, J. C., Brodsky, J. L., Guerriero, V., Hartl, F. U., and Bracher, A. (2005) Regulation of Hsp70 function by HspBP1: Structural analysis reveals an alternate mechanism for Hsp70 nucleotide exchange. *Mol Cell* 17:367–79.

Skowronek, M. H., Hendershot, L. M., and Haas, I. G. (1998) The variable domain of nonassembled Ig light chains determines both their half-life and binding to the chaperone BiP. *Proc Natl Acad Sci U S A* 95:1574–8.

Smith D. F. (2004) Tetratricopeptide repeat cochaperones in steroid receptor complexes. *Cell Stress Chap* 9(2):109–21.

Smith, D. F., Whitesell, L., Nair, S. C., Chen, S., Prapapanich, V., and Rimerman, R. A. (1995) Progesterone receptor structure and function altered by geldanamycin, an hsp90-binding agent. *Mol Cell Biol* 15:6804–12.

Sondermann, H., Scheufler, C., Schneider, C., Hohfeld, J., Hartl, F. U., and Moarefi, I. (2001) Structure of a Bag/Hsc70 complex: Convergent functional evolution of Hsp70 nucleotide exchange factors. *Science* 291:1553–7.

Sondermann, H., Ho, A. K., Listenberger, L. L., Siegers, K,. Moarefi, I., Wente, S. R., Hartl, F. U., and Young, J. C. (2002) Prediction of novel Bag-1 homologs based on structure/function analysis identifies Snl1p as an Hsp70 co-chaperone in *Saccharomyces cerevisiae*. *J Biol Chem* 277:33220–7.

Song, J., Takeda, M., and Morimoto, R. I. (2001) Bag1-Hsp70 mediates a physiological stress signalling pathway that regulates Raf-1/ERK and cell growth. *Nat Cell Biol* 3:276–82.

Steel, G. J., Fullerton, D. M., Tyson, J. R., and Stirling, C. J. (2004) Coordinated action of Hsp70 chaperones. *Science* 303:98–101.

Suh, W. C., Burkholder, W. F., Lu, C. Z., Zhao, X., Gottesman, M. E., and Gross, C. A. (1998) Interaction of the Hsp70 molecular chaperone, DnaK, with its cochaperone DnaJ. *Proc Natl Acad Sci U S A* 95:15223–8.

Szabo, A., Korszun, R., Hartl, F. U., and Flanagan, J. (1996) A zinc finger-like domain of the molecular chaperone DnaJ is involved in binding to denatured protein substrates. *EMBO J* 15:408–17.

Szabo, A., Langer, T., Schroder, H., Flanagan, J., Bukau, B., and Hartl, F. U. (1994) The ATP hydrolysis-dependent reaction cycle of the *Escherichia coli* Hsp70 system DnaK, DnaJ, and GrpE. *Proc Natl Acad Sci U S A* 91:10345–9.

Szyperski, T., Pellecchia, M., Wall, D., Georgopoulos, C., and Wuthrich, K. (1994) NMR structure determination of the *Escherichia coli* DnaJ molecular chaperone: Secondary structure and backbone fold of the N-terminal region (residues 2–108) containing the highly conserved J domain. *Proc Natl Acad Sci U S A* 91:11343–7.

Takayama, S., Bimston, D. N., Matsuzawa, S., Freeman, B. C., Aime-Sempe, C., Xie, Z., Morimoto, R. I., and Reed, J. C. (1997) BAG-1 modulates the chaperone activity of Hsp70/Hsc70. *EMBO J* 16:4887–96.

Takayama, S., Krajewski, S., Krajewska, M., Kitada, S., Zapata, J. M., Kochel, K., Knee, D., Scudiero, D., Tudor, G., Miller, G. J., Miyashita, T., Yamada, M., and Reed, J. C. (1998) Expression and location of Hsp70/Hsc-binding anti-apoptotic protein BAG-1 and its variants in normal tissues and tumor cell lines. *Cancer Res* 58:3116–31.

Takayama, S., and Reed, J. C. (2001) Molecular chaperone targeting and regulation by BAG family proteins. *Nat Cell Biol* 3:E237–41.

Takayama, S., Sato, T., Krajewski, S., Kochel, K., Irie, S., Millan, J. A., and Reed, J. C. (1995) Cloning and functional analysis of BAG-1: A novel Bcl-2-binding protein with anti-cell death activity. *Cell* 80:279–84.

Takayama, S., Xie, Z., and Reed, J. C. (1999) An evolutionarily conserved family of Hsp70/Hsc70 molecular chaperone regulators. *J Biol Chem* 274:781–6.

Thress, K., Song, J., Morimoto, R. I., and Kornbluth, S. (2001) Reversible inhibition of Hsp70 chaperone function by Scythe and Reaper. *EMBO J* 20(5):1033–41.

Tyson, J. R., and Stirling, C. J. (2000) LHS1 and SIL1 provide a lumenal function that is essential for protein translocation into the endoplasmic reticulum. *EMBO J* 19:6440–52.

Wall, D., Zylicz, M., and Georgopoulos, C. (1995) The conserved G/F motif of the DnaJ chaperone is necessary for the activation of the substrate binding properties of the DnaK chaperone. *J Biol Chem* 270:2139–44.

Walsh, P., Bursac, D., Law, Y. C., Cyr, D., and Lithgow, T. (2004) The J-protein family: Modulating protein assembly, disassembly and translocation. *EMBO Rep* 5:567–71.

Walter, J., Urban, J., Volkwein, C., and Sommer, T. (2001) Sec61p-independent degradation of the tail-anchored ER membrane protein Ubc6p. *EMBO J* 20:3124–31.

Xu, W., Marcu, M., Yuan, X., Mimnaugh, E., Patterson, C., and Neckers, L. (2002) Chaperone-dependent E3 ubiquitin ligase CHIP mediates a degradative pathway for c-ErbB2/Neu. *Proc Natl Acad Sci U S A* 99:12847–52.

Yam, A. Y. W., Albanese, V., Lin, H. T. J., and Frydman, J. (2005) Hsp110 cooperates with different cytosolic Hsp70 systems in a pathway for de novo folding. *J Biol Chem* 280(50):41252–61.

Yang, X., Chernenko, G., Hao, Y., Ding, Z., Pater, M. M., Pater, A., and Tang, S. C. (1998) Human BAG-1/RAP46 protein is generated as four isoforms by alternative translation initiation and overexpressed in cancer cells. *Oncogene* 17:981–9.

Youker, R. T., Walsh, P., Beilharz, T., Lithgow, T., and Brodsky, J. L. (2004) Distinct roles for the Hsp40 and Hsp90 molecular chaperones during cystic fibrosis transmembrane conductance regulator degradation in yeast. *Mol Biol Cell* 15:4787–97.

Young, P., Deveraux, Q., Beal, R. E., Pickart, C. M., and Rechsteiner, M. (1998) Characterization of two polyubiquitin binding sites in the 26 S protease subunit 5a. *J Biol Chem* 273:5461–7.

Zeiner, M., Gebauer, M., and Gehring, U. (1997) Mammalian protein RAP46: An interaction partner and modulator of 70 kDa heat shock proteins. *EMBO J* 16:5483–90.

Zeiner, M., and Gehring, U. (1995) A protein that interacts with members of the nuclear hormone receptor family: Identification and cDNA cloning. *Proc Natl Acad Sci U S A* 92:11465–9.

Zhang, Y., Nijbroek, G., Sullivan, M. L., McCracken, A. A., Watkins, S. C., Michaelis, S., and Brodsky, J. L. (2001) Hsp70 molecular chaperone facilitates endoplasmic reticulum-associated protein degradation of cystic fibrosis transmembrane conductance regulator in yeast. *Mol Biol Cell* 12:1303–14.

Zhou, M., and Schekman, R. (1999) The engagement of Sec61p in the ER dislocation process. *Mol Cell* 4:925–34.

Zhu, X., Zhao, X., Burkholder, W. F., Gragerov, A., Ogata, C. M., Gottesman, M. E., and Hendrickson, W. A. (1996) Structural analysis of substrate binding by the molecular chaperone DnaK. *Science* 272:1606–14.

Ziegelhoffer, T., Johnson, J. L., and Craig, E. A. (1996) Chaperones get Hip. Protein folding. *Curr Biol* 6:272–5.

Zimmerman, S. B., and Trach, S. O. (1991) Estimation of macromolecule concentrations and excluded volume effects for the cytoplasm of *Escherichia coli*. *J Mol Biol* 222:599–620.

Zylicz, M., LeBowitz, J. H., McMacken, R., and Georgopoulos, C. (1983) The dnaK protein of *Escherichia coli* possesses an ATPase and autophosphorylating activity and is essential in an *in vitro* DNA replication system. *Proc Natl Acad Sci U S A* 80:6431–5.

10
Protein Disassembly by Hsp40–Hsp70

SAMUEL J. LANDRY
Department of Biochemistry, New Orleans, LA 70112

1. Introduction

This chapter surveys evidence that Hsp40 and Hsp70 act as a molecular machine to disassemble protein complexes. Recent papers have discussed whether Hsp70 actively unfolds proteins during translocation across membranes (Matouschek et al., 2000), whether it fragments aggregates or extracts polypeptides (Zietkiewicz et al., 2006), and whether it acts a "holdase" or a "foldase" (Slepenkov and Witt, 2002a,b). These are related questions because they ask whether Hsp70 exerts a force on its clients during these processes. The first two sections of the chapter discuss the structures of Hsp70 and Hsp40 with an eye toward their interactions and conformational changes, that is, the "moving parts". Hsp40–Hsp70 chaperone machines have three levels of functional organization: (i) the articulation of Hsp70 domains, (ii) the interactions with J-domain-containing protein(s), and (iii) interactions with additional protein factors for targeting (e.g., TPR proteins) and regulation (e.g., GrpE). These additional factors are diverse and probably not fundamental to a disassembly activity, and thus they generally will not be discussed. Subsequent sections discuss the machine-like role of Hsp40–Hsp70 in several major classes of biochemical systems: the disassembly of protein complexes, protein degradation, protein translocation across membranes, and protein folding.

2. Hsp70 Structure

Hsp70s can be divided into two major domains—the N-terminal, 40-kDa ATPase domain (AD) and the C-terminal, 30-kDa peptide-binding domain (PBD). The AD is a member of a family of ancient and versatile bilobal kinases that includes hexokinase and actin (Bork et al., 1992). The common structure-function relationship in all family members is the ATP-dependent closing and hydrolysis-dependent opening of the two lobes. In the case of Hsp70, ATP binding also drives opening of the PBD, and ATP hydrolysis drives closing of the PBD. The PBD is less well conserved than the AD, and it includes subdomains that may have distinct functions.

2.1. ATPase Domain

All of the available crystal structures of the Hsp70 AD show the "open" form, by comparison to the open and closed structures that have been observed for hexokinase (Bennett and Steitz, 1978). The open form of bovine Hsc70 was obtained even when the structure included a bound nucleotide (Obrien et al., 1996). The *E. coli* Hsp70/DnaK AD was crystallized in the complex with its nucleotide-exchange factor, GrpE, which appears to be holding the two lobes apart to allow the nucleotide easy access to its binding site (Harrison et al., 1997). Hsc70 also was crystallized with nucleotide-exchange factor Bag1 (Sondermann et al., 2001). The bacterial and mammalian nucleotide-exchange factors do not appear to be homologous in spite of their similar function and mode of binding to the respective Hsp70s.

2.2. Peptide-Binding Domain

The Hsp70/DnaK PBD can be divided into the 20-kDa β-sandwich (PBDβ), 10kDa α-helical subdomain (PBDα), and a 33-residue C-terminal tail. Two α-helices (A and B) connect the N-terminal PBDβ to C-terminal PBDα (Fig. 1). Helix B and the PBDα, together, constitute the "lid" because they appear to close the PBD and block client peptide dissociation. The structure of the PBDβ was described in the crystal structure of the entire PBD and in several NMR structures that included PBDβ and various C-terminal segments (Pellecchia et al., 2000; Stevens et al., 2003; Wang et al., 1998; Zhu et al., 1996). The PBDβ resembles a curved palm as it grips an extended peptide representing the client protein. The C-terminal tail is disordered in both the crystal structure of the PBD and in the NMR structure of the PBDα (Bertelsen et al., 1999; Zhu et al., 1996).

The PBDα is less conserved than the PBDβ (Gupta and Singh, 1994), and it is unclear whether this is due to sequence drift or functional specialization. The sequence-similar domains of *E. coli* and rat Hsc70 adopt different arrangements of the helices (Bertelsen et al., 1999; Chou et al., 2003; Zhu et al., 1996). On the one hand, PBDα forms part of the lid, and thus constraints on its structure could

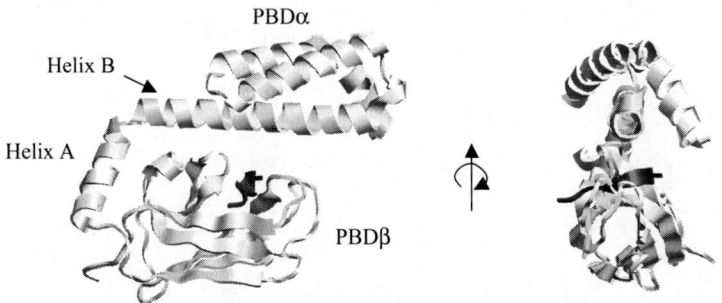

FIGURE 1. Ribbon diagram of the *E. coli* DnaK PBD (PDB: 1DKX). A synthetic peptide (NRLLLTG) that was co-crystallized in the PBD is illustrated in dark gray.

be limited to interactions with PBDβ. Deletion of the entire lid or deletion of only PBDα accelerates peptide binding and release (Mayer et al., 2000; Pellecchia et al., 2000; Slepenkov and Witt, 2002a,b). On the other hand, the main function of PBDα may involve interactions with components of the translation, translocation, or degradation machinery. For example, the PBD is necessary for BiP to seal the ER translocation channel, presumably so that the contents of the ER lumen do not leak into the cytosol (Alder et al., 2005). Removal of the lid from yeast mitochondrial Hsp70/Ssc1 rendered the cells inviable (Strub et al., 2003). As suggested by the ER example, perhaps the PBDα is important for sealing the mitochondrial translocation channel so that matrix contents do not leak into the cytosol.

2.3. Minimum Functional Unit

A minimum functional unit of Hsp70 could be said to be composed of the AD and the PBDβ, i.e., "lidless" Hsp70, because it is able to couple cycles of ATP hydrolysis and ADP/ATP exchange with cycles of peptide binding and releasing (Mayer et al., 2000; Pellecchia et al., 2000; Slepenkov and Witt, 2002a,b). This view of Hsp70 functionality highlights the importance of both the conserved linker peptide and tertiary structural contacts between the AD and PBDβ. For the first time in 20 years of intense study, a snapshot was taken of the two domains together in a crystal structure of lidless Hsc70 (Jiang et al., 2005). The polypeptide linker between the AD and PBDβ is visible, but it is not making much contact with the rest of the protein, and it is not revealing much about its role in the articulation of the two domains (Fig. 2). Tertiary contacts connect two of the most interesting surfaces of the respective domains, a cleft between the lobes of the AD and helix A of the PBD. This cleft on the AD is also thought to be the site for J-domain binding. The cleft was implicated in J-domain binding because it is the location of a mutation that suppresses a J-domain mutant (Suh et al., 1998), and the suppressor mutation affects NMR signals in the J-domain-AD interface (Landry, 2003). Numerous mutations in the area of the cleft cause defects in J-domain binding (Davis et al., 1999; Gassler et al., 1998), but the recent structure renews the question of whether the mutations affect J-domain binding directly or by their effects on tertiary contacts with the PBD.

2.4. Conformational Change

Conformational changes in Hsp70 associated with ATP binding and hydrolysis probably involve large-scale domain movements. A solution small-angle X-ray scattering study by McKay and co-workers found that ATP-bound Hsc70 was compact and that ADP-bound Hsc70 was elongated (Wilbanks et al., 1995). Mutations in the AD that affected interdomain communication shifted the ATP-bound protein to a more elongated conformation, and a second mutation in the PBD could shift it back toward a more compact conformation (Ha et al., 1997; Johnson and McKay, 1999; Sousa and McKay, 1998). The dimensions of nucleotide-free

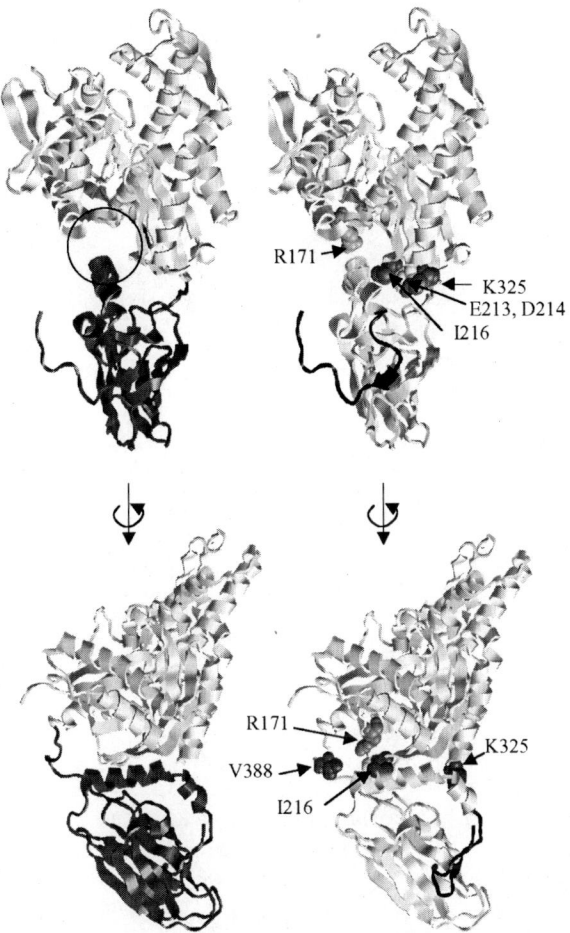

FIGURE 2. Ribbon diagram of nucleotide-free bovine Hsc70 residues 1-554 (PDB: 1YUW), highlighting the AD-PBD interface. The PBD is illustrated in dark gray in panels at left. Most of helix B is unfolded (residues 534–554, illustrated in black in the panels at right), and a segment is bound in the peptide-binding site. Helix A of the PBD interacts with an acidic cleft (circled) in the AD. This same cleft has been implicated in J-domain binding. A mutation at R167 in DnaK (R171 in Hsc70) suppressed a mutation in the HPD tripeptide of DnaJ (Suh et al., 1998). Mutations (e.g., at I216) in the cleft of Hsc70 disrupted auxilin binding. However, other mutations in the Hsc70 AD-PBD interface, but which lie outside of the cleft (e.g., at K325 or V388), also affected auxilin binding. Thus, it is unclear whether any of these mutations directly affected J-domain binding or whether they affected AD-PBD interactions, which in turn affected J-domain binding. The ATP-bound form of Hsc70 is much less elongated (Wilbanks et al., 1995), suggesting that ATP-binding causes the AD and PBD to dramatically rearrange their interaction. Crystallization was aided by the introduction of mutations in the AD-PBD interface (E213A and D214A).

Hsc70 in the recent crystal structure were consistent with the size and shape of the ADP-form in solution. Several studies have shown that the ADP-form is more protease-sensitive, which is consistent with the exposure of the interdomain linker observed in the crystal structure (Buchberger et al., 1995; Jiang et al., 2005; Liberek et al., 1991). Taken together, these studies suggest that ATP-form of the protein may look quite different from the recent crystallographic snapshot. In the ATP-form, the PBD may have a very different orientation and interaction with the AD. Such large domain-wise displacements may be employed by the Hsp40–Hsp70 machine to generate force that acts on other macromolecules, in the manner of molecular motors such as actomyosin and kinesin.

3. Hsp40 Structure

The classical Hsp40s are a small group of molecular chaperones in a very large family of J-domain-containing proteins. Classical Hsp40s are homologous to *E. coli* DnaJ, which was divided into four domains on the basis of sequence comparisons: the N-terminal J-domain, a flexible gly/phe-rich region, a Zinc-binding domain, a "C-terminal" domain (Bork et al., 1992). Classical Hsp40s are dimers, stabilized by a coiled-coil motif and sometimes other interactions near the C terminus (Wu et al., 2005).

3.1. J-domain

The number of NMR and crystal structures determined of isolated J-domains emphasizes their importance and diversity (Berjanskii et al., 2000; Gruschus et al., 2004; Huang et al., 1999; Jiang et al., 2003; Pellecchia et al., 1996; Qian et al., 1996). However, comparatively little is known about J-domain interactions with other Hsp40 domains or other proteins (Fig. 3). One crystal structure shows the SV40 T-antigen J-domain complexed with the retinoblastoma pocket domain (Kim et al., 2001) (Fig. 4). Presumably, this complex represents the end-product of the Hsc70-dependent disassembly of the Rb-E2F complex (see below). Another

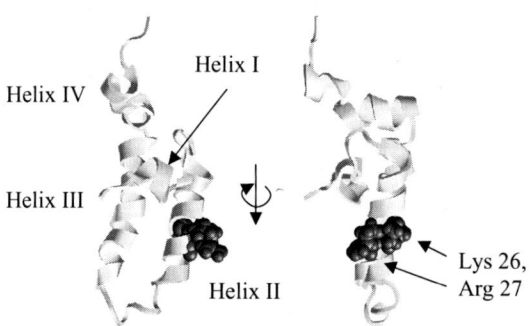

FIGURE 3. Ribbon diagram of the J-domain from *E. coli* DnaJ (PDB: 1BQZ). According to NMR studies, the binding site for the Hsp70 AD is centered on residues K26–R27 and covers helix II and the H33-P34-D35 tripeptide (Greene et al., 1998).

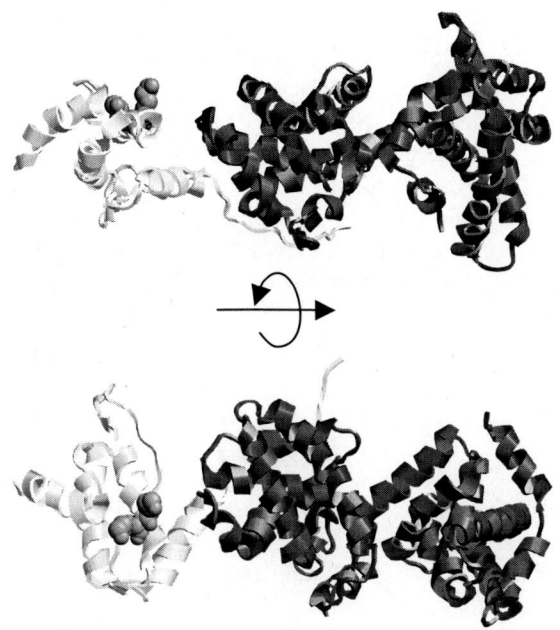

FIGURE 4. Ribbon diagram of the J-domain from the polyomavirus T antigen complexed with the retinoblastoma (Rb) pocket domain (dark gray). Space-filled atoms (K36, K37) correspond to residues at the center of the binding site for the Hsp70 AD in the *E. coli* J-domain (see Fig. 3). Rb interacts with J-domain helix IV and a C-terminal stretch of irregularly structured polypeptide, where it should not interfere with Hsp70 binding (Kim et al., 2001).

crystal structure shows the complete E. coli HscB, a J-domain protein involved in the assembly of iron-sulfur cofactors (Cupp-Vickery and Vickery, 2000). However, HscB interactions with Hsp70/HscA and the IscU client protein are poorly defined (Andrew et al., 2006); and thus the HscB structure so far explains little of how HscB-Hsp70/HscA acts on IscU.

The J-domain is a bundle of four α-helices. Helices II and III form a coiled-coil joined at one end by a flexible, but highly conserved loop that includes a universally conserved tripeptide, HPD. Mutation of any of these residues destroys Hsp40–Hsp70 cooperation (Kelley and Georgopoulos, 1997; Scidmore et al., 1993; Wall et al., 1995). On the basis of NMR signal perturbations, helix II binds to the Hsp70 AD (Greene et al., 1998). Subsequent studies with mutant DnaJ and auxilin showed that disruptions in the Hsp70-binding site have consequences for Hsp40–Hsp70 cooperation (Genevaux et al., 2002; Jiang et al., 2003).

3.2. Gly/Phe-Rich Region

The gly/phe-rich region seems to have great importance to Hsp40–Hsp70 function. It is flexible but not completely disordered (Greene et al., 2000; Pellecchia et al., 1996). *S. cerevisiae* has a paralogous Hsp40, Sis1, in which the gly/phe-rich region extends into a gly/met-rich region (Luke et al., 1991). Domain-swapping studies between Sis1 and the yeast's classical Hsp40, Ydj1, found that this flexible domain determined functional specialization of the Hsp40–Hsp70 machine (Fan et al., 2004; Lopez et al., 2003; Yan and Craig, 1999). Several "DIF" motifs in the gly/phe-rich region of DnaJ are critical for function (Cogelja Cajo et al., 2006). Nevertheless, the specific role of the gly/phe-rich region remains unknown. Investigators have suggested that it serves as a pseudo-client peptide that binds with very low affinity in the Hsp70 peptide-binding site, where it readily exchanges with authentic client peptides (Karzai and McMacken, 1996). Others have suggested that it acts as a linker between domains that must move during the interaction with Hsp70 (Aron et al., 2005).

3.3. Zinc-Binding and C-Terminal Domains

In the crystal structure of Ydj1, the monomer the zinc-binding and C-terminal domains form an "L"-shaped monomer (Li et al., 2003) that assembles into a large clasp-like dimer (Fig. 5). The Zn-binding and C-terminal domains are organized into three structural domains, Domains I-III, which are numbered N-terminal to C-terminal. Domains I and III form the long arm of the "L." The first of two Zn-binding motifs, ZnI, is integrated into Domain I. ZnII comprises Domain II, and it forms the short arm of the "L." The zinc-binding domains have been implicated

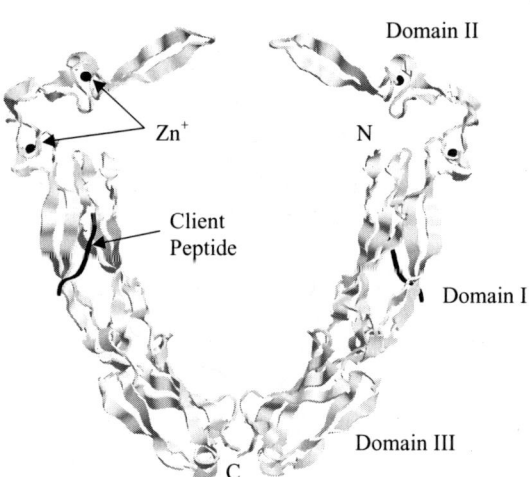

FIGURE 5. Ribbon diagram of the zinc-binding domain and C-terminal domain of *S. cerevisiae* Ydj1. The cleft between monomers is large enough to accommodate the PBD of Hsp70, which could bind an extended client polypeptide that is bound to both monomers of Ydj1 (Qian et al., 2002). A model for the dimer was constructed by superimposing Domain III of each monomer (PDB: 1NLT) with Domains III in the crystal structure of the Ydj1 dimerization motif (PDB: 1XAO) (Wu et al., 2005).

in client protein binding, redox activity, and interactions with Hsp70s (Banecki et al., 1996; Choi et al., 2006; Linke et al., 2003; Shi et al., 2005; Szabo et al., 1996; Tang and Wang, 2001).

Domain I of Ydj1 contains a small client-binding site that accepts a peptide as an anti-parallel β-strand at the edge of a β-sheet, and it has a pocket that accommodates the hydrophobic sidechain of a single amino acid residue. Such a modest interaction is consistent with several studies showing that this domain is dispensable for a minimal Hsp40 function (Lu and Cyr, 1998; Wall et al., 1994). In the context of the Ydj1 dimer, the two client-binding sites could engage a single polypeptide spanning across the large cleft. The cleft is large enough to accommodate the Hsp70 peptide-binding domain (Qian et al., 2002; Wu et al., 2005), which could simultaneously bind the client polypeptide between the two Ydj1 sites.

3.4. Models for the Hsp40–Hsp70 Mechanism

In a minimal model for the Hsp40–Hsp70 mechanism, the role of the J-domain is to stabilize peptide-binding to the Hsp70 by stimulating the Hsp70 ATPase activity. The J-domain protein probably binds independently to the client protein, and then it stimulates tight binding of the Hsp70. In the case of Hsp40/DnaJ, the J-domain's affinity for Hsp70/DnaK is very low, and thus the J-domain probably is not responsible for recruiting DnaK to client proteins (Wittung-Stafshede et al., 2003). Moreover, simple binding of the J-domain to Hsp70 could not be sufficient for J-domain function because at least one mutation in the J-domain causes a modest increase in its affinity for the Hsp70 AD and yet destroys Hsp40/DnaJ function (Landry, 2003a). The cellular concentration of Hsp40/DnaJ is substantially lower than the concentration of Hsp70/DnaK, which is consistent with Hsp40/DnaJ acting catalytically (Neidhardt and VanBogelen, 1987). Experiments support a catalytic role for J-domain proteins. A truncated DnaJ catalytically stimulated loading of DnaK onto the sigma-32 transcription factor (Liberek et al., 1995). Sub-stoichiometric amounts of Auxilin 1 optimally support clathrin uncoating (Ma et al., 2002).

The Hsp70 may be recruited to the client by interactions outside of the J-domain, such as have been described for Hsp40/Sis1-Hsp70/Ssa1 (Aron et al., 2005; Qian et al., 2002), or the Hsp70 may encounter the client protein independently (Knieszner et al., 2005; Russell et al., 1999). In higher eukaryotes, the Hsp70 may be recruited to client proteins through a tetratricopeptide repeat (TPR) protein (Blatch and Lassle, 1999), at least one of which contains a J-domain (Brychzy et al., 2003).

Most models for Hsp40–Hsp70 action on client proteins have emphasized how cycles of Hsp70 binding and releasing could accomplish the biological activity. However, recent studies have begun to enlarge the role of Hsp40–Hsp70, giving this machine the ability to generate force, such as in "pulling" precursors into the mitochondrion (Chauwin et al., 1998; Elston, 2000) or "prying apart" segments of clathrin from each other (Heymann et al., 2005). Hsp70 could exert a force on the client peptide by applying a "power stroke" in the manner of a lever or pump (Chauwin et al., 1998; Elston, 2000; Pierpaoli et al., 1997).

Force-generation models would require a combination of static and dynamic interactions with the client protein and the translocation channel or other molecules in the protein complex (Landry, 2003a,b). The requirement for PAM16 and PAM17 to organize the J-domain function of PAM18 suggests that the mitochondrial import motor has enough sophistication to provide force generation (Li et al., 2004; van der Laan et al., 2005). Alternatively, the Hsp70 might exert a force on the client protein by harnessing Brownian motions while locked onto to the client protein until ADP/ATP exchange releases the Hsp70 (Ben-Zvi et al., 2004; Okamoto et al., 2002).

3.5. Non-Classical J-Domain Proteins

Non-classical J-domain proteins exhibit extreme specialization. Some of the most divergent J-domain proteins include polyomavirus T-antigens, auxilin, Sec63, and cysteine-string protein (CSP). On the one hand, in proteins like Sec63 and CSP, the J-domain appears to have been joined with domains whose sole purpose is to recruit Hsp70 function to a specific cellular process. In Sec63, the J-domain lies between membrane-spanning segments that position the J-domain in the lumen of the endoplasmic reticulum, where it engages Hsc70 during protein translocation (Feldheim et al., 1992). The lipidated cysteine-rich sequences in CSP localize the J-domain to secretory membranes, where CSP is thought to help recycle the machinery of vesicle fusion (Schmitz et al., 2006). On the other hand, the J-domain may have been joined with other domains that, transiently, must converge on the same client protein complex. For example, large T-antigen binds the viral DNA, acts as an ATP-dependent helicase, and sequesters tumor suppressors using distinct domains (Ahuja et al., 2005; Campbell et al., 1997). While the J-domain is necessary to initiate DNA replication, its function is separable from the helicase activity (Campbell et al., 1997). Auxilin's J-domain is necessary for clathrin uncoating, but other domains in auxilin have (possibly independent) roles in the formation of clathrin-coated vesicles (Ma et al., 2002). In these examples, it is unclear whether the J-domain "hitched a ride" to the complex with the other domains, or whether the other domains "hitched a ride" to the complex with a primordial J-domain protein.

4. The Hsp40–Hsp70 Disassembler

Various protein systems, ranging from indeterminant aggregates to highly evolved multi-subunit complexes are disassembled by the Hsp40–Hsp70.

4.1. Nascent and Denatured Protein Aggregates

The role of the Hsp40–Hsp70 chaperone machine in the stress response is thought to involve the ATP-dependent disassembly of protein aggregates. The earliest indication of this activity was the observation that Hsp70 accumulated in the nucleolus of stressed cells, a site where highly concentrated ribosome-assembly intermediates are prone to aggregation (Pelham, 1984). Later on, protein aggregates were found to accumulate to higher levels in heat-stressed cells of yeast and E. coli that expressed low levels of Hsp40 or Hsp70 (Glover and Lindquist, 1998; Gragerov et al., 1992).

The possibility that Hsp40–Hsp70 assists nascent protein folding has been studied intensely. Many early experiments suggested that nascent polypeptides interact with Hsp40–Hsp70 (Beckmann et al., 1990; Hendrick et al., 1993; Nelson et al., 1992). Studies with purified proteins showed that Hsp40–Hsp70 could dramatically increase the efficiency of refolding firefly luciferase (Schroder et al., 1993). However, more recent studies lead to the conclusion that other chaperones, most notably Hsp60-Hsp10, have a general role in folding of newly synthesized proteins (Young et al., 2004). In contrast, the ability of Hsp40–Hsp70 to facilitate protein folding is largely restricted to the refolding of denatured and/or aggregated proteins (Hesterkamp and Bukau, 1998). To the extent that Hsp40–Hsp70 interacts with nascent proteins, it may be dedicated to passing the client proteins along to other chaperones or a membrane-translocation apparatus.

4.2. Origin DNA-Binding Complexes

The Hsp40/DnaJ-Hsp70/DnaK disassembler has an essential role in the initiation of bacteriophage λ DNA replication, hence the names DnaJ and DnaK (Georgopoulos and Eisen, 1974). Together with the nucleotide-exchange factor, GrpE, they partially disassemble the protein complex composed of the bacteriophage O and P and the *E. coli* DnaB at the bacteriophage origin of DNA replication (Liberek et al., 1988). P recruits the DnaB helicase by binding more tightly than the host DnaC and thus redirects the replication machinery to the viral DNA (Mallory et al., 1990). The O-P-DnaB complex is extremely stable, and therefore it must be partially disassembled by DnaJ/Hsp40-DnaK/Hsp70 so that DnaB can begin unwinding the DNA.

Hsp40/DnaJ-Hsp70/DnaK disassembles dimers of RepA, whose monomers bind to the plasmid P1 origin of replication (Wickner et al., 1991). The DnaJ-binding site on RepA was localized by mutagenesis to residues 185–200, which contains a stretch of hydrophobic residues that would be suitable for interaction with a peptide-binding site on DnaJ (Kim et al., 2002). Wickner and co-workers constructed an homology model of the RepA monomer on the basis of a similar replication initiator protein, RepE (Sharma et al., 2004). In the model, the DnaJ-binding sequence appears in an ordered, but irregularly structured segment connecting two α-helices. The model may be valid for the RepA dimer because DnaJ binds both dimers and monomers. It is interesting that the DnaJ-binding site on RepA is ordered because chaperones are generally thought to recognize features typical of unfolded proteins (Landry and Gierasch, 1994). Nevertheless, the DnaJ-binding site is exposed on the molecular surface, and it is hydrophobic. The DnaK-binding site in RepA was localized to residues 36–49 (Sharma et al., 2004). As yet, there is no structural information available for this segment of RepA because it has no counterpart in the structures of similar replication initiators. The monomerization of RepA can also be accomplished by the AAA-family chaperone, ClpA, which suggests that Hsp40–Hsp70 and AAA-family chaperones have similar activities and overlapping functions (Wickner et al., 1994).

4.3. Tumor-Suppressor Complex

The J-domains of polyomavirus T/t antigens have multiple roles in the virus life cycle. SV40 is a polyomavirus-family member that makes a large T antigen and a small t antigen. The two proteins share a "common domain," which is a J-domain (Kelley and Landry, 1994). Mutations in the J-domain cause defects in the control of the host cell cycle, viral DNA replication, and viral capsid assembly (Campbell et al., 1997; Chromy et al., 2003; Fewell et al., 2002; Whalen et al., 2005). Several of large T's myriad roles in the viral life cycle involve the disassembly of protein complexes. In some cases, these activities are clearly associated with the ability of large T's J-domain to engage Hsc70. For example, SV40 uses Hsc70 to disassemble complexes of retinoblastoma tumor suppressor Rb with the transcription factor E2F (Sullivan et al., 2000). Released E2F drives expression of numerous genes required for viral DNA replication. In vitro, large T associates with Rb-E2F complexes but does not cause disassembly until Hsc70 and ATP are added.

4.4. Clathrin Cage

The clathrin cage is one of the most well-established Hsp40–Hsp70 client protein complexes. Clathrin cages assemble on coated vesicles of donor membranes and must be disassembled prior to fusion of the vesicle with the target membrane. Hsc70 was originally identified as a clathrin-uncoating ATPase (Chappell et al., 1986; Schlossman et al., 1984). Specialized J-domain proteins, known as auxilins, provide the Hsp40 function by stably associating with clathrin cages and stimulating ATP-dependent disassembly by Hsc70 (Ungewickell et al., 1995). Auxilins are different from classical Hsp40s in that the J-domain is located near the C-terminus. Toward the N-terminus, mammalian auxilins (Auxilin 1 and Auxilin 2) have a clathrin-binding domain and a PTEN/tensin tyrosine phosphatase domain (Lemmon, 2001). Auxilin 2, also known as the cyclin-G-associated kinase (GAK), has an additional serine/threonine kinase domain. Auxilin 1 is selectively expressed in neurons, whereas Auxilin 2 is predominant in non-neuronal cells. The phosphatase and kinase domains of Auxilin 2 may be involved in binding and sorting of cargo during the formation of clathrin-coated vesicles (Zhang et al., 2005). Neither the phosphatase nor the kinase domain is required for clathrin uncoating (Ma et al., 2002). The uncoating reaction has been studied using clathrin coats assembled in vitro from clathrin and adaptin subunits and subsequently mixed with auxilin, Hsc70, ATP, and an ATP-regenerating system (Ma et al., 2002). The uncoating reaction may be arrested after binding of the chaperones by mixing the components at pH 6. Disassembly proceeds when the reaction is shifted to pH 7.

A mechanism for disassembly of clathrin by Hsc70 has been proposed on the basis of atomic structures obtained by cryo-electron microscopy. The J-domain of a functional auxilin fragment has been identified in the 12-Å structure of a clathrin coat (Fotin et al., 2004). The auxilin fragment contains the J-domain and a minimal clathrin-binding segment. Binding of the fragment distorts the coat in comparison to the unbound coat. The auxilin J-domains are located in vertices beneath the clathrin hubs (Fig. 6). The J-domain fits at the intersection of three

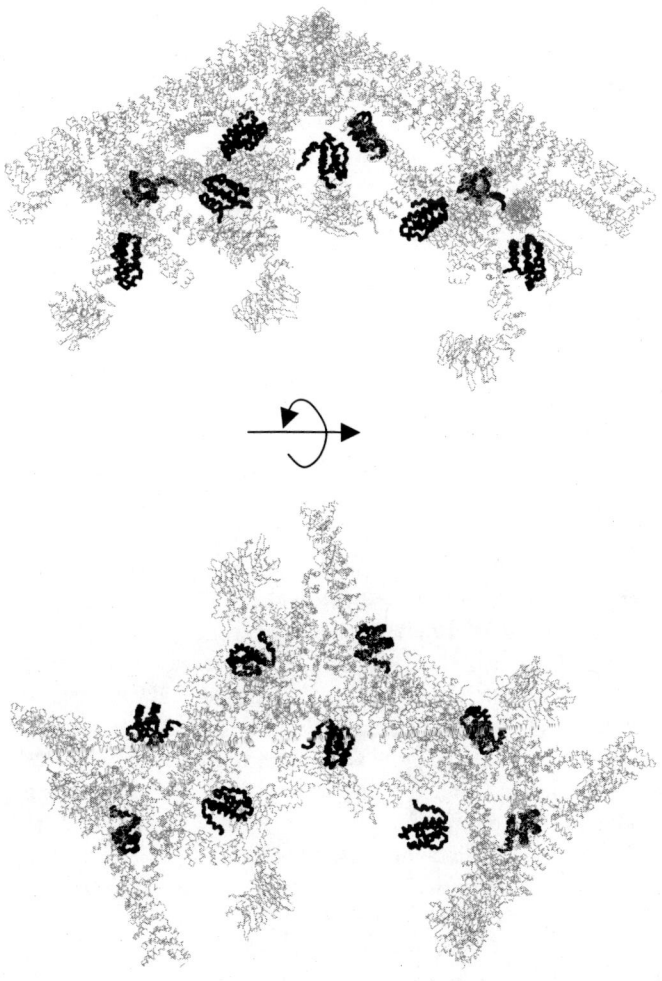

FIGURE 6. Polypeptide backbone illustrating a portion of a clathrin cage saturated with an auxilin fragment, composed of residues 797-910, that includes the J-domain (PDB: 1XI5). The auxilin fragments are highlighted in dark gray.

crossed clathrin-heavy chain "ankles," and it contacts the terminal domain of a fourth heavy chain. The Hsc70-binding surface of the J-domain faces toward the inside of the coat, where there is space to accommodate the Hsc70 near the vertex. The only part of the heavy chains that may be sufficiently disordered for Hsc70 to bind is the C-terminal 40-residue segment, which is thought to "brace" the corners formed by the crossing ankle regions. It is proposed that the coat distortion caused by auxilin releases the C-terminal segments, and then Hsc70 sequesters them, thereby destabilizing the vertex.

An alternative model system for clathrin uncoating suggests that the J-domain can cooperate with Hsc70 irrespective of its location on the coat. In this model system, the auxilin J-domain is incorporated into the coats by co-assembly of a fusion protein (C58J) composed of the J-domain and the clathrin-binding domain of adaptin AP80 (Ma et al., 2002). Using cryo-electron microscopy, C58J-coats were compared with C58J-coats plus Hsc70 (Heymann et al., 2005). The Hsc70 was found near the surface of the coat, alongside the clathrin hubs. The investigators found that Hsc70 caused a distortion of the coat that resembled the distortion caused by auxilin described above. It is unclear whether Hsc70 binds to the same site on the clathrin heavy chains when cooperating with C58J versus authentic auxilin.

5. Disassembly for or Against Protein Degradation

5.1. Proteolysis of Aggregates

There is little evidence for a direct connection between Hsp40–Hsp70 and proteolytic enzymes in spite of strong indications that these two types of machinery act on converging pathways of protein traffic. Defects in Hsp40/DnaJ-Hsp70/DnaK induce the expression of proteases, and defects in proteolysis induce the expression of Hsp40–Hsp70 (Bukau, 1993; Bush et al., 1997; Georgopoulos et al., 1994; Neidhardt and VanBogelen, 1987). Several chaperones of the AAA-family, some of which are thought to act either in parallel or sequentially with Hsp40–Hsp70, are closely associated with proteases. The *E. coli* ClpA and ClpX chaperones form complexes with the ClpP protease, and the ClpY(HslU) chaperone forms a similar complex with the ClpQ (HslV) protease (Gottesman et al., 1997). The *E. coli* lon protease has probably assembled both the chaperone and protease functions into a single polypeptide (Smith et al., 1999). These machines are similar to the eukaryotic 26S proteasome, which has subunits that are homologous to ClpA/ClpX. The chaperones are responsible for unfolding and transfer of client proteins into the degradation chamber of the protease.

In spite of similarities between protease-associated chaperones and Hsp40–Hsp70, such as ATP dependence and the ability to bind common clients/substrates proteins, Hsp40–Hsp70 does not appear to be responsible for transfer of clients/substrates to proteases. *E. coli* strains that are depleted of DnaJ and/or

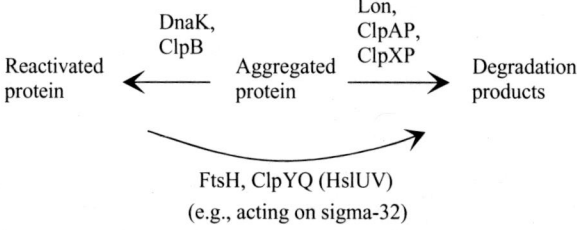

FIGURE 7. Alternate fates for aggregated proteins.

DnaK can still degrade proteins using proteases such as lon and ClpAP (Jubete et al., 1996; Tomoyasu et al., 2001). The role of Hsp40–Hsp70 may be to protect proteins from aggregation or to restore them from the aggregated state, without regard to their subsequent fate (Fig. 7). Interestingly, DnaJ-DnaK may be aided in this role by the chaperone activity of ClpB, a AAA-family chaperone that is not associated with a protease (Goloubinoff et al., 1999; Zietkiewicz et al., 2006; Zolkiewski, 1999).

5.2. Proteolysis of Sigma-32

Efficient degradation of at least one client protein depends on the presence of Hsp40/DnaJ-Hsp70/DnaK. One of the best-studied client proteins for Hsp40/DnaJ and Hsp70/DnaK is the *E. coli* heat shock transcription factor sigma-32, which directs the synthesis of heat-shock protein genes. The turnover of sigma-32 is regulated by a feedback loop that senses the availability of the DnaJ and DnaK chaperones (Georgopoulos et al., 1994). In unstressed cells, DnaJ and DnaK bind to sigma-32 and somehow facilitate its rapid degradation. During heat shock, the chaperones become sequestered by other client proteins, and the half-life of sigma-32 increases by more than 10-fold (Morita et al., 2000). The resulting increase in sigma-32 stability and concentration allows sigma-32 to compete effectively with the constitutively expressed sigma-70 for binding to the RNA polymerase core. When the stress is over, sigma-32 re-associates with DnaJ–DnaK and resumes rapid turnover.

The protease most responsible for sigma-32 degradation is FtsH, an AAA-family protease that has only a weak ability to unfold substrate proteins (Herman et al., 2003). Perhaps this explains the dependence of sigma-32 turnover on DnaJ–DnaK. Sigma-32 may be resistant to proteolysis by FtsH unless it has been unfolded or disaggregated by DnaJ–DnaK. However, sigma-32 must be released from DnaJ–DnaK before it associates with FtsH. Repeated attempts to show transfer of proteins from DnaJ–DnaK to FtsH in vitro have been unsuccessful (Blaszczak et al., 1999; Herman et al., 2003; Tatsuta et al., 2000). DnaJ–DnaK apparently is not required for the dissociation of sigma-32 from the RNA polymerase core. Sigma-32 mutants that are defective in core binding still require DnaJ–DnaK for rapid degradation in vivo (Tatsuta et al., 2000).

5.3. Disassembly in Support of Protein Translocation Across Membranes

The necessity that proteins be unfolded prior to and during membrane translocation creates an obvious need for molecular chaperones. In this process Hsp40–Hsp70 may be utilized for disassembly of tertiary structure, rather than of quaternary structure or aggregates. Cells use Hsp40–Hsp70 on the cis-side of the membrane to stabilize some precursors. However, the more important roles of Hsp40–Hsp70 appear to be on the trans-side. The binding of Hsp70 to precursors on the trans-side of the membrane gives direction to translocation, and the ATP-dependent Hsp70 conformational change drives the their unfolding on the cis-side of the membrane.

5.4. Hsp70 on the Cis-Side of the Membrane

One of the first constitutive roles to be ascribed to Hsp70 is the maintenance of precursor proteins in a translocation-competent state prior to transport into the endoplasmic reticulum. Hsp70 was purified by its ability to stimulate protein translocation into the ER (Chirico et al., 1988). Moreover, the down-regulation of Hsp70 in yeast cells caused a subset of ER and mitochondrial precursors to accumulate in the cytosol (Deshaies et al., 1988). The requirement for Hsp70 could be rationalized by a necessity for precursors to be unfolded during translocation. Prior experiments studying bacterial protein export and import into mitochondria had demonstrated that stable structure in precursors could block their translocation (Eilers and Schatz, 1986; Randall and Hardy, 1986). Most ER proteins are translocated co-translationally, and thus the precursors never have a chance to fold before they arrive in the ER lumen. In bacteria, Hsp40/DnaJ-Hsp70/DnaK facilitates protein export, but this machine is not essential, probably because the SecB chaperone and the SecA translocation motor are dedicated to the task (Wild et al., 1992).

A general role for cytosolic Hsp70 in preventing the folding of mitochondrial protein precursors is not so well established, possibly because Hsp70 on the cis-side of the membrane drives unfolding (see below). ATP and Hsp70 from a reticulocyte lysate are required for import of some precursors into yeast mitochondria (Asai et al., 2004; Young et al., 2003), but the ATP may be necessary only to recycle Hsp70 after it targets the precursors to a receptor on the mitochondrial outer membrane (Young et al., 2003).

5.5. Hsp70 on the Trans-Side of the Membrane

The involvement of Hsp40–Hsp70 on the trans-side of the membrane was first demonstrated for translocation into the ER of yeast, where Hsp70/BiP binds to precursors during translocation and interacts with the membrane-bound J-domain protein, Sec63 (Brodsky and Schekman, 1993). Mutations in either of these proteins could block import, and a mutation in Sec63 could be suppressed by a mutation in Hsp70/Bip (Scidmore et al., 1993). However, the fact that most translocation was co-translational made it impossible to establish whether the chaperones provided a driving force for translocation, and clearly there was no need for protein unfolding on the cis-side of the membrane. There is no obvious Sec63-homolog in mammalian cells. Nevertheless, Hsp70/BiP can interact with any of a number of J-domain proteins that have strikingly diverse structures and cellular roles that probably go well beyond the modulation of the Hsp70/BiP ATPase (Chevalier et al., 2000; Cunnea et al., 2003; Shen et al., 2002; Yu et al., 2000).

Compelling evidence that Hsp40–Hsp70 drives translocation and unfolds precursors emerged for protein transport into yeast mitochondria, and the story continues to develop rapidly. An Hsp70/Ssc1 mutant that cannot hydrolyze ATP is able to provide the driving force for translocation, but only if unfolding is not required on the cis-side of the membrane (Kang et al., 1990). This is explained in a "Brownian-ratchet" model for translocation in which Hsp70 molecules prevent the precursor from diffusing back across the membrane. As segments of the protein spontaneously move across the membrane, molecules of Hsp70 continue to bind until the entire polypeptide is in the mitochondrial matrix. However, if the precursor has the ability to fold into a stable structure on the cis-side of the membrane, then the Hsp70/Ssc1 mutant cannot drive import. Presumably, ATP hydrolysis by Hsp70/Ssc1 is required to exert a pulling force from the trans-side of the membrane, and the pulling force squeezes the precursor into the translocation channel, forcing the precursor to unfold. In order for Hsp70/Ssc1 to unfold the precursor, the precursor must have an unfolded segment (including the mitochondrial targeting sequence) that is long enough to reach into the matrix (Matouschek et al., 1997). The idea that precursor unfolding limits translocation is supported by studies showing inverse correlations of translocation rate with precursor stability (Gaume et al., 1998; Huang et al., 1999, 2002), which suggests that translocation is delayed until the precursor spontaneously unfolds. A recent study found that translocation rate correlated better with the rate of forced unfolding as measured by atomic force microscopy, which may more accurately mimic unfolding induced by Hsp70/Ssc1 (Sato et al., 2005).

The mechanics of the mitochondrial import motor are beginning to take shape in the structural details of Hsp70/Ssc1's interaction with the translocation channel. Hsp70-ATP docks at the inactive translocation channel by binding to TIM44 through bivalent interactions involving the Hsp70/Ssc1 AD and PBD (D'Silva et al., 2004). When the precursor reaches the motor, Hsp70-ATP dissociates from TIM44 and binds loosely to the precursor. PAM18 uses its J-domain to stimulate ATP hydrolysis, which triggers Hsp70 to bind tightly to the precursor (D'Silva et al., 2003; Truscott et al., 2003). PAM16 is thought to control the activity of PAM18 by an unprecedented mechanism (D'Silva et al., 2005). PAM16 has a domain that resembles a J-domain, but it lacks the characteristic HPD tripeptide and cannot stimulate Hsp70/Ssc1. The J-like domain of PAM16 binds to the J-domain of PAM18, which inhibits the ability of PAM18 to stimulate Hsp70/Ssc1. PAM17 stabilizes the association of PAM18 and PAM16, and the three proteins are anchored by transmembrane spans in PAM18 and PAM17 (van der Laan et al.,

2005). The motor uses a GrpE-like nucleotide-exchange factor (Mge1) to recycle Hsp70/Ssc1-ADP into the ATP form (Sakuragi et al., 1999).

6. Disassembly in Support of Folding or Assembly

6.1. Transcription Factors and Signaling Proteins

Hsp40–Hsp70 is most likely required for the maturation of a large number of transcription factors and other proteins involved in signal-transduction pathways (Freeman and Yamamoto, 2002). The prototypical client is the glucocorticoid receptor, which follows a multistep pathway, initially through interactions with Hsp40–Hsp70, and then with Hsp90, other chaperones, and immunophilins before acquiring its activity as a hormone-sensitive transcription factor (Hernandez et al., 2002). Certain protein tyrosine kinases may follow a similar path to the active state (Pratt and Toft, 2003). The precise role of Hsp40–Hsp70 is not known. Since there is no evidence that the client proteins are aggregated, it seems likely that Hsp40–Hsp70 uses its peptide-binding and force-generating activities to partially unfold the receptor and then hand it off to the Hsp90.

6.2. SV40 Capsid

An ability of Hsp40–Hsp70 to unfold proteins could explain Hsp40–Hsp70 action in the assembly of viral particles. Hsp70 associates with the major capsid protein of SV40 in infected mammalian cells, as well as in recombinant *E. coli* (Cripe et al., 1995). Since mutations in the J-domain of SV40 large T antigen block virus assembly (Brodsky and Pipas, 1998; Kelley and Landry, 1994; Peden and Pipas, 1992; Spence and Pipas, 1994), it is possible that the viral J-domain and host Hsp70 work together in the capsid assembly. Chaperone-mediated assembly of SV40 capsids was demonstrated in vitro (Chromy et al., 2003). Capsids can be assembled in vitro from recombinant capsid protein in calcium or high ionic strength, but the structures of the resulting particles are much less homogeneous than authentic capsids. In contrast, assembly reactions carried out in the presence of Hsp40/DnaJ-Hsp70/DnaK yield much more uniform, authentic-looking capsids. Similar results were obtained when Hsp40/DnaJ was substituted with SV40 large T antigen. Enhanced uniformity in the capsid structures could result from cyclic disassembly and reassembly of the structures. This activity would be analogous to the "iterative annealing" mechanism that has been used to explain assisted refolding by Hsp60-Hsp10 (Todd et al., 1996). The idea is that multiple conformations (assemblies) have free energies or kinetic probabilities that are only slightly less favorable than the native conformation (assembly), and thus misfolds (or incorrect assemblies) are highly probable. Misfolded proteins and incorrect assemblies may be preferred clients for chaperones because they expose segments of hydrophobic polypeptide. By constantly unraveling and disassembling incorrect structures, Hsp40–Hsp70 could favor the assembly of the correct structure.

7. Concluding Remarks

The Hsp40–Hsp70 machine has been deployed by nature on a great variety of cellular processes that involve protein disassembly or unfolding. Hsp70 can deliver this function because it undergoes structural transformations that radically change its affinity for client proteins and possibly allow it to generate force. The diverse Hsp40/J-domain proteins activate the Hsp70 wherever the disassembly function is required. Future studies will reveal whether Hsp40–Hsp70 operates simply as a Brownian ratchet or in a more sophisticated manner, such as a molecular motor.

References

Ahuja, D., Saenz-Robles, M. T., and Pipas, J. M. (2005) SV40 large T antigen targets multiple cellular pathways to elicit cellular transformation. *Oncogene* 24:7729–45.

Alder, N. N., Shen, Y., Brodsky, J. L., Hendershot, L. M., and Johnson, A. E. (2005) The molecular mechanisms underlying BiP-mediated gating of the Sec61 translocon of the endoplasmic reticulum. *J Cell Biol* 168:389–99.

Andrew, A. J., Dutkiewicz, R., Knieszner, H., Craig, E. A., and Marszalek, J. (2006) Characterization of the interaction between the J-protein Jac1 and Isu1, the scaffold for Fe–S cluster biogenesis. *J Biol Chem*.

Aron, R., Lopez, N., Walter, W., Craig, E. A., and Johnson, J. (2005) *In vivo* bipartite interaction between the Hsp40 Sis1 and Hsp70 in *Saccharomyces cerevisiae*. *Genetics* 169:1873–82.

Asai, T., Takahashi, T., Esaki, M., Nishikawa, S., Ohtsuka, K., Nakai, M., and Endo, T. (2004) Reinvestigation of the requirement of cytosolic ATP for mitochondrial protein import. *J Biol Chem* 279:19464–70.

Banecki, B., Liberek, K., Wall, D., Wawrzynow, A., Georgopoulos, C., Bertoli, E., Tanfani, F., and Zylicz, M. (1996) Structure–function analysis of the zinc finger region of the DnaJ molecular chaperone. *J Biol Chem* 271:14840–8.

Beckmann, R. P., Mizzen, L. A., and Welch, W. J. (1990) Interaction of Hsp70 with newly synthesized proteins: Implications for protein folding and assembly. *Science* 248:850–4.

Bennett, W. S., Jr., and Steitz, T. A. (1978) Glucose-induced conformational change in yeast hexokinase. *Proc Natl Acad Sci U S A* 75:4848–52.

Ben-Zvi, A., De Los Rios, P., Dietler, G., and Goloubinoff, P. (2004) Active solubilization and refolding of stable protein aggregates by cooperative unfolding action of individual hsp70 chaperones. *J Biol Chem* 279:37298–303.

Berjanskii, M. V., Riley, M. I. Xie, A., Semenchenko, V., Folk, W. R., and Van Doren, S. R. (2000) NMR structure of the N-terminal J domain of murine polyomavirus T antigens. Implications for DnaJ-like domains and for mutations of T antigens. *J Biol Chem* 275:36094–103.

Bertelsen, E. B., Zhou, H. J., Lowry, D. F., Flynn, G. C., and Dahlquist, F. W. (1999) Topology and dynamics of the 10 kDa C-terminal domain of DnaK in solution. *Prot Sci* 8:343–54.

Blaszczak, A., Georgopoulos, C., and Liberek, K. (1999) On the mechanism of FtsH-dependent degradation of the sigma(32) transcriptional regulator of *Escherichia coli* and the role of the DnaK chaperone machine. *Mol Microbiol* 31:157–66.

Blatch, G. L., and Lassle, M. (1999) The tetratricopeptide repeat: A structural motif mediating protein–protein interactions. *Bioessays* 21:932–9.

Bork, P., Sander, C., Valencia, A., and Bukau, B. (1992) A module of the DnaJ heat shock proteins found in malaria parasites. *Trends Biochem Sci* 17:129.

Brodsky, J. L., and Pipas, J. M. (1998) Polyomavirus T antigens: Molecular chaperones for multiprotein complexes. *J Virol* 72:5329–34.

Brodsky, J. L., and Schekman, R. (1993) A sec63p–BiP complex from yeast is required for protein translocation in a reconstituted proteoliposome. *J Cell Biol* 123:1355–63.

Brychzy, A., Rein, T., Winklhofer, K. F., Hartl, F. U., Young, J. C., and Obermann, W. M. (2003). Cofactor Tpr2 combines two TPR domains and a J domain to regulate the Hsp70/Hsp90 chaperone system. *EMBO J* 22:3613–23.

Buchberger, A., Theyssen, H., Schroder, H., McCarty, J. S., Virgallita, G., Milkereit, P., Reinstein, J., and Bukau. B. (1995) Nucleotide-induced conformational changes in the ATPase and substrate binding domains of the DnaK chaperone provide evidence for interdomain communication. *J Biol Chem* 270:16903–10.

Bukau, B. (1993) Regulation of the *Escherichia coli* heat–shock response. *Mol Microbiol* 9:671–80.

Bush, K. T., Goldberg, A. L., and Nigam, S. K. (1997). Proteasome inhibition leads to a heat–shock response, induction of endoplasmic reticulum chaperones, and thermotolerance. *J Biol Chem* 272:9086–92.

Campbell, K. S., Mullane, K. P., Aksoy, I. A., Stubdal, H., Zalvide, J., Pipas, J. M., Silver, P. A., Roberts, T. M., Schaffhausen, B. S., and DeCaprio. J. A. (1997) DnaJ/hsp40 chaperone domain of SV40 large T antigen promotes efficient viral DNA replication. *Genes Dev* 11:1098–110.

Chappell, T. G., Welch, W. J., Schlossman, D. M., Palter, K. B., Schlesinger, M. J., and Rothman, J. E. (1986) Uncoating ATPase is a member of the 70 kilodalton family of stress proteins. *Cell* 45:3–13.

Chauwin, J. F., Oster, G., and Glick, B. S. (1998) Strong precursor–pore interactions constrain models for mitochondrial protein import. *Biophys J* 74:1732–43.

Chevalier, M., Rhee, H., Elguindi, E. C., and Blond, S. Y. (2000) Interaction of murine BiP/GRP78 with the DnaJ homologue MTJ1. *J Biol Chem* 275:19620–7.

Chirico, W. J., Waters, M. G., and Blobel, G. (1988) 70K heat shock related proteins stimulate protein translocation into microsomes. *Nature* 332:805–10.

Choi, H. I., Lee, S. P., Kim, K. S., Hwang, C. Y., Lee, Y R, Chae, S. K.,Kim, Y. S., Chae, H. Z., and Kwon, K. S. (2006) Redox-regulated cochaperone activity of the human DnaJ homolog Hdj2. *Free Radic Biol Med* 40:651–9.

Chou, C. C., Forouhar, F., Yeh, Y. H., Shr, H. L., Wang, C., and Hsiao, C. D. (2003) Crystal structure of the C-terminal 10-kDa subdomain of Hsc70. *J Biol Chem* 278:30311–6.

Chromy, L. R., Pipas, J. M., and Garcea, R. L. (2003) Chaperone-mediated *in vitro* assembly of *Polyomavirus capsids*. *Proc Natl Acad Sci U S A* 100:10477–82.

Cogelja Cajo, G., Horne, B. E., Kelley, W. L., Schwager, F., Georgopoulos, C., and Genevaux, P. (2006) The role of the DIF motif of the DnaJ (Hsp40) co-chaperone in the regulation of the DnaK (Hsp70) chaperone cycle. *J Biol Chem*.

Cripe, T. P., Delos, S. E., Estes, P. A., and Garcea, R. L. (1995) *In vivo* and *in vitro* association of hsc70 with polyomavirus capsid proteins. *J Virol* 69:7807–13.

Cunnea, P. M., Miranda-Vizuete, A., Bertoli, G., Simmen, T., Damdimopoulos, A. E., Hermann, S., Leinonen, S., Huikko, M. P., Gustafsson, J. A., Sitia, R., and Spyrou, G. (2003) ERdj5, an endoplasmic reticulum (ER)-resident protein containing DnaJ and thioredoxin domains, is expressed in secretory cells or following ER stress. *J Biol Chem* 278:1059–66.

Cupp-Vickery, J. R., and Vickery, L. E. (2000) Crystal structure of Hsc20, a J-type Cochaperone from *Escherichia coli*. *J Mol Biol* 304:835–45.

Davis, J. E., Voisine, C., and Craig, E. A. (1999) Intragenic suppressors of Hsp70 mutants: Interplay between the ATPase- and peptide-binding domains. *Proc Natl Acad Sci U S A* 96:9269–76.

Deshaies, R. J., Koch, B. D., Werner-Washburne, M., Craig, E. A., and Schekman, R. (1988) A subfamily of stress proteins facilitates translocation of secretory and mitochondrial precursor polypeptides. *Nature* 332:800–5.

D'Silva, P., Liu, Q., Walter, W., and Craig, E. A. (2004) Regulated interactions of mtHsp70 with Tim44 at the translocon in the mitochondrial inner membrane. *Nat Struct Mol Biol* 11:1084–91.

D'Silva, P. D., Schilke, B., Walter, W., Andrew, A., and Craig, E. A. (2003) J protein cochaperone of the mitochondrial inner membrane required for protein import into the mitochondrial matrix. *Proc Natl Acad Sci U S A* 100:13839–44.

D'Silva, P. R., Schilke, B., Walter, W., and Craig, E. A. (2005) Role of Pam16's degenerate J domain in protein import across the mitochondrial inner membrane. *Proc Natl Acad Sci U S A* 102:12419–24.

Eilers, M., and Schatz, G. (1986) Binding of a specific ligand inhibits import of a purified precursor protein into mitochondria. *Nature* 322:228–32.

Elston, T. C. (2000) Models of post-translational protein translocation. *Biophys J* 79:2235–51.

Fan, C. Y., Lee, S.,Ren, H. Y., and Cyr, D. M. (2004) Exchangeable chaperone modules contribute to specification of type I and type II Hsp40 cellular function. *Mol Biol Cell* 15:761–73.

Feldheim, D., Rothblatt, J., and Schekman, R. (1992) Topology and functional domains of Sec63p, an endoplasmic reticulum membrane protein required for secretory protein translocation. *Mol Cell Biol* 12:3288–96.

Fewell, S. W., Pipas, J. M, and Brodsky, J. L. (2002) Mutagenesis of a functional chimeric gene in yeast identifies mutations in the simian virus 40 large T antigen J domain. *Proc Natl Acad Sci U S A* 99:2002–7.

Fotin, A., Cheng, Y., Grigorieff, N., Walz, T., Harrison, S. C., and Kirchhausen, T. (2004) Structure of an auxilin-bound clathrin coat and its implications for the mechanism of uncoating. *Nature* 432:649–53.

Freeman, B. C., and Yamamoto, K. R. (2002) Disassembly of transcriptional regulatory complexes by molecular chaperones. *Science* 296:2232–5.

Gassler, C. S., Buchberger, A., Laufen, T., Mayer, M. P., Schroder, H., Valencia, A., and Bukau, B. (1998) Mutations in the DnaK chaperone affecting interaction with the DnaJ cochaperone. *Proc Natl Acad Sci U S A* 95:15229–34.

Gaume, B., Klaus, C. Ungermann, C., Guiard, B., Neupert, W., and Brunner, M. (1998) Unfolding of preproteins upon import into mitochondria. *EMBO J* 17:6497–507.

Genevaux, P., Schwager, F., Georgopoulos, C., and Kelley, W. L. (2002) Scanning mutagenesis identifies amino acid residues essential for the *in vivo* activity of the *Escherichia coli* DnaJ (Hsp40) J-domain. *Genetics* 162:1045–53.

Georgopoulos, C., Liberek, K., Zylicz, M., and Ang, D. (1994) Properties of the heat shock proteins of *Escherichia coli* and the autoregulation of the heat shock response. *In* Morimoto, R. I., Tissières, A., and Georgopoulos, C. (eds.): *The Biology of Heat Shock*

Proteins and Molecular Chaperones. Cold Spring Harbor Laboratory Press, Cold Spring Harbor, NY, p. 209–49.

Georgopoulos, C. P., and Eisen, H. (1974) Bacterial mutants which block phage assembly. *J Supramol Struct* 2:349–59.

Glover, J. R., and Lindquist, S. (1998) Hsp104, Hsp70, and Hsp40: A novel chaperone system that rescues previously aggregated proteins. *Cell* 94:73–82.

Goloubinoff, P., Mogk, A., Zvi, A. P., Tomoyasu, T., and Bukau, B. (1999) Sequential mechanism of solubilization and refolding of stable protein aggregates by a bichaperone network. *Proc Natl Acad Sci U S A* 96:13732–7.

Gottesman, S., Maurizi, M. R., and Wickner, S. (1997) Regulatory subunits of energy-dependent proteases. *Cell* 91:435–8.

Gragerov, A., Nudler, E., Komissarova, N., and Gaitanaris, G. A. (1992) Cooperation of GroEL/GroES and DnaK/DnaJ heat shock proteins in preventing protein misfolding in *Escherichia coli*. *Proc Natl Acad Sci U S A* 89:10341–4.

Greene, M. K., Maskos, K., and Landry, S. J. (1998) Role of the J-domain in the cooperation of Hsp40 with Hsp70. *Proc Natl Acad Sci U S A* 95:6108–13.

Greene, M. K., Steede, N. K., and Landry, S. J. (2000) Domain-specific spectroscopy of 5-hydroxytryptophan-containing variants of *Escherichia coli* dnaJ [In Process Citation]. *Biochim Biophys Acta* 1480:267–77.

Gruschus, J. M., Han, C. J., Greener, T., Ferretti, J. A., Greene, L. E., and Eisenberg, E. (2004). Structure of the functional fragment of auxilin required for catalytic uncoating of clathrin-coated vesicles. *Biochemistry* 43:3111–9.

Gupta, R. S., and Singh, B. (1994) Phylogenetic analysis of 70 kD heat shock protein sequences suggests a chimeric origin for the eukaryotic cell nucleus. *Curr Biol* 4:1104–14.

Ha, J. H., Hellman, U., Johnson, E. R., Li, L. S., McKay, D. B., Sousa, M. C., Takeda, S., Wernstedt, C., and Wilbanks, S. M. (1997) Destabilization of peptide binding and interdomain communication by an E543K mutation in the bovine 70-kDa heat shock cognate protein, a molecular chaperone. *J Biol Chem* 272:27796–803.

Harrison, C. J., HayerHartl, M., DiLiberto, M., Hartl, F. U., and Kuriyan, J. (1997) Crystal structure of the nucleotide exchange factor GrpE bound to the ATPase domain of the molecular chaperone DnaK. *Science* 276:431–5.

Hendrick, J. P., Langer, T., Davis, T. A., Hartl, F. U., and Wiedmann, M. (1993) Control of folding and membrane translocation by binding of the chaperone DnaJ to nascent polypeptides. *Proc Natl Acad Sci U S A* 90:10216–20.

Herman, C., Prakash, S., Lu, C. Z., Matouschek, A., and Gross, C. A. (2003) Lack of a robust unfoldase activity confers a unique level of substrate specificity to the universal AAA protease FtsH. *Mol Cell* 11:659–69.

Hernandez, M. P., Chadli, A., and Toft, D. O. (2002) HSP40 binding is the first step in the HSP90 chaperoning pathway for the progesterone receptor. *J Biol Chem* 277:11873–81.

Hesterkamp, T., and Bukau, B. (1998) Role of the DnaK and HscA homologs of Hsp70 chaperones in protein folding in *E-coli*. *EMBO J* 17:4818–28.

Heymann, J. B., Iwasaki, K., Yim, Y. I., Cheng, N., Belnap, D. M., Greene, L. E., Eisenberg, E., and Steven, A. C. (2005) Visualization of the binding of Hsc70 ATPase to clathrin baskets: Implications for an uncoating mechanism. *J Biol Chem* 280:7156–61.

Huang, K., Flanagan, J. M., and Prestegard, J. H. (1999) The influence of C-terminal extension on the structure of the "J-domain" in *E-coli* DnaJ. *Prot Sci* 8:203–14.

Huang, S., Ratliff, K. S., and Matouschek, A. (2002) Protein unfolding by the mitochondrial membrane potential. *Nat Struct Biol* 9:301–7.

Huang, S., Ratliff, K. S., Schwartz, M. P., Spenner, J. M., and Matouschek, A. (1999) Mitochondria unfold precursor proteins by unraveling them from their N-termini. *Nat Struct Biol* 6:1132–8.

Jiang, J., Prasad, K., Lafer, E. M., and Sousa, R. (2005) Structural basis of interdomain communication in the Hsc70 chaperone. *Mol Cell* 20:513–24.

Jiang, J., Taylor, A. B., Prasad, K., Ishikawa-Brush, Y., Hart, P. J., Lafer, E. M., and Sousa, R. (2003) Structure–function analysis of the auxilin J-domain reveals an extended Hsc70 interaction interface. *Biochemistry* 42:5748–53.

Johnson, E. R., and McKay, D. B. (1999) Mapping the role of active site residues for transducing an ATP-induced conformational change in the bovine 70-kDa heat shock cognate protein. *Biochemistry* 38:10823–30.

Jubete, Y., Maurizi, M. R., and Gottesman, S. (1996) Role of the heat shock protein DnaJ in the Lon-dependent degradation of naturally unstable proteins. *J Biol Chem* 271:30798–803.

Kang, P. J., Ostermann, J., Shilling, J., Neupert, W., Craig, E. A., and Pfanner, N. (1990) Requirement for hsp70 in the mitochondrial matrix for translocation and folding of precursor proteins. *Nature* 348:137–43.

Karzai, A. W., and McMacken, R. (1996) A bipartite signaling mechanism involved in DnaJ-mediated activation of the *Escherichia coli* DnaK protein. *J Biol Chem* 271:11236–46.

Kelley, W. L., and Georgopoulos, C. (1997) Positive control of the two-component RcsC/B signal transduction network by DjlA: A member of the DnaJ family of molecular chaperones in *Escherichia coli*. *Mol Microbiol* 25:913–31.

Kelley, W. L., and Landry, S. J. (1994) Chaperone power in a virus? *Trends Biochem Sci* 19:277–8.

Kim, H. Y., Ahn, B. Y., and Cho, Y. (2001) Structural basis for the inactivation of retinoblastoma tumor suppressor by SV40 large T antigen. *EMBO J* 20:295–304.

Kim, S. Y., Sharma, S., Hoskins, J. R., and Wickner, S. (2002) Interaction of the DnaK and DnaJ chaperone system with a native substrate, P1 RepA. *J Biol Chem* 277:44778–83.

Knieszner, H., Schilke, B., Dutkiewicz, R., D'Silva, P., Cheng, S., Ohlson, M., Craig, E. A., and Marszalek, J. (2005) Compensation for a defective interaction of the hsp70 ssq1 with the mitochondrial Fe–S cluster scaffold isu. *J Biol Chem* 280:28966–72.

Landry, S. J. (2003a) Structure and energetics of an allele-specific genetic interaction between dnaJ and dnaK: correlation of nuclear magnetic resonance chemical shift perturbations in the J-domain of Hsp40/DnaJ with binding affinity for the ATPase domain of Hsp70/DnaK. *Biochemistry* 42:4926–36.

Landry, S. J. (2003b) Swivels and stators in the Hsp40-Hsp70 chaperone machine. *Structure* 11:1465–6.

Landry, S. J., and Gierasch, L. M. (1994) Polypeptide interactions with molecular chaperones and the relationship to *in vivo* protein folding. *Ann Rev Biophys Biomol Struct* 23:645–69.

Lemmon, S. K. (2001) Clathrin uncoating: Auxilin comes to life. *Curr Biol* 11:R49–52.

Li, J., Qian, X., and Sha, B. (2003) The crystal structure of the yeast Hsp40 Ydj1 complexed with its peptide substrate. *Structure* 11:1475–83.

Li, Y., Dudek, J., Guiard, B., Pfanner, N., Rehling, P., and Voos, W. (2004). The presequence translocase-associated protein import motor of mitochondria. Pam16 functions in an antagonistic manner to Pam18. *J Biol Chem* 279:38047–54.

Liberek, K., Georgopoulos, C., and Zylicz, M. (1988) Role of the *Escherichia coli* dnaK and dnaJ heat shock proteins in the initiation of bacteriophage lambda. DNA replication. *Proc Natl Acad Sci U S A* 85:6632–6.

Liberek, K., Skowyra, D., Zylicz, M., Johnson, C., and Georgopoulos. C. (1991). The *Escherichia-coli* DnaK Chaperone, the 70-kDa heat shock protein eukaryotic equivalent, changes conformation upon ATP hydrolysis, thus triggering its dissociation from a bound target protein. *J Biol Chem* 266:14491–6.

Liberek, K., Wall, D., and Georgopoulos, C. (1995) The DnaJ chaperone catalytically activates the DnaK chaperone to preferentially bind the sigma(32) heat shock transcriptional regulator. *Proc Natl Acad Sci U S A* 92:6224–8.

Linke, K., Wolfram, T., Bussemer, J., and Jakob, U. (2003) The roles of the two zinc binding sites in DnaJ. *J Biol Chem* 278:44457–66.

Lopez, N., Aron, R., and Craig, E. A. (2003) Specificity of class II Hsp40 Sis1 in maintenance of yeast prion [RNQ(+)]. *Mol Biol Cell* 14:1172–81.

Lu, Z., and Cyr, D. M. (1998) The conserved carboxyl terminus and zinc finger-like domain of the co-chaperone Ydj1 assist Hsp70 in protein folding. *J Biol Chem* 273: 5970–8.

Luke, M. M., Sutton, A., and Arndt, K. T. (1991) Characterization of SIS1, a *Saccharomyces cerevisiae* homologue of bacterial dnaJ proteins. *J Cell Biol* 114:623–38.

Ma, Y., Greener, T., Pacold, M. E., Kaushal, S., Greene, L. E., and Eisenberg, E. (2002) Identification of domain required for catalytic activity of auxilin in supporting clathrin uncoating by Hsc70. *J Biol Chem* 107.

Mallory, J. B., Alfano, C., and McMacken, R. (1990) Host virus interactions in the initiation of bacteriophage-lambda DNA replication—Recruitment of Escherichia-Coli Dnab Helicase by Lambda-P Replication Protein. *J Biol Chem* 265:13297–13307.

Matouschek, A., Azem, A., Ratliff, K., Glick, B. S., Schmid, K., and Schatz, G. (1997). Active unfolding of precursor proteins during mitochondrial protein import. *EMBO J.* 16:6727–6736.

Matouschek, A., Pfanner, N, and Voos, W. (2000). Protein unfolding by mitochondria. The Hsp70 import motor. *EMBO Rep* 1:404–10.

Mayer, M. P., Schroder, H., Rudiger, S., Paal, K., Laufen, T., and Bukau, B. (2000) Multistep mechanism of substrate binding determines chaperone activity of Hsp70. *Nat Struct Biol* 7:586–93.

Morita, M. T., Kanemori, M., Yanagi, H., and Yura, T. (2000) Dynamic interplay between antagonistic pathways controlling the sigma 32 level in *Escherichia coli*. *Proc Natl Acad Sci U S A* 97:5860–5.

Neidhardt, F. C., and VanBogelen, R. A., (1987) Heat shock response. In Neidhardt, F. C., Ingraham, J. L., Low, K. B., Magasanik, B., Schaechter, M., and Umbarger, H. E. (eds.): *Escherichia Coli* and *Salmonella Typhimurium*. Cellular and Molecular Biology. American Society of Microbiology, Washington, D.C., p. 1334–45.

Nelson, R. J., Ziegelhoffer, T., Nicolet, C., Wernerwashburne, M., and Craig, E. A. (1992) The translation machinery and 70 kd heat shock protein cooperate in protein synthesis. *Cell* 71:97–105.

Obrien, M. C., Flaherty, K. M., and McKay, D. B. (1996) Lysine 71 of the chaperone protein Hsc70 is essential for ATP hydrolysis. *J Biol Chem* 271:15874–8.

Okamoto, K., Brinker, A., Paschen, S. A., Moarefi, I., Hayer-Hartl, M., Neupert, W., and Brunner, M. (2002) The protein import motor of mitochondria: A targeted molecular ratchet driving unfolding and translocation. *EMBO J* 21:3659–71.

Peden, K. W., and Pipas, J. M. (1992) Simian virus 40 mutants with amino-acid substitutions near the amino terminus of large T antigen. *Virus Genes* 6:107–18.

Pelham, H. R. B. (1984) Hsp70 accelerates the recovery of nucleolar morphology after heat shock. *EMBO J.* 3:3095–100.

Pellecchia, M., Montgomery, D. L., Stevens, S. Y., Vander Kooi, C. W., Feng, H. P., Gierasch, L. M., and Zuiderweg, E. R. (2000) Structural insights into substrate binding by the molecular chaperone DnaK. *Nat Struct Biol* 7:298–303.

Pellecchia, M., Szyperski, T., Wall, D., Georgopoulos, C., and Wuthrich. K. (1996) NMR structure of the J-domain and the Gly/Phe-rich region of the *Escherichia coli* DnaJ chaperone. *J Mol Biol* 260:236–50.

Pierpaoli, E. V., Sandmeier, E., Baici, A., Schonfeld, H. J., Gisler, S., and Christen, P. (1997) The power stroke of the DnaK/DnaJ/GrpE molecular chaperone system. *J Mol Biol* 269:757–68.

Pratt, W. B., and Toft, D. O. (2003) Regulation of signaling protein function and trafficking by the hsp90/hsp70-based chaperone machinery. *Exp Biol Med (Maywood)* 228: 111–33.

Qian, X., Hou, W., Zhengang, L., and Sha, B. (2002) Direct interactions between molecular chaperones heat-shock protein (Hsp) 70 and Hsp40: Yeast Hsp70 Ssa1 binds the extreme C-terminal region of yeast Hsp40 Sis1. *Biochem J* 361:27–34.

Qian, Y. Q., Patel, D., Hartl, F. U., and McColl, D. J. (1996) Nuclear magnetic resonance solution structure of the human Hsp40 (HDJ-1) J-domain. *J Mol Biol* 260: 224–35.

Randall, L. L., and Hardy, S. J. (1986) Correlation of competence for export with lack of tertiary structure of the mature species: A study *in vivo* of maltose-binding protein in *E. coli*. *Cell* 46:921–8.

Russell, R., Karzai, A. W., Mehl, A. F., and McMacken, R. (1999) DnaJ dramatically stimulates ATP hydrolysis by DnaK: Insight into targeting of Hsp70 proteins to polypeptide substrates. *Biochemistry* 38:4165–76.

Sakuragi, S., Liu, Q. L., and Craig, E. (1999) Interaction between the nucleotide exchange factor Mge1 and the mitochondrial Hsp70 Ssc1. *J Biol Chem* 274:11275–82.

Sato, T., Esaki, M., Fernandez, J. M., and Endo, T. (2005) Comparison of the protein-unfolding pathways between mitochondrial protein import and atomic-force microscopy measurements. *Proc Natl Acad Sci U S A* 102:17999–8004.

Schlossman, D. M., Schmid, S. L., Braell, W. A., and Rothman, J. E. (1984) An enzyme that removes clathrin coats: Purification of an uncoating ATPase. *J Cell Biol* 99: 723–33.

Schmitz, F., Tabares, L., Khimich, D., Strenzke, N., de la Villa-Polo, P., Castellano-Munoz, M., Bulankina, A., Moser, T., Fernandez-Chacon, R., and Sudhof, T. C. (2006) CSPα-deficiency causes massive and rapid photoreceptor degeneration. *Proc Natl Acad Sci U S A*.

Schroder, H., Langer, T., Hartl, F. U., and Bukau, B. (1993) DnaK, DnaJ and GrpE form a cellular chaperone machinery capable of repairing heat-induced protein damage. *EMBO J* 12:4137–44.

Scidmore, M. A., Okamura, H. H., and Rose, M. D. (1993) Genetic interactions between Kar2 and Sec63, encoding eukaryotic homologues of DnaK and DnaJ in the *Endoplasmic reticulum*. *Mol Biol Cell* 4:1145–59.

Sharma, S., Sathyanarayana, B. K., Bird, J. G., Hoskins, J. R., Lee, B., and Wickner, S. (2004) Plasmid P1 RepA is homologous to the F plasmid RepE class of initiators. *J Biol Chem* 279:6027–34.

Shen, Y., Meunier, L., and Hendershot, L. M. (2002) Identification and characterization of a novel endoplasmic reticulum (ER) DnaJ homologue, which stimulates ATPase activity of BiP *in vitro* and is induced by ER stress. *J Biol Chem* 277:15947–56.

Shi, Y. Y., Tang, W., Hao, S. F., and Wang, C. C. (2005) Contributions of cysteine residues in Zn2 to zinc fingers and thiol-disulfide oxidoreductase activities of chaperone DnaJ. *Biochemistry* 44:1683–9.

Slepenkov, S. V., and Witt, S. N. (2002a) Kinetic analysis of interdomain coupling in a lidless variant of the molecular chaperone DnaK: DnaK's lid inhibits transition to the low affinity state. *Biochemistry* 41:12224–35.

Slepenkov, S. V., and Witt, S. N. (2002b) The unfolding story of the *Escherichia coli* Hsp70 DnaK: Is DnaK a holdase or an unfoldase? *Mol Microbiol* 45:1197–206.

Smith, C. K., Baker, T. A., and Sauer, R. T. (1999) Lon and Clp family proteases and chaperones share homologous substrate-recognition domains. *Proc Natl Acad Sci U S A* 96:6678–82.

Sondermann, H., Scheufler, C., Schneider, C., Hohfeld, J., Hartl, F. U., and Moarefi, I. (2001) Structure of a Bag/Hsc70 complex: Convergent functional evolution of Hsp70 nucleotide exchange factors. *Science* 291:1553–7.

Sousa, M. C., and McKay, D. B. (1998) The hydroxyl of threonine 13 of the bovine 70-kDa heat shock cognate protein is essential for transducing the ATP-induced conformational change. *Biochemistry* 37:15392–9.

Spence, S. L., and Pipas, J. M. (1994) SV40 large T antigen functions at two distinct steps in virion assembly. *Virology* 204:200–9.

Stevens, S. Y., Cai, S., Pellecchia, M., and Zuiderweg, E. R. (2003) The solution structure of the bacterial HSP70 chaperone protein domain DnaK(393–507) in complex with the peptide NRLLLTG. *Prot Sci* 12:2588–96.

Strub, A., Zufall, N., and Voos, W. (2003) The putative helical lid of the Hsp70 peptide-binding domain is required for efficient preprotein translocation into mitochondria. *J Mol Biol* 334:1087–99.

Suh, W. C., Burkholder, W. F., Lu, C. Z., Zhao, X., Gottesman, M. E., and Gross, C. A. (1998) Interaction of the Hsp70 molecular chaperone, DnaK, with its cochaperone DnaJ. *Proc Natl Acad Sci U S A* 95:15223–8.

Sullivan, C. S., Cantalupo, P., and Pipas, J. M. (2000) The molecular chaperone activity of simian virus 40 large T antigen is required to disrupt Rb-E2F family complexes by an ATP-dependent mechanism. *Mol Cell Biol* 20:6233–43.

Szabo, A., Korszun, R., Hartl, F.-U., and Flanagan, J. (1996) A zinc finger-like domain of the molecular chaperone DnaJ is involved in binding to denatured protein substrates. *EMBO J* 15:408–17.

Tang, W., and Wang, C. C. (2001) Zinc fingers and thiol-disulfide oxidoreductase activities of chaperone DnaJ. *Biochemistry* 40:14985–94.

Tatsuta, T., Joob, D. M., Calendar, R., Akiyama, Y., and Ogura, T. (2000) Evidence for an active role of the DnaK chaperone system in the degradation of sigma(32). *FEBS Lett* 478:271–5.

Todd, M. J., Lorimer, G. H., and Thirumalai, D. (1996) Chaperonin-facilitated protein folding: Optimization of rate and yield by an iterative annealing mechanism. *Proc Natl Acad Sci U S A* 93:4030–5.

Tomoyasu, T., Mogk, A., Langen, H., Goloubinoff, P., and Bukau, B. (2001) Genetic dissection of the roles of chaperones and proteases in protein folding and degradation in the *Escherichia coli* cytosol. *Mol Microbiol* 40:397–413.

Truscott, K. N., Voos, W., Frazier, A. E., Lind, M., Li, Y., Geissler, A., Dudek, J., Muller, H., Sickmann, A., Meyer, H. E., Meisinger, C., Guiard, B., Rehling, P., and Pfanner, N. (2003) A J-protein is an essential subunit of the presequence translocase-associated protein import motor of mitochondria. *J Cell Biol* 163:707–13.

Ungewickell, E., Ungewickell, H., Holstein, S. E., Lindner, R., Prasad, K., Barouch, W., Martin, B., Greene, L. E., and Eisenberg, E. (1995) Role of auxilin in uncoating clathrin-coated vesicles. *Nature* 378:632–5.

van der Laan, M., Chacinska, A., Lind, M., Perschil, I., Sickmann, A., Meyer, H. E., Guiard, B., Meisinger, C., Pfanner, N., and Rehling, P. (2005) Pam17 is required for architecture and translocation activity of the mitochondrial protein import motor. *Mol Cell Biol* 25:7449–58.

Wall, D., Zylicz, M., and Georgopoulos, C. (1995) The conserved G/F motif of the DnaJ chaperone is necessary for the activation of the substrate binding properties of the DnaK chaperone. *J Biol Chem* 270:2139–44.

Wall, D., Zylicz, M., and Georgopoulos, C. (1994) The NH2-Terminal 108 amino acids of the *Escherichia coli* DnaJ protein stimulate the ATPase activity of Dnak and are sufficient for lambda replication. *J Biol Chem*. 269:5446–51.

Wang, H., Kurochkin, A. V., Pang, Y., Hu, W. D., Flynn, G. C., and Zuiderweg, E. R. P. (1998) NMR solution structure of the 21 kDa chaperone protein DnaK substrate binding domain: A preview of chaperone–protein interaction. *Biochemistry* 37:7929–40.

Whalen, K. A., de Jesus, R., Kean, J. A., and Schaffhausen, B. S. (2005) Genetic analysis of the polyomavirus DnaJ domain. *J Virol* 79:9982–90.

Wickner, S., Gottesman, S., Skowyra, D., Hoskins, J., McKenney, K., and Maurizi, M. R. (1994) A molecular chaperone, ClpA, functions like DnaK and DnaJ. *Proc Natl Acad Sci U S A* 91:12218—22.

Wickner, S., Hoskins, J., and McKenney, K. (1991) Monomerization of RepA dimers by heat shock proteins activates binding to DNA replication origin. *Proc Natl Acad Sci U S A* 88:7903—7.

Wilbanks, S. M., Chen, L. L., Tsuruta, H., Hodgson, K. O., and McKay. D. B. (1995) Solution small-angle X-ray scattering study of the molecular chaperone Hsc70 and its subfragments. *Biochemisty* 34:12095–106.

Wild, J., Altman, E., Yura, T., and Gross. C. A.,(1992) DnaK and DnaJ heat shock proteins participate in protein export in *Escherichia-coli*. *Genes Dev* 6:1165–72.

Wittung-Stafshede, P., Guidry, J., Horne, B. E., and Landry. S. J. (2003) The J-domain of Hsp40 couples ATP hydrolysis to substrate capture in Hsp70. *Biochemistry* 42:4937–44.

Wu, Y., Li, J., Jin, Z., Fu, Z., and Sha, B. (2005) The crystal structure of the C-terminal fragment of yeast Hsp40 Ydj1 reveals novel dimerization motif for Hsp40. *J Mol Biol* 346:1005–11.

Yan, W., and Craig, E. A. (1999) The glycine-phenylalanine-rich region determines the specificity of the yeast Hsp40 Sis1. *Mol Cell Biol* 19:7751–8.

Young, J. C., Agashe, V. R., Siegers, K., and Hartl F. U., (2004) Pathways of chaperone-mediated protein folding in the cytosol. *Nat Rev Mol Cell Biol* 5:781–91.

Young, J. C., Hoogenraad, N. J., and Hartl, F. U. (2003) Molecular chaperones Hsp90 and Hsp70 deliver preproteins to the mitochondrial import receptor Tom70. *Cell* 112: 41–50.

Yu, M., Haslam, R. H., and Haslam, D. B. (2000) HEDJ, an Hsp40 co-chaperone localized to the endoplasmic reticulum of human cells. *J Biol Chem* 275:24984–92.

Zhang, C. X., Engqvist-Goldstein, A. E., Carreno, S., Owen, D. J., Smythe, E., and Drubin, D. G. (2005) Multiple roles for cyclin G-associated kinase in clathrin-mediated sorting events. *Traffic* 6:1103–13.

Zhu, X. T., Zhao, X., Burkholder, W. F., Gragerov, A., Ogata, C. M., Gottesman, M. E., and Hendrickson, W. A. (1996) Structural analysis of substrate binding by the molecular chaperone DnaK. *Science* 272:1606–14.

Zietkiewicz, S., Lewandowska, A., Stocki, P., and Liberek, K. (2006) HSP70 chaperone machine remodels protein aggregates at the initial step of HSP40–HSP100 dependent disaggregation. *J Biol Chem.*

Zolkiewski, M. (1999) ClpB cooperates with DnaK, DnaJ, and GrpE in suppressing protein aggregation—A novel multi-chaperone system from *Escherichia coli*. *J Biol Chem* 274:28083–6.

11
Mammalian HSP40/DnaJ Chaperone Proteins in Cytosol

KAZUTOYO TERADA AND MASATAKA MORI
Department of Molecular Genetics, Graduate School of Medical Sciences, Kumamoto University, Kumamoto, Kumamoto 860-8556, Japan

1. Introduction

The HSP70-based molecular chaperone system in cytosol mediates various biological processes such as folding of newly synthesized proteins, refolding of misfolded proteins, translocation of proteins into organelles, assembly of proteins, and their degradation (Gething and Sambrook, 1992; Frydman, 2001; Young et al., 2004; Mayer and Bukau, 2005). HSP70 chaperone proteins catalyze these biological processes with the aid of HSP40/DnaJ proteins and co-chaperones (Fig. 1A). HSP40/DnaJ proteins initially recognize substrate polypeptides, and transfer them to the substrate-binding domain of HSP70 proteins. During or just after transfer of the substrate polypeptide to HSP70, the J-domain of HSP40/DnaJ proteins accelerates hydrolysis of bound ATP in the nucleotide-binding domain of HSP70. The hydrolysis of ATP in the nucleotide-binding domain leads a conformational change of the substrate-binding domain to bind the substrate tightly. After binding of a substrate polypeptide to HSP70, various co-chaperones interact with the HSP70–polypeptide binary complex. Many co-chaperones in mammals contain a tetratricopeptide repeat (TPR) domain(s), and this domain recognizes the EEVD sequence at the carboxyl-terminal of cytosolic HSP70 proteins in eukaryotes. The combination of these three components of the HSP70-based chaperone system is important to determine the proper destination of the substrate polypeptide.

All chaperone activities of the HSP70-based system appear to be based on the property of HSP70 to interact with hydrophobic segments of substrate polypeptide in an ATP-controlled fashion (Mayer and Bukau, 2005). Chaperone activities of HSP70 can be classified as (1) folding of non-native intermediate to the native state (folding activity), (2) holding of non-native proteins through association with hydrophobic patches of the substrate molecules to prevent aggregation (holding activity), and (3) disassembly of protein complexes such as clathrin, viral capsid coat, and nucleoprotein complex (disassembling activity). Among these chaperone functions of the HSP70-based chaperone system, protein-folding activity is the most prominent (Fig. 1B,C). HSP70, HSP40, and nucleotide exchange factor (NEF) drive folding of substrate polypeptides and are well-characterized in

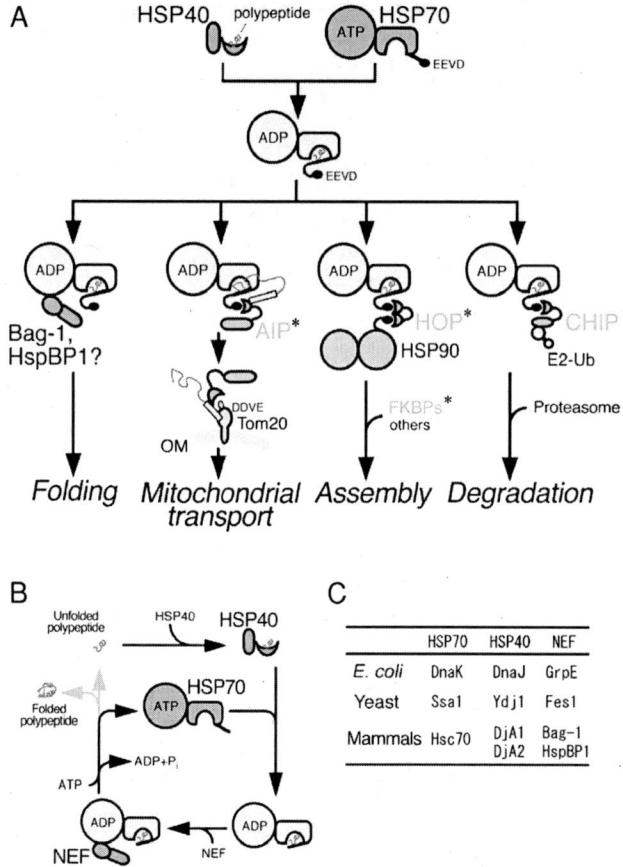

FIGURE 1. HSP70–HSP40 co-chaperone systems. A: Quality control model of the HSP70-based chaperone systems. A substrate polypeptide is initially recognized by HSP40/DnaJ and HSP70 proteins. A co-chaperone subsequently interacts with the polypeptide-HSP70 binary complex, and the polypeptide substrate is destined for either protein folding, mitochondrial protein transport, assembly, or degradation. The carboxyl-terminal of animal Tom20 is ended with the DDVE sequence and is recognized by the AIP co-chaperone (Yano et al., 2003). The assembly pathway is shown for steroid receptors. The pathways are shown uni-directionally for simplicity. The action points and functions of co-chaperones may be overlapping and interrelated. The co-chaperones with TPR domains are shown with asterisks. OM; outer membrane of mitochondria, E2; ubiquitin ligase, Ub; ubiquitin. B: Folding cycle by HSP70–HSP40–nucleotide exchange factor (NEF). Modulation of the cycle by CHIP and HspBP1 co-chaperones may be present. C: Components of HSP70-type I HSP40-NEF for the folding cycle from *E. coli,* yeast, and mammals. Modified from Terada (2005) Seikagaku (in Japanese) 77:101–112.

E. coli. Folding activity is also prominent in the chaperonin system, but appears to be limited to substrate polypeptides of small size (Bukau and Horwich, 1998). All classes of chaperone proteins have holding activity. Disassembling activity of the HSP70-based chaperone system requires other classes of chaperones. It is exerted with HSP100 (ClpP/Hsp104) family proteins in cytosol of bacteria and yeast (Glover and Lindquist, 1998; Motohashi et al., 1999; Goloubinoff et al., 1999).

2. Components of HSP70–HSP40–Co-Chaperone System

2.1. HSP70 Family

Mammalian HSP70 family proteins in cytosol have amino-terminal ATPase domain, substrate-binding domain, and carboxyl-terminal domain that ends by the EEVD sequence (Fig. 2A). HSP70 family proteins are present in all terrestrial organisms except some species of extreme thermophiles (Pahl et al., 1997; Tatusov et al., 2001). Representative cytosolic HSP70 family protein is DnaK in *E. coli,* while constitutive Hsc70 (p73) and inducible Hsp70 (p72) in mammals. Hsc70 constitutes about 1% of total cellular proteins. Inducible Hsp70 is expressed under stress conditions including heat shock (Morimoto, 1998; Kedersha and Anderson, 2002). Under normal conditions, Hsp70 is expressed in some tissues (Huang et al., 2001) and cultured cells. Besides these two HSP70 proteins, Hsp70-2 and Hsc70t are highly expressed in testis germ cells. Specific members of HSP70 family protein are present in matrix of mitochondria and lumen of ER (Fig. 2B).

Members of cytosolic HSP70 proteins have overlapping chaperone functions. For instance, Hsc70, Hsp70, and Hsp70-2 showed similar folding and holding activities against a model protein substrate in vitro (Hafizur et al., 2004). However, nonoverlapping functions among these three cytosolic HSP70 proteins were reported quite recently based on RNA interference experiments (Rohde et al., 2005). Hsc70 was essential for survival of both tumorigenic and nontumorigenic cells, whereas Hsp70 and Hsp70-2 were required for cancer cell growth. Targeted knockdown of Hsp70 and Hsp70-2 showed different morphologies, cell cycle distributions, and gene expression profiles.

Structure of the nucleotide-binding domain (ATPase domain) of HSP70 protein resembles those of actin and hexokinase (Fig. 2C). Structure of the ATPase domain from bovine Hsc70 is similar to that of bacterial DnaK with some exception in the subdomain 3 (Brehmer et al., 2001). Structure of the substrate-binding domain with a model polypeptide was solved in bacterial DnaK (Fig. 2D). The amino-terminal portion of the substrate-binding domain is structured mainly with β-sheets, and forms a cleft for substrate polypeptide binding. The carboxyl-terminal portion is structured mainly with α-helices, and acts like a "lid" to bind the substrate tightly (Bukau and Horwich, 1998). The preferable motif of the substrate polypeptide is seven consecutive residues that have central hydrophobic core of aliphatic and

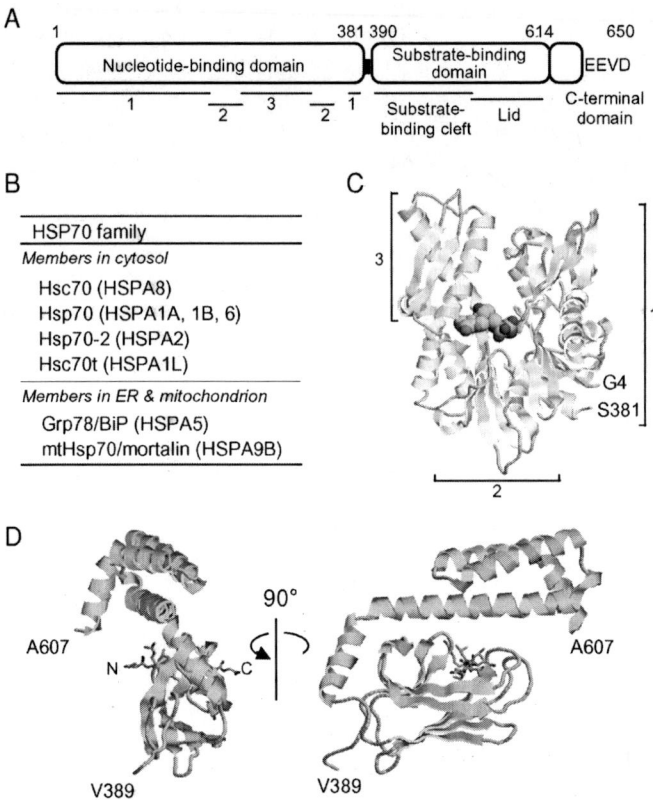

FIGURE 2. HSP70 family proteins. A: Schematic model for human Hsc70. The structures of the nucleotide-binding domain (C) and the substrate-binding domain (D) are shown. Numbers on the top denotes residue numbers. The carboxyl-terminal EEVD sequence is present in the cytosolic members of HSP70 proteins in (B). B: Major human HSP70 family proteins. HSP110 family is not included in the table. Grp78/BiP resides in the lumen of ER and mtHSP70/mortalin locates in the matrix of mitochondria. Registered names in the UniGene database are shown in parenthesis. C: Structure of the nucleotide-binding domain (ribbons) from bovine Hsc70 with ATP (spacefill) (Flaherty et al., 1994; Wilbanks and McKay, 1995). The α- and β-phosphates are presented on the right hand. The γ-phosphate is hidden behind the β-phosphate. One Mg^{2+} and three K^+ that coordinate to the β- and γ-phosphates are not shown. The subdomains 1–3 correspond to the schematic diagram in (A). The structure was generated using a RasMol software. PDB accession; 1KAX. D: Structure of the substrate-binding domain (ribbons) from bacterial DnaK with a model polypeptide substrate (sticks, NRLLLTG) (Zhu et al., 1996). PDB accession; 1DKX. Modified from Terada (2005) Seikagaku (in Japanese) 77:101–112.

aromatic residues and two flanking regions with basic residues (Rudiger et al., 1997a,b). Interaction between the ATPase domain and substrate-binding domain is essential for chaperone function of HSP70. However, no precise information on spatial arrangement of the two domains is currently available.

2.2. HSP40/DnaJ Protein Family

Members of the HSP40/DnaJ protein family are classified into three subfamilies (Fig. 3A) (Cheetham and Caplan, 1998; Ohtsuka and Hata, 2000). There are four members of type I HSP40/DnaJ protein subfamily in mammals, and they are the authentic orthologs of bacterial DnaJ. Mammalian DjA1 (DnajA1, HomoloGene #20385) and DjA2 (DnajA2, #21193) proteins locate in cytosol and are expressed ubiquitously. DjA1 and DjA2 efficiently refold a denatured model substrate in cooperation with cytosolic HSP70 proteins (Terada et al., 1997; Terada and Mori, 2000; Hafizur et al., 2004). DjA3 (DnajA3, #36170) locates in mitochondrial matrix, and DjA4 (DnajA4, #23110) is highly expressed in the cytosol of selective tissues such as heart and testis (Abdul et al., 2002). Members of type II HSP40/DnaJ protein subfamily lack zinc-finger region but have glycine/methionine-rich domain. Since DjB1 (hsp40/hdj-1) was first discovered within the members of mammalian HSP40 family (Hattori et al., 1993), it has been widely used for research of mammalian HSP70-based chaperone system. Members of type III HSP40/DnaJ protein subfamily lack glycine/phenylalanine-rich domain, and location of their J-domains are not always amino-termini. Members of type III HSP40/DnaJ protein act on specific cellular events. For instance, DjC6 (auxilin) engages in disassembly of clathrin-coat (Gall et al., 2000; Pishvaee et al., 2000; Greener et al., 2001) and DjC5 (cysteine string proteins) releases neurotransmitters from synaptic vesicles (Chamberlain and Burgoyne, 2000). There are 45 human HSP40 proteins and 22 yeast HSP40 proteins in the Unigene database (Walsh et al., 2004). All these HSP40/DnaJ proteins have J-domains around 70 residues in length (Fig. 3B,C) (Fan et al., 2003). Characteristic HPD sequence in the J-domain is essential for accelerating ATPase activity of HSP70 proteins (Wall et al., 1994; Tsai and Douglas, 1996).

Li et al. (2003) solved the structure of the peptide-binding fragment of yeast type I HSP40 Ydj1 with a model polypeptide substrate. The peptide-binding fragment of Ydj1 is divided by three domains and forms a L-shaped structure. The model peptide binds to a hydrophobic pocket of domain I (Fig. 3D). Characteristics of the polypeptide substrate used for the structural analysis closely resemble those identified by peptide-screening against bacterial DnaJ and by phage-display against Ydj1 (Rudiger et al., 2001; Fan et al., 2004). The polypeptide substrate binds to the domain I by forming an extra β strand. The leucine residue in the middle of the polypeptide substrate GWLYEIS inserts its side chain deeply into the hydrophobic pocket. Although hydrophobic nature of the polypeptide substrates is common to the substrate motifs required for binding to HSP70 and type I HSP40 proteins, the binding motifs of the polypeptide substrates to the respective chaperone proteins are not totally similar. The domain I is divided by the insertion of the domain II that coordinates two zinc atoms. The Zn 2 region of the domain II is important for the bacterial DnaK to bind the substrate polypeptide tightly (Linke et al., 2003). Landry (2003) proposed a "swivels and stators" model for the folding process by HSP70 and type I HSP40. The Zn 2 region of type I HSP40 serves as a stator for HSP70 rotor.

260   Kazutoyo Terada and Masataka Mori

The structure of the peptide-binding fragment of yeast type II HSP40 Sis1 (CTD 1 and 2) resembles that of type I HSP40 (domain I and III) (Fig. 3E) (Sha et al., 2000). A polypeptide substrate may bind to a hydrophobic pocket in the CTD 1 of Sis1 protein, but characteristics of the screened polypeptide substrates for Sis1 are somewhat different from those for Ydj1 (Fan et al., 2004). Sis1 forms a dimer at their carboxyl-terminal dimerization domain (DD), and Ydj1 is also supposed to form a dimer similarly (Li et al., 2003). Recently, biophysical analysis of a type I

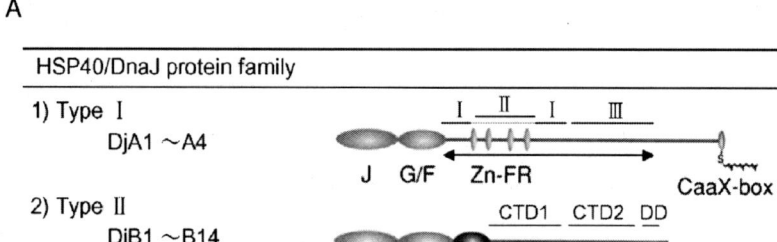

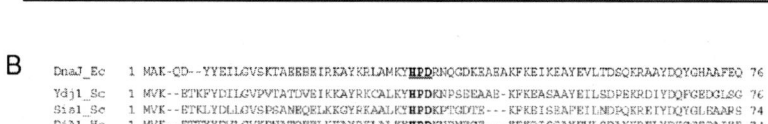

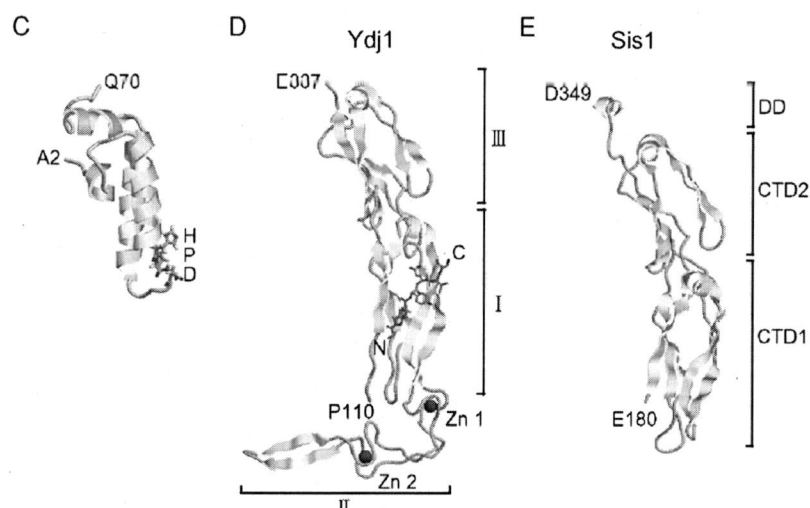

HSP40 DjA1 and a type II HSP40 DjB4 (Hlj1) revealed that both HSP40/DnaJ proteins form homodimers (Borges et al., 2005).

2.3. Co-Chaperone Proteins

An efficient folding reaction by the HSP70-based chaperone system requires nucleotide exchange factor (NEF). Although no homolog of bacterial GrpE has been found in eukaryotic cytosol, Bag-1 and HspBP1/Fes1 were identified as NEFs in the cytosol. Bag-1 was initially identified as an anti-apoptotic factor and was subsequently revealed to have a nucleotide exchange activity for Hsc70 (Takayama et al., 1997; Hohfeld and Jentsch, 1997). Bag-1 inhibited the refolding reaction by HSP70–DjB1 chaperone pair (Bimston et al., 1998), but accelerated refolding by HSP70–DjA1/A2 chaperone pairs (Terada and Mori, 2000). This discrepancy may arise from the different subfamily of HSP40/DnaJ proteins used for in vitro assay reactions. Altenatively, Bag-1 can act in concentration-dependent manner (Gassler et al., 2001). The carboxyl-terminal half of Bag-1 protein constitutes the BAG domain and has nucleotide exchange activity. Involvement of Bag-1 in proteasomal degradation was reported (Alberti et al., 2003). The structure of Hsc70–Bag-1 complex was somewhat different from bacterial DnaK–(GrpE)^2 complex (Harrison et al., 1997; Sondermann et al., 2001). Six members of the BAG family are present in mammals and are proposed to be involved with HSP70 in various biological processes including apoptosis, signal transduction, proliferation, transcription, and immortalization (Takayama and Reed, 2001; Townsend

←

FIGURE 3. HSP40/DnaJ family proteins. A: The members of human HSP40/DnaJ protein subfamilies and their representative schematic models. Some unclassified HSP40/DnaJ proteins in the UniGene database are omitted, including mammalian counterpart of Sec63 translocon. J; J-domain, G/F; glycine/phenylalanine-rich domain, Zn-FR; zinc-finger region, G/M; glycicne/methionine-rich domain, CTD1 and CTD2; carboxyl-terminal domain 1 and 2, DD; dimerization domain. B: Sequence alignment of the J-domain from selected species. Type I HSP40 proteins are DnaJ from *E. coli*, Ydj1 from *S. cerevisiae*, DjA1 and DjA2 from *H. sapiens*. Type II HSP40 proteins are Sis1 from *S. cerevisiae* and DjB1 from *H. sapiens*. The HPD sequences in bold are indispensable sequence of the J-domain. C: The molecular structure of the J-domain from bacterial DnaJ (Szyperski et al., 1994; Pellecchia et al., 1996). A representative structure that was selected from 20 structures determined by NMR analysis is shown. The HPD residues are shown by sticks. PDB accession; 1XBL. D: The structure of the peptide-binding fragment from yeast Ydj1 (ribbons) and a model polypeptide substrate (sticks, GWLYEIS) (Li et al., 2003). Two zinc ions are shown by balls. The peptide-binding fragment corresponds to the region with arrows in the schematic model for the type I HSP40 subfamily in (A). The domain I is divided by the insertion of the domain II. PDB accession; 1NLT. E: The structure of the peptide-binding fragment from yeast Sis1 (ribbons) (Sha et al., 2000). One monomer molecule of the Sis1 dimer is shown. The peptide-binding fragment corresponds to the region with arrows in the schematic model for the type II HSP40 subfamily in (A). PDB acccession; 1C3G. Modified from Terada (2005) Seikagaku (in Japanese) 77:101–112.

et al., 2003). Indeed, targeted disruption of Bag-1 in mice resulted in apoptotic cell death in differentiating motoneurons and hematopoietic cells (Gotz et al., 2005). Absence of Bag-1 affected anti-apoptotic signaling pathway by inhibiting Bad phosphorylation with the disturbance of a tripartite complex formed by Akt, B-Raf, and Bag-1.

HspBP1 was initially identified as an HSP70-interacting protein that inhibits ATP binding to HSP70 (Raynes and Guerriero, 1998). From the study of yeast lumenal Sls1 in ER (Kabani et al., 2000), homologs from yeast Fes1 and mammalian HspBP1 were identified as NEFs for cytosolic HSP70 proteins (Kabani et al., 2002a,b). Fes1 promoted protein folding in yeast cytosol and the structure of Hsp70–HspBP1 complex was quite different from that of Hsc70–Bag-1 complex (Shomura et al., 2005).

The carboxyl-termini of the cytosolic members of eukaryotic HSP70 and HSP90 proteins end with the EEVD sequence, and co-chaperones with a TPR domain(s) recognize and bind to this sequence motif. Hop is a co-chaperone with three TPR domains and binds to HSP70 and HSP90 simultaneously (Scheufler et al., 2000). Hop was initially identified as a co-chaperone for HSP90 proteins. Steroid receptors and kinases of the signal transduction pathway are "client" proteins for the HSP90-based chaperone system. The client protein remains bound to HSP90 protein in an inactive state. When the ligand hormone or upstream signal reaches the client protein, it is released from HSP90, and exerts its function (Pratt and Toft, 2003). The HSP70-based chaperone system initially recognizes a client protein, and Hop mediates transfer of the client protein from an HSP70- to the HSP90-based chaperone system. Initial recognition of a client steroid receptor by the HSP70 chaperone system is mediated by type I HSP40 proteins (Kimura et al., 1995; Kanelakis and Pratt, 2003), but type III DjC7 (Tpr2) was reported to mediate the retrograde transfer of glucocorticoid receptor from Hsp90 onto Hsp70 (Brychzy et al., 2003). Type III DjC7 has a TPR domain together with a J-domain.

Chip has not only a TPR domain but a U-box that has E3-ubiquitin ligase activity (Murata et al., 2001). Overexpression of Chip increased ubiquitination of a polypeptide bound to HSP70 and HSP90, and promoted proteasomal degradation (Connell et al., 2001; Meacham et al., 2001), while overexpression of Chip in fibroblasts enhanced the protein-folding capacity of HSP70 (Kampinga et al., 2003). Thus Chip is involved in quality control of polypeptides recognized by the HSP70 and HSP90 chaperone systems (Cyr et al., 2002; Murata et al., 2003). Chip-targeted mouse was viable but induction of inducible Hsp70 after thermal challenge was impaired (Dai et al., 2003). The impairment was due to the activation process of HSF1 transcription factor. Degradation of proteins by Chip may be compensated for by other E3 ligases (Hershko and Ciechanover, 1998) and the aggresome pathway (Kawaguchi et al., 2003).

A group of the metabolic carrier proteins are translated in the cytosol and targeted into the inner membrane of mitochondria. The carrier proteins in the cytosol are kept in a translocation-competent state by the HSP70 and HSP90 proteins (Young et al., 2003). Tom70 in the outer membrane of mitochondria serves as a receptor for these carrier proteins. Tom70 has three TPR domains and binds to the EEVD sequence of HSP70 and HSP90 proteins for efficient protein transport. Most pre-proteins that have a cleavable amino-terminal pre-sequence are targeted into the matrix of mitochondria. These pre-proteins are usually recognized directly by Tom20 receptor in the outer membrane of mitochondria. We found that the TPR domain of arylhydrocarbon receptor-interacting protein (AIP) recognizes the carboxyl-terminal DDVE sequence of mammalian Tom20 and EEVD sequence of Hsc70 (Fig. 1A) (Yano et al., 2003). AIP prevented a pre-protein from losing translocation competency in vitro. AIP also showed a holdase activity to prevent substrate proteins from aggregation. Overexpression of AIP enhanced import of a pre-protein into mitochondria, and depletion of AIP by RNA interference impaired the import.

In apoptosis, we found that HSP70–HSP40 chaperone pairs prevent ER stress-induced apoptosis by inhibiting translocation of a pro-apoptotic protein Bax in the cytosol to the outer membrane of mitochondria (Gotoh et al., 2004). Although both DjA1 (a type I HSP40) and DjB1 (a type II HSP40) were effective, the EEVD sequence of Hsp70 was required for prevention of the translocation. This implies that a co-chaperone(s) with TPR domain is required for this prevention.

The action points and functions of co-chaperones may be distinct among co-chaperones but partially overlapping and interrelated. For instance, the Chip ubiquitin ligase activity is attenuated specifically by HspBP1 (Alberti et al., 2004) and Bag-2 (Dai et al., 2005a). The overlapping and interrelated functions of HSP70 and HSP90 chaperone proteins are also suggested (Ahner et al., 2005).

3. In Vitro Folding with Mammalian Type I HSP40/DNAJ Proteins

As a biological source of molecular chaperones, rabbit reticulocyte lysate has been used for various in vitro studies, including protein translocation, folding, assembly, and degradation (Nimmesgern and Hartl, 1993). This chaperone source contains a Hsc70-based chaperone system which can maintain the translocation competency of pre-proteins into mitochondria and refold a chemically denatured protein (Terada et al., 1997; Terada and Mori, 2000). The chaperone activity required for luciferase refolding was abolished by either single depletion of Hsc70 or simultaneous depletion of DjA1 and DjA2, but not by depletion of DjB1 (Fig. 4). These results demonstrate that type I HSP40 proteins are efficient partner chaperones of Hsc70 for the folding activity. Similar results were reported in yeast cytosolic HSP70-based chaperone system in vitro (Levy et al., 1995) and in vivo (Kim et al., 1998).

We further tested the folding activity of the purified Hsc70, Hsp70, and Hsp70-2 proteins. These HSP70 proteins showed a similar folding activity but require specific type I HSP40/DnaJ proteins (Hafizur et al., 2004). The folding activity was prominent in HSP70–DjA1 and HSP70–DjA2 pairs, moderately in HSP70–DjB1 pairs, but not in HSP70–DjA4 pairs (Fig. 5). These results demonstrate that DjA1 and DjA2 are functionally equivalent and potent HSP40 proteins in luciferase

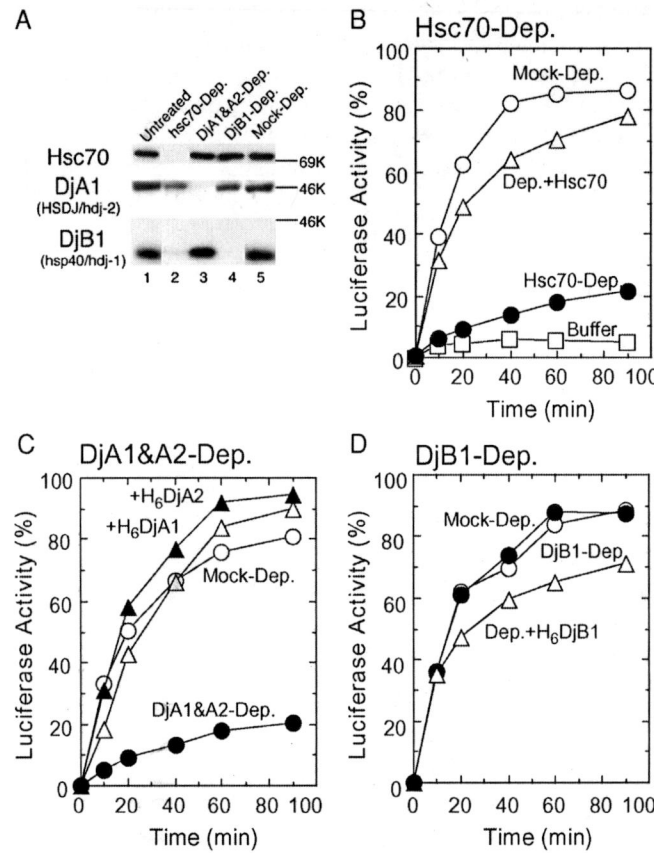

FIGURE 4. Requirement of Hsc70 and type I HSP40 proteins for the refolding of denatured luciferase. A: Immunodepletion of DjA1 and DjA2 from rabbit reticulocyte lysate. The HSP40 proteins were depleted from reticulocyte lysate by corresponding antibodies. The antibody against DjA1 crossreacted with DjA2 proteins in the lysate, and DjA2 protein was efficiently depleted from the lysate together with DjA1 protein (data not shown). B–D: Refolding of chemically denatured luciferase in Hsc70- (B), DjA1-, and DjA2- (C), and DjB1- (D) depleted lysates. Firefly luciferase was denatured in 8 M guanidinum hydrochloride and diluted into the depleted lysate by 1,000-fold dilution. The refolded luciferase activity was monitored at the indicated periods. Enzymatic activity of native luciferase was set as 100%. Chaperone proteins added to the depleted lysates are: Hsc70; 1.8 µM, H6DjA1; 0.4 µM, H6DjA2; 0.4 µM, and H6DjB1; 0.5 µM. Panels A, B, and D are reproduced from The Journal of Cell Biology, 1997, 139,1089–1095 by copyright permission of The Rockefeller University Press. Panel C is adapted from Terada and Mori (2000).

refolding. Tissue-specific DjA4 was not effective in the refolding reaction and was distinct from other members of cytosolic type I HSP40 proteins. The refolding reaction by HSP70–DjA1 and HSP70–DjA2 pairs proceeded linearly with initial lag. Further addition of Bag-1 accelerated the refolding reaction, and the kinetics

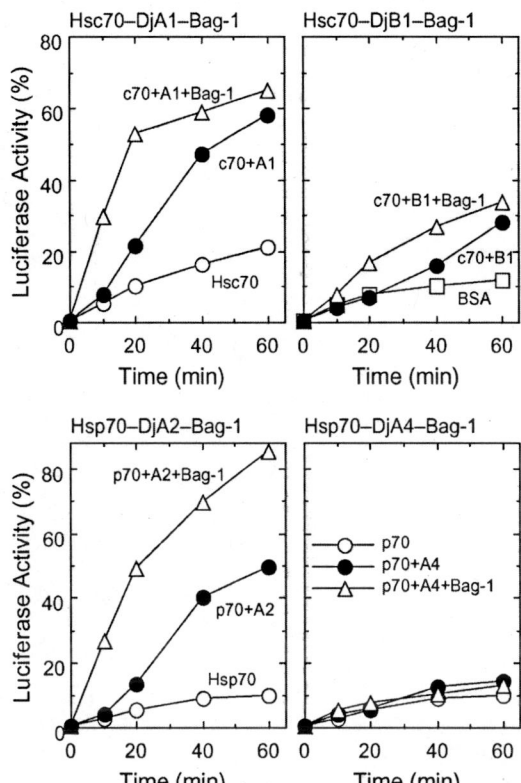

FIGURE 5. Refolding of luciferase by reconstituted HSP70–HSP40–Bag-1 chaperone systems. Firefly luciferase was denatured in 8 M guanidinum hydrochloride, and diluted into the reaction mixture containing the purified chaperones. Enzymatic activity of native luciferase was set as 100%. Similar results were obtained with Hsp70-2 protein (data not shown). Chaperone proteins added to the reaction mixture are: Hsc70; 1.8 μM, Hsp70; 1.8 μM, H6DjA1; 0.4 μM, H6DjA2; 0.4 μM, H6DjB1; 0.5 μM, and H6Bag-1; 0.4 μM. Figures are adapted from Terada and Mori (2000) and Hafizur et al. (2004).

of the refolding was similar to that observed in the lysate (Fig. 4). This result suggests that a nucleotide exchange factor(s), such as Bag-1, is important for efficient refolding.

The ATPase activity of mammalian HSP70 proteins is quite low (Minami et al., 1996; Hohfeld and Jentsch, 1997). The intrinsic ATPase activities of purified Hsc70, Hsp70, and Hsp70-2 were 0.056 min^{-1}, 0.019 min^{-1}, and 0.040 min^{-1}, respectively (Fig. 6, and Hafizur et al. (2004)). The ATPase activities of HSP70 proteins were increased by severalfold in the presence of substoichiometric amount of HSP40 proteins. Turnover of the ATP hydrolysis is proportional to that of folding cycle, and this may explain the initial lag of the folding reaction observed for HSP70–DjA1 and HSP70–DjA2 pairs (Fig. 5). Further addition of a nucleotide exchange factor Bag-1 stimulated the ATPase activities of HSP70–DjA1, HSP70–DjA2, and HSP–DjB1, but not of HSP70–DjA4. The reason why Bag-1 cannot accelerate the ATPase activity of HSP70–DjA4 pair is currently elusive. Thus DjA4 protein is functionally distinct from DjA1 and DjA2 proteins, although they belong to the same type I HSP40/DnaJ protein subfamily.

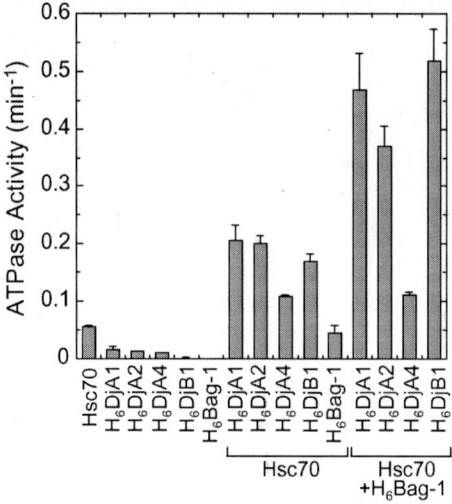

FIGURE 6. Effects of HSP40 and Bag-1 proteins on ATPase activity of Hsc70. ATPase activity was assayed by extracting the liberated inorganic phosphate from [γ-^{32}P]ATP with ethyl acetate. Similar results were obtained with Hsp70 and Hsp70-2 proteins (data not shown). Chaperone proteins added to the reaction mixture are: Hsc70; 1.8 μM, H6DjA1; 0.4 μM, H6DjA2; 0.4 μM, H6DjB1; 0.5 μM, and H6Bag-1; 0.4 μM. Values are adapted from Terada and Mori (2000) and Hafizur et al. (2004).

4. In Vivo Functions of Mammalian HSP40/DNAJ Proteins

Recently, several HSP40/DnaJ genes were disrupted to analyze chaperone functions in mice (Table 1). These genetic approaches provided surprisingly distinct functions for respective members of HSP40/DnaJ proteins in vivo. In this section, we will overview cellular functions of some members of HSP40/DnaJ proteins, including those disrupted in mice.

4.1. Type I HSP40/DnaJ Proteins

The male DjA1-null mice were infertile due to a defect in the late stages of spermatogenesis (Fig. 7) (Terada et al., 2005). It is surprising that single disruption

TABLE 1. A list of HSP40/DnaJ genes disrupted in mice.

HSP40/DnaJ gene	Phenotype	Reference
DjA1(HSDJ/Hdj2)	Male infertility, defect in spermiogenesis by aberrant androgen receptor signaling	Terada et al., 2005
DjA3(hTid1)	1) Embryonic lethal in early stage (E4.5 to 7.5)	Lo et al., 2004
	2) Lethal with dilated cardiomyopathy by selective targeting in heart	Hayashi et al., 2005
DjB6(Mrj)	Embryonic lethal, defect in placental development (E 8.5)	Hunter et al., 1999
DjC3(P58IPK)	Pancreatic β-cell apoptosis and diabetes	Ladiges et al., 2005
DjC5(CSPα)	Progressive, fatal sensorimotor disorder	Fernandez-Chacon et al., 2004

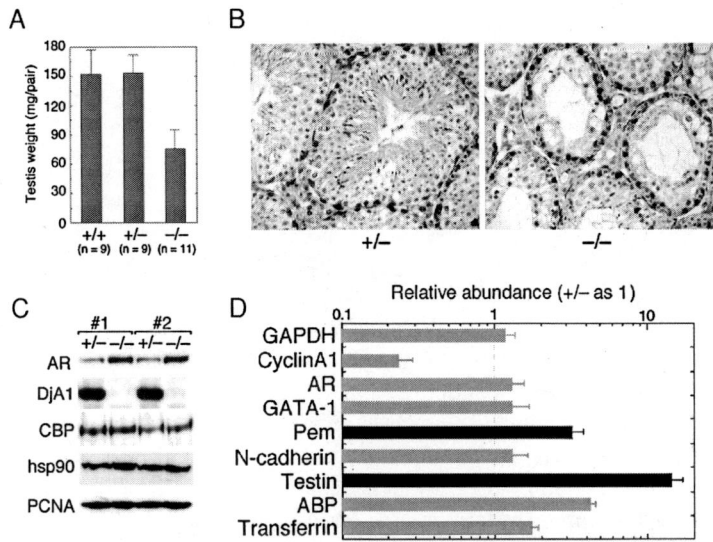

FIGURE 7. Defect of spermatogenesis in DjA1-null mice. A: Testis weight of each phenotype at 4 months of age. B: Representative section of testis probed with andorogen receptor antibody. C: Western blot analysis of total protein from testis. Protein samples were prepared from two pairs of littermates (3 months of age). D: Real time PCR analysis. Values were normalized for levels of β-actin expression (n = 5 pairs of littermates), and mean value for respective gene expression for DjA1+/− mice was set as 1. Modified form Terada et al. (2005). A type I DnaJ homolog, DjA1, regulates androgen receptor signaling and spermatogenesis. EMBO J. 24:611–622.

of the DjA1 gene resulted in an abnormality in mice. The DjA2 protein level was increased in the DjA1-null mice, and the remaining DjA2 was expected to compensate for the chaperone activity of DjA1. Transplantation experiments revealed a critical defect of the Sertoli cells in maintaining spermiogenesis at steps 8 to 9. These steps overlap to the steps in which androgen receptor in Sertoli cells dominantly promotes gene expression for maintaining the spermiogenesis (De Gendt et al., 2004). However the androgen receptor protein was increased post-transcriptionally in the DjA1-null testis. Androgen receptor protein was functional since it robustly enhanced transcription of several androgen-responsive genes in Sertoli cells. Disruption of adherens junction between Sertoli cell and germ cell was also evident. These results reveal a critical and distinct role of DjA1 in spermiogenesis through androgen receptor-mediated signaling in Sertoli cells, and raise a question whether excess androgen signaling is deleterious for spermatogenesis.

DjA3 (hTid1) is a mammalian homolog of yeast Mdj1. Yeast Mdj1 localizes in mitochondria, is engaged in translocation of proteins across inner membrane of mitochondria, and folding of proteins in the matrix (Neupert, 1997; Voos and Rottgers, 2002). DjA3 in the mitochondrial matrix was identified as a mammalian counterpart of Drosophila tumor suppressor, lethal(2) tumorous imaginal discs, and was a regulator of apoptosis (Syken et al., 1999) DjA3 was also regarded as a regulatory protein for a number of signaling molecules in the cytosol (Trentin et al., 2001 ; Kim et al., 2004). The mechanism for the dual intracellular localization of DjA3 remains unsolved. DjA3 targeting in mice resulted in embryonic death just after implantation (Lo et al., 2004). Recent research with this DjA3-floxed mice showed dilated cardiomyopathy in heart by a defect of mitochondrial biogenesis (Hayashi et al., 2005).

4.2. Type II HSP40/DnaJ Proteins

DjB1 (Hsp40/hdj1) is the most well-known mammalian HSP40/DnaJ protein in cytosol. DjB1 was reported to be engaged in many cellular functions with HSP70 proteins, including de novo folding of nascent polypeptides (Frydman, 2001; Young et al., 2004). Involvement of DjB1 in de novo folding in reticulocyte lysate (Frydman et al., 1994) and in refolding of a model polypeptide after thermal stress within the cell (Michels et al., 1999) was demonstrated.

DjB2 (Hsj1) is preferentially expressed in neuronal cells and is engaged in protein degradation by binding to ubiquitilated client proteins with its ubiquitin interaction motifs (Westhoff et al., 2005).

DjB4 (Hlj1) is a typical chaperone induced by heat stress (Hoe et al., 1998), and interacts with the DNA-damage protein GADD34 (Hasegawa et al., 2000). Upregulation of DjB4 expression by a transcription factor YY1 was reported to suppress cancer cell invasive capability through increase in E-cadherin expression (Wang et al., 2005).

DjB6 (Mrj) gene disruption resulted in embryonic death in mice by a defect in placental development (Hunter et al., 1999). Significant reduction was observed in expression of the chorionic trophoblast-specific transcription factor genes Err2 and Gcm1. DjB6 specifically binds to intermediate filament keratins 8 and 18 by its carboxyl-terminal domain and to cytosolic Hsc70 and Hsp70 by its amino-terminal J-domain in HeLa cells (Izawa et al., 2000). Antibodies against DjB6 disrupted organization of keratin 8/18 filaments without affecting the actin filaments and microtubules. The assembly of keratin filament cytoskeleton may be regulated by HSP70–DjB6 pair. Recently, DjB6 was reported to bind to nuclear factor of activated T cells c3 and mediated transcriptional repression through class II histone deacetylase recruitment (Dai et al., 2005b). A specific member of HSP40/DnaJ chaperone proteins may recruit cytosolic HSP70 protein to disassemble the transcriptional complexes and serve as an important regulator of gene expression.

DjB9 (MDG1/ERdj4) and DjB11 (HEDJ/Erdj3) are lumenal members of HSP40/DnaJ proteins, and assist BiP/Grp78 HSP70 protein in the translocation, folding, and ER-associated degradation (Shen et al., 2002; Meunier et al., 2002).

4.3. Type III HSP40/DnaJ Proteins

DjC1 (Mtj1/ERdj1) is an ER lumenal HSP40/DnaJ protein that contains a large carboxyl-terminal cytosolic extension composed of two SANT domains

(Kroczynska et al., 2004). The SANT2 domain of DjC1 binds to a serpin family protein and may be related to the ER-stress response.

DjC2 (MPP11/MIDA1) has yeast zuotin-homology domain in its aminoterminal and two-repeat Myb domain in its carboxyl-terminal. In yeast, zuotin (Zuo1) forms ribosome-associated complex (RAC) with Ssz1 (Pdr13), and acts as a chaperone for nascent polypeptide chains (Huang et al., 2005). DjC2 associates with ribosomes and forms a stable complex with Hsp70L1 (HSPA14), a mammalian counterpart of yeast Ssz1 (Otto et al., 2005). DjC2 has binding activity against a specific sequence of DNA by its Myb domain and stimulates transcription by its zuotin-homology domain (Inoue et al., 2000; Yoshida et al., 2004).

DjC3 (P58IPK) has a TPR domain in its amino-terminal and J-domain in its carboxyl-terminal. DjC3 is induced by influenza virus infection, and inhibits the interferon-induced, double-stranded RNA-activated, eukaryotic initiation factor 2α protein kinase R (PKR) (Melville et al., 2000). DjC3 was subsequently revealed to be a novel ER stress-induced gene and attenuates the unfolded protein response (Yan et al., 2002; van Huizen et al., 2003). DjC3 gene disruption in mice resulted in pancreatic β-cell failure and diabetes, and provided in vivo evidence for DjC3 functions in the ER stress response (Ladiges et al., 2005).

DjC5 (CSPα) is a palmitoylated secretory vesicle membrane protein and recruits Hsc70 to Ca^{2+}-channels and synaptic vesicles (Evans et al., 2003). DjC5 (CSPα) gene disruption in mice exhibited a progressive lethal phenotype that starts from sensorimotor disorder around weaning period (Fernandez-Chacon et al., 2004). This result suggests that DjC5 is indispensable for the normal operation of Ca^{2+}-channels or exocytosis but acts as a chaperone for maintenance of continued synaptic function. However, Chandra et al. (Chandra et al., 2005) recently reported that α-synuclein abrogated the lethal effects of the DjC5 deletion in mice and partially rescued a loss of proper SNARE complex assembly.

DjC6 (auxilin 1) is a neuronal-specific J-protein, and cyclin-G-associated kinase (GAK/auxilin 2) is a ubiquitous form of DjC6. GAK and Hsc70 proteins cooperatively disassemble clathrin from the cathrin-coated vesicle, exchange clathrin from the clathrin-coated pits, and mediate binding of clathrin and adaptors to the plasma membrane and the trans-Golgi network (Lee et al., 2005). The structure of an auxilin-bound clathrin coat was determined (Fotin et al., 2004). Binding of auxilin produced a local change in the clathrin heavy-chain contact, and created a significant distortion on the clathrin coat. Subsequent recruitment of Hsc70 protein may promote general uncoating.

DjC19 (DNADJ1) is a mammalian counterpart of yeast Tim14/Pam18. Another homologous gene is also expressed in human, and their functional redundancy remains to be elucidated. Recently, an autosomal recessive hereditary disease, dilated cardiomyopathy with ataxia (DCMA) syndrome, was mapped to the DjC19 gene (Davey et al., 2005). Yeast Tim14 is an integral protein of the inner membrane and an essential component of the TIM23 preprotein translocase (Mokranjac et al., 2003; Truscott et al., 2003). A J-domain of Tim14 is exposed to the matrix space and required for the interaction of mtHsp70 with Tim44 for translocation.

The large and small tumor antigens (T-antigens) of SV40 polyoma virus are essential HSP40/DnaJ proteins for replication (Sullivan and Pipas, 2002). They have the same J-domain in their amino-termini, and correspond to type III HSP40/DnaJ proteins. They bind stoichiometrically to the cytosolic HSP70 protein of the infected cells. The large T-antigen also binds the tumor suppressors, pRB and p53. Mutations in the J-domain of T-antigens significantly reduce efficacy of not only replication, but transcription, tumorigenesis, and virion assembly.

5. Concluding Remarks

The HSP40/DnaJ proteins manage HSP70 proteins to fold, hold, and disassemble specific target protein substrates properly. The list of specific protein substrates for the HSP40/DnaJ proteins in biological processes continues to grow, and indicates that HSP40/DnaJ protein is a key determinant for the HSP70-based chaperone system in intracellular protein society. Some target proteins for the HSP70-based chaperone system may be managed by the redundant functions of the respective members of HSP70, HSP40/DnaJ, and co-chaperone proteins in cytosol, but others require specific sets of the components of the HSP70 system. Various combinations of the components of the HSP70 system may impart diversity to the system. A mechanism(s) by which members of the HSP40/DnaJ proteins cooperate with HSP70 and co-chaperone proteins in specific chaperone activities may be eventually elucidated.

References

Abdul, K. M., Terada, K., Gotoh, T., Hafizur, R. M., and Mori, M. (2002) Characterization and functional analysis of a heart-enriched DnaJ/Hsp40 homolog dj4/DjA4. *Cell Stress Chap* 7:156–66.

Ahner, A., Whyte, F. M., and Brodsky, J. L. (2005) Distinct but overlapping functions of Hsp70, Hsp90, and an Hsp70 nucleotide exchange factor during protein biogenesis in yeast. *Arch Biochem Biophys* 435:32–41.

Alberti, S., Esser, C., and Hohfeld, J. (2003) BAG-1–a nucleotide exchange factor of Hsc70 with multiple cellular functions. *Cell Stress Chap* 8:225–31.

Alberti, S., Bohse, K., Arndt, V., Schmitz, A., and Hohfeld, J. (2004) The co-chaperone HspBP1 inhibits the CHIP ubiquitin ligase and stimulates the maturation of the cystic fibrosis transmembrane conductance regulator. *Mol Biol Cell* 15:4003–10.

Bimston, D., Song, J., Winchester, D., Takayama, S., Reed, J. C., and Morimoto, R. I. (1998) BAG-1, a negative regulator of Hsp70 chaperone activity, uncouples nucleotide hydrolysis from substrate release. *EMBO J* 17:6871–8.

Borges, J. C., Fischer, H., Craievich, A. F., and Ramos, C. H. (2005) Low resolution structural study of two human HSP40 chaperones in solution. DJA1 from subfamily A and DJB4 from subfamily B have different quaternary structures. *J Biol Chem* 280:13671–81.

Brehmer, D., Rudiger, S., Gassler, C. S., Klostermeier, D., Packschies, L., Reinstein, J., Mayer, M. P., and Bukau, B. (2001) Tuning of chaperone activity of Hsp70 proteins by modulation of nucleotide exchange. *Nat Struct Biol* 8:427–32.

Brychzy, A., Rein, T., Winklhofer, K. F., Hartl, F. U., Young, J. C., and Obermann, W. M. (2003) Cofactor Tpr2 combines two TPR domains and a J domain to regulate the Hsp70/Hsp90 chaperone system. *EMBO J.* 22:3613–23.

Bukau, B., and Horwich, A. L. (1998) The Hsp70 and Hsp60 chaperone machines. *Cell* 92:351–66.

Chamberlain, L. H., and Burgoyne, R. D. (2000) Cysteine-string protein: The chaperone at the synapse. *J Neurochem* 74:1781–9.

Chandra, S., Gallardo, G., Fernandez-Chacon, R., Schluter, O. M., and Sudhof, T. C. (2005) Alpha-synuclein cooperates with CSPalpha in preventing neurodegeneration. *Cell* 123:383–96.

Cheetham, M. E., and Caplan, A. J. (1998) Structure, function and evolution of DnaJ: Conservation and adaptation of chaperone function. *Cell Stress Chap* 3:28–36.

Connell, P., Ballinger, C. A., Jiang, J., Wu, Y., Thompson, L. J., Hohfeld, J., and Patterson, C. (2001) The co-chaperone CHIP regulates protein triage decisions mediated by heat-shock proteins. *Nat Cell Biol* 3:93–6.

Cyr, D. M., Hohfeld, J., and Patterson, C. (2002) Protein quality control: U-box-containing E3 ubiquitin ligases join the fold. *Trends Biochem Sci* 27:368–75.

Dai, Q., Zhang, C., Wu, Y., McDonough, H., Whaley, R. A., Godfrey, V., Li, H. H., Madamanchi, N., Xu, W., Neckers, L., Cyr, D., and Patterson, C. (2003) CHIP activates HSF1 and confers protection against apoptosis and cellular stress. *EMBO J* 22:5446–58.

Dai, Q., Qian, S. B., Li, H. H., McDonough, H., Borchers, C., Huang, D., Takayama, S., Younger, J. M., Ren, H. Y., Cyr, D. M., and Patterson, C. (2005a) Regulation of the cytoplasmic quality control protein degradation pathway by BAG2. *J Biol Chem* 280:38673–81.

Dai, Y. S., Xu, J., and Molkentin, J. D. (2005b) The DnaJ-related factor Mrj interacts with nuclear factor of activated T cells c3 and mediates transcriptional repression through class II histone deacetylase recruitment. *Mol Cell Biol* 25:9936–48.

Davey, K. M., Parboosingh, J. S., McLeod, D. R., Chan, A., Casey, R., Ferreira, P., Snyder, F. F., Bridge, P. J., and Bernier, F. P. (2005) Mutation of DNAJC19, a human homolog of yeast inner mitochondrial membrane co-chaperones, causes DCMA syndrome, a novel autosomal recessive Barth syndrome-like condition. *J Med Genet* 3:3.

De Gendt, K., Swinnen, J. V., Saunders, P. T., Schoonjans, L., Dewerchin, M., Devos, A., Tan, K., Atanassova, N., Claessens, F., Lecureuil, C., Heyns, W., Carmeliet, P., Guillou, F., Sharpe, R. M., and Verhoeven, G. (2004) A Sertoli cell-selective knockout of the androgen receptor causes spermatogenic arrest in meiosis. *Proc Natl Acad Sci USA* 101: 1327–32.

Evans, G. J., Morgan, A., and Burgoyne, R. D. (2003) Tying everything together: The multiple roles of cysteine string protein (CSP) in regulated exocytosis. *Traffic* 4:653–9.

Fan, C. Y., Lee, S., and Cyr, D. M. (2003) Mechanisms for regulation of Hsp70 function by Hsp40. *Cell Stress Chap* 8:309–16.

Fan, C. Y., Lee, S., Ren, H. Y., and Cyr, D. M. (2004) Exchangeable chaperone modules contribute to specification of type I and type II Hsp40 cellular function. *Mol Biol Cell* 15:761–73.

Fernandez-Chacon, R., Wolfel, M., Nishimune, H., Tabares, L., Schmitz, F., Castellano-Munoz, M., Rosenmund, C., Montesinos, M. L., Sanes, J. R., Schneggenburger, R., and Sudhof, T. C. (2004) The synaptic vesicle protein CSP alpha prevents presynaptic degeneration. *Neuron* 42:237–51.

Flaherty, K. M., Wilbanks, S. M., DeLuca-Flaherty, C., and McKay, D. B. (1994) Structural basis of the 70-kilodalton heat shock cognate protein ATP hydrolytic activity. II. Structure of the active site with ADP or ATP bound to wild type and mutant ATPase fragment. *J Biol Chem* 269:12899–907.

Fotin, A., Cheng, Y., Grigorieff, N., Walz, T., Harrison, S. C., and Kirchhausen, T. (2004) Structure of an auxilin-bound clathrin coat and its implications for the mechanism of uncoating. *Nature* 432:649–53.

Frydman, J., Nimmesgern, E., Ohtsuka, K., and Hartl, F. U. (1994) Folding of nascent polypeptide chains in a high molecular mass assembly with molecular chaperones. *Nature* 370:111–7.

Frydman, J. (2001) Folding of newly translated proteins *in vivo*: The role of molecular chaperones. *Annu Rev Biochem* 70:603–47.

Gall, W. E., Higginbotham, M. A., Chen, C., Ingram, M. F., Cyr, D. M., and Graham, T. R. (2000) The auxilin-like phosphoprotein Swa2p is required for clathrin function in yeast. *Curr Biol* 10:1349–58.

Gassler, C. S., Wiederkehr, T., Brehmer, D., Bukau, B., and Mayer, M. P. (2001) Bag-1M accelerates nucleotide release for human Hsc70 and Hsp70 and can act concentration-dependent as positive and negative cofactor. *J Biol Chem* 276:32538–44.

Gething, M. J., and Sambrook, J. (1992) Protein folding in the cell. *Nature* 355:33–45.

Glover, J. R., and Lindquist, S. (1998) Hsp104, Hsp70, and Hsp40: A novel chaperone system that rescues previously aggregated proteins. *Cell* 94:73–82.

Goloubinoff, P., Mogk, A., Zvi, A. P., Tomoyasu, T., and Bukau, B. (1999) Sequential mechanism of solubilization and refolding of stable protein aggregates by a bichaperone network. *Proc Natl Acad Sci USA* 96:13732–7.

Gotoh, T., Terada, K., Oyadomari, S., and Mori, M. (2004) hsp70–DnaJ chaperone pair prevents nitric oxide- and CHOP-induced apoptosis by inhibiting translocation of Bax to mitochondria. *Cell Death Differ* 11:390–402.

Gotz, R., Wiese, S., Takayama, S., Camarero, G. C., Rossoll, W., Schweizer, U., Troppmair, J., Jablonka, S., Holtmann, B., Reed, J. C., Rapp, U. R., and Sendtner, M. (2005) Bag1 is essential for differentiation and survival of hematopoietic and neuronal cells. *Nat Neurosci* 8:1169–78.

Greener, T., Grant, B., Zhang, Y., Wu, X., Greene, L. E., Hirsh, D., and Eisenberg, E. (2001) *Caenorhabditis elegans* auxilin: A J-domain protein essential for clathrin-mediated endocytosis *in vivo*. *Nat Cell Biol* 3:215–9.

Hafizur, R. M., Yano, M., Gotoh, T., Mori, M., and Terada, K. (2004) Modulation of chaperone activities of Hsp70 and Hsp70-2 by a mammalian DnaJ/Hsp40 homolog, DjA4. *J Biochem* 135:193–200.

Harrison, C. J., Hayer-Hartl, M., Di Liberto, M., Hartl, F., and Kuriyan, J. (1997) Crystal structure of the nucleotide exchange factor GrpE bound to the ATPase domain of the molecular chaperone DnaK. *Science* 276:431–5.

Hasegawa, T., Xiao, H., Hamajima, F., and Isobe, K. (2000) Interaction between DNA-damage protein GADD34 and a new member of the Hsp40 family of heat shock proteins that is induced by a DNA-damaging reagent. *Biochem J* 352:795–800.

Hattori, H., Kaneda, T., Lokeshwar, B., Laszlo, A., and Ohtsuka, K. (1993) A stress-inducible 40 kDa protein (hsp40): Purification by modified two-dimensional gel electrophoresis and co-localization with hsc70(p73) in heat-shocked HeLa cells. *J Cell Sci* 104:629–38.

Hayashi, M., Imanaka-Yoshida, K., Yoshida, T., Wood, M., Fearns, C., Tatake, R. J., and Lee, J. D. (2006) A crucial role of mitochondrial Hsp40 in preventing dilated cardiomyopathy. *Nat Med* 12:128–32.

Hershko, A., and Ciechanover, A. (1998) The ubiquitin system. *Annu Rev Biochem* 67:425–79.

Hoe, K. L., Won, M., Chung, K. S., Jang, Y. J., Lee, S. B., Kim, D. U., Lee, J. W., Yun, J. H., and Yoo, H. S. (1998) Isolation of a new member of DnaJ-like heat shock protein 40 (Hsp40) from human liver. *Biochim Biophys Acta* 1383:4–8.

Hohfeld, J., and Jentsch, S. (1997) GrpE-like regulation of the hsc70 chaperone by the anti-apoptotic protein BAG-1. *EMBO J* 16:6209–16.

Huang, L., Mivechi, N. F., and Moskophidis, D. (2001) Insights into regulation and function of the major stress-induced hsp70 molecular chaperone *in vivo*: Analysis of mice with targeted gene disruption of the hsp70.1 or hsp70.3 gene. *Mol Cell Biol* 21:8575–91.

Huang, P., Gautschi, M., Walter, W., Rospert, S., and Craig, E. A. (2005) The Hsp70 Ssz1 modulates the function of the ribosome-associated J-protein Zuo1. *Nat Struct Mol Biol* 12:497–504.

Hunter, P. J., Swanson, B. J., Haendel, M. A., Lyons, G. E., and Cross, J. C. (1999) Mrj encodes a DnaJ-related co-chaperone that is essential for murine placental development. *Development* 126:1247–58.

Inoue, T., Shoji, W., and Obinata, M. (2000) MIDA1 is a sequence specific DNA binding protein with novel DNA binding properties. *Genes Cells* 5:699–709.

Izawa, I., Nishizawa, M., Ohtakara, K., Ohtsuka, K., Inada, H., and Inagaki, M. (2000) Identification of Mrj, a DnaJ/Hsp40 family protein, as a keratin 8/18 filament regulatory protein. *J Biol Chem* 275:34521–7.

Kabani, M., Beckerich, J. M., and Gaillardin, C. (2000) Sls1p stimulates Sec63p-mediated activation of Kar2p in a conformation-dependent manner in the yeast endoplasmic reticulum. *Mol Cell Biol* 20:6923–34.

Kabani, M., Beckerich, J. M., and Brodsky, J. L. (2002a) Nucleotide exchange factor for the yeast Hsp70 molecular chaperone Ssa1p. *Mol Cell Biol* 22:4677–89.

Kabani, M., McLellan, C., Raynes, D. A., Guerriero, V., and Brodsky, J. L. (2002b) HspBP1, a homologue of the yeast Fes1 and Sls1 proteins, is an Hsc70 nucleotide exchange factor. *FEBS Lett* 531:339–42.

Kampinga, H. H., Kanon, B., Salomons, F. A., Kabakov, A. E., and Patterson, C. (2003) Overexpression of the co-chaperone CHIP enhances Hsp70-dependent folding activity in mammalian cells. *Mol Cell Biol* 23:4948–58.

Kanelakis, K. C., and Pratt, W. B. (2003) Regulation of glucocorticoid receptor ligand-binding activity by the hsp90/hsp70-based chaperone machinery. *Methods Enzymol* 364:159–73.

Kawaguchi, Y., Kovacs, J. J., McLaurin, A., Vance, J. M., Ito, A., and Yao, T. P. (2003) The deacetylase HDAC6 regulates aggresome formation and cell viability in response to misfolded protein stress. *Cell* 115:727–38.

Kedersha, N., and Anderson, P. (2002) Stress granules: Sites of mRNA triage that regulate mRNA stability and translatability. *Biochem Soc Trans* 30:963–9.

Kim, S., Schilke, B., Craig, E. A., and Horwich, A. L. (1998) Folding *in vivo* of a newly translated yeast cytosolic enzyme is mediated by the SSA class of cytosolic yeast Hsp70 proteins. *Proc Natl Acad Sci USA* 95:12860–5.

Kim, S. W., Chao, T. H., Xiang, R., Lo, J. F., Campbell, M. J., Fearns, C., and Lee, J. D. (2004) Tid1, the human homologue of a *Drosophila* tumor suppressor, reduces the malignant activity of ErbB-2 in carcinoma cells. *Cancer Res* 64:7732–9.

Kimura, Y., Yahara, I., and Lindquist, S. (1995) Role of the protein chaperone YDJ1 in establishing Hsp90-mediated signal transduction pathways. *Science* 268:1362–5.

Kroczynska, B., Evangelista, C. M., Samant, S. S., Elguindi, E. C., and Blond, S. Y. (2004) The SANT2 domain of the murine tumor cell DnaJ-like protein 1 human homologue interacts with alpha1-antichymotrypsin and kinetically interferes with its serpin inhibitory activity. *J Biol Chem* 279:11432–43.

Ladiges, W. C., Knoblaugh, S. E., Morton, J. F., Korth, M. J., Sopher, B. L., Baskin, C. R., MacAuley, A., Goodman, A. G., LeBoeuf, R. C., and Katze, M. G. (2005) Pancreatic beta-cell failure and diabetes in mice with a deletion mutation of the endoplasmic reticulum molecular chaperone gene P58IPK. *Diabetes* 54:1074–81.

Landry, S. J. (2003) Swivels and stators in the Hsp40–Hsp70 chaperone machine. *Structure* 11:1465–6.

Lee, D. W., Zhao, X., Zhang, F., Eisenberg, E., and Greene, L. E. (2005) Depletion of GAK/auxilin 2 inhibits receptor-mediated endocytosis and recruitment of both clathrin and clathrin adaptors. *J Cell Sci* 118:4311–21.

Levy, E. J., McCarty, J., Bukau, B., and Chirico, W. J. (1995) Conserved ATPase and luciferase refolding activities between bacteria and yeast Hsp70 chaperones and modulators. *FEBS Lett* 368:435–40.

Li, J., Qian, X., and Sha, B. (2003) The crystal structure of the yeast Hsp40 Ydj1 complexed with its peptide substrate. *Structure* 11:1475–83.

Linke, K., Wolfram, T., Bussemer, J., and Jakob, U. (2003) The roles of the two zinc binding sites in DnaJ. *J Biol Chem* 278:44457–66.

Lo, J. F., Hayashi, M., Woo-Kim, S., Tian, B., Huang, J. F., Fearns, C., Takayama, S., Zapata, J. M., Yang, Y., and Lee, J. D. (2004) Tid1, a co-chaperone of the heat shock 70 protein and the mammalian counterpart of the *Drosophila* tumor suppressor l(2)tid, is critical for early embryonic development and cell survival. *Mol Cell Biol* 24:2226–36.

Mayer, M. P., and Bukau, B. (2005) Hsp70 chaperones: Cellular functions and molecular mechanism. *Cell Mol Life Sci* 62:670–84.

Meacham, G. C., Patterson, C., Zhang, W., Younger, J. M., and Cyr, D. M. (2001) The Hsc70 co-chaperone CHIP targets immature CFTR for proteasomal degradation. *Nat Cell Biol* 3:100–5.

Melville, M. W., Katze, M. G., and Tan, S. L. (2000) P58IPK, a novel co-chaperone containing tetratricopeptide repeats and a J-domain with oncogenic potential. *Cell Mol Life Sci* 57:311–22.

Meunier, L., Usherwood, Y. K., Chung, K. T., and Hendershot, L. M. (2002) A subset of chaperones and folding enzymes form multiprotein complexes in endoplasmic reticulum to bind nascent proteins. *Mol Biol Cell* 13:4456–69.

Michels, A. A., Kanon, B., Bensaude, O., and Kampinga, H. H. (1999) Heat shock protein (Hsp) 40 mutants inhibit Hsp70 in mammalian cells. *J Biol Chem* 274:36757–63.

Minami, Y., Hohfeld, J., Ohtsuka, K., and Hartl, F. U. (1996) Regulation of the heat-shock protein 70 reaction cycle by the mammalian DnaJ homolog, Hsp40. *J Biol Chem* 271:19617–24.

Mokranjac, D., Sichting, M., Neupert, W., and Hell, K. (2003) Tim14, a novel key component of the import motor of the TIM23 protein translocase of mitochondria. *EMBO J* 22:4945–56.

Morimoto, R. I. (1998) Regulation of the heat shock transcriptional response: Cross talk between a family of heat shock factors, molecular chaperones, and negative regulators. *Genes Dev* 12:3788–96.

Motohashi, K., Watanabe, Y., Yohda, M., and Yoshida, M. (1999) Heat-inactivated proteins are rescued by the DnaK.J-GrpE set and ClpB chaperones. *Proc Natl Acad Sci USA* 96:7184–9.

Murata, S., Minami, Y., Minami, M., Chiba, T., and Tanaka, K. (2001) CHIP is a chaperone-dependent E3 ligase that ubiquitylates unfolded protein. *EMBO Rep* 2:1133–8.

Murata, S., Chiba, T., and Tanaka, K. (2003) CHIP: A quality-control E3 ligase collaborating with molecular chaperones. *Int J Biochem Cell Biol* 35:572–8.

Neupert, W. (1997) Protein import into mitochondria. *Annu Rev Biochem* 66:863–917.

Nimmesgern, E., and Hartl, F. U. (1993) ATP-dependent protein refolding activity in reticulocyte lysate. Evidence for the participation of different chaperone components. *FEBS Lett* 331:25–30.

Ohtsuka, K., and Hata, M. (2000) Mammalian HSP40/DNAJ homologs: Cloning of novel cDNAs and a proposal for their classification and nomenclature. *Cell Stress Chap* 5:98–112.

Otto, H., Conz, C., Maier, P., Wolfle, T., Suzuki, C. K., Jeno, P., Rucknagel, P., Stahl, J., and Rospert, S. (2005) The chaperones MPP11 and Hsp70L1 form the mammalian ribosome-associated complex. *Proc Natl Acad Sci USA* 102:10064–9.

Pahl, A., Brune, K., and Bang, H. (1997) Fit for life? Evolution of chaperones and folding catalysts parallels the development of complex organisms. *Cell Stress Chap* 2:78–86.

Pellecchia, M., Szyperski, T., Wall, D., Georgopoulos, C., and Wuthrich, K. (1996) NMR structure of the J-domain and the Gly/Phe-rich region of the *Escherichia coli* DnaJ chaperone. *J Mol Biol* 260:236–50.

Pishvaee, B., Costaguta, G., Yeung, B. G., Ryazantsev, S., Greener, T., Greene, L. E., Eisenberg, E., McCaffery, J. M., and Payne, G. S. (2000) A yeast DNA J protein required for uncoating of clathrin-coated vesicles *in vivo*. *Nat Cell Biol* 2:958–63.

Pratt, W. B., and Toft, D. O. (2003) Regulation of signaling protein function and trafficking by the hsp90/hsp70-based chaperone machinery. *Exp Biol Med* 228:111–33.

Raynes, D. A., and Guerriero, V., Jr. (1998) Inhibition of Hsp70 ATPase activity and protein renaturation by a novel Hsp70-binding protein. *J Biol Chem* 273:32883–8.

Rohde, M., Daugaard, M., Jensen, M. H., Helin, K., Nylandsted, J., and Jaattela, M. (2005) Members of the heat-shock protein 70 family promote cancer cell growth by distinct mechanisms. *Genes Dev* 19:570–82.

Rudiger, S., Buchberger, A., and Bukau, B. (1997a) Interaction of Hsp70 chaperones with substrates. *Nat Struct Biol* 4:342–9.

Rudiger, S., Germeroth, L., Schneider-Mergener, J., and Bukau, B. (1997b) Substrate specificity of the DnaK chaperone determined by screening cellulose-bound peptide libraries. *EMBO J* 16:1501–7.

Rudiger, S., Schneider-Mergener, J., and Bukau, B. (2001) Its substrate specificity characterizes the DnaJ co-chaperone as a scanning factor for the DnaK chaperone. *EMBO J* 20:1042–50.

Scheufler, C., Brinker, A., Bourenkov, G., Pegoraro, S., Moroder, L., Bartunik, H., Hartl, F. U., and Moarefi, I. (2000) Structure of TPR domain-peptide complexes: Critical elements in the assembly of the Hsp70–Hsp90 multichaperone machine. *Cell* 101:199–210.

Sha, B., Lee, S., and Cyr, D. M. (2000) The crystal structure of the peptide-binding fragment from the yeast Hsp40 protein Sis1. *Structure* 8:799–807.

Shen, Y., Meunier, L., and Hendershot, L. M. (2002) Identification and characterization of a novel endoplasmic reticulum (ER) DnaJ homologue, which stimulates ATPase activity of BiP *in vitro* and is induced by ER stress. *J Biol Chem* 277:15947–56.

Shomura, Y., Dragovic, Z., Chang, H. C., Tzvetkov, N., Young, J. C., Brodsky, J. L., Guerriero, V., Hartl, F. U., and Bracher, A. (2005) Regulation of Hsp70 function by HspBP1: Structural analysis reveals an alternate mechanism for Hsp70 nucleotide exchange. *Mol Cell* 17:367–79.

Sondermann, H., Scheufler, C., Schneider, C., Hohfeld, J., Hartl, F. U., and Moarefi, I. (2001) Structure of a Bag/Hsc70 complex: Convergent functional evolution of Hsp70 nucleotide exchange factors. *Science* 291:1553–7.

Sullivan, C. S., and Pipas, J. M. (2002) T antigens of simian virus 40: Molecular chaperones for viral replication and tumorigenesis. *Microbiol Mol Biol Rev* 66:179–202.

Syken, J., De-Medina, T., and Munger, K. (1999) TID1, a human homolog of the *Drosophila* tumor suppressor l(2)tid, encodes two mitochondrial modulators of apoptosis with opposing functions. *Proc Natl Acad Sci USA* 96:8499–504.

Szyperski, T., Pellecchia, M., Wall, D., Georgopoulos, C., and Wuthrich, K. (1994) NMR structure determination of the *Escherichia coli* DnaJ molecular chaperone: Secondary structure and backbone fold of the N-terminal region (residues 2-108) containing the highly conserved J domain. *Proc Natl Acad Sci USA* 91:11343–7.

Takayama, S., Bimston, D. N., Matsuzawa, S., Freeman, B. C., Aime-Sempe, C., Xie, Z., Morimoto, R. I., and Reed, J. C. (1997) BAG-1 modulates the chaperone activity of Hsp70/Hsc70. *EMBO J* 16:4887–96.

Takayama, S., and Reed, J. C. (2001) Molecular chaperone targeting and regulation by BAG family proteins. *Nat Cell Biol* 3:E237–41.

Tatusov, R. L., Natale, D. A., Garkavtsev, I. V., Tatusova, T. A., Shankavaram, U. T., Rao, B. S., Kiryutin, B., Galperin, M. Y., Fedorova, N. D., and Koonin, E. V. (2001) The COG database: New developments in phylogenetic classification of proteins from complete genomes. *Nucleic Acids Res* 29:22–8.

Terada, K., Kanazawa, M., Bukau, B., and Mori, M. (1997) The human DnaJ homologue dj2 facilitates mitochondrial protein import and luciferase refolding. *J Cell Biol* 139:1089–95.

Terada, K., and Mori, M. (2000) Human DnaJ homologs dj2 and dj3, and bag-1 are positive co-chaperones of hsc70. *J Biol Chem* 275:24728–34.

Terada, K., Yomogida, K., Imai, T., Kiyonari, H., Takeda, N., Kadomatsu, T., Yano, M., Aizawa, S., and Mori, M. (2005) A type I DnaJ homolog, DjA1, regulates androgen receptor signaling and spermatogenesis. *EMBO J* 24:611–22.

Townsend, P. A., Cutress, R. I., Sharp, A., Brimmell, M., and Packham, G. (2003) BAG-1: A multifunctional regulator of cell growth and survival. *Biochim Biophys Acta* 1603:83–98.

Trentin, G. A., Yin, X., Tahir, S., Lhotak, S., Farhang-Fallah, J., Li, Y., and Rozakis-Adcock, M. (2001) A mouse homologue of the *Drosophila* tumor suppressor l(2)tid gene defines a novel Ras GTPase-activating protein (RasGAP) binding protein. *J Biol Chem* 276:13087–95.

Truscott, K. N., Voos, W., Frazier, A. E., Lind, M., Li, Y., Geissler, A., Dudek, J., Muller, H., Sickmann, A., Meyer, H. E., Meisinger, C., Guiard, B., Rehling, P., and Pfanner, N. (2003) A J-protein is an essential subunit of the presequence translocase-associated protein import motor of mitochondria. *J Cell Biol* 163:707–13.

Tsai, J., and Douglas, M. G. (1996) A conserved HPD sequence of the J-domain is necessary for YDJ1 stimulation of Hsp70 ATPase activity at a site distinct from substrate binding. *J Biol Chem* 271:9347–54.

van Huizen, R., Martindale, J. L., Gorospe, M., and Holbrook, N. J. (2003) P58IPK, a novel endoplasmic reticulum stress-inducible protein and potential negative regulator of eIF2alpha signaling. *J Biol Chem* 278:15558–64.

Voos, W., and Rottgers, K. (2002) Molecular chaperones as essential mediators of mitochondrial biogenesis. *Biochim Biophys Acta* 1592:51–62.

Wall, D., Zylicz, M., and Georgopoulos, C. (1994) The NH2-terminal 108 amino acids of the *Escherichia coli* DnaJ protein stimulate the ATPase activity of DnaK and are sufficient for lambda replication. *J Biol Chem* 269:5446–51.

Walsh, P., Bursac, D., Law, Y. C., Cyr, D., and Lithgow, T. (2004) The J-protein family: Modulating protein assembly, disassembly and translocation. *EMBO Rep* 5:567–71.

Wang, C. C., Tsai, M. F., Hong, T. M., Chang, G. C., Chen, C. Y., Yang, W. M., Chen, J. J., and Yang, P. C. (2005) The transcriptional factor YY1 upregulates the novel invasion suppressor HLJ1 expression and inhibits cancer cell invasion. *Oncogene* 24:4081–93.

Westhoff, B., Chapple, J. P., van der Spuy, J., Hohfeld, J., and Cheetham, M. E. (2005) HSJ1 is a neuronal shuttling factor for the sorting of chaperone clients to the proteasome. *Curr Biol* 15:1058–64.

Wilbanks, S. M., and McKay, D. B. (1995) How potassium affects the activity of the molecular chaperone Hsc70. II. Potassium binds specifically in the ATPase active site. *J Biol Chem* 270:2251–7.

Yan, W., Frank, C. L., Korth, M. J., Sopher, B. L., Novoa, I., Ron, D., and Katze, M. G. (2002) Control of PERK eIF2alpha kinase activity by the endoplasmic reticulum stress-induced molecular chaperone P58IPK. *Proc Natl Acad Sci USA* 99:15920–5.

Yano, M., Terada, K., and Mori, M. (2003) AIP is a mitochondrial import mediator that binds to both import receptor Tom20 and pre-proteins. *J Cell Biol* 163:45–56.

Yoshida, M., Inoue, T., Shoji, W., Ikawa, S., and Obinata, M. (2004) Reporter gene stimulation by MIDA1 through its DnaJ homology region. *Biochem Biophys Res Commun* 324:326–32.

Young, J. C., Hoogenraad, N. J., and Hartl, F. U. (2003) Molecular chaperones Hsp90 and Hsp70 deliver preproteins to the mitochondrial import receptor Tom70. *Cell* 112:41–50.

Young, J. C., Agashe, V. R., Siegers, K., and Hartl, F. U. (2004) Pathways of chaperone-mediated protein folding in the cytosol. *Nat Rev Mol Cell Biol* 5:781–91.

Zhu, X., Zhao, X., Burkholder, W. F., Gragerov, A., Ogata, C. M., Gottesman, M. E., and Hendrickson, W. A. (1996) Structural analysis of substrate binding by the molecular chaperone DnaK. *Science* 272:1606–14.

V
Role of Molecular Chaperones in Cell Regulation

12
FKBP Co-Chaperones in Steroid Receptor Complexes

JOYCE CHEUNG-FLYNN, SEAN P. PLACE, MARC B. COX, VIRAVAN PRAPAPANICH, AND DAVID F. SMITH

Department of Biochemistry & Molecular Biology, Mayo Clinic, Scottsdale, AZ 85259

1. Introduction

Molecular chaperones are critical for cell survival by assuring proper protein folding in general, but chaperones, in particular the Hsp90 machinery, are also important for the activity of multiple specific client proteins involved in cellular signal transduction pathways. One class of extensively studied Hsp90 client is the steroid receptor subfamily of nuclear receptors. Chaperones are required for folding and stabilizing steroid receptors in a functionally competent state for hormone binding, and chaperones can also modulate steroid receptor responsiveness to hormone binding. In this chapter, we review recent advances in understanding the biochemical and physiological functions of a class of co-chaperones, the Hsp90-binding peptidylprolyl isomerases, that populate steroid receptor complexes.

1.1. Steroid Hormone Action and Physiology

Steroid hormones regulate numerous developmental and physiological processes in vertebrates. For example, estrogens, progesterone, and androgens control development and function of reproductive tissues; mineralocorticoids such as aldosterone regulate water and electrolyte balance; and glucocorticoids regulate a wide range of processes, including cellular metabolism, stress responses, and the immune system. Steroid hormone actions are mediated by hormone specific receptor proteins. The classic steroid receptors are intracellular, ligand-dependent transcription factors that have three major structural domains (Fig. 1A). The N-terminal domain, which is highly variable among steroid receptors, is the least structurally characterized but plays important roles in interactions with transcriptional factors. The DNA-binding domain targets receptor to specific DNA sequences in the promoter regions of target genes. Finally, the ligand binding domain (LBD) is the site for specific hormone binding and is the receptor region to which chaperones bind; subsequent to hormone binding, the LBD also participates in receptor dimerization and interactions with transcriptional factors. Recent studies have also suggested non-transcriptional activities for classic receptors localized to the plasma membrane, and unique membrane steroid receptors that are structurally distinct from the

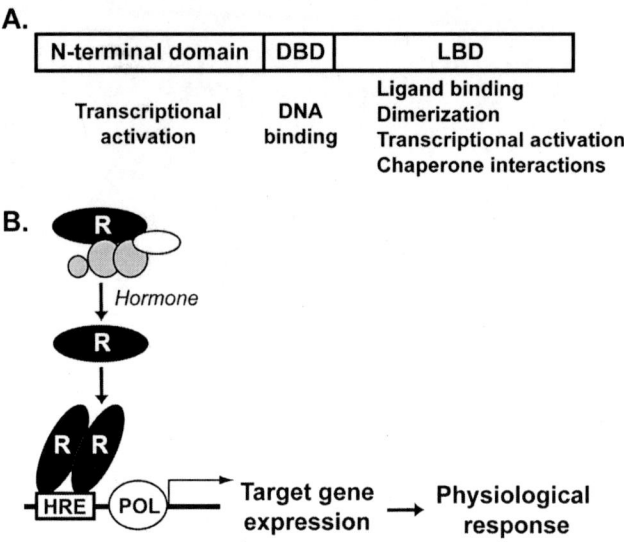

FIGURE 1. Basics of steroid receptor structure and function. A: Classic steroid receptors are comprised of three structural domains, each with distinct functions. These are the N-terminal domain, the DNA binding domain (DBD), and the ligand binding domain (LBD) to which hormone binds. B: In the absence of hormone, receptor (R) is associated with a combination of molecular chaperones whose release is stimulated by hormone binding. The hormone-activated receptor forms a homodimer bound to specific hormone response elements (HRE), and, through association with various co-transcriptional factors, receptor promotes transcription of target genes by the DNA polymerase complex (POL).

classic receptors have also been identified (Simoncini and Genazzani, 2003). As roles for molecular chaperones in membrane-localized and non-classic steroid receptors are largely unexplored, the remainder of this chapter will focus on chaperone interactions with classic steroid receptors that respond to hormone by altering transcription of specific target genes.

Steroid actions are tissue-specific, due in part to local expression of the cognate receptor but also due to expression of a range of protein effectors. The molecular chaperones are effectors that typically bind receptor prior to hormone binding, whereas a diverse assortment of co-transcriptional factors interact with hormone-bound receptor. As illustrated in Fig. 1B, steroid receptor that lacks hormone is typically found in dynamic complexes with Hsp90 and other chaperone proteins. Upon binding hormone, receptor interactions with chaperones are interrupted, presumably resulting from a receptor conformational change. Hormone-bound receptor typically homodimerizes, undergoes any of several covalent modifications, binds to specific DNA response elements on target genes, and establishes interactions with protein components of the chromatin remodeling machinery and basal transcription machinery. The consequence of these molecular rearrangements is up- or down-regulation of target genes. Mechanistic details of transcriptional events mediated by steroid receptors are not within the scope of this chapter, and the reader is referred to recent articles for a more in-depth review (Aranda and Pascual, 2001). Suffice it to say that numerous cellular factors—chaperones, kinases and other modifying enzymes, chromatin remodeling proteins, and an assortment of other transcription factors—dictate the cell-specific consequence of hormone binding to receptor.

1.2. Assembly of the Chaperone-Receptor Heterocomplex

Little is known about chaperone interactions with the nascent receptor polypeptide as it emerges from the ribosome, but it is safe to assume that, as with most eukaryotic, multi-domain proteins (Fink, 1999), chaperones participate in the initial folding of receptor. Steroid receptors, however, persist in a conformation that remains attractive to chaperones until hormone binding. In fact, steroid receptors are typically dependent on continued chaperone interactions under physiological conditions to maintain the receptor in a functionally mature form that is competent for hormone binding. Much has been learned about the ordered nature of chaperone interactions with receptors in their persistent state of attraction. Following the initial discovery that steroid receptor complexes can be faithfully assembled in vitro using rabbit reticulocyte lysate (Smith et al., 1990b), cell-free assays have been exploited to identify chaperones critical for complex assembly, to describe the ordering of assembly steps, and to understand the biochemical role of individual chaperones in assembly of functionally mature steroid receptors. There is general agreement from minimal assembly systems developed for GR by the Pratt laboratory (Hutchison et al., 1994) and for PR by the Toft laboratory (Kosano et al., 1998) that five chaperones are required to efficiently assemble functionally mature receptor complexes (Fig. 2). Hsp40 is the initial chaperone that binds receptor, and its binding is obligatory for subsequent assembly steps to proceed efficiently. Hsp40 appears to bind a single site in the LBD, which is the receptor domain that all chaperone interactions localize to, but the exact binding site has not been mapped more precisely. Hsp40 recruits Hsp70 to the receptor, and Hsp70 attracts the adaptor co-chaperone Hop, which simultaneously binds Hsp90. In a process that remains poorly understood, Hsp90 becomes directly associated with the receptor LBD, Hsp40, Hsp70, and Hop exit the receptor complex, and the co-chaperone p23 binds Hsp90 and stabilizes Hsp90 binding to receptor. Only when Hsp90 and p23 are properly assembled on the LBD does the receptor bind hormone efficiently, but the conformational change that presumably occurs to the LBD to facilitate hormone binding has not been identified. Hsp90 dissociates from receptor in a spontaneous manner that depends on the Hsp90 ATPase cycle. In the absence of hormone, receptor again attracts Hsp40 to reinitiate assembly and maturation. On the other hand, if hormone is bound when Hsp90 dissociates, it appears that Hsp40 is no longer attracted to the LBD; receptor effectively escapes renewed chaperone interactions and undergoes the multiple events that lead to transcriptional changes at target genes.

Although a purified system of five chaperones is sufficient to generate receptor complexes that are competent for hormone binding, the resulting receptor complex

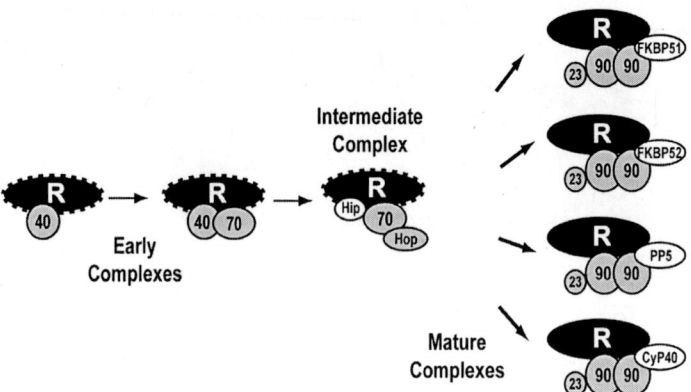

FIGURE 2. Ordered assembly of chaperones with steroid receptor. Receptor (R) initially is bound by Hsp40 (40), which recruits Hsp70 (70) and co-chaperones Hip and Hop. In the final assembly stage, Hsp70 is replaced by an Hsp90 (90) dimer plus the co-chaperone p23 (23) and one of the TPR co-chaperones FKBP51, FKBP52, PP5, or CyP40. Receptor in early and intermediate complexes lacks hormone-binding ability (indicated by dashed border). The five chaperone components that are minimally required for maturation of receptor hormone binding ability are Hsp40, Hsp70, Hop, Hsp90, and p23 (shaded in dark gray).

fails to dissociate in response to hormone binding. Thus, there are additional factors in reticulocyte lysate and in cells that are required for normal response of the receptor complex to hormone binding. Efforts have so far failed to identify the missing factor or, more likely, combination of factors that restore full responsiveness to reconstituted receptor complexes. Several additional co-chaperones have been identified as participants in native receptor complexes. Among these factors are Hsp70 co-chaperones Hip (Smith, 1993) and BAG1 (Froesch et al., 1998), the latter of which exists in several isoforms. Additional Hsp90 co-chaperones observed in native receptor complexes include members of the immunophilin protein families and the protein phosphatase PP5 (Cox and Smith, 2005). Although none of these Hsp90 co-chaperones is necessary for the basic assembly and maturation of receptor complexes, recent studies have begun to uncover physiologically relevant ways in which they alter receptor activity.

2. Receptor-Associated PPIases

2.1. PPIases, Drugs, and Immunosuppression

Immunophilins are proteins that bind certain immunosuppressant drugs and are divided into two, largely unrelated gene families: the FK506 binding proteins (FKBP) that bind FK506, rapamycin and their derivatives, and cyclophilins (CyP) that bind cyclosporin A (CsA) and related compounds. FK506 (tacrolimus, ProGraf), rapamycin (sirolimus, Rapimmune), and CsA are widely used in the clinic for immunosuppression following transplantation procedures or in autoimmune disease. Multiple members of the FKBP and CyP gene families are found in all eukaryotes. CyP and FKBP members characteristically have an active peptidylprolyl *cis/trans* isomerase (PPIase) domain, although the FKBP PPIase domain is unrelated to the CyP PPIase domain. PPIases catalyze the isomerization of proline in a peptide backbone between the more favorable *trans* isomer and the alternate *cis* isomer; this activity is important since proline isomerization can be the rate-limiting step in protein folding and can alter native protein conformation and function. A third structural class of PPIase is represented by the vertebrate protein Pin1, whose substrate proline must be preceded by phospho-serine or phospho-threonine (Yaffe et al., 1997). Pin1 has numerous protein substrates and is an important regulator of cell cycle and other cellular pathways, although it is not a target for immunosuppressive drugs, and thus not an immunophilin, and is not a component of the molecular chaperone machinery.

The relevant immunosuppressive drugs bind an FKBP or CyP PPIase active site and block PPIase activity, yet PPIase inhibition is not the mechanism for immunosuppression since there are drug analogs that efficiently inhibit PPIase activity yet are non-immunosuppressive. The immunosuppressive ligands contain effector domains that combine with portions of the cognate immunophilin to form specific binding sites for proteins involved in T-cell activation or proliferation. For example, FK506 in complex with FKBP12, the prototypical FKBP, binds tightly to and inhibits calcineurin, a calcium-dependent Ser-Thr phosphatase that is critical for T-cell activation and proliferation (Bierer, 1994). CsA-CyPA complexes act in a similar manner to inhibit calcineurin. On the other hand, the rapamycin-FKBP12 complex does not bind calcineurin, but rather binds and inhibits TOR (target of rapamycin), a protein that stimulates cell cycle progression in T-cells (Heitman et al., 1991).

Understanding the physiological roles for PPIase immunophilins and identification of relevant PPIase substrates has progressed slowly, although recent findings with mouse gene knockout models for individual immunophilins have greatly enhanced evaluation of PPIase immunophilin function in vertebrates. For example, mice lacking FKBP12 display defects related to hyperactivity of TGF-β receptor and to altered activity of skeletal muscle ryanodine receptors (Shou et al., 1998). Mice lacking a related gene, FKBP12.6, have normal TGF-β pathways but die due to dysfunction of cardiac, not skeletal muscle ryanodine receptors (Xin et al., 2002). Finally, mice lacking CyPA have heightened allergic responses due to defects in Itk, a tyrosine kinase in Th2 class T-cells (Colgan et al., 2004). More recent findings with mice lacking steroid receptor-associated FKBP are discussed later in this chapter.

2.2. *PPIase Co-Chaperone Structure*

FKBP51 and FKBP52 share approximately 70% amino acid sequence similarity and have similar domain arrangements, Hsp90-binding ability, and PPIase activity

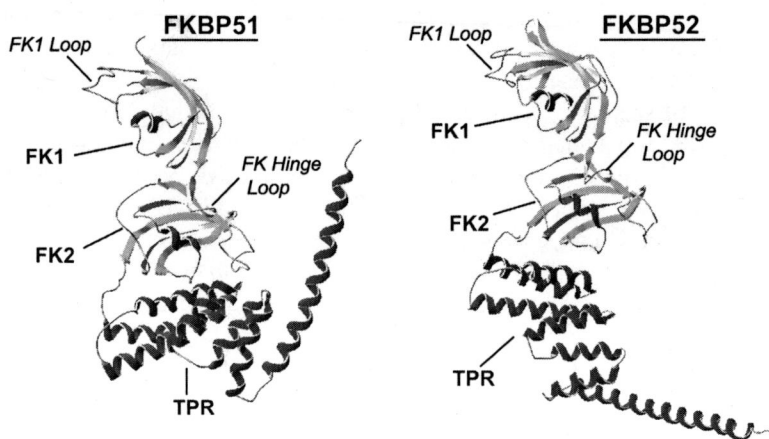

FIGURE 3. X-ray crystallographic structures of FKBP51 and FKBP52. Each FKBP is composed of two FKBP12-like domains (FK1 and FK2) and an Hsp90-binding tetratricopeptide repeat domain (TPR). The FK1 Loop and FK Hinge Loop structures have been shown to have functional significance in steroid receptor interactions. Original structures described by Sinars et al. (2003) and Wu et al. (2004).

toward small peptide substrates (Pirkl and Buchner, 2001). Nonetheless, these two FKBP have dissimilar actions in steroid receptor complexes, as will be detailed below. Three-dimensional X-ray crystallographic structures have been obtained for full-length FKBP51 (Sinars et al., 2003) and for overlapping fragments of FKBP52 (Wu et al., 2004) that can be merged to model a full-length structure.

Both FKBP51 and FKBP52 are three-domain proteins, as depicted in Fig. 3. The N-terminal domain is structurally similar to FKBP12 (Van Duyne et al., 1991), the archetypal, single domain member of the FKBP family, and this FKBP12-like domain (FK1) confers PPIase activity and FK506 binding ability. There is a second FKBP12-like domain (FK2) that lacks PPIase and drug binding due to several amino acid differences with FK1. FK2 domain function is poorly understood, although one report noted that mutations in FK2 of FKBP51 reduced assembly with steroid receptor complexes despite normal binding to Hsp90 (Sinars et al., 2003), which suggests potential FK2 interaction with receptor in the heteromeric complex. The third domain shared by FKBP51 and FKBP52 is composed of tetratricopeptide repeat motifs (TPR) and forms the Hsp90-binding site. A structurally similar Hsp90-binding TPR domain is found in the co-chaperones PP5 (Das et al., 1998), CyP40 (Taylor et al., 2001), and Hop (Scheufler et al., 2000).

The FK1 domains of FKBP51 and FKBP52 are identical at several amino acids known to be critical for PPIase activity and exhibit similar enzymatic properties toward small peptide substrates (Pirkl and Buchner, 2001). Nonetheless, there are few differences at positions surrounding the PPIase pocket that become important when considering interactions with full protein substrates (addressed further in a later section). Of particular interest is a loop that typically overhangs the FKBP

PPIase pocket; others have previously referred to this as the 80s loop in reference to numbering of amino acids in FKBP12 (Sinars et al., 2003), but we prefer the more generic term FK1 loop (see Fig. 3). Forming the bases of the FK1 loop in FKBP52 are prolines at positions 119 and 124; the corresponding positions in FKBP51 are leucine and serine. As will be discussed later, these differences in the FK1 loop relate directly to functional differences observed for FKBP52 and FKBP51 in steroid receptor complexes. Another difference between FKBP51 and FKBP52 is within the loop that forms the hinge between FK1 and FK2 domains (FK hinge loop; Fig. 3). Capping the FK loop in FKBP52 is the sequence TEEED, which is a consensus Casein Kinase 2 (CK2) phosphorylation site. Baulieu and colleagues reported that CK2-mediated threonine phosphorylation occurs at this site in vitro and further provided evidence that phosphorylated FKBP52 has reduced binding to Hsp90 (Miyata et al., 1997). The potential physiological significance of FKBP52 phosphorylation needs to be further explored. In FKBP51 the corresponding sequence in the FK hinge loop is FED and thus would not be amenable to phosphorylation.

Whereas the individual domains of FKBP51 and FKBP52 have similar conformations, there is an apparent distinction in the relative orientation of the FK domain couplet with the TPR domain. In FKBP51 there are unique ionic interactions between FK2 and TPR side-chains that stabilize orientation of the FK couplet and TPR domain. In contrast, FKBP52 differs at relevant charged amino acids in FK2, precluding the stabilizing interactions observed with FKBP51. The FKBP52 crystal structure has a more extended domain orientation, but this might reflect a preferred conformation for crystal formation in a molecule that otherwise has greater flexibility in solution than FKBP51. Conceivably, global conformational constraints unique to FKBP51 could contribute to functional differences in FKBP51 and FKBP52.

3. Role of FKBPs in Steroid Receptor Signaling

3.1. Biochemical Studies

3.1.1. FKBP Chaperone Activity

In addition to the PPIase activity discussed above, the large FKBP immunophilins also have been shown to possess chaperoning activity that is independent of Hsp90. In vitro aggregation of thermally denatured citrate synthase can be inhibited by FKBP52 (Bose et al., 1996) and even more efficiently by FKBP51 (Pirkl and Buchner, 2001). The holding activity localizes to the C-terminal TPR/Hsp90-binding domain (Pirkl et al., 2001) and is accordingly independent of PPIase activity or treatment with immunosuppressive ligands. Competition experiments demonstrated that, within the TPR domain, the Hsp90-binding site is distinct from the binding site for non-native proteins. To date, however, no studies have directly linked the inherent chaperone activity of receptor-associated FKBPs to Hsp90 client protein regulation. Since holding activity in vitro is displayed by multiple chaperone components and does not appear to be particularly substrate-specific, it seems unlikely that FKBP holding activity is central to functional changes elicited in a specific manner by FKBP in steroid receptor complexes.

3.1.2. FKBP Binding to Hsp90

The receptor-associated immunophilins through their respective TPR domains compete for binding the EEVD motif that terminates Hsp90 (Ratajczak et al., 2003; Smith, 2004), but sequences downstream from the core TPR domain of FKBP co-chaperones also influence Hsp90 binding (Cheung-Flynn et al., 2003). As measured by isothermal titration calorimetry, it appears that one PPIase monomer binds per Hsp90 dimer with FKBP52 displaying approximately 3–4-fold higher affinity for Hsp90 than either FKBP51 or CyP40 (Pirkl and Buchner, 2001); however, the particular Hsp90 client protein can greatly influence the relative abundance of PPIase co-chaperones in heteromeric complexes. When progesterone receptor (PR), glucocorticoid receptor (GR), and estrogen receptor (ER) complexes were assembled in parallel in rabbit reticulocyte lysate, FKBP51 was clearly preferred in PR complexes, FKBP51 and PP5 in GR complexes, and CyP40 was more prominent in ER complexes (Barent et al., 1998). How the receptor client discriminates among Hsp90 co-chaperones is not yet clear; one possibility is that the co-chaperone alters Hsp90 binding properties in a client-specific manner; another is that the co-chaperone directly and specifically interacts with receptor while bound to Hsp90 in the heterocomplex.

Hsp90-mediated client protein maturation requires ATP turnover at the N-terminal ATPase domain of Hsp90 (Panaretou et al., 1998), and regulation of Hsp90 ATPase activity has been examined as a mechanism by which co-chaperones can affect client protein activity. The TPR co-chaperone Hop, which competes with PPIase co-chaperones for Hsp90 binding, inhibits Hsp90 ATPase activity (Richter et al., 2003) and could retain Hsp90 dimer in an open conformation that facilitates loading of client protein. For vertebrate Hsp90, client binding stimulates ATP hydrolysis and would thus favor rapid client release. However, the Hsp90 co-chaperone p23 suppresses ATP hydrolysis and enhances interaction between the N-terminal regions of Hsp90 monomers (Cox and Miller, 2004; McLaughlin et al., 2002; Richter et al., 2004). According to the clamp model for Hsp90 binding to client proteins (Prodromou et al., 1997, 2000), this explains how p23 can stabilize and prolong mature receptor complexes (Dittmar et al., 1997; Smith et al., 1995). A study conducted by McLaughlin et al. suggests that FKBP52 has no effect on the basal rate of ATP hydrolysis by Hsp90 but enhances the client-stimulated ATPase (McLaughlin et al., 2002). It would seem, then, that FKBP52 and p23 could act at odds in the stabilization of client complexes. Whether FKBP51 directly affects Hsp90 ATPase has not been explored; however, since the FKBPs competitively displace Hop from Hsp90, they could passively alter Hsp90 ATPase properties. In support of this idea, the yeast CyP40 ortholog Cpr6 does not directly affect Hsp90 ATPase activity (Siligardi et al., 2002), but it can reactivate ATP turnover on Hsp90 by displacing Sti1, the yeast Hop ortholog. Whatever the influence of

FKBP on Hsp90 function in general, it is not clear how this could differentially affect the function of one steroid receptor but not another that similarly assembles with FKBPs.

3.1.3. Dynamics and Nature of FKBP Binding in Receptor Complexes

Formation and maintenance of the steroid receptor heterocomplex is a dynamic process. Kinetic studies demonstrated that under optimum conditions for receptor assembly in a cell-free system, mature complexes form in approximately 2 minutes and Hsp90 dissociates with a half-life of approximately 5 minutes (Smith, 1993). However, immunophilin association and dissociation on mature PR-Hsp90 complexes is up to tenfold faster (Barent et al., 1998). The receptor-Hsp90 complex continuously samples the environment for available immunophilins, and, as noted above, the preference given to an immunophilin is determined by the particular receptor present in the complex. Immunophilin binding also differs between hormone-free and hormone-bound receptor. For example, PR complexes that are stabilized in vitro against Hsp90 dissociation show a rapid loss of FKBP51, but not FKBP52, upon addition of hormone (Smith et al., 1990a, 1993b). When cells expressing GR were maintained on ice to retard Hsp90 dissociation, hormone addition similarly resulted in loss of FKBP51 and concomitant gain of FKBP52 in GR complexes (Davies et al., 2002).

The extent to which FKBPs have direct interactions with steroid receptors and the nature of such interactions are unclear. Pratt and colleagues detected Hsp90-independent interaction between GR and FKBP52 that involved the N-terminal FK domains of FKBP52 (Silverstein et al., 1999). Evidence also suggests that FK domains of FKBP51 are required for preferred binding to GR (Denny et al., 2005; Sinars et al., 2003; Wochnik et al., 2005). As discussed above, crystal structures suggest that FKBP52 and FKBP51 differ in the orientation between FK domains and the Hsp90-binding TPR domain. Furthermore, when complexed with Hsp90, FKBP51, and FKBP52 differ in their ability to bind an FK506 affinity resin, consistent with distinct accessibilities of respective FK1 domains (Smith et al., 1993a). Therefore, in addition to amino acid sequence differences that might exist at a site or sites of receptor binding, differences in how the FK domains interface with receptor might also contribute to distinct receptor binding properties of FKBP52 and FKBP51.

3.2. Cellular Studies

3.2.1. Effects of FK506 and Cyclosporin A on Steroid Receptor Activity

Immunosuppressive ligands can potentially influence steroid receptor function by binding to immunophilins in receptor heterocomplexes. Several studies have addressed this possibility, but findings have been inconsistent and defy any single mechanistic explanation.

GR complexes were isolated on an FK506 affinity resin through association with FKBP52 (Tai et al., 1992). Addition of FK506 to rabbit reticulocyte lysate had no effect on GR assembly or maturation of hormone binding ability (Hutchison et al., 1993). In contract, FK506 treatment of L929 mouse fibroblasts resulted in opposing, hormone dose-dependent effects on GR function (Ning and Sanchez, 1993). FK506 enhanced GR-mediated reporter gene expression at a low hormone concentration but inhibited reporter expression at a high concentration. Confounding interpretation of cellular findings with immunosuppressant drugs is the presence of multiple immunophilins in addition to those associated with a particular steroid receptor and the actions of drugs at non-immunophilin targets. For instance, Kralli et al. discovered that FK506 could enhance hormone-induced GR function by an indirect mechanism involving membrane efflux pumps in the family of ABC transporters (Kralli et al., 1995; Kralli and Yamamoto, 1996). Some glucocorticoid hormones were found to be substrates pumped out of cells by efflux pumps, and FK506, a known inhibitor of these pumps (Arceci et al., 1992; Naito et al., 1992), was found to increase intracellular hormone concentrations.

Tai and co-workers demonstrated that FK506 enhanced progesterone-induced transcription in a yeast PR model (Tai et al., 1994), but this did not involve PR-associated FKBP. Instead, FK506 elevated hormone-dependent phosphorylation and transcriptional activity of PR by inhibiting the phosphatase calcineurin.

The cyclosporin CsA, a ligand for receptor-associated CyP40, has been compared with the FKBP ligands FK506 or rapamycin in several reports. For example, CsA had no effect on PR in T47D cells (Le Bihan et al., 1998). Likewise, FK506 and rapamycin, much more than CsA, inhibited GR function in A6 cells (Edinger et al., 2002), but this inhibition was only observed after an 8-hour drug treatment, which suggests remodeling of pathways that may not directly relate to receptor-associated immunophilins. Both FK506 and CsA were found to enhance nuclear translocation of GR (Prima et al., 2000); on the other hand, CsA but neither FK506 nor rapamycin, stimulated nuclear translocation of a mutant PR lacking an endogenous nuclear localization signal (Jung-Testas et al., 1995; Lebeau et al., 1999). As with other reports, drug effects related to receptor nuclear translocation were not directly linked to receptor-associated immunophilin.

FK506 can also have direct effects on receptor heterocomplexes. In a study involving T47D breast cancer cells, which express endogenous PR, rapamycin and FK506 were found to inhibit hormone-dependent activation of a reporter gene (Le Bihan et al., 1998), and some evidence was presented that the inhibition might relate to receptor-associated FKBP. FK506 induces dissociation of FKBP51 from GR complexes (Reynolds et al., 1999) and disrupts functional interaction of FKBP52 with GR (Riggs et al., 2003) and androgen receptor (AR) (Cheung-Flynn et al., 2005). Cellular administration of FK506 prior to hormone was shown to cause replacement of FKBP51 by PP5 in GR complexes and to block hormone-induced recruitment of FKBP52 (Davies et al., 2005). As described below, FKBP52 and FKBP51 have opposite influences on the sensitivity of some steroid receptors to hormone, so FK506-mediated inhibition of one, the other, or both immunophilins can change receptor activity in complex ways.

3.2.2. New World Primates

The first physiological evidence that receptor-associated FKBP can alter steroid receptor activity came from studies of cortisol resistance in New World primates. New World primates such as squirrel monkey (SM), owl monkey, and cotton-top tamarin have markedly elevated circulating cortisol levels. Interestingly, New World primates also have elevated serum levels of progesterone, androgens, and other hormones, hinting at a general mechanism for desensitizing steroid response pathways (Scammell, 2000). Endogenous GR in SM cells has reduced affinity for hormone, which could account for glucocorticoid resistance. However, Scammell and colleagues demonstrated that cortisol insensitivity in New World primates is not a consequence of GR sequence divergence as SM-GR exhibits in vitro hormone binding affinity that is high and similar to human GR (Reynolds et al., 1997). From this observation, Scammell concluded that a cellular factor specific to or missing from SM cells might account for differences in hormone binding affinity. When cellular levels of receptor-associated chaperones were compared in SM and human lymphocytes, FKBP51 was greater than fivefold more abundant in SM lysates (Reynolds et al., 1999). In addition, FKBP52 levels in monkey cells were only half the level in human cells, but all other chaperones examined were present at similar levels. The discovery of FKBP51 overexpression, which is consistently observed in all New World primates but not in Old World primates (Scammell et al., 2001), led to experiments that clearly demonstrate FKBP51 reduces GR hormone binding affinity (Denny et al., 2000; Reynolds et al., 1999) and similarly reduces PR and AR response to hormone (Cheung-Flynn et al., 2005; Hubler et al., 2003; Tranguch et al., 2005). SM-FKBP51 is more potent in repressing GR function than its human counterpart (Denny et al., 2000; Reynolds et al., 1999), and the difference traces to a combination of amino acid changes in both FK domains and the TPR domain (Denny et al., 2005). Still, human FKBP51 can significantly inhibit steroid receptor function.

3.2.3. Yeast-Model for FKBP Function

Saccharomyces cerevisiae, the yeast of brewers and bakers, has long been a valued experimental model system, and yeast have been adapted to provide a useful background for studying steroid receptor function. The yeast genome lacks steroid receptor genes but contains chaperone and transcription factor genes necessary to support steroid receptor function; therefore, yeast transformed with a receptor cDNA and an appropriate reporter gene will induce reporter gene expression in a hormone-dependent manner (Schena and Yamamoto, 1988). Recently, our lab has explored the function of FKBP51 and FKBP52 in steroid receptor complexes by exploiting the fact that there are no orthologs for these particular immunophilins in yeast. Based on findings in New World primates and confirmed in human cells, we anticipated that human FKBP51 would reduce GR function in yeast; this was not what we found. Instead, we were surprised to discover that FKBP52 dramatically potentiates GR activity in yeast (Riggs et al., 2004). FKBP52 also potentiates AR activity in yeast, but no effect is seen toward ER or mineralocorticoid receptor (MR) activity, despite the ability of FKBP52 to assemble in all steroid receptor complexes. Chimeric receptors were generated by exchanging GR and ER domains to demonstrate that potentiation localizes to the LBD of GR, which is consistent with the general localization of chaperone interactions in steroid receptors. Overexpression of human FKBP51, PP5, CyP40, or endogenous yeast cyclophilins did not duplicate the potentiation of receptor activity seen with FKBP52. However, we observed that co-expression of FKBP51, but not PP5, with FKBP52 returned GR activity toward the lower baseline level in yeast. From these experiments we concluded FKBP52 specifically potentiates steroid receptor function and that FKBP51 inhibits steroid receptor function by antagonizing the actions of FKBP52.

Functional mapping with point mutants determined that FKBP52-mediated promotion of receptor activity requires PPIase and Hsp90-binding activities (Riggs et al., 2003). Our current working model is that FKBP52 piggybacks on Hsp90 into receptor complexes where the FKBP52 PPIase domain is positioned to act on the receptor LBD; a resulting change in receptor conformation leads to enhanced activity. At this time, however, we have not identified specific substrate sites on the receptor for FKBP52 PPIase and cannot exclude other mechanisms by which receptor activity is altered.

Antagonism of FKBP52 by FKBP51 is dependent on FKBP51's ability to bind Hsp90 (Wochnik et al., 2005). More, however, appears to be involved than simple competitive displacement of FKBP52 from Hsp90 and receptor complexes. First, PP5, which assembles into GR complexes at least as well as FKBP51 (Silverstein et al., 1997), failed to block FKBP52-mediated potentiation of GR activity in yeast (Riggs et al., 2003). Next, despite the irrelevance of FKBP51 PPIase activity, it does appear the FK domains are necessary for efficient antagonism (Denny et al., 2005; Wochnik et al., 2005). The observation that preferential binding of FKBP51 to receptor complexes is blocked by FK2 mutation (Sinars et al., 2003) is consistent with a model in which FK sites unique to FKBP51 interact with receptor and efficiently block the actions of FKBP52.

3.2.4. Mammalian Cellular Models of FKBP Action

The enhancement of steroid receptor function by FKBP52 and antagonism by FKBP51, as first observed in the yeast model, have since been confirmed in various mammalian cell backgrounds (Cheung-Flynn et al., 2005; Davies et al., 2005; Hubler et al., 2003; Tranguch et al., 2005; Wochnik et al., 2005). In cells overexpressing FKBP51 from a transfected plasmid vector, hormone-induced transcriptional activity is reduced for GR (Davies et al., 2005; Denny et al., 2000; Wochnik et al., 2005), PR (Hubler et al., 2003), and AR (Tranguch et al., 2005). In contrast overexpression of an exogenous FKBP52 can mitigate the inhibitory effect of FKBP51 (Wochnik et al., 2005) and enhance receptor activity (Davies et al., 2005). As would be expected from these findings, knockdown of FKBP52 expression by an siRNA approach (Cheung-Flynn et al., 2005) was shown to reduce activity of AR, and AR activity could be restored by expression of exogenous FKBP52.

Recently, we have generated permanent mouse embryonic fibroblast (MEF) lines from mice devoid of FKBP52 (52KO), devoid of FKBP51 (51KO), and from wildtype littermates (WT). As predicted from yeast and other cellular studies, AR, PR, and GR activities are reduced in 52KO MEF as compared to WT MEF (Tranguch et al., 2005, and unpublished observations from the authors); in all cases, receptor activity is restored by expression of exogenous FKBP52 but not by FKBP52 point mutants that lack PPIase or Hsp90 binding activity. In contrast to the 52KO MEF background, receptor assays in 51KO MEF displayed maximal receptor activity, but this activity was reduced by expression of exogenous FKBP51 (Tranguch et al., 2005). Taken together, cellular studies support the possibility that receptor-associated FKBP contribute to the physiological actions of glucocorticoids, androgens, and progesterone in vertebrates; this has recently been confirmed with mouse gene knockout models.

3.3. Phenotypes of FKBP Gene Knockout Mice

Standard gene targeting approaches were used to disrupt genes for FKBP52 or FKBP51 in mouse embryonic stem cells, from which homozygous mutant mouse strains were developed (Cheung-Flynn et al., 2005; Tranguch et al., 2005). As described above, MEF lines generated from 52KO and 51KO embryos are proving very useful for cellular assessment of FKBP function. Adult mice, on the other hand, provide information of physiological processes that depend on FKBP52 or FKBP51 function. Of particular interest so far are the dramatic reproductive phenotypes in either sex of 52KO mice.

3.3.1. Reproductive Phenotype of the 52KO Male

In humans and vertebrates, androgen plays a central role in organogenesis of the male reproductive tract. Expression of genes crucial for sexual development depends on AR and its sensitivity toward hormonal signals. Androgen insensitivity syndrome (AIS) is a fairly common congenital defect in males and is most often due to mutation in the X-linked AR gene, which is a single-copy gene in males (McPhaul, 2002). Complete AIS is often characterized by a lack of virilization of external genitalia, post-pubertal breast development, the presence of undescended, cryptorchid testes, a blind vagina, and an absence of ovary and uterus. Partial AIS, which frequently cannot be linked to AR mutation, encompasses a broad spectrum of phenotypes. At the mild end may be infertility due to low sperm number, but the spectrum extends through all degrees of ambiguous genital development. 52KO male mice have variable phenotypes, but all exhibit features that resemble AIS, consistent with defects in androgen signaling (Cheung-Flynn et al., 2005). Importantly, though, the 52KO phenotypes differ in several regards from mice completely lacking AR (Table 1). Defects in genital development of 52KO males ranges from mild hypospadias to complete feminization of external genitalia; the degree of nipple retention into adulthood parallels the degree of genital defect. Internally, defects range from partial dysgenesis of seminal vesicle and anterior

TABLE 1. Male Sexual Differentiation and Development in Mouse Models*

Reproductive organs/functions	52KO	Tfm/ARKO	5α-reductase KO
Fertility	−	−	+
External Genitalia	Ambiguous	Feminine	Masculine
Accessory sex organs	Dysgenic	Agenic	Reduced size
Testes	Mostly normal	Cryptorchid	Normal
Spermatogenesis	Normal; reduced sperm motility	Incomplete	Normal
Testosterone Level	Normal to elevated	Low	Normal to elevated in serum; elevated in target tissues

52KO = FKBP52 gene knockout (Cheung-Flynn et al., 2005); Tfm = testicular feminization mouse, an AR mutant (Charest et al., 1991; Couse and Korach, 1998); ARKO = androgen receptor knockout (Yeh et al., 2002).

prostate to essential absence of these organs. In contrast to AR mutant mouse models (Charest et al., 1991; Yeh et al., 2002), 52KO males have fully formed testes, epididymis, and vas deferens; spermatogenesis proceeds normally in 52KO males, although there is reduced sperm motility and fertilization efficiency in vitro. The collection of defects observed in 52KO male mice resembles those seen in humans with mutations in 5α-reductase, an enzyme found in many secondary sex organs that converts testosterone to the more potent dihydrotestosterone (Imperato-McGinley et al., 1974). Unlike the situation in humans, deletion of 5α-reductase genes in mice does not have a major impact on virilization (Mahendroo et al., 2001; Table 1). There is no decrease in circulating testosterone levels in 52KO males, and AR protein is detected in target tissues at levels and localizations comparable to those of WT mice. Since 52KO defects are not due to absence of hormone or AR protein, there instead appears to be a resistance of AR response to hormone in a subset of tissues in the 52KO mouse. Such resistance is consistent with observations in cellular models presented above.

FKBP52 is critical for male sexual development and fertility, which relates to AR function, but some androgen target tissues in the 52KO mouse appear to develop and function normally. Reasons for the tissue-selective requirement for FKBP52 are not understood. One relevant factor may be local concentration of hormone. Since testis is the site of most androgen production, the relatively high levels of hormone here could be sufficient to drive AR function, whereas hormone levels in peripheral tissues may be inadequate to support full AR activity. Another possibility is that testis and a subset of androgen target tissues express an unknown factor that replicates FKBP52 function.

3.3.2. Reproductive Phenotype of the 52KO Female

As with males, 52KO females are infertile, yet no morphological or histological defects are observed in the virgin female reproductive tract (Tranguch et al., 2005). Female ovaries develop normally, folliculogenesis is unperturbed, follicles

TABLE 2. Female Reproductive Functions in Mouse Models*

	52KO	PRKO	PRBKO	PRAKO	αβERKO
Fertility	−	−	−	+	∼∼ −
Mammary gland development	Normal; pregnancy-related development unknown	Impaired	Normal	Impaired	Impaired
Follicolugenesis/ ovulation	Normal/normal	Impaired/ anovulatory	Reduced oocytes/ anovulatory	Normal/ normal	Sex-reversed follicles/ anovulatory
Implantation	Impaired	Impaired	Impaired	Normal	Impaired

*52KO = FKBP52 gene knockout (Tranguch et al., 2005); PRKO = PR gene knockout (Lydon et al., 1995); PRAKO = PR A selective gene knockout (Mulac-Jericevic et al., 2000); PRBKO = PR-B selective gene knockout (Mulac-Jericevic et al., 2003); αβERKO = double knockout of ER-α and ER-β genes (Couse et al., 1999).

rupture in response to exogenous hormone, and oocytes can be fertilized in vitro. The critical defect is an inability of the 52KO uterus to accept blastocysts for implantation. Since implantation is known to be progesterone-dependent (Dey et al., 2004), it was reasonable to assess whether PR in the uterus is hormone-resistant in the absence of FKBP52. Cellular studies are consistent with this model, and molecular analyses confirm that progesterone-induced genes are underexpressed in the 52KO uterus (Tranguch et al., 2005).

Female mice lacking the PR gene (PRKO) display multiple reproductive abnormalities, including ovulation failure, implantation failure, uterine hyperplasia, and defective mammary development (Lydon et al., 1995). 52KO mice share only a few phenotypic features with PRKO females (Table 2). First, it should be noted that PR is expressed from a single gene in two isoforms, PR-A and PR-B, the latter of which contains 120 N-terminal amino acids unique to PR-B (Conneely et al., 1987; Gronemeyer et al., 1987). Both PR isoforms form similar complexes with FKBP52 and other chaperones (Smith et al., 1990a), which is expected since they share identical LBD. Each isoform regulates overlapping and unique gene sets (Conneely et al., 2003), and selective mouse knockout models (Mulac-Jericevic et al., 2000, 2003) have identified tissue-specific roles for either isoform (Table 2). Since PR-A is critical for uterine function, it appears that FKBP52 is necessary for adequate response of PR-A to hormone in the uterus. PR-A is also required for follicular rupture and oocyte release, but the typical number of oocytes were recovered from the uterus of normally cycling 52KO females or mice stimulated with a superovulatory dose of hormone. Perhaps since progesterone is synthesized in the ovary, local concentrations of hormone are sufficient to support PR function and follicular rupture but insufficient for uterine implantation. PR-B is uniquely responsible for progesterone-stimulated changes in the mammary during pregnancy, but it has not been straightforward to assess these mammary changes in 52KO females since they do not become pregnant. It was shown that a PR-B responsive gene in the uterus was underexpressed in 52KO females and PR-B function is impaired in cellular models lacking FKBP52 (Tranguch et al., 2005). To assess PR-B function in the 52KO mammary more directly will require tissue transplantation into a wild-type female that is subsequently impregnated.

3.3.3. Phenotypes of 51KO and 52KO+51KO Mice

Physiological roles for FKBP51, as opposed to FKBP52, are less apparent since 51KO mice display no apparent growth or reproductive anomalies. FKBP51 is expressed in a variety of human and mouse tissues (Baughman et al., 1997; Nair et al., 1997; Yeh et al., 1995), including reproductive tissues affected by FKBP52 deletion. Clearly, FKBP51 and FKBP52 function uniquely in some physiological settings. On the other hand, there is complete embryonic lethality in mice lacking genes for both FKBP52 and FKBP51 (authors' unpublished observations). More analysis will be required to identify whether the two FKBP have critical redundant functions that only become apparent with loss of both gene products or whether the two FKBP separately contribute to a critical physiological process.

3.3.4. Glucocorticoid Signaling in Knockout Models

Since receptor-associated FKBP52 enhances GR activity in cellular models, phenotypes associated with a reduction in GR activity would be expected in 52KO mice. GR knockout mice die perinatally due to defective lung function (Cole et al., 1995), a phenotype we do not observe with 52KO mice. Yet based on observations with AR and PR, one would not expect a complete loss of GR function in 52KO mice. The strongest current evidence for a change in GR function is the elevated level of circulating corticosterone in 52KO mice. Endocrine feedback mechanisms might be sufficient to overcome GR resistance in 52KO, much as feedback mechanisms minimize overt evidence of glucocorticoid resistance in New World monkeys. Still, it is reasonable to expect that 52KO mice will have alterations in immunological or stress responses that relate to depressed GR function.

4. Interaction of FKBP with Other Cellular Factors

4.1. FKBP51 Interactions

FKBP51 has been identified in a variety of Hsp90 client protein complexes, including kinase complexes, but the functional significance of FKBP51 in non-steroid receptor complexes has not been assessed in most cases. Recently, FKBP51 was identified through a proteomic approach as an interaction partner with the IκB kinase IKKα (Bouwmeester et al., 2004), as well as several additional kinases. Previously, Hsp90 was shown to be important for IKK biogenesis and activity (Broemer et al., 2004), so FKBP51 might be participating in IKK complexes in association with Hsp90. IKKα is a heterodimeric partner with IKKβ in a complex that phosphorylates IκB, the inhibitory subunit of NF-κB complexes; phosphorylation of IκB stimulates its degradation and thus frees NF-κB for nuclear localization and transcriptional activity (Karin and Ben-Neriah). Knockdown of FKBP51 expression by an siRNA approach reduced NF-κB-mediated expression of a reporter gene (Bouwmeester et al., 2004). IKKα is independently involved in alternate NF-κB pathways (Senftleben et al., 2001), and IKKα mutant mice display defects in B-cell, mammary, skin, and skeletal development (Cao et al., 2001; Hu et al., 1999; Senftleben et al., 2001). We have not observed corresponding defects in 51KO mice and have had difficulty replicating functional measures of IKKα/FKBP51 interactions in cellular models, so the physiological importance of FKBP51 in IKKα complexes remains an open issue.

4.2. FKBP52 Interactions

FKBP52 has been identified in a wide array of complexes, not only with alternate Hsp90 clients, but also in Hsp90-independent interactions (Table 3). Since FKBP52 is selective in altering steroid receptor function yet can be identified in association with all steroid receptors, the functional significance of FKBP52 in non-receptor client complexes will need to be assessed individually. Although major features of the 52KO mouse phenotype are attributable to resistance of

TABLE 3. Association of FKBPs with Other Cellular Factors

Interactors	Experimental Approach	Physiological Implication	References
FKBP51			
IKKα	Tandem affinity purification	Subset of NFκB signaling pathways	(Bouwmeester et al., 2004)
FKBP52			
Dynein	Co-immunoprecipitation	Intracellular trafficking of steroid receptor complexes	(Silverstein et al., 1999)
p53	Co- immunoprecipitation	Cancer	(Galigniana et al., 2004)
HSF-1	Co- immunoprecipitation	Cellular stress	(Bharadwaj et al., 1999)
TRPCs	Co- immunoprecipitation	B- and T-cell activation; neuronal survival and growth	(Sinkins et al., 2004)
FAP48	Yeast two-hybrid	T-cell activation	(Chambraud et al., 1996; Krummrei et al., 2003; Neye, 2001)
PHAX	Yeast two-hybrid	Refsum disease, lupus	(Chambraud et al., 1999)
IRF-4	Yeast two-hybrid	Immune regulation	(Mamane et al., 2000)
Atox1	Yeast two-hybrid	Copper transport	(Sanokawa-Akakura et al., 2004)
AAV DNA	EMSA	Gene therapy	(Qing et al., 2001; Qing et al., 1997; Zhong et al., 2004a)

steroid receptors to hormone, it seems unlikely that FKBP52 does not play an active role in additional Hsp90 client complexes.

Pratt and colleagues have advocated a role for FKBP52 in the intracellular transport of receptor complexes (Galigniana et al., 2001; Silverstein et al., 1999). They showed that FKBP52 binds a subunit of dynein complexes and propose that this interaction can link GR complexes to the cytoskeletal machinery for rapid translocation from cytoplasm to nucleus. However, PP5 and CyP40 also bind dynein (Galigniana et al., 2002), and should therefore duplicate the actions of FKBP52, but 52KO mice clearly show that FKBP52 acts in a non-redundant manner to support steroid receptor function. In the yeast model, FKBP52 enhances GR function independent of dynein (Riggs et al., 2003) and results from other cellular studies purporting to causally link FKBP52, dynein, and enhanced GR function (Davies et al., 2002; Wochnik et al., 2005) could be alternatively explained on the basis of FKBP52 acting directly on steroid receptor.

Yeast two-hybrid screens have identified FKBP52 as an interaction partner of the copper transport protein Atox1, with a subset of TRPC-class Ca^{2+} channels, and the transcription factor interferon regulatory factor 4 (IRF4). The PPIase region of FKBP52 was found to interact directly with Atox1 in a copper-sensitive manner (Sanokawa-Akakura et al., 2004); overexpression of FKBP52 increased copper efflux from cells, suggesting that FKBP52 positively regulates Atox1. Following up a prior observation that insect FKBP59 associates with a related channel protein in *Drosophila* (Goel et al., 2001), Sinkins et al. discovered that FKBP52 and FKBP12 interact with distinct subsets of the mammalian TRPC family of calcium channel and enhance channel function in a PPIase-dependent manner (Sinkins et al., 2004). They further characterized proline motifs that differed between FKBP12 and FKBP52 substrates to address the specificity observed between FKBP and particular TRPC. In yet another study, FKBP52 was shown to alter IRF4 conformation and inhibit IRF4 transcriptional activity in a PPIase-dependent manner; moreover, a proline-rich region of IRF4 was identified as a site of interaction for FKBP52 (Mamane et al., 2000). A link between FKBP52 and cardiac hypertrophy has also been proposed.

A unique, non-protein partner for FKBP52 is adeno-associated virus (AAV) DNA. Srivastava and colleagues found that FKBP52 was the previously unidentified cellular factor that binds AAV DNA and regulates replication of the viral genome (Qing et al., 2001). FKBP52 binds to a specific single-stranded segment of the viral genome and inhibits second-strand DNA synthesis. Importantly, FKBP52 binds AAV DNA and inhibits expression only when tyrosine phosphorylated, as can be stimulated by epidermal growth factor pathways. It has been demonstrated that Hsp90 is not involved in FKBP52 binding to AAV DNA (Zhong et al., 2004b), so it appears that FKBP52 acts on its own. The physiological importance of phospho-FKBP52 in limiting AAV transduction of hepatocytes was confirmed using independently generated FKBP52 gene knockout mice and transgenic mice overexpressing a tyrosine phosphatase (Zhong et al., 2004a). As yet, the specific DNA binding site of FKBP52 has not been identified, and its mechanism of phosphorylation-dependent binding to DNA remains a mystery. The availability of FKBP52 mutants that disrupt AAV DNA binding would be very useful in assessing whether FKBP52 relies on similar mechanisms in interactions with normal cellular partners.

5. Regulation of FKBP Expression and Function

Based on the protein sequence similarity and identical intron/exon boundaries within the coding sequences of FKBP51 and FKBP52 genes, it is clear that a gene duplication event generated the related genes. Incidentally, only a single gene, which encodes an FK1-FK2-TPR product that equally resembles either FKBP51 or FKBP52, has been identified in insects (Zaffran, 2000); conversely, all vertebrate genomes reported to date contain two genes easily discernible as FKBP52 or FKBP51. Therefore, it is possible that the seminal gene duplication event occurred around the time vertebrate precursors evolved. Despite their common origin and structural similarities, several lines of evidence have been discussed for unique functions of FKBP52 and FKBP51. Expression and activity of the two FKBP are also uniquely regulated, as befits proteins with distinct functions. These patterns of regulation likely influence steroid receptor function in physiologically relevant ways.

5.1. Transcriptional Regulation of FKBP51

Transcriptional expression of the FKBP51 gene is highly stimulated by certain steroid hormones, as first described by Baughman et al. using glucocorticoids (Baughman et al., 1995 and references therein). Since, it has been shown that progesterone (Hubler et al., 2003) and androgens (Febbo et al., 2005; Zhu et al., 2001) induce FKBP51 expression, and numerous gene expression profiles have identified FKBP51 as up-regulated by these steroids. Hormone-induced expression of FKBP51, which is an inhibitor of steroid receptor function, can serve as a negative feedback mechanism to limit cellular responsiveness to secondary exposures to hormone.

Estrogen does not induce FKBP51 expression, which suggests that the other steroids act through consensus hormone responsive elements that are common transcriptional targets for GR, PR, and AR. Hubler and Scammell have recently reported identification of such elements that are conserved among human, mouse, and rat genes for FKBP51 (Hubler and Scammell, 2004), but absent from genes encoding FKBP52. Interestingly, these elements are not located in the more traditional upstream promoter region, but instead are located in the fifth intron, at least 50 kilobases from the transcriptional start site. They confirmed that these elements confer glucocorticoid- and progestin-inducibility to a heterologous reporter, that GR and PR bind these elements, and that mutation of these elements block hormone-induced transcription. However, these elements did not confer androgen inducibility.

5.2. Transcriptional Regulation of FKBP52

Sanchez and colleagues discovered that FKBP52 expression can be induced by heat shock or chemical stress (Sanchez, 1990). Based on what we now understand about FKBP52 ability to enhance GR activity, stress-induced expression of FKBP52 might contribute to glucocorticoid roles in physiological responses to stress. The cytokine cardiotrophin-1 (CT-1) can induce cardiac hypertrophy, and this has been causally linked to CT-1-induced expression of FKBP52 (Jamshidi et al., 2004; Railson et al., 2001). Overexpression of FKBP52 in cardiac cells was found to stimulate signaling cascades involving several kinases, but the mechanism for kinase stimulation by FKBP52 was not explored.

FKBP52 expression in the pregnant mouse uterus is stimulated around the time of implantation by the transcription factor Hoxa10 (Daikoku et al., 2005), whose expression is in turn induced by progesterone and PR (Lim et al., 1999). Preexisting FKBP52 is necessary for uterine PR to induce Hoxa10 expression (Tranguch et al., 2005), but additional FKBP52 might be required to maintain uterine sensitivity to progesterone in the face of increased FKBP51 expression.

5.3. Post-Transcriptional Regulation of FKBP52

It was noted previously that tyrosine phosphorylation of FKBP52 is necessary for AAV DNA binding. Phosphorylation by serine/threonine kinases can also alter FKBP52 function. The Baulieu lab first noted a consensus CK2 site in the FK hinge loop of FKBP52 (Miyata et al., 1997); as discussed previously, phosphorylation of this site appeared to reduce binding to Hsp90. We have generated the phosphomimetic mutant T143E and find that Hsp90 binding is intact, as is assembly with receptor complexes, but find a loss of FKBP52 ability to enhance receptor function (M. Cox and D.F. Smith, unpublished). These preliminary observations are being pursued to determine if kinase-mediated signaling events can indirectly regulate steroid receptor function through FKBP52 phosphorylation.

6. Clinical Implications for FKBP Function

The physiological importance of FKBP52 has been established in a mouse model, and the potential importance of FKBP51, is suggested by New World primates and by cellular models. Currently, neither FKBP has been identified as a causative agent in human disease or as a specific target for therapeutic drugs, but there are several connections to human development and physiology worthy of further exploration, as well as conceivable applications for diagnosis or therapeutic intervention.

6.1. Androgen Insensitivity and Prostate Cancer

An enlightening case of complete AIS (CAIS) has been a recent focus of research in our lab. We surveyed the human AR mutant database (http://www.androgendb.mcgill.ca/) for LBD mutations in or near proline sites that might relate to FKBP function and identified the mutation P723S for further study. This AR mutation was identified in a patient with CAIS (Ahmed et al., 2000), and hormone-binding measurements of genital skin fibroblasts derived from this patient were consistent with a low affinity AR. We generated the AR-P723S mutant and tested activity in the yeast model (Cheung-Flynn et al., 2005). Hormone-dependent reporter gene expression was minimal in the yeast reporter strain transformed with AR-P723S as compared to WT-AR; however, co-expression of FKBP52 with AR-P723S rescued reporter expression to the fully enhanced level observed with WT-AR plus FKBP52. Neither FKBP51 nor other chaperones were able to rescue AR-P723S activity, and both PPIase and Hsp90-binding domains of FKBP52 were required for rescue. All of these observations have since been duplicated in the mouse 52KO MEF background. The findings from these experiments demonstrate that FKBP52 can not only enhance WT-AR activity but can also chaperone proper folding of the AR-P723S mutant. Yet the findings also begged the question of why the CAIS patient cannot, apparently, overcome the deleterious effect of AR mutation; perhaps there is a concomitant defect in this patient's FKBP52. In a cohort of 111 patients with partial AIS (PAIS), only 24% were found to contain mutations in the AR gene (Deeb et al., 2005), which implicates other cellular factors in the majority of PAIS cases. The variety of observations that implicate FKBP in AR function make these genes good candidates for analysis in PAIS lacking AR mutation.

Receptor-associated FKBP could also play a role in prostate cancer. By cDNA microarray analysis, FKBP51 is one of the genes that is upregulated in prostate cancer specimens (Febbo et al., 2005). This is not surprising since FKBP51 expression, much like prostate specific antigen, is known to be androgen-inducible; moreover, one might expect that overexpression of FKBP51 would dampen AR activity and thus tend to ameliorate androgen-dependent proliferation of cancer cells. To the contrary, it was found that FKBP51 overexpression somehow augmented androgen-inducible reporter activity in the LNCaP prostate cancer cell line (Febbo et al., 2005). Apart from this study, one could predict that underexpression or mutation of FKBP51 or overexpression of FKBP52 could constitutively enhance AR sensitivity to hormone, and thus promote androgen-dependent hyperplasia or prostate carcinogenesis. Another consideration is that mutation of AR could mimic the enhanced-activity state induced by FKBP52 and generate an AR with abnormally high constitutive activity. Each of these possibilities merits further study.

6.2. Breast and Uterine Cancer

PR and ER are both critical for normal development and pregnancy-related morphogenetic changes in the human mammary gland and uterus. Estrogen and progesterone also play roles in carcinogenesis of these tissues; anti-estrogens are one of the primary treatment options for breast cancer and progestins are commonly used for early stage uterine cancer. 52KO mice confirm that FKBP52 is important for PR function in the uterus and there is the untested likelihood that FKBP52 is relevant to PR activity in the mammary as well. FKBP52 is not directly important for ER activity, but many of the proliferative actions of estrogens are held in check by the opposition of progesterone and PR (Clarke and Sutherland, 1990). Thus, a decrease in PR activity could promote the proliferative effects of estrogen; this possibility is supported by the observation of epithelial hyperplasia in the uterus of 52KO mice (Tranguch et al., 2005). To date, little has been done to directly assess FKBP expression and function in relation to breast or uterine cancer.

6.3. Male and Female Fertility

Male and female fertility represents an area of significant clinical and societal interest. Hormone therapy is widely employed adult females for reproductive control. On the flip side, one in ten couples seeking to have children experience some degree of fertility impediment (Gurmankin et al., 2005). As touched upon previously, defects in male sexual development and fertility are more common than generally appreciated, and these problems often relate to some aspect of androgen production or androgen sensitivity of target tissues. On the female side, idiopathic failures of uterine implantation and/or embryonic maintenance are frequently associated with infertility; deficits in FKBP52 function or, conceivably, excessive FKBP51 activity could underlie some of these problems.

6.4. Stress and Immunology

Acting through GR, glucocorticoids (GC) regulate diverse physiological functions, having major actions in the central nervous system, cardiovascular maintenance, metabolic homeostasis, growth, stress, and immune responses. GC are also commonly employed as immunosuppressive and anti-inflammatory drugs. GC are primarily produced in the adrenal cortex under positive and negative control by the hypothalamus-pituitary-adrenal axis. Cellular studies indicate that FKBP52 and FKBP51 can also regulate GC actions through changes in GR responsiveness to hormone. Elevated serum corticosterone in 52KO mice is consistent with a role for FKBP52 in maintaining maximum GR responsiveness to hormone. Physiological support for FKBP51 involvement is provided by apparent GC resistance in New World monkeys that overexpress FKBP51. There has not yet been a careful characterization of GR-related processes in 51KO mice, but loss of the GC-inducible down-modulator of GR activity could result in immunological or behavioral changes associated with chronically elevated response to GC. Evidence that FKBP51 is clinically relevant to human GR function comes from a study of single-nucleotide polymorphisms in FKBP5, the human gene for FKBP51, in patients with clinical depression (Binder et al., 2004). Certain FKBP5 genotypes were found to be associated with increased expression of FKBP51, a greater recurrence of depressive episodes, and a heightened response to antidepressant drugs..

7. Pharmacological Prospects and Toxicological Concerns

As do most members of the FKBP family, FKBP52 and FKBP51 bind FK506 and rapamycin, which inhibit PPIase and perturb interactions with steroid receptors. The immunosuppressive actions of these drugs is thought to be largely FKBP12-dependent, although some of the unwanted and toxic effects of these compounds are likely mediated by other FKBP family members. There is a need for selective FKBP inhibitors whose actions are restricted to a single FKBP. Unfortunately, specific inhibitors of the receptor-associated FKBP are not currently available, but one could imagine useful clinical properties of specific inhibitors. For instance, a specific FKBP52 inhibitor could be an effective female contraceptive. Based on FKBP52-dependence of implantation in the mouse uterus, efficient inhibition of FKBP52 function would be expected to block blastocyst implantation. Another attractive prospect is the potential for an FKBP52 inhibitor to dampen AR activity in the prostate as an alternative to anti-androgen therapy for prostate cancer. Conversely, a specific inhibitor of FKBP51 might enhance AR function in select PAIS individuals and help restore androgen sensitivity and fertility; likewise, an FKBP51 inhibitor might be effective in boosting uterine PR function and treating some types of female infertility. Additionally, specific FKBP inhibitors could be used to up- or down-modulate GR activity and therapeutically alter GC-sensitive pathologies.

A sobering consideration, given the critical role of FKBP52 in both male a female reproductive processes, is the potential that a chemical inhibitor of FKBP52, either a natural product or perhaps a synthetic byproduct, could act as an endocrine disruptor to impact the fitness of vertebrate organisms in the environment. There is great concern about environmental endocrine disruptors that can impact humans as well as non-human vertebrate populations. Although some disruptors are known to act directly on AR, ER, or PR, other suspected disruptors do not appear to bind these receptors with relevant affinities (Fisher, 2004). FKBP52 could be an unrecognized target of action for compounds that impair reproductive development and fertility.

8. Future Prospects

The functional involvement of FKBP co-chaperones in steroid receptor complexes is just beginning to be appreciated and understood. While existing mouse gene knockout models will continue to be useful in addressing the physiological importance of each FKBP, the development of additional transgenic models with tissue-specific or inducible disruption of FKBP expression, or knockin models where endogenous FKBP is substituted with a defined mutant, will refine our understanding of FKBP function and our rationales for targeting FKBP actions. Yeast, human cell, and animal models will all contribute to the identification and characterization of new pharmacological agents that alter FKBP function. Finally, we feel the work of many that has contributed to our current appreciation that the FKBP co-chaperones functionally impact only a subset of Hsp90 client proteins provides impetus for similar experimental analysis of other co-chaperones that are likely to have unique functions of their own.

Acknowledgments. Relevant studies in the authors' laboratory were supported by NIH R01-DK48218 (D.F.S.), NIH F32 DK068983 (M.B.C.), and the Mayo Foundation.

References

Ahmed, S. F., Cheng, A., Dovey, L., Hawkins, J. R., Martin, H., Rowland, J., Shimura, N., Tait, A. D., and Hughes, I. A. (2000) Phenotypic features, androgen receptor binding, and mutational analysis in 278 clinical cases reported as androgen insensitivity syndrome. *J Clin Endocrinol Metab* 85:658–65.

Aranda, A., and Pascual, A. (2001) Nuclear hormone receptors and gene expression. *Physiol Rev* 81:1269–304.

Arceci, R. J., Stieglitz, K., and Bierer, B. E. (1992) Immunosuppressants FK506 and rapamycin function as reversal agents of the multidrug resistance phenotype. *Blood* 80:1528–36.

Barent, R. L., Nair, S. C., Carr, D. C., Ruan, Y., Rimerman, R. A., Fulton, J., Zhang, Y., and Smith, D. F. (1998) Analysis of FKBP51/FKBP52 chimeras and mutants for Hsp90 binding and association with progesterone receptor complexes. *Mol Endocrinol* 12:342–54.

Baughman, G., Wiederrecht, G. J., Campbell, N. F., Martin, M. M., and Bourgeois, S. (1995) FKBP51, a novel T-cell-specific immunophilin capable of calcineurin inhibition. *Mol Cell Biol* 15:4395–402.

Baughman, G., Wiederrecht, G. J., Chang, F., Martin, M. M., and Bourgeois, S. (1997) Tissue distribution and abundance of human FKBP51, an FK506-binding protein that can mediate calcineurin inhibition. *Biochem Biophys Res Commun* 232:437–43.

Bharadwaj, S., Ali, A., and Ovsenek, N. (1999) Multiple components of the HSP90 chaperone complex function in regulation of heat shock factor 1 *in vivo*. *Mol Cell Biol* 19:8033–41.

Bierer, B. E. (1994) Cyclosporin A, FK506, and rapamycin: Binding to immunophilins and biological action. *Chem Immunol* 59:128–55.

Binder, E. B., Salyakina, D., Lichtner, P., Wochnik, G. M., Ising, M., Putz, B., Papiol, S., Seaman, S., Lucae, S., Kohli, M. A. et al. (2004) Polymorphisms in FKBP5 are associated with increased recurrence of depressive episodes and rapid response to antidepressant treatment. *Nat Genet* 36:1319–25.

Bose, S., Weikl, T., Bugl, H., and Buchner, J. (1996) Chaperone function of Hsp90-associated proteins. *Science* 274:1715–7.

Bouwmeester, T., Bauch, A., Ruffner, H., Angrand, P. O., Bergamini, G., Croughton, K., Cruciat, C., Eberhard, D., Gagneur, J., Ghidelli, S. et al. (2004) A physical and functional map of the human TNF-alpha/NF-kappa B signal transduction pathway. *Nat Cell Biol* 6:97–105.

Broemer, M., Krappmann, D., and Scheidereit, C. (2004) Requirement of Hsp90 activity for IkappaB kinase (IKK) biosynthesis and for constitutive and inducible IKK and NF-kappaB activation. *Oncogene* 23:5378–86.

Cao, Y., Bonizzi, G., Seagroves, T. N., Greten, F. R., Johnson, R., Schmidt, E. V., and Karin, M. (2001) IKKalpha provides an essential link between RANK signaling and cyclin D1 expression during mammary gland development. *Cell* 107:763–75.

Chambraud, B., Radanyi, C., Camonis, J. H., Rajkowski, K., Schumacher, M., and Baulieu, E. E. (1999) Immunophilins, Refsum disease, and lupus nephritis: The peroxisomal enzyme phytanoyl-COA alpha-hydroxylase is a new FKBP-associated protein. *Proc Natl Acad Sci USA* 96:2104–9.

Chambraud, B., Radanyi, C., Camonis, J. H., Shazand, K., Rajkowski, K., and Baulieu, E. E. (1996) FAP48, a new protein that forms specific complexes with both immunophilins FKBP59 and FKBP12. Prevention by the immunosuppressant drugs FK506 and rapamycin. *J Biol Chem* 271:32923–9.

Charest, N. J., Zhou, Z. X., Lubahn, D. B., Olsen, K. L., Wilson, E. M., and French, F. S. (1991) A frameshift mutation destabilizes androgen receptor messenger RNA in the Tfm mouse. *Mol Endocrinol* 5:573–81.

Cheung-Flynn, J., Prapapanich, V., Cox, M. B., Riggs, D. L., Suarez-Quian, C., and Smith, D. F. (2005) Physiological role for the cochaperone FKBP52 in androgen receptor signaling. *Mol Endocrinol* 19:1654–66.

Cheung-Flynn, J., Roberts, P. J., Riggs, D. L., and Smith, D. F. (2003) C-terminal sequences outside the tetratricopeptide repeat domain of FKBP51 and FKBP52 cause differential binding to Hsp90. *J Biol Chem* 278:17388–94.

Clarke, C. L., and Sutherland, R. L. (1990) Progestin regulation of cellular proliferation. *Endocr Rev* 11:266–301.

Cole, T. J., Blendy, J. A., Monaghan, A. P., Krieglstein, K., Schmid, W., Aguzzi, A., Fantuzzi, G., Hummler, E., Unsicker, K., and Schutz, G. (1995) Targeted disruption of the glucocorticoid receptor gene blocks adrenergic chromaffin cell development and severely retards lung maturation. *Genes Dev* 9:1608–21.

Colgan, J., Asmal, M., Neagu, M., Yu, B., Schneidkraut, J., Lee, Y., Sokolskaja, E., Andreotti, A., and Luban, J. (2004) Cyclophilin A regulates TCR signal strength in CD4+ T cells via a proline-directed conformational switch in Itk. *Immunity* 21:189–201.

Conneely, O. M., Dobson, A. D., Tsai, M. J., Beattie, W. G., Toft, D. O., Huckaby, C. S., Zarucki, T., Schrader, W. T., and O'Malley, B. W. (1987) Sequence and expression of a functional chicken progesterone receptor. *Mol Endocrinol* 1:517–25.

Conneely, O. M., Mulac-Jericevic, B., and Lydon, J. P. (2003) Progesterone-dependent regulation of female reproductive activity by two distinct progesterone receptor isoforms. *Steroids* 68:771–8.

Couse, J. F., Hewitt, S. C., Bunch, D. O., Sar, M., Walker, V. R., Davis, B. J., and Korach, K. S. (1999) Postnatal sex reversal of the ovaries in mice lacking estrogen receptors alpha and beta. *Science* 286:2328–31.

Couse, J. F., and Korach, K. S. (1998) Exploring the role of sex steroids through studies of receptor deficient mice. *J Mol Med* 76:497–511.

Cox, M. B., and Miller, C. A., III (2004) Cooperation of heat shock protein 90 and p23 in aryl hydrocarbon receptor signaling. *Cell Stress Chap* 9:4–20.

Cox, M. B., and Smith, D. F. (2005) Functions of the Hsp90-Binding FKBP immunophilins. In G. Blatch (ed.): *Networking of Chaperones by Cochaperones*. Landes Bioscience, Georgetown, TX, pp. 1–13.

Daikoku, T., Tranguch, S., Friedman, D. B., Das, S. K., Smith, D. F., and Dey, S. K. (2005) Proteomic analysis identifies immunophilin FK506 binding protein 4 (FKBP52) as a downstream target of Hoxa10 in the periimplantation mouse uterus. *Mol Endocrinol* 19:683–97.

Das, A. K., Cohen, P. W., and Barford, D. (1998) The structure of the tetratricopeptide repeats of protein phosphatase 5: Implications for TPR-mediated protein–protein interactions. *EMBO J* 17:1192–9.

Davies, T. H., Ning, Y. M., and Sanchez, E. R. (2002) A new first step in activation of steroid receptors: Hormone-induced switching of FKBP51 and FKBP52 immunophilins. *J Biol Chem* 277:4597–600.

Davies, T. H., Ning, Y. M., and Sanchez, E. R. (2005) Differential control of glucocorticoid receptor hormone-binding function by tetratricopeptide repeat (TPR) proteins and the immunosuppressive ligand FK506. *Biochemistry* 44:2030–8.

Deeb, A., Mason, C., Lee, Y. S., and Hughes, I. A. (2005) Correlation between genotype, phenotype and sex of rearing in 111 patients with partial androgen insensitivity syndrome. *Clin Endocrinol (Oxf)* 63:56–62.

Denny, W. B., Prapapanich, V., Smith, D. F., and Scammell, J. G. (2005) Structure–function analysis of squirrel monkey FK506-binding protein 51, a potent inhibitor of glucocorticoid receptor activity. *Endocrinology* 146:3194–201.

Denny, W. B., Valentine, D. L., Reynolds, P. D., Smith, D. F., and Scammell, J. G. (2000) Squirrel monkey immunophilin FKBP51 is a potent inhibitor of glucocorticoid receptor binding. *Endocrinology* 141:4107–13.

Dey, S. K., Lim, H., Das, S. K., Reese, J., Paria, B. C., Daikoku, T., and Wang, H. (2004) Molecular cues to implantation. *Endocr Rev* 25:341–73.

Dittmar, K. D., Demady, D. R., Stancato, L. F., Krishna, P., and Pratt, W. B. (1997) Folding of the glucocorticoid receptor by the heat shock protein (hsp) 90-based chaperone machinery. The role of p23 is to stabilize receptor.hsp90 heterocomplexes formed by hsp90.p60.hsp70. *J Biol Chem* 272:21213–20.

Edinger, R. S., Watkins, S. C., Pearce, D., and Johnson, J. P. (2002) Effect of immunosuppressive agents on glucocorticoid receptor function in A6 cells. *Am J Physiol Renal Physiol* 283:F254–61.

Febbo, P. G., Lowenberg, M., Thorner, A. R., Brown, M., Loda, M., and Golub, T. R. (2005) Androgen mediated regulation and functional implications of fkbp51 expression in prostate cancer. *J Urol* 173:1772–7.

Fink, A. L. (1999) Chaperone-mediated protein folding. *Physiol Rev* 79:425–49.

Fisher, J. S. (2004) Are all EDC effects mediated via steroid hormone receptors? *Toxicology* 205:33–41.

Froesch, B. A., Takayama, S., and Reed, J. C. (1998) BAG-1L protein enhances androgen receptor function. *J Biol Chem* 273:11660–6.

Galigniana, M. D., Harrell, J. M., Murphy, P. J., Chinkers, M., Radanyi, C., Renoir, J. M., Zhang, M., and Pratt, W. B. (2002) Binding of hsp90-associated immunophilins to cytoplasmic dynein: Direct binding and *in vivo* evidence that the peptidylprolyl isomerase domain is a dynein interaction domain. *Biochemistry* 41:13602–10.

Galigniana, M. D., Harrell, J. M., O'Hagen, H. M., Ljungman, M., and Pratt, W. B. (2004) Hsp90-binding immunophilins link p53 to dynein during p53 transport to the nucleus. *J Biol Chem* 279:22483–9. Epub 22004 Mar 22485.

Galigniana, M. D., Radanyi, C., Renoir, J. M., Housley, P. R., and Pratt, W. B. (2001) Evidence that the peptidylprolyl isomerase domain of the hsp90-binding immunophilin FKBP52 is involved in both dynein interaction and glucocorticoid receptor movement to the nucleus. *J Biol Chem* 276:14884–9.

Goel, M., Garcia, R., Estacion, M., and Schilling, W. P. (2001) Regulation of *Drosophila* trpl channels by immunophilin fkbp59. *J Biol Chem* 276:38762–73.

Gronemeyer, H., Turcotte, B., Quirin-Stricker, C., Bocquel, M. T., Meyer, M. E., Krozowski, Z., Jeltsch, J. M., Lerouge, T., Garnier, J. M., and Chambon, P. (1987) The chicken progesterone receptor: Sequence, expression and functional analysis. *EMBO J* 6:3985–94.

Gurmankin, A. D., Caplan, A. L., and Braverman, A. M. (2005) Screening practices and beliefs of assisted reproductive technology programs. *Fertil Steril* 83:61–7.

Heitman, J., Movva, N. R., and Hall, M. N. (1991) Targets for cell cycle arrest by the immunosuppressant rapamycin in yeast. *Science* 253:905–9.

Hu, Y., Baud, V., Delhase, M., Zhang, P., Deerinck, T., Ellisman, M., Johnson, R., and Karin, M. (1999) Abnormal morphogenesis but intact IKK activation in mice lacking the IKKalpha subunit of IkappaB kinase. *Science* 284:316–20.

Hubler, T. R., Denny, W. B., Valentine, D. L., Cheung-Flynn, J., Smith, D. F., and Scammell, J. G. (2003) The FK506-binding immunophilin FKBP51 is transcriptionally regulated by progestin and attenuates progestin responsiveness. *Endocrinology* 144:2380–7.

Hubler, T. R., and Scammell, J. G. (2004) Intronic hormone response elements mediate regulation of FKBP5 by progestins and glucocorticoids. *Cell Stress Chap* 9:243–52.

Hutchison, K. A., Dittmar, K. D., and Pratt, W. B. (1994) All of the factors required for assembly of the glucocorticoid receptor into a functional heterocomplex with heat shock protein 90 are preassociated in a self-sufficient protein folding structure, a "foldosome". *J Biol Chem* 269:27894–9.

Hutchison, K. A., Scherrer, L. C., Czar, M. J., Ning, Y., Sanchez, E. R., Leach, K. L., Deibel, M. R., Jr., and Pratt, W. B. (1993) FK506 binding to the 56-kilodalton immunophilin (Hsp56) in the glucocorticoid receptor heterocomplex has no effect on receptor folding or function. *Biochemistry* 32:3953–7.

Imperato-McGinley, J., Guerrero, L., Gautier, T., and Peterson, R. E. (1974) Steroid 5alpha-reductase deficiency in man: An inherited form of male pseudohermaphroditism. *Science* 186:1213–5.

Jamshidi, Y., Zourlidou, A., Carroll, C. J., Sinclair, J., and Latchman, D. S. (2004) Signal-transduction pathways involved in the hypertrophic effect of hsp56 in neonatal cardiomyocytes. *J Mol Cell Cardiol* 36:381–92.

Jung-Testas, I., Lebeau, M. C., Catelli, M. G., and Baulieu, E. E. (1995) Cyclosporin A promotes nuclear transfer of a cytoplasmic progesterone receptor mutant. *C R Acad Sci III* 318:873–8.

Karin, M., and Ben-Neriah, Y. Phosphorylation meets ubiquitination: The control of NF-[kappa]B activity. 621–663.

Kosano, H., Stensgard, B., Charlesworth, M. C., McMahon, N., and Toft, D. (1998) The assembly of progesterone receptor–hsp90 complexes using purified proteins. *J Biol Chem* 273:32973–9.

Kralli, A., Bohen, S. P., and Yamamoto, K. R. (1995) LEM1, an ATP-binding-cassette transporter, selectively modulates the biological potency of steroid hormones. *Proc Natl Acad Sci USA* 92:4701–5.

Kralli, A., and Yamamoto, K. R. (1996) An FK506-sensitive transporter selectively decreases intracellular levels and potency of steroid hormones. *J Biol Chem* 271:17152–6.

Krummrei, U., Baulieu, E. E., and Chambraud, B. (2003) The FKBP-associated protein FAP48 is an antiproliferative molecule and a player in T cell activation that increases IL2 synthesis. *Proc Natl Acad Sci USA* 100:2444–9.

Le Bihan, S., Marsaud, V., Mercier-Bodard, C., Baulieu, E. E., Mader, S., White, J. H., and Renoir, J. M. (1998) Calcium/calmodulin kinase inhibitors and immunosuppressant macrolides rapamycin and FK506 inhibit progestin- and glucocorticosteroid receptor-mediated transcription in human breast cancer T47D cells. *Mol Endocrinol* 12:986–1001.

Lebeau, M. C., Jung-Testas, I., and Baulieu, E. E. (1999) Intracellular distribution of a cytoplasmic progesterone receptor mutant and of immunophilins cyclophilin 40 and FKBP59: Effects of cyclosporin A, of various metabolic inhibitors and of several culture conditions. *J Steroid Biochem Mol Biol* 70:219–28.

Lim, H., Ma, L., Ma, W. G., Maas, R. L., and Dey, S. K. (1999) Hoxa-10 regulates uterine stromal cell responsiveness to progesterone during implantation and decidualization in the mouse. *Mol Endocrinol* 13:1005–17.

Lydon, J. P., DeMayo, F. J., Funk, C. R., Mani, S. K., Hughes, A. R., Montgomery, C. A., Jr., Shyamala, G., Conneely, O. M., and O'Malley, B. W. (1995) Mice lacking progesterone receptor exhibit pleiotropic reproductive abnormalities. *Genes Dev* 9:2266–78.

Mahendroo, M. S., Cala, K. M., Hess, D. L., and Russell, D. W. (2001) Unexpected virilization in male mice lacking steroid 5 alpha-reductase enzymes. *Endocrinology* 142:4652–62.

Mamane, Y., Sharma, S., Petropoulos, L., Lin, R., and Hiscott, J. (2000) Posttranslational regulation of IRF-4 activity by the immunophilin FKBP52. *Immunity* 12:129–40.

McLaughlin, S. H., Smith, H. W., and Jackson, S. E. (2002) Stimulation of the weak ATPase activity of human hsp90 by a client protein. *J Mol Biol* 315:787–98.

McPhaul, M. J. (2002) Androgen receptor mutations and androgen insensitivity. *Mol Cell Endocrinol* 198:61–7.

Miyata, Y., Chambraud, B., Radanyi, C., Leclerc, J., Lebeau, M. C., Renoir, J. M., Shirai, R., Catelli, M. G., Yahara, I., and Baulieu, E. E. (1997) Phosphorylation of the immunosuppressant FK506-binding protein FKBP52 by casein kinase II: Regulation of HSP90-binding activity of FKBP52. *Proc Natl Acad Sci USA* 94:14500–5.

Mulac-Jericevic, B., Lydon, J. P., DeMayo, F. J., and Conneely, O. M. (2003) Defective mammary gland morphogenesis in mice lacking the progesterone receptor B isoform. *Proc Natl Acad Sci USA* 100:9744–9.

Mulac-Jericevic, B., Mullinax, R. A., DeMayo, F. J., Lydon, J. P., and Conneely, O. M. (2000) Subgroup of reproductive functions of progesterone mediated by progesterone receptor-B isoform. *Science* 289:1751–4.

Nair, S. C., Rimerman, R. A., Toran, E. J., Chen, S., Prapapanich, V., Butts, R. N., and Smith, D. F. (1997) Molecular cloning of human FKBP51 and comparisons of immunophilin interactions with Hsp90 and progesterone receptor. *Mol Cell Biol* 17:594–603.

Naito, M., Oh-hara, T., Yamazaki, A., Danki, T., and Tsuruo, T. (1992) Reversal of multidrug resistance by an immunosuppressive agent FK-506. *Cancer Chemother Pharmacol* 29:195–200.

Neye, H. (2001) Mutation of FKBP associated protein 48 (FAP48) at proline 219 disrupts the interaction with FKBP12 and FKBP52. *Regul Pept* 97:147–52.

Ning, Y. M., and Sanchez, E. R. (1993) Potentiation of glucocorticoid receptor-mediated gene expression by the immunophilin ligands FK506 and rapamycin. *J Biol Chem* 268:6073–6.

Panaretou, B., Prodromou, C., Roe, S. M., O'Brien, R., Ladbury, J. E., Piper, P. W., and Pearl, L. H. (1998) ATP binding and hydrolysis are essential to the function of the Hsp90 molecular chaperone *in vivo*. *EMBO J* 17:4829–36.

Pirkl, F., and Buchner, J. (2001) Functional analysis of the Hsp90-associated human peptidyl prolyl *cis/trans* isomerases FKBP51, FKBP52 and Cyp40. *J Mol Biol* 308:795–806.

Pirkl, F., Fischer, E., Modrow, S., and Buchner, J. (2001) Localization of the chaperone domain of FKBP52. *J Biol Chem* 276:37034–41.

Prima, V., Depoix, C., Masselot, B., Formstecher, P., and Lefebvre, P. (2000) Alteration of the glucocorticoid receptor subcellular localization by non steroidal compounds. *J Steroid Biochem Mol Biol* 72:1–12.

Prodromou, C., Panaretou, B., Chohan, S., Siligardi, G., O'Brien, R., Ladbury, J. E., Roe, S. M., Piper, P. W., and Pearl, L. H. (2000) The ATPase cycle of Hsp90 drives a molecular 'clamp' via transient dimerization of the N-terminal domains. *EMBO J* 19:4383–92.

Prodromou, C., Roe, S. M., Piper, P. W., and Pearl, L. H. (1997) A molecular clamp in the crystal structure of the N-terminal domain of the yeast Hsp90 chaperone [see comments]. *Nat Struct Biol* 4:477–82.

Qing, K., Hansen, J., Weigel-Kelley, K. A., Tan, M., Zhou, S., and Srivastava, A. (2001) Adeno-associated virus type 2-mediated gene transfer: Role of cellular FKBP52 protein in transgene expression. *J Virol* 75:8968–76.

Qing, K., Wang, X. S., Kube, D. M., Ponnazhagan, S., Bajpai, A., and Srivastava, A. (1997) Role of tyrosine phosphorylation of a cellular protein in adeno-associated virus 2-mediated transgene expression. *Proc Natl Acad Sci USA* 94:10879–84.

Railson, J. E., Lawrence, K., Buddle, J. C., Pennica, D., and Latchman, D. S. (2001) Heat shock protein-56 is induced by cardiotrophin-1 and mediates its hypertrophic effect. *J Mol Cell Cardiol* 33:1209–21.

Ratajczak, T., Ward, B. K., and Minchin, R. F. (2003) Immunophilin chaperones in steroid receptor signalling. *Curr Top Med Chem* 3:1348–57.

Reynolds, P. D., Pittler, S. J., and Scammell, J. G. (1997) Cloning and expression of the glucocorticoid receptor from the squirrel monkey (*Saimiri boliviensis boliviensis*), a glucocorticoid-resistant primate. *J Clin Endocrinol Metab* 82:465–72.

Reynolds, P. D., Ruan, Y., Smith, D. F., and Scammell, J. G. (1999) Glucocorticoid resistance in the squirrel monkey is associated with overexpression of the immunophilin FKBP51. *J Clin Endocrinol Metab* 84:663–9.

Richter, K., Muschler, P., Hainzl, O., Reinstein, J., and Buchner, J. (2003) Sti1 is a noncompetitive inhibitor of the Hsp90 ATPase. Binding prevents the N-terminal dimerization reaction during the atpase cycle. *J Biol Chem* 278:10328–33.

Richter, K., Walter, S., and Buchner, J. (2004) The co-chaperone Sba1 connects the ATPase reaction of Hsp90 to the progression of the chaperone cycle. *J Mol Biol* 342:1403–13.

Riggs, D. L., Cox, M. B., Cheung-Flynn, J., Prapapanich, V., Carrigan, P. E., and Smith, D. F. (2004) Functional specificity of co-chaperone interactions with Hsp90 client proteins. *Crit Rev Biochem Mol Biol* 39:279–95.

Riggs, D. L., Roberts, P. J., Chirillo, S. C., Cheung-Flynn, J., Prapapanich, V., Ratajczak, T., Gaber, R., Picard, D., and Smith, D. F. (2003) The Hsp90-binding peptidylprolyl isomerase FKBP52 potentiates glucocorticoid signaling *in vivo*. *EMBO J* 22:1158–67.

Sanchez, E. R. (1990) Hsp56: A novel heat shock protein associated with untransformed steroid receptor complexes. *J Biol Chem* 265:22067–70.

Sanokawa-Akakura, R., Dai, H., Akakura, S., Weinstein, D., Fajardo, J. E., Lang, S. E., Wadsworth, S., Siekierka, J., and Birge, R. B. (2004) A novel role for the immunophilin FKBP52 in copper transport. *J Biol Chem* 279:27845–8.

Scammell, J. G. (2000) Steroid resistance in the squirrel monkey: An old subject revisited. *Ilar J* 41:19–25.

Scammell, J. G., Denny, W. B., Valentine, D. L., and Smith, D. F. (2001) Overexpression of the FK506-binding immunophilin FKBP51 is the common cause of glucocorticoid resistance in three new world primates. *Gen Comp Endocrinol* 124:152–65.

Schena, M., and Yamamoto, K. R. (1988) Mammalian glucocorticoid receptor derivatives enhance transcription in yeast. *Science* 241:965–7.

Scheufler, C., Brinker, A., Bourenkov, G., Pegoraro, S., Moroder, L., Bartunik, H., Hartl, F. U., and Moarefi, I. (2000) Structure of TPR domain–peptide complexes: Critical elements in the assembly of the Hsp70–Hsp90 multichaperone machine. *Cell* 101:199–210.

Senftleben, U., Cao, Y., Xiao, G., Greten, F. R., Krahn, G., Bonizzi, G., Chen, Y., Hu, Y., Fong, A., Sun, S. C., and Karin, M. (2001) Activation by IKKalpha of a second, evolutionary conserved, NF-kappa B signaling pathway. *Science* 293:1495–9.

Shou, W., Aghdasi, B., Armstrong, D. L., Guo, Q., Bao, S., Charng, M. J., Mathews, L. M., Schneider, M. D., Hamilton, S. L., and Matzuk, M. M. (1998) Cardiac defects and altered ryanodine receptor function in mice lacking FKBP12. *Nature* 391:489–92.

Siligardi, G., Panaretou, B., Meyer, P., Singh, S., Woolfson, D. N., Piper, P. W., Pearl, L. H., and Prodromou, C. (2002) Regulation of Hsp90 ATPase activity by the co-chaperone Cdc37p/p50cdc37. *J Biol Chem* 277:20151–9. Epub 22002 Mar 20126.

Silverstein, A. M., Galigniana, M. D., Chen, M. S., Owens-Grillo, J. K., Chinkers, M., and Pratt, W. B. (1997) Protein phosphatase 5 is a major component of glucocorticoid receptor.hsp90 complexes with properties of an FK506-binding immunophilin. *J Biol Chem* 272:16224–30.

Silverstein, A. M., Galigniana, M. D., Kanelakis, K. C., Radanyi, C., Renoir, J. M., and Pratt, W. B. (1999) Different regions of the immunophilin FKBP52 determine its association with the glucocorticoid receptor, hsp90, and cytoplasmic dynein [In Process Citation]. *J Biol Chem* 274:36980–6.

Simoncini, T., and Genazzani, A. R. (2003) Non-genomic actions of sex steroid hormones. *Eur J Endocrinol* 148:281–92.

Sinars, C. R., Cheung-Flynn, J., Rimerman, R. A., Scammell, J. G., Smith, D. F., and Clardy, J. (2003) Structure of the large FK506-binding protein FKBP51, an Hsp90-binding protein and a component of steroid receptor complexes. *Proc Natl Acad Sci USA* 100:868–73.

Sinkins, W. G., Goel, M., Estacion, M., and Schilling, W. P. (2004) Association of immunophilins with mammalian TRPC channels. *J Biol Chem* 279:34521–9.

Smith, D. F. (1993) Dynamics of heat shock protein 90-progesterone receptor binding and the disactivation loop model for steroid receptor complexes. *Mol Endocrinol* 7:1418–29.

Smith, D. F. (2004) Tetratricopeptide repeat cochaperones in steroid receptor complexes. *Cell Stress Chap* 9:109–21.

Smith, D. F., Albers, M. W., Schreiber, S. L., Leach, K. L., and Deibel, M. R., Jr. (1993a) FKBP54, a novel FK506-binding protein in avian progesterone receptor complexes and HeLa extracts. *J Biol Chem* 268:24270–3.

Smith, D. F., Baggenstoss, B. A., Marion, T. N., and Rimerman, R. A. (1993b) Two FKBP-related proteins are associated with progesterone receptor complexes. *J Biol Chem* 268:18365–71.

Smith, D. F., Faber, L. E., and Toft, D. O. (1990a) Purification of unactivated progesterone receptor and identification of novel receptor-associated proteins. *J Biol Chem* 265:3996–4003.

Smith, D. F., Schowalter, D. B., Kost, S. L., and Toft, D. O. (1990b) Reconstitution of progesterone receptor with heat shock proteins. *Mol Endocrinol* 4:1704–11.

Smith, D. F., Whitesell, L., Nair, S. C., Chen, S., Prapapanich, V., and Rimerman, R. A. (1995) Progesterone receptor structure and function altered by geldanamycin, an hsp90-binding agent. *Mol Cell Biol* 15:6804–12.

Tai, P. K., Albers, M. W., Chang, H., Faber, L. E., and Schreiber, S. L. (1992) Association of a 59-kilodalton immunophilin with the glucocorticoid receptor complex. *Science* 256:1315–8.

Tai, P. K., Albers, M. W., McDonnell, D. P., Chang, H., Schreiber, S. L., and Faber, L. E. (1994) Potentiation of progesterone receptor-mediated transcription by the immunosuppressant FK506. *Biochemistry* 33:10666–71.

Taylor, P., Dornan, J., Carrello, A., Minchin, R. F., Ratajczak, T., and Walkinshaw, M. D. (2001) Two structures of cyclophilin 40: Folding and fidelity in the TPR domains. *Structure (Camb)* 9:431–8.

Tranguch, S., Cheung-Flynn, J., Daikoku, T., Prapapanich, V., Cox, M. B., Xie, H., Wang, H., Das, S. K., Smith, D. F., and Dey, S. K. (2005) Co-chaperone immunophilin FKBP52 is critical to uterine receptivity for embryo implantation. *Proc Natl Acad Sci USA* 102:14326–31.

Van Duyne, G. D., Standaert, R. F., Karplus, P. A., Schreiber, S. L., and Clardy, J. (1991) Atomic structure of FKBP-FK506, an immunophilin–immunosuppressant complex. *Science* 252:839–42.

Wochnik, G. M., Ruegg, J., Abel, G. A., Schmidt, U., Holsboer, F., and Rein, T. (2005) FK506-binding proteins 51 and 52 differentially regulate dynein interaction and nuclear translocation of the glucocorticoid receptor in mammalian cells. *J Biol Chem* 280:4609–16.

Wu, B., Li, P., Liu, Y., Lou, Z., Ding, Y., Shu, C., Ye, S., Bartlam, M., Shen, B., and Rao, Z. (2004) 3D structure of human FK506-binding protein 52: Implications for the assembly of the glucocorticoid receptor/Hsp90/immunophilin heterocomplex. *Proc Natl Acad Sci USA* 101:8348–53.

Xin, H. B., Senbonmatsu, T., Cheng, D. S., Wang, Y. X., Copello, J. A., Ji, G. J., Collier, M. L., Deng, K. Y., Jeyakumar, L. H., Magnuson, M. A. et al. (2002) Oestrogen protects FKBP12.6 null mice from cardiac hypertrophy. *Nature* 416:334–8.

Yaffe, M. B., Schutkowski, M., Shen, M., Zhou, X. Z., Stukenberg, P. T., Rahfeld, J. U., Xu, J., Kuang, J., Kirschner, M. W., Fischer, G. et al. (1997) Sequence-specific and phosphorylation-dependent proline isomerization: A potential mitotic regulatory mechanism. *Science* 278:1957–60.

Yeh, S., Tsai, M. Y., Xu, Q., Mu, X. M., Lardy, H., Huang, K. E., Lin, H., Yeh, S. D., Altuwaijri, S., Zhou, X. et al. (2002) Generation and characterization of androgen receptor knockout (ARKO) mice: An *in vivo* model for the study of androgen functions in selective tissues. *Proc Natl Acad Sci USA* 99:13498–503.

Yeh, W. C., Li, T. K., Bierer, B. E., and McKnight, S. L. (1995) Identification and characterization of an immunophilin expressed during the clonal expansion phase of adipocyte differentiation. *Proc Natl Acad Sci USA* 92:11081–5.

Zaffran, S. (2000) Molecular cloning and embryonic expression of dFKBP59, a novel *Drosophila* FK506-binding protein. *Gene* 246:103–9.

Zhong, L., Li, W., Yang, Z., Chen, L., Li, Y., Qing, K., Weigel-Kelley, K. A., Yoder, M. C., Shou, W., and Srivastava, A. (2004a) Improved transduction of primary murine hepatocytes by recombinant adeno-associated virus 2 vectors *in vivo*. *Gene Ther* 11:1165–9.

Zhong, L., Qing, K., Si, Y., Chen, L., Tan, M., and Srivastava, A. (2004b) Heat-shock treatment-mediated increase in transduction by recombinant adeno-associated virus 2 vectors is independent of the cellular heat-shock protein 90. *J Biol Chem.*

Zhu, W., Zhang, J. S., and Young, C. Y. (2001) Silymarin inhibits function of the androgen receptor by reducing nuclear localization of the receptor in the human prostate cancer cell line LNCaP. *Carcinogenesis* 22:1399–403.

13
Up and Down
Regulation of the Stress Response by the Co-Chaperone Ubiquitin Ligase CHIP

SHU-BING QIAN AND CAM PATTERSON

Carolina Cardiovascular Biology Center, School of Medicine, University of North Carolina, Chapel Hill, NC 27599

1. Introduction

Following exposure to environmental insults, the cells in most tissues dramatically increase the production of a group of proteins that are collectively known as "heat shock" or stress proteins (Parsell and Lindquist, 1993). This set of proteins functions as the major cellular defense against the accumulation of damaged proteins. One large group of heat shock proteins are molecular chaperones, including members of the Hsp70, Hsp90, Hsp104, Hsp40 (DnaJ), and small Hsp families (Hsp27, α-crystallins) (Feder and Hofmann, 1999). Chaperones retard protein denaturation and aggregation by selectively binding unfolded domains in polypeptides (Frydman and Hartl, 1996). Most of these chaperones are major cell constituents under normal conditions, where they are essential to ensure the proper folding and intracellular localization of newly synthesized polypeptides (Frydman, 2001). Their synthesis is further increased upon exposure to various proteotoxic stressors (including heat, heavy metals, hypoxia, and acidosis). The heat shock response is an evolutionary conserved mechanism that enables cells to better withstand these diverse physical and chemical insults (Lindquist and Craig, 1988).

Two main protective strategies, the repair of damaged proteins and their selective degradation, are complementary mechanisms to eliminate aberrant cellular proteins to ensure protein homeostasis in cells (Sherman and Goldberg, 2001). The ubiquitin/ proteasome system is the major proteolytic machinery in eukaryotic cells (Coux et al., 1996). There is ample evidence that both the molecular chaperones and ubiquitin/proteasome systems are required for maintaining intracellular homeostasis (Wickner et al., 1999). However, refolding and degradation are not independent processes involving distinct cellular components, but appear to be linked at several levels (Cyr et al., 2002). Of particular importance in such linkages is the function of CHIP, carboxy-terminus of Hsp70 interacting protein (McDonough and Patterson, 2003). Its interaction with molecular chaperones (Hsc70, Hsp70, and Hsp90) results in ubiquitylation of client substrates followed by proteasomal degradation. Accumulating evidence indicates that CHIP acts to tilt the folding/refolding machinery toward the degradative pathway (McDonough and Patterson, 2003). In addition to facilitating the degradation of misfolded proteins, CHIP is able to regulate the stress response at multiple levels (Dai et al., 2003).

This review focuses on the role of CHIP in the regulation of stress responses and provides an overview of both induction and restitution processes of the stress response. We also describe emerging insights into the ways that co-factors regulate CHIP activity. Since the other chapters in this book provide an in-depth treatment of stress proteins and the mechanisms of heat shock response, the following sections provide only a brief summary of the ubiquitin/proteasome system before proceeding to discuss recent advances in our understanding of CHIP functions.

2. Ubiquitin/Proteasome System

Elaborate enzymatic mechanisms have evolved to ensure that the degradation of cellular proteins is a highly selective process that eliminates certain proteins rapidly and does not lead to nonspecific destruction of other cell constituents. To achieve this selectivity, the majority of proteins to be degraded are first marked by covalent linkage to ubiquitin (Hershko and Ciechanover, 1998). The linkage of a chain of four or more ubiquitins to a lysine on the substrate marks it for rapid breakdown by a multisubunit ATP-dependent protease known as the 26S proteasome (Pickart, 2001). The 26S proteasome consists of two 19S regulatory complexes situated at either end of the 20S core proteasome, a hollow, barrel-shaped structure, within which proteins are degraded (Hochstrasser, 1996). In eukaryotes, selective protein degradation by the 26S proteasome is pivotal for a host of cellular processes, such as cell-cycle regulation, signal transduction, and development (Glickman and Ciechanover, 2002).

The exquisite specificity of the degradative process results from a cascade of enzymes: E1 (ubiquitin-activating enzyme) hydrolyzes ATP and forms a high-energy thioester bond between an internal cysteine residue and the COOH terminus of ubiquitin. E2 (ubiquitin-conjugating enzyme) receives ubiquitin from E1 and forms a similar thiolester-linked intermediate with ubiquitin. E3 (ubiquitin ligase) finally binds both E2 and a substrate and catalyzes the transfer of ubiquitin to the substrate (Pickart and Eddins, 2004; Hershko and Ciechanover, 1998). The organization of the enzymatic conjugating cascade is hierarchical: there is one E1, a significant but limited number of E2s, and a much larger number of E3s (hundreds). The pairing of specific enzymes with cognate substrates allows for exquisite specificity in regulating substrate ubiquitylation (Glickman and Ciechanover, 2002).

The known E3s belong to just three protein families: homologous to E6AP carboxy-terminus (HECT), really interesting new gene (RING), and UFD2 homology (U-box) proteins (Pickart and Eddins, 2004). While HECT E3s form a thiolester with ubiquitin prior to modification of a target protein, RING and U-box domain E3s mediate the direct transfer of ubiquitin from an activated E2~Ub complex to the protein substrate (Deshaies, 1999). The RING domain consists of a short motif rich in cysteine and histidine residues, which coordinate two zinc ions

(Joazeiro and Weissman, 2000). The U-box domain adopts a RING domain-like conformation in which electrostatic interactions, rather than metal binding, provide the organizing principle (Ohi et al., 2003). Both domains feature a central α-helix and several β-strands that are separated by variable loops. Like conventional RING domain E3s, U-box E3s bind E2 and facilitate ubiquitin conjugation (Hatakeyama et al., 2001). The best-studied member of the U-box E3 family is CHIP (Patterson, 2002).

3. CHIP: Co-Chaperone Ubiquitin Ligase

Though the first hint that U-box proteins might act as conventional E3s came from bioinformatics (Aravind and Koonin, 2000), CHIP was first identified as a 35 kDa co-chaperone by screening a cDNA library using a probe containing tetratricopeptide repeat (TPR) motif (Ballinger et al., 1999). Further characterization led to the discovery that CHIP is not only a bona fide interaction partner with the molecular chaperones Hsc70, Hsp70, and Hsp90, but a prototypical U-box E3 as well (Jiang et al., 2001). CHIP is thus the first described co-chaperone E3.

3.1. TPR-Containing Co-Chaperone

The TPR domain mediates protein–protein interactions, including those of Hsp70 and Hsp90 chaperones with some of their co-chaperones (D'Andrea and Regan, 2003). The TPR domains of phosphatase 5, cyclophilin 40, and FKBP52 bind to Hsp90, assisting in folding hormone receptors and kinases (Young et al., 2001). Hip (Hsc70-interacting protein) binds to the ATPase domain of Hsc/Hsp70 by its TPR domain and increases the affinity for substrates by stabilizing the ADP-bound state of Hsc/Hsp70 (Hohfeld et al., 1995). HOP (Hsc70–Hsp90 organizing protein) connects Hsc70 with Hsp90, facilitating the cooperation between these two chaperones (Frydman and Hohfeld, 1997). As for CHIP, it is associated with the carboxy-terminus of Hsc70, Hsp70, and Hsp90 through its TPR domain (Ballinger et al., 1999). CHIP attenuates the Hsp40-timulated ATPase activity of Hsc70, and diminishes the refolding activity of Hsc70-Hsp40 complex for denatured substrates (Ballinger et al., 1999). In addition, CHIP has a profound effect on the composition of Hsp90 heterocomplexes (Connell et al., 2001). By occupying a docking site on Hsp90 for TPR domain-containing co-chaperones, CHIP reduces the amount of HOP associated with Hsp90 and completely blocked the interaction of Hsp90 with p23. Therefore, CHIP shifts the mode of action of Hsp70 and Hsp90 from protein folding to protein degradation (McDonough and Patterson, 2003).

3.2. U-Box E3

Another important and unique domain that is not found in other TPR containing proteins is the U-box domain located at the carboxy-terminal region of CHIP. Indeed, CHIP utilizes its U-box for binding to E2s of the Ubc4/5 family and acts as an E3 during the ubiquitylation of known chaperone substrates (Jiang et al., 2001). As a consequence, elevating the cellular concentration of CHIP results in increased degradation of chaperone substrates by the proteasome. These substrates include glucocorticoid receptor (GR) (Connell et al., 2001), cystic fibrosis transmembrane-conductance regulator (CFTR) (Meacham et al., 2001), and ErbB-2 (Xu et al., 2002). The chaperones apparently select substrates for CHIP-mediated ubiquitylation. Consistent with this idea, in vitro reconstitution ubiquitylation assays confirm that efficient ubiquitylation of heat-denatured firefly luciferase by CHIP is dependent on the presence of Hsp70 or Hsp90, whereas native luciferase is not a substrate for CHIP (Murata et al., 2001). In vivo evidence further indicates that phosphorylated Tau, an aggregation-prone protein implicated in neurodegenerative diseases, can be efficiently ubiquitylated by CHIP-Hsc70 complex (Shimura et al., 2004).

The CHIP/chaperone complex can be viewed as a multi-subunit E3 that contains either Hsc/Hsp70 or Hsp90 as the main substrate recognition factor. The complex thus resembles the Skp1/cullin/F-box E3 complex (SCF), in which the F-box is a dedicated substrate-binding partner (Jackson et al., 2000). By relegating recognition to a chaperone, CHIP can target diverse proteins that resemble one another only by virtue of their unfolded states. It is noteworthy that CHIP is able to ubiquitylate some protein substrates in the absence of chaperones if there is a direct interaction between CHIP and substrates (Demand et al., 2001). By the same token, CHIP ubiquitylates its associated molecular chaperones in vivo and in vitro (Jiang et al., 2001).

3.3. Asymmetric Homodimer

There is a mixed charged segment between the NH_2-terminal TPR domain and COOH-terminal U-box domain of CHIP. Interestingly, this predicted coiled-coil region is necessary and sufficient for the dimerization of CHIP (Nikolay et al., 2004). Under physiological conditions, no CHIP monomer can be detected. More important, deletion of this coiled-coil region resultes in the loss of CHIP ubiquitylation activity (Nikolay et al., 2004). There are several possible explanations for the biological significance of the dimeric state of CHIP. First, dimerization of CHIP could be essential for efficient binding to the chaperones. For example, the co-chaperone HOP exists as a dimer and binds the same carboxy-terminal EEVD motif of chaperones. There might be a competition between CHIP and HOP for chaperone binding. Second, the ubiquitylation activity of CHIP could be controlled by regulation of dimer formation, though little is known about the underlying mechanism.

Recently, the crystal structure of CHIP was determined at 3.2 Å (Zhang et al., 2005). Surprisingly, the CHIP homodimer exhibits a radical asymmetric structure. The major dimer interface is provided by a hydrophobic patch on the surface of helical hairpins, forming a four-helix bundle. However, the long helix in one protomer is broken and causes the dislocation of the TPR domain. Consequently, the asymmetric arrangement of TPR domains in the CHIP dimer occludes one UbC binding site of U-box, displaying "half-sites" activity. It remains to be elucidated how CHIP asymmetry regulates its E3 ligase activity, and the significance of coupling a dimeric chaperone to a single ubiquitylation system.

4. CHIP and Stress Response

The heat shock response is an ordered genetic response to diverse environmental and physiological stressors. The heat shock response is regulated at the transcriptional level by the activities of the heat shock transcription factors (HSF), of which the principle one in vertebrates is HSF1 (Wu, 1995). The DNA binding and transcriptional activities of HSF1 are stress inducibly regulated by a multistep activation pathway (Ahn and Thiele 2003). HSF1 exists normally in a negatively regulated state as an inert monomer in either the cytoplasmic or nuclear compartments. Upon exposure to a variety of stresses, HSF1 is derepressed, trimerizes, and accumulates in the nucleus. HSF1 trimers bind with high affinity to heat shock elements (HSEs) consisting of multiple contiguous inverted repeats of the pentamer sequence nGAAn located in the promoter regions of targets genes (Morimoto, 1998).

4.1. Capacitor of Stress

As a co-chaperone E3, CHIP regulates the heat shock response via multiple pathways. The overwhelming data in the literature support the proximal stress signal for the heat shock response as the appearance of misfolded proteins and an imbalance of protein homeostasis (Ananthan et al., 1986). For example, a large induction of heat shock response occurs upon treatment of normal cells with inhibitors of the proteasome, which prevents the breakdown of abnormal proteins and leads to their accumulation (Lee and Goldberg, 1998). In unstressed cells, the expression of HSP genes is normally inhibited by the presence of molecular chaperones, and HSF1 is maintained as an inactive monomer. The heat shock response is triggered when abnormal proteins accumulate and preferentially bind to Hsp90 and Hsp70, thus preventing these chaperones from inhibiting HSF1 activity (Zou et al., 1998; Shi et al., 1998). Therefore, HSF1 activation status is controlled by the availability of intracellular HSPs which is in turn dependent on the amount of substrates.

The important role of CHIP in maintaining the intracellular homeostasis is well illustrated in CHIP(-/-) mice, which develop normally but are temperature-sensitive and develop apoptosis in multiple organs after environmental challenge (Dai et al., 2003). This phenotype indicates that CHIP plays a pivotal role in cell survival under stress conditions. Accumulating evidence indicates that CHIP is central in the protein quality control by facilitating the clearance of misfolded proteins (Murata et al., 2003). Indeed, CHIP(-/-) cells demonstrate deficiency in degrading some chaperone substrates (Xu et al., 2002). CHIP thus acts as a protein quality control E3 that selectively leads abnormal proteins recognized by molecular chaperones to degradation by the proteasome. This activity is essential, because if the degradative mechanism is insufficient, the chaperone's ability to refold denatured proteins will be quickly exceeded. Consequently, these potentially toxic misfolded proteins accumulate and tend to aggregate. Therefore, CHIP acts as a stress capacitor by cooperation with molecular chaperones to alleviate the load of stress-induced misfolded proteins in the cell.

4.2. HSF1 Activator

The efficient induction of HSP gene expression is often required for cells to withstand various harmful conditions, since both removal and refolding of misfolded proteins need molecular chaperones. Unfortunately, the multiple activation steps of HSF1 take some time and the maximal induction of HSP gene expression occurs at a relatively late stage. The gap of several hours leaves the cell more susceptible to acute stress. Indeed, exposure of cells to a mild insult (i.e., a nonlethal buildup of unfolded proteins) enables the cells to withstand more toxic conditions that would otherwise be lethal (Rokutan et al., 1998). Interest in this acquired stress tolerance has spurred investigations on regulators of HSF1 (Westerheide and Morimoto, 2005). Therapeutically active small molecules that regulate HSF1 or modulate chaperones activities could benefit diseases that have in common alterations in protein conformation that cause an imbalance in protein homeostasis.

Surprisingly, CHIP also functions as an activator of HSF1 (Dai et al., 2003). CHIP overexpression resultes in higher basal level of HSF1 activity and lower HSF1 activity is observed in CHIP(-/-) cells. The physical interaction of CHIP with HSF1 and Hsp70 indicates that CHIP might directly trigger HSF1 trimerization followed by the activation of transcriptional activity (Dai et al., 2003). Moreover, CHIP is able to shuttle into the nucleus shortly after heat shock. Therefore, CHIP is able to regulate the heat shock response by mechanisms independent of protein quality control. Though the underlying mechanism is still elusive, it appears that both activities of CHIP lead to the increased availability of molecular chaperones via distinct pathways. As mentioned above, the rapid induction of HSP gene expression is critical for cell viability under stress conditions. CHIP-mediated HSF1 activation not only increases the basal level of molecular chaperones, but supplements protein quality control as well.

4.3. Stress Restitution Mediator

Under normal growth conditions, HSPs make up to ~5% of the total protein content of a cell. After stress-mediated induction, these proteins can reach up to 15% of the total cellular protein content (Pockley, 2001). The generation of HSPs must be transient, as a continued presence of HSPs would adversely influence protein homeostasis and a variety of intracellular functions. For example, sustained expression of Hsp70 in *Drosophila* is detrimental to growth at normal temperature (Feder and Hofmann, 1999). The HSF1 regulatory mechanisms not only act to provide rapid induction of Hsp70 during stress conditions, but also to prevent expression at normal conditions. However, the excess HSP molecules synthesized during stress must also be removed in order to maintain intracellular homeostasis.

Interestingly, CHIP is capable of ubiquitylating its associated chaperone partners in vivo and in vitro, although it has not been apparent that this leads to their degradation, nor have physiological roles for CHIP-dependent chaperone ubiquitylation been assigned (Jiang et al., 2001). During the course of studying CHIP effects on the stress response, we found that CHIP targets its associated HSPs for degradation, with a hierarchy that favors degradation of inducible over constitutive chaperones (Qian et al., 2006). The unexpected effect of CHIP on Hsp70 does not counteract its active role in the clearance of misfolded proteins. Instead, both functions are integrally related during the stress response and recovery process. Supporting this idea, we observed that CHIP-mediated Hsp70 turnover occurs in a substrate-dependent manner. The hierarchic catalysis of the CHIP-associated chaperone adaptors and its bound substrate provides an elegant mechanism for maintaining homeostasis by tuning chaperone levels appropriately to reflect the status of protein folding within the cytoplasm.

5. Regulators of CHIP Activity

Association of CHIP with Hsc/Hsp70 and Hsp90 appears to enable CHIP to gain access to a broad spectrum of chaperone-bound substrates for subsequent ubiquitylation and targeting to the proteasome. This unidirectional process often needs fine regulation in order to avoid uncontrolled degradation of chaperone substrates. A cooperation of diverse co-chaperones apparently provides additional levels of regulation to alter chaperone-assisted folding and degradation pathways. To date, several cofactors have been characterized to affect CHIP functions either positively or negatively (Figure 1). The identification of more regulators of CHIP may open new roads to modulate pathologically relevant processes that involve chaperone action.

5.1. BAG-1

BAG-1 was first identified as a suppressor of apoptosis. The protein exists in several translation initiation forms, and a family of mammalian paralogues has also been found. All family members possess as a common signature motif the carboxy-terminal BAG domain (Takayama and Reed, 2001). This domain interacts with the ATPase domain of Hsp70, thereby accelerating exchange of ADP for ATP and stimulating substrate release (Sondermann et al., 2001). Intriguingly, BAG-1 has an amino-terminal ubiquitin-like domain (UBL). BAG-1 has been shown to associate with the 26S proteasome via its UBL domain, and has been proposed to recruit Hsp70 to the proteasome (Lindquist and Craig, 1988). At the proteasome, BAG-1 may act as unloading factor that promotes direct delivery of Hsp70 substrates for degradation.

Both BAG-1 and CHIP are capable of binding to Hsp70 simultaneously. This cooperation leads to speculation that BAG-1 could act in concert with, or downstream of, CHIP to target ubiquitylated substrates to the proteasome. Consistent with this idea, the overexpression of BAG-1 and CHIP together in cell culture resultes in a stronger reduction of glucocorticoid receptor (GR) levels than does expression of CHIP alone (Demand et al., 2001). Interestingly, BAG-1 is efficiently ubiquitylated by the CHIP conjugation machinery, which further promotes the binding of BAG-1 to the proteasome complex (Alberti et al., 2002).

5.2. BAG-2

In searching for components of endogenous CHIP complexes, another BAG member, BAG-2, has been identified as a prominent component of CHIP-containing holocomplexes (Dai et al., 2005; Arndt et al., 2005). In contrast to BAG-1, which facilitates the transfer of substrates ubiquitylated by CHIP to the proteasome, BAG-2 inhibits CHIP ubiquitylation activity. Consequently, BAG-2 overexpression stimulates the chaperone-assisted maturation of CFTR. It appears that BAG-2 is able to suppress the interaction of CHIP with its cognate E2, possibility by sterically hindering their association or conformational remodeling of Hsp70-CHIP complexes as proposed for the BAG domain. The underlying molecular mechanism remains to be elucidated. Taken together, the disparate effects of BAG family members provide a plausible mechanism to regulate the activity of Hsp70-CHIP complexes and therefore to govern the balance between folding and degradation of misfolded proteins.

5.3. XAP2

XAP2 (hepatitis B virus X-associated protein), which is also known as Ara9 (AhR-associated protein 9) or AIP2 (AhR-interacting protein 2), is a component of the dioxin receptor (DR) chaperone complex (Kuzhandaivelu et al., 1996). In addition to XAP2, this complex comprises the molecular chaperone Hsp90 and co-chaperone p23. Overexpression of CHIP induces the degradation of the DR, whereas overexpressing of XAP2 inhibits this action of CHIP, suggesting that XAP2 may directly suppress the ubiquitin ligase of CHIP (Lees et al., 2003). The protection of CHIP-mediated DR degradation is dependent on the TPR domain of XAP2, suggesting a mechanism whereby competition for the binding site of Hsp90 guides the protein triage decision.

5.4. HspBP1

Like BAG-1, HspBP1 interacts with the ATPase domain of Hsc70 and is another nucleotide release factor (Kabani et al., 2002). However, HspBP1 acts as an inhibitor of CHIP ubiquitin ligase activity and stimulates the maturation of CFTR at the ER membrane (Alberti et al., 2004). It is speculated that the association of HspBP1 with the chaperone complex seems to shield Hsc70 and the bound client against CHIP-mediated ubiquitylation. However, not all the chaperone substrates

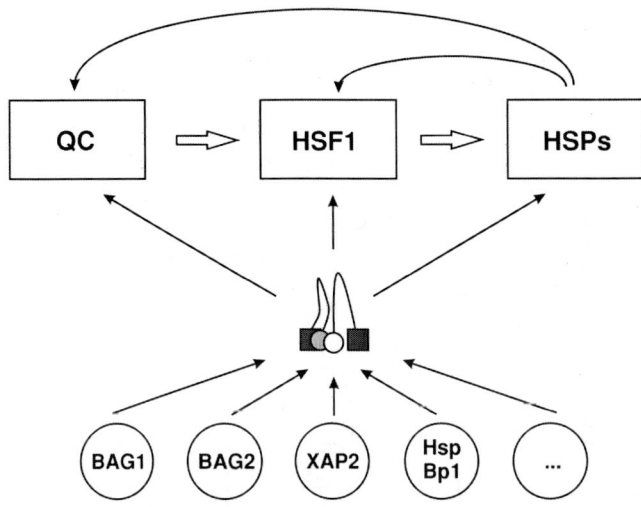

FIGURE 1. Regulation of the stress response by the co-chaperone ubiquitin ligase CHIP. Illustrated as an asymmetric dimer, CHIP plays a critical role in protein quality control (QC) by facilitating degradation of misfolded proteins; stimulates HSF1 activity; and targets excess HSPs for degradation. The activity of CHIP is tightly regulated by cofactors like BAG1, BAG2, XAP2, and HspBP1.

are affected by this mechanism. It remains an open question why the Hsp90-associated ubiquitylation is unaffected under the overexpression of HspBP1.

6. Concluding Remarks

The appreciation for the role of CHIP in chaperone-dependent protein degradation has grown rapidly in recent years. Though the concept of protein quality control was raised several decades ago, the identification of CHIP represents a great stride, in which the two "counteracting" systems (chaperones and proteasomes) have been functionally linked. To maintain intracellular homeostasis, the whole chaperone network needs to be finely tuned. The complexity of protein quality control in vivo raises many exciting and important questions. It remains unanswered how decisions are made by a chaperone complex to either refold or degrade a protein, a process referred as "protein triage" (Wickner et al.,1999). Furthermore, the physiological substrates of CHIP need to be further explored. It is almost certain that new regulators of CHIP have yet to be discovered.

In terms of the stress response, it becomes clearer that CHIP acts as a "stress capacitor" via multiple pathways. However, the expression of *CHIP* is not stress-inducible. Given the extreme importance of protein quality control under normal and stress condition, it is speculated that CHIP has a global regulatory function for chaperone activities. On the other hand, the CHIP expression profile indeed differs in different tissues, with highest expression levels in muscle and heart (Ballinger et al., 1999). This pattern might reflect the "wear and tear" on cardiac tissue that is constant and for which protein turnover and degradation is a continual requirement. Given the pleiotropic functions of CHIP, a more profound challenge is to define the appropriate way to fine tune the CHIP activity with some degree of tissue specificity. The identification of new regulators of CHIP activity will prove invaluable as tools to dissect further the multistep pathway of stress response.

References

Ahn, S. G., and Thiele, D. J. (2003) Redox regulation of mammalian heat shock factor 1 is essential for Hsp gene activation and protection from stress. *Genes Dev* 17:516–28.

Alberti, S., Bohse, K., Arndt, V., Schmitz, A., and Hohfeld, J. (2004) The cochaperone HspBP1 inhibits the CHIP ubiquitin ligase and stimulates the maturation of the cystic fibrosis transmembrane conductance regulator. *Mol Biol Cell* 15:4003–10.

Alberti, S., Demand, J., Esser, C., Emmerich, N., Schild, H., and Hohfeld, J. (2002) Ubiquitylation of BAG-1 suggests a novel regulatory mechanism during the sorting of chaperone substrates to the proteasome. *J Biol Chem* 277:45920–7.

Ananthan, J., Goldberg, A. L., and Voellmy, R. (1986) Abnormal proteins serve as eukaryotic stress signals and trigger the activation of heat shock genes. *Science* 232:522–4.

Aravind, L., and Koonin, E. V. (2000) The U box is a modified RING finger—a common domain in ubiquitination. *Curr Biol* 10:R132–4.

Arndt, V., Daniel, C., Nastainczyk, W., Alberti, S., and Hohfeld, J. (2005) BAG-2 acts as an Inhibitor of the chaperone-associated ubiquitin ligase CHIP. *Mol Biol Cell* 16:5891–5900.

Ballinger, C. A., Connell, P., Wu, Y., Hu, Z., Thompson, L. J., Yin, L. Y., and Patterson, C. (1999) Identification of CHIP, a novel tetratricopeptide repeat-containing protein that interacts with heat shock proteins and negatively regulates chaperone functions. *Mol Cell Biol* 19:4535–45.

Connell, P., Ballinger, C. A., Jiang, J., Wu, Y., Thompson, L. J., Hohfeld, J., and Patterson, C. (2001) The co-chaperone CHIP regulates protein triage decisions mediated by heat-shock proteins. *Nat Cell Biol* 3:93–6.

Coux, O., Tanaka, K., and Goldberg, A. L. (1996) Structure and functions of the 20S and 26S proteasomes. *Annu Rev Biochem* 65.801–47.

Cyr, D. M., Hohfeld, J., and Patterson, C. (2002) Protein quality control: U-box-containing E3 ubiquitin ligases join the fold. *Trends Biochem Sci* 27:368–75.

D'Andrea, L. D., and Regan, L. (2003) TPR proteins: The versatile helix. *Trends Biochem Sci* 28:655–62.

Dai, Q., Qian, S. B., Li, H. H., McDonough, H., Borchers, C., Huang, D., Takayama, S., Younger, J. M., Ren, H. Y., Cyr, D. M., and Patterson, C. (2005) Regulation of the cytoplasmic quality control protein degradation pathway by BAG2. *J Biol Chem* 280:38673–81.

Dai, Q., Zhang, C., Wu, Y., McDonough, H., Whaley, R. A., Godfrey, V., Li, H. H., Madamanchi, N., Xu, W., Neckers, L., Cyr, D., and Patterson, C. (2003) CHIP activates HSF1 and confers protection against apoptosis and cellular stress. *EMBO J* 22:5446–58.

Demand, J., Alberti, S., Patterson, C., and Hohfeld, J. (2001) Cooperation of a ubiquitin domain protein and an E3 ubiquitin ligase during chaperone/proteasome coupling. *Curr Biol* 11:1569–77.

Deshaies, R. J. (1999) SCF and Cullin/Ring H2-based ubiquitin ligases. *Annu Rev Cell Dev Biol* 15:435–67.

Feder, J. H., Rossi, J. M., Solomon, J., Solomon, N., and Lindquist, S. (1992) The consequences of expressing hsp70 in *Drosophila* cells at normal temperatures. *Genes Dev* 6:1402–13.

Feder, M. E., and Hofmann, G. E. (1999) Heat-shock proteins, molecular chaperones, and the stress response: Evolutionary and ecological physiology. *Annu Rev Physiol* 61:243–82.

Frydman, J. (2001) Folding of newly translated proteins *in vivo*: The role of molecular chaperones. *Annu Rev Biochem* 70:603–47.

Frydman, J., and Hartl, F. U. (1996) Principles of chaperone-assisted protein folding: Differences between *in vitro* and *in vivo* mechanisms. *Science* 272:1497–502.

Frydman, J., and Hohfeld, J. (1997) Chaperones get in touch: The Hip–Hop connection. *Trends Biochem Sci* 22:87–92.

Glickman, M. H., and Ciechanover, A. (2002) The ubiquitin-proteasome proteolytic pathway: Destruction for the sake of construction. *Physiol Rev* 82:373–428.

Hatakeyama, S., Yada, M., Matsumoto, M., Ishida, N., and Nakayama, K. I. (2001) U box proteins as a new family of ubiquitin–protein ligases. *J Biol Chem* 276:33111–20.

Hershko, A., and Ciechanover, A. (1998) The ubiquitin system. *Annu Rev Biochem* 67:425–79.

Hochstrasser, M. (1996) Ubiquitin-dependent protein degradation. *Annu Rev Genet* 30:405–39.

Hohfeld, J., Cyr, D. M., and Patterson, C. (2001) From the cradle to the grave: Molecular chaperones that may choose between folding and degradation. *EMBO Rep* 2:885–90.

Hohfeld, J., Minami, Y., and Hartl, F. U. (1995) Hip, a novel cochaperone involved in the eukaryotic Hsc70/Hsp40 reaction cycle. *Cell* 83:589–98.

Jackson, P. K., Eldridge, A. G., Freed, E., Furstenthal, L., Hsu, J. Y., Kaiser, B. K., and Reimann, J. D. (2000) The lore of the RINGs: Substrate recognition and catalysis by ubiquitin ligases. *Trends Cell Biol* 10:429–39.

Jiang, J., Ballinger, C. A., Wu, Y., Dai, Q., Cyr, D. M., Hohfeld, J., and Patterson, C. (2001) CHIP is a U-box-dependent E3 ubiquitin ligase: Identification of Hsc70 as a target for ubiquitylation. *J Biol Chem* 276:42938–44.

Joazeiro, C. A., and Weissman, A. M. (2000) RING finger proteins: Mediators of ubiquitin ligase activity. *Cell* 102:549–52.

Kabani, M., McLellan, C., Raynes, D. A., Guerriero, V., and Brodsky, J. L. (2002) HspBP1, a homologue of the yeast Fes1 and Sls1 proteins, is an Hsc70 nucleotide exchange factor. *FEBS Lett* 531:339–42.

Kuzhandaivelu, N., Cong, Y. S., Inouye, C., Yang, W. M., and Seto, E. (1996) XAP2, a novel hepatitis B virus X-associated protein that inhibits X transactivation. *Nucleic Acids Res* 24:4741–50.

Lee, D. H., and Goldberg, A. L. (1998) Proteasome inhibitors cause induction of heat shock proteins and trehalose, which together confer thermotolerance in *Saccharomyces cerevisiae*. *Mol Cell Biol* 18:30–8.

Lees, M. J., Peet, D. J., and Whitelaw, M. L. (2003) Defining the role for XAP2 in stabilization of the dioxin receptor. *J Biol Chem* 278:35878–88.

Lindquist, S., and Craig, E. A. (1988) The heat–shock proteins. *Annu Rev Genet* 22:631–77.

Luders, J., Demand, J., and Hohfeld, J. (2000) The ubiquitin-related BAG-1 provides a link between the molecular chaperones Hsc70/Hsp70 and the proteasome. *J Biol Chem* 275:4613–7.

McDonough, H., and Patterson, C. (2003) CHIP: A link between the chaperone and proteasome systems. *Cell Stress Chap* 8:303–8.

Meacham, G. C., Patterson, C., Zhang, W., Younger, J. M., and Cyr, D. M. (2001) The Hsc70 co-chaperone CHIP targets immature CFTR for proteasomal degradation. *Nat Cell Biol* 3:100–5.

Morimoto, R. I. (1998) Regulation of the heat shock transcriptional response: Cross talk between a family of heat shock factors, molecular chaperones, and negative regulators. *Genes Dev* 12:3788–96.

Murata, S., Chiba, T., and Tanaka, K. (2003) CHIP: A quality-control E3 ligase collaborating with molecular chaperones. *Int J Biochem Cell Biol* 35:572–8.

Murata, S., Minami, Y., Minami, M., Chiba, T., and Tanaka, K. (2001) CHIP is a chaperone-dependent E3 ligase that ubiquitylates unfolded protein. *EMBO Rep* 2:1133–8.

Nikolay, R., Wiederkehr, T., Rist, W., Kramer, G., Mayer, M. P., and Bukau, B. (2004) Dimerization of the human E3 ligase CHIP via a coiled-coil domain is essential for its activity. *J Biol Chem* 279:2673–8.

Ohi, M. D., Vander Kooi, C. W., Rosenberg, J. A., Chazin, W. J., and Gould, K. L. (2003) Structural insights into the U-box, a domain associated with multi-ubiquitination. *Nat Struct Biol* 10:250–5.

Parsell, D. A., and Lindquist, S. (1993) The function of heat–shock proteins in stress tolerance: Degradation and reactivation of damaged proteins. *Annu Rev Genet* 27:437–96.

Patterson, C. (2002) A new gun in town: The U box is a ubiquitin ligase domain. *Sci STKE* 2002:E4.

Pickart, C. M. (2001) Mechanisms underlying ubiquitination. *Annu Rev Biochem* 70:503–33.

Pickart, C. M., and Eddins, M. J. (2004) Ubiquitin: Structures, functions, mechanisms. *Biochim Biophys Acta* 1695:55–72.

Pockley, A. G. (2001) Heat–shock proteins in health and disease: Therapeutic targets or therapeutic agents? *Expert Rev Mol Med* 2001:1–21.

Qian, S. B., McDonough, H., Boellmann, F., Cyr, D. M., and Patterson, C. (2006) CHIP-mediated stress recovery by sequential ubiquitination of substrates and Hsp70. *Nature* 440:551–5.

Rokutan, K., Hirakawa, T., Teshima, S., Nakano, Y., Miyoshi, M., Kawai, T., Konda, E., Morinaga, H., Nikawa, T., and Kishi, K. (1998) Implications of heat shock/stress proteins for medicine and disease. *J Med Invest* 44:137–47.

Sherman, M. Y., and Goldberg, A. L. (2001) Cellular defenses against unfolded proteins: A cell biologist thinks about neurodegenerative diseases. *Neuron* 29:15–32.

Shi, Y., Mosser, D. D., and Morimoto, R. I. (1998) Molecular chaperones as HSF1-specific transcriptional repressors. *Genes Dev* 12:654–66.

Shimura, H., Schwartz, D., Gygi, S. P., and Kosik, K. S. (2004) CHIP-Hsc70 complex ubiquitinates phosphorylated tau and enhances cell survival. *J Biol Chem* 279:4869–76.

Sondermann, H., Scheufler, C., Schneider, C., Hohfeld, J., Hartl, F. U., and Moarefi, I. (2001) Structure of a Bag/Hsc70 complex: Convergent functional evolution of Hsp70 nucleotide exchange factors. *Science* 291:1553–7.

Takayama, S., and Reed, J. C. (2001) Molecular chaperone targeting and regulation by BAG family proteins. *Nat Cell Biol* 3:E237–41.

Westerheide, S. D., and Morimoto, R. I. (2005) Heat shock response modulators as therapeutic tools for diseases of protein conformation. *J Biol Chem* 280:33097–100.

Wickner, S., Maurizi, M. R., and Gottesman, S. (1999) Posttranslational quality control: Folding, refolding, and degrading proteins. *Science* 286:1888–93.

Wu, C. (1995) Heat shock transcription factors: Structure and regulation. *Annu Rev Cell Dev Biol* 11:441–69.

Xu, W., Marcu, M., Yuan, X., Mimnaugh, E., Patterson, C., and Neckers, L. (2002) Chaperone-dependent E3 ubiquitin ligase CHIP mediates a degradative pathway for c-ErbB2/Neu. *Proc Natl Acad Sci U S A* 99:12847–52.

Young, J. C., Moarefi, I., and Hartl, F. U. (2001) Hsp90: A specialized but essential protein-folding tool. *J Cell Biol* 154:267–73.

Zhang, M., Windheim, M., Roe, S. M., Peggie, M., Cohen, P., Prodromou, C., and Pearl, L. H. (2005) Chaperoned ubiquitylation-crystal Structures of the CHIP U box E3 ubiquitin ligase and a chip-Ubc13-Uev1a complex. *Mol Cell* 20:525–38.

Zou, J., Guo, Y., Guettouche, T., Smith, D. F., and Voellmy, R. (1998) Repression of heat shock transcription factor HSF1 activation by HSP90 (HSP90 complex) that forms a stress–sensitive complex with HSF1. *Cell* 94:471–80.

14
Role of Cdc37 in Protein Kinase Folding

ATIN K. MANDAL, DEVI M. NAIR, AND AVROM J. CAPLAN
Department of Pharmacology and Biological Chemistry, Mount Sinai School of Medicine, New York, NY 10029

1. Introduction

As nascent chains emerge from the ribosome, they interact with molecular chaperone proteins that prevent aggregation and promote protein folding. Chaperones such as Hsp70 and Hsp40 function together to protect nascent chains while still ribosome bound, and function with little if any specificity for the unfolded polypeptide. These interactions may be sufficient to promote folding, but in many cases they are not, and further binding between the nascent chain and more specific chaperones is needed. Two well-characterized chaperones that function downstream of Hsp70/Hsp40 are the chaperonins and Hsp90. Chaperonins are barrel-shaped multi-subunit proteins that have a chamber in which the unfolded polypeptide can attempt to fold. Hsp90 is known for its large number of co-chaperones or helper chaperones, and for folding of transcription factors and protein kinases among many other client types (Frydman, 2001) (Prodromou and Pearl, 2003).

Cdc37 is a molecular chaperone that functions with Hsp90 and is required for protein kinase folding and maturation (MacLean and Picard, 2003; Pearl, 2005). Both chaperones interact with each other and with nascent kinase chains during the folding reaction. And yet, little is understood regarding Cdc37 mechanism of action during kinase folding. Cdc37 was first discovered in yeast as a cell division cycle gene (Reed, 1980). Subsequent yeast genetic studies found that Cdc37 was involved in Cdc28 cyclin-dependent kinase function, including stabilizing the kinase itself and promoting cyclin binding (Gerber et al., 1995). In addition, mutation of *CDC37* suppressed the lethal growth phenotype when the oncogene v-Src was expressed in yeast (Dey et al., 1996). These studies provided evidence that Cdc37 promoted kinase maturation or activity, but did not demonstrate what that function was. Kimura et al. (1997) was the first to show that Cdc37 had a chaperone activity of its own by demonstrating that the purified protein could prevent aggregation of ß-galactosidase (Kimura et al., 1997). Other studies showed that Cdc37 was a homologue of the mammalian Hsp90 binding protein called p50, and appeared to recruit client kinases to Hsp90 (Stepanova et al., 1996). These early studies clarified the role of Cdc37 as a co-chaperone of Hsp90 that might function in the early stages of kinase recruitment to the Hsp90 complex. The interaction of cell cycle regulating kinases, such as Cdk4, with Cdc37 prompted investigations of the role Cdc37 plays in tumor growth (Dai et al., 1996; Stepanova et al., 1996). Several studies have now reported that Cdc37 levels are increased in tumor cells suggesting that the chaperone may be an important target for future cancer therapeutics (Pearl, 2005).

In the following sections we will discuss Cdc37 structure and its binding to Hsp90 and other co-chaperones. This is followed by a discussion of Cdc37 interaction with kinase and non-kinase clients. Finally, we will discuss what is known about how these interactions promote kinase folding and speculate on the mechanism of Cdc37 function.

2. Cdc37 Structure

Cdc37 is an essential cytosolic protein of eukaryotic cells. The protein sequence is moderately conserved among higher eukaryotes and becomes divergent in fungi (see Fig. 1 for alignment). Mammalian Cdc37 is 44.5 kDa and exists as a dimer in solution. It comprises three distinct domains: N-terminal, middle, and C-terminal (Fig. 2A). The N-terminal domain, approximately 126 amino acids, contains highly conserved residues among Cdc37 from different organisms and binds to protein kinases. This conservation is highest over the first 30 amino acids and contains residues that are critical for Cdc37's kinase-binding activity. *S. cerevisiae* Cdc37 shares only 20% sequence identity with its mammalian orthologue and it has a large insertion in the N-domain compared to Cdc37 from metazoan phyla and it is 58.4 kDa. The structure of this region is unknown and seems to be a less compact, protease sensitive coil-coiled structure (Zhang et al., 2004).

In contrast to the N-terminal kinase-binding domain, the middle domain of Cdc37 is well characterized at a structural level. The sequence in this region is reasonably conserved in vertebrates and becomes more diverged in invertebrates and fungi. This domain comprises of protease stable α-helical structure of approximately 150 amino acids that binds to Hsp90. The crystal structure of the Hsp90-Cdc37 core complex, lacking the N-terminal region of Cdc37 reveals that this domain is composed of large six-helix bundle of unusual topology containing approximately 90 amino acids that binds to N-terminal ATP binding domain of Hsp90 (Roe et al., 2004). C-terminal to this bundle is a long single α-helix that connects the middle domain and the rest of C-terminal portion of Cdc37 (see Fig. 2). This long helix is a part of a protease-stable portion of 150 amino acids and belongs to the middle domain. The middle domain contains the residues that are involved in dimerization of Cdc37. These residues (amino acids 240-254) are located in the first turn of the long connecting helix as showed in the crystal structure as well as by comparison to a related protein, Harc (an Hsp90-associating relative of Cdc37 found in mammalian cells) (Scholz et al., 2001). The middle domain of Harc shares 62% identity with Cdc37 and bears an Hsp90-binding site. Unlike Cdc37 however, Harc does not bind to protein kinases. The C-terminal portion of Cdc37 is a less well ordered structure and consists of small three-helix bundle.

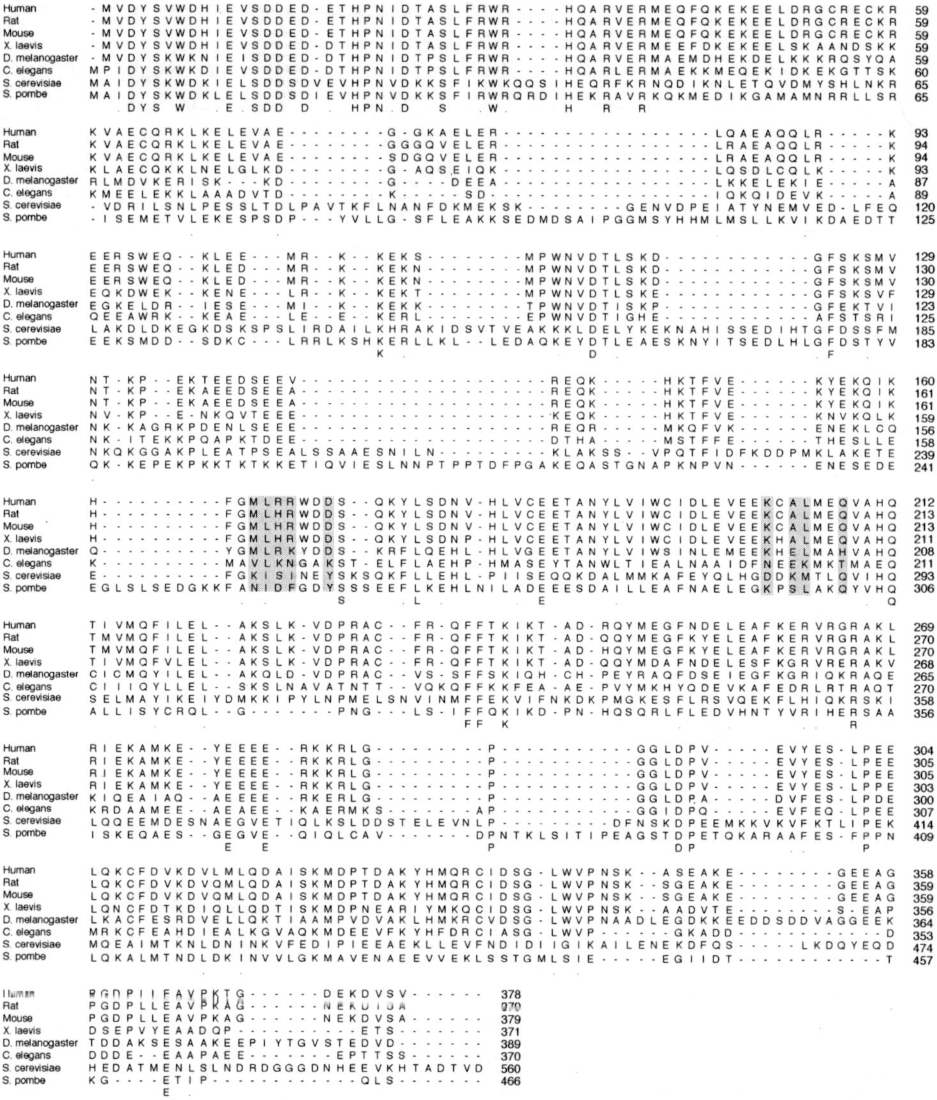

FIGURE 1. Alignment of the sequence of human Cdc37 with the amino acid sequences of homologous Cdc37 from rat, mouse, xenopus, drosophila, worms, and yeast. Identical residues between Cdc37 from different phyla are indicated and similar residues are depicted as (-). Hsp90 interacting residues are highlighted in cyan.

The function of the C-terminal region is not clear and is dispensable for normal and stressful growth in yeast (Lee et al., 2002).

Phosphorylation of Cdc37 is an essential requirement for its chaperoning activity. This phosphorylation is mediated by casein kinase II (CK2) on a highly conserved N-terminal Ser residue (Bandhakavi et al., 2003; Miyata and Nishida,

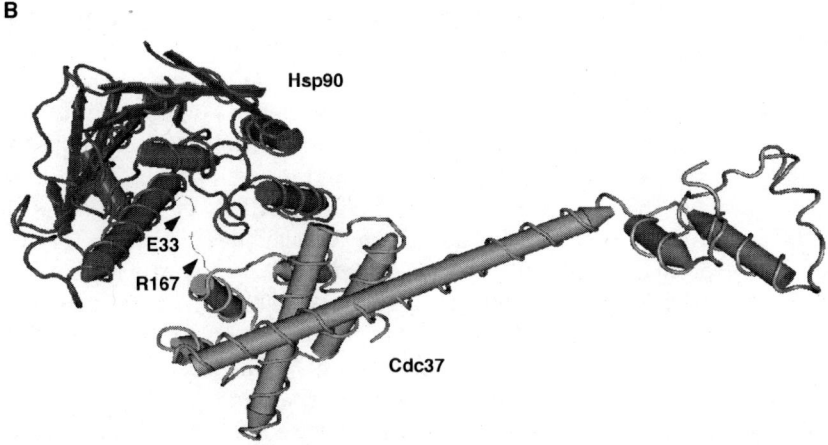

FIGURE 2. Structure of Cdc37: A: Schematic representation of domain organization of Cdc37. N-terminus, middle domain, and C-terminus is shown in maroon, pink, and gray, respectively. The conserved phosphorylated serine residue is indicated. B: Structure of Cdc37 (residues 138–378) with the isolated N-domain of yeast Hsp90 (residues 1–208) as described by Roe at el. (2004); PDB 1US7. Yeast Hsp90 is shown in blue color. The hydrogen bonds between ATP hydrolyzing residue E33 of Hsp90 and carboxyl side chain of R167 of Cdc37 is depicted. Cn3D was used to render the structure (http://www.biosino.org/mirror/www.ncbi.nhl.nih.gov/Structure/cn3d/cn3d.html).

2004; Shao et al., 2003b). Yeast Cdc37 is phophorylated on both Ser14 and Ser17, however, Ser14 is more important for its function compared to Ser17. On the other hand, mammalian Cdc37 is phophorylated on Ser13 (equivalent of Ser14 of yeast Cdc37). Phophorylation at Ser13 or Ser14 is crucial for its stable interaction with protein kinases, as unphosphorylated Cdc37 is unable to interact with client kinases and these kinases often degrade or have reduced function/activity. Therefore, CK2 regulates Cdc37 activity and subsequent protein kinase folding broadly. Additionally, Cdc37 physically interacts with CK2 and enhances its activity; thus CK2 and Cdc37 constitute a positive feedback loop to promote the function of multiple protein kinases (Bandhakavi et al., 2003; Miyata and Nishida, 2004).

3. Cdc37 as a Co-Chaperone of Hsp90

Cdc37 was first identified as p50, an Hsp90-binding protein, in association with Rous sarcoma virus-encoded oncogenic protein kinase pp60$^{v\text{-}src}$ (Brugge, 1986). Subsequent cloning and biochemical analysis of mammalian Cdc37 described its function as a protein-kinase-targeting co-chaperone of Hsp90. This was based on the observation that overexpression of mammalian Cdc37 in insect cells assembles with Cdk4 in high molecular weight complexes that also contain Hsp90 (Stepanova et al., 1996). Moreover, Cdc37 interacts with Hsp90 in the absence of client kinases (Whitelaw et al., 1991).

In vitro interaction studies of Cdc37 show a species-variant binding affinity with Hsp90. Binding of mammalian Cdc37 to Hsp90 is significantly tighter than that of its yeast orthologue (Siligardi et al., 2002). The dissociation constant (K_d) value for the interaction of mammalian Cdc37 with Hsp90 is 2.5–4.0 µM in the absence of a kinase (Siligardi et al., 2002; Zhang et al., 2004). In contrast, the K_d value for yeast Cdc37 with Hsp90 is approximately 100 µM (Siligardi et al., 2002). Both mammalian and yeast Cdc37 can inhibit the ATPase activity of Hsp90 although to different extents: Mammalian Cdc37 is a potent inhibitor of Hsp90's ATPase activity compared to yeast Cdc37. This might be due to the presence of mammalian Cdc37 Arg167 side chain which hydrogen bonds to the catalytic residue Glu33 of Hsp90, which is thought to be involved in ATP hydrolysis (see Fig. 2B). This interaction of Cdc37 with Hsp90 is able to arrest the Hsp90 ATPase cycle and prevents its ATP-driven conformational change. However, this might not be the case in yeast since yCdc37 contains isoleucine at the equivalent position.

In vitro and in vivo interaction studies provide evidence that the middle domain of Cdc37 binds to the N-domain of Hsp90 and the kinase-binding N-domain of Cdc37 modulates this interaction. However, the precise interaction between Hsp90 and Cdc37 is not understood clearly. The complex between human Cdc37 with the isolated yeast N-domain of Hsp90 described the interacting residues 164–170 and 204–208 of Cdc37 that form a hydrophobic patch with the ATP binding lid segment (residues 100–121) of Hsp90. The structure explained the inhibitory role of Cdc37 on the ATPase activity of Hsp90 (Roe et al., 2004). Moreover, the dimeric conformation of Cdc37 in the complex correlates with the previous studies. However, this structure is not a complete picture of the interaction between Hsp90 and Cdc37 since the kinase-binding N-terminal domain of Cdc37 was not present. Furthermore, full-length Cdc37 was unable to interact with the isolated N-domain of Hsp90 (Zhang et al., 2004). This is also supported by the evidence that mutation of Trp7 in the N-terminus of Cdc37 reduces complex formation with Hsp90 present in cell lysates (Shao et al., 2003a).

The minimal region of Hsp90 that is required for stable complex formation with full-length Cdc37 comprises the N-terminal domain and the flexible charged linker region (Zhang et al., 2004). Furthermore, the complex is more stable in the presence of the N-domain, flexible linker, middle, and C-terminal domains of Hsp90. On the other hand, truncated Cdc37 lacking the kinase-binding domain can form stable complexes with the isolated N-terminal part of Hsp90. This is based on the assumption that the N-terminus of Cdc37 can interact with Hsp90, perhaps in the flexible linker; however, no direct evidence has been reported so far.

Hsp90 is the core component of a chaperone complex that assembles other co-chaperones and facilitates maturation of client proteins (Pratt and Toft, 2003). These co-chaperones interact with Hsp90 at different stages of client protein folding. Cdc37 was found in large chaperone complexes with Hsp90 during client protein folding along with some other co-chaperones, but not all of them. Aha1, for example, binds to the middle domain of Hsp90 and stimulates its ATPase activity (Lotz et al., 2003; Panaretou et al., 2002). Binding of Aha1 induces conformational changes of the catalytic loop of Hsp90 that facilitates interaction of catalytic Arg380 with the ATP in the N-terminal nucleotide binding domain of the Hsp90. Aha1 competes with Cdc37 for binding with Hsp90 during client protein folding (Harst et al., 2005). However, N-terminal truncated Cdc37 does not interfere with Aha1 binding to Hsp90 (Siligardi et al., 2004). This suggests that N-terminal Cdc37 can interact with Hsp90 or it sterically hinders binding of Aha1. On the other hand, Cdc37 was also found in the heterocomplex of hsp90 and p23/Sba1 during late stages of client protein maturation (Hartson et al., 2000). p23 inhibits the ATPase activity of Hsp90 by stabilizing the ATP bound form of Hsp90. By contrast, truncated Cdc37 could not form a complex with Hsp90/p23 (Harst et al., 2005; Siligardi et al., 2004).

The best-characterized interactions that Cdc37 has with other co-chaperones is with Sti1, the yeast homologue of Hop, that bridges Hsp70 with Hsp90 (Abbas-Terki et al., 2002; Lee et al., 2002; 2004). Deletion of Sti1 has very little growth effects on yeast, although when combined with a mutant form of *CDC37* results in a synthetic growth defect. This has been interpreted to mean that Cdc37 and Sti1 function together. Indeed, physical interaction between Sti1 and Cdc37 has been described and *CDC37* overexpression can suppress defects in kinase activity in a *sti1*Δ mutant. Importantly, *CDC37* overexpression suppressed loss of binding between the protein kinase Ste11 and Hsp90 that occurred in a *sti1*Δ mutant. The need for Sti1 in kinase biogenesis suggests that nascent kinases interact with Hsp70 and Hsp40 proteins shortly after synthesis. Recruitment of Hsp90 is likely to occur via interaction with Sti1, which binds to both chaperones. Cdc37 is likely to enter the complex at this point based on its interaction with Sti1. However, since Cdc37 can function independently of Sti1 (since it can suppress the effects of *STI1* deletion), there may be Hsp70/Sti1-independent mechanisms by which Cdc37 can promote client kinase binding to Hsp90.

4. Interaction of Cdc37 with Kinase and Non-Kinase Clients

Cdc37, or p50 as it was originally called, was first discovered as a v-Src interacting protein (Brugge, 1986). Since then, many kinases have been shown to interact with Cdc37, with ILK (Aoyagi et al., 2005), and JAK 1/2 (Shang and Tomasi, 2006) discovered recently as interactors. Akt, Cdk4, Lck, v-Src, and Raf-1 are some of the well-studied Cdc37 client kinases and a more comprehensive list of Cdc37 interacting client proteins can be found at the following website: Cdc37interactors.pdf at http://www.picard.ch/. The requirement for Cdc37 by many kinases makes this molecule an important regulator for major cellular events such as cell cycle, signal transduction, and transcription.

Several proteins apart from kinases have been discovered to interact with Cdc37. The first non-kinase client to be discovered was the human androgen receptor (AR). A comparative study of AR with glucocorticoid receptor (GR), heterologously expressed in yeast, revealed that unlike GR, AR binds to Cdc37 both in vitro and in vivo (Fliss et al., 1997; Rao et al., 2001). This interaction is via the ligand-binding domain of AR, and was sensitive to geldanamycin treatment. Transcription factor MyoD, a major regulator of myoblast differentiation, was also shown to interact with Cdc37 (Yun and Matts, 2005). MyoD was previously shown to interact with Hsp90 (Shaknovich et al., 1992), and geldanamycin also disrupted its interaction with Cdc37 (Yun and Matts, 2005). These results suggest that Cdc37 action in chaperoning non-kinase clients also involves Hsp90. Reverse transcriptase from hepadnaviruses and endothelial nitric oxide synthase are two other non-kinase proteins whose interactions with Cdc37 are essential for their important roles in viral assembly and cardiovascular functioning, respectively (Harris et al., 2006; Wang et al., 2002). Whether Cdc37 has distinct domains or utilizes its kinase-binding domain for interactions with non-kinase clients is unknown. Also unknown is the amino acid sequence or structural motif(s) recognized by Cdc37 in these diverse proteins and if there are any common properties that they share with kinases.

Despite limited understanding of how non-kinase clients interact with Cdc37, much more is known about its interaction with protein kinases. Eukaryotic protein kinases have a highly conserved bi-lobal structure with ATP binding between the lobes. The N-domain is predominantly β-sheet, while the C-domain is largely α-helical in structure (Hanks and Hunter, 1995). Initial studies described both Cdc37 and Hsp90 as interacting directly with the catalytic domain of protein kinases (Silverstein et al., 1998). Furthermore, more recent studies have shown that the N-lobe is the direct target of Cdc37 (Zhao et al., 2004; Prince and Matts, 2004). Studies with Cdk4 showed that a region that included the glycine rich loop, the α-C helix and the loop between the α-C helix and ß4 strand were important for Cdc37 binding (Zhao et al., 2004). Studies of Lck showed that the α-C helix, ß-4 and ß-5 strands were important for interaction with Hsp90 and Cdc37 (Prince and Matts, 2004). Hsp90 appears to interact with both N- and C-lobes of the kinase. Subsequent studies with Cdk2 were consistent with the view that Cdc37 bound to the N-lobe via the glycine-rich loop while Hsp90 bound to motifs in both lobes (Prince et al., 2005).

The domain of Cdc37 that interacts with protein kinases was originally described at the N-terminus of the chaperone. Initial studies used truncation analysis to define the N-terminal domain as binding directly to Raf-1, HRI and v-Src (Grammatikakis et al., 1999; Lee et al., 2002; Shao et al., 2003a). The shortest construct used in these studies was the first 126 amino acids of Cdc37 which bound directly to HRI (Shao et al., 2003a). Strong support for a role of this domain in kinase binding was shown by mutagenesis studies demonstrating that Trp 7 of Cdc37 was important for kinase binding (Shao et al., 2003a). Furthermore, several studies, as noted above, demonstrated that phosphorylation of Ser13 in mammalian Cdc37 and Ser14/17 in yeast Cdc37 were important for kinase binding. Importance of Cdc37 phosphorylation was also recently confirmed in an elegant in vitro system, using purified chaperones and co-chaperone proteins and the results clearly show that Casein kinase II dependent activation of Cdc37 is a pre-requisite for folding and activity of Chk1 kinase (Arlander et al., 2006).

In combination, the above studies show that the N-terminal domain of the chaperone is required for kinase binding. However, two recent reports have shown that a second kinase binding site is situated in a 20 amino acid segment contained within the Hsp90 binding middle domain of Cdc37. This was first discovered in truncation studies where deletion of the N-domain actually induced binding of some kinases to Cdc37 (Terasawa and Minami, 2005). Indeed, this second site appeared to bind to some kinases that failed to interact with the N-terminal kinase-binding domain, such as Cdk2. Furthermore, this second site appears to bind to the glycine-rich loop of Raf1 (Terasawa et al., 2006). These two kinase-binding sites may bind to different kinases and the significance of these two sites to the chaperoning function of Cdc37 is not clear at this time.

Although the significance of the second kinase-binding site is unknown, it is clear from genetic studies that the essential domain of Cdc37 is at the N-terminus. Deletion mutants of Cdc37 that lacked the complete C-terminal domain and Hsp90 interacting middle domain could support cell growth in *S. cerevisiae* (Lee et al., 2002). Similar observations were made with Cdc37 deletion mutant studies in *S. pombe*. Also, truncations from the N-terminal domain by as little as 20 amino acids abolished cell growth in this yeast (Turnbull et al., 2005). Since the N-domain of Cdc37 binds to protein kinases, these data suggest that kinase binding is the essential function of the chaperone rather than Hsp90 binding. Indeed, overexpression of *CDC37* can suppress defects associated with mutation of Hsp90 and the co-chaperone Sti1 which functions in client recruitment to the Hsp90 complex (Kimura et al., 1997; Lee et al., 2004).

Further studies have characterized the N-terminal domain of Cdc37 as being dominant negative for activation of Raf1 in mammalian cells (Grammatikakis et al., 1999). Overerexpression of the N-domain also leads to growth arrest in human prostate epithelial cells (Schwarze et al., 2003). These results are consistent with the view that Cdc37 binding to kinases needs to be coupled to Hsp90 binding for proper folding. However, in one documented case, overexpression of the N-domain of Cdc37 activated a mutant form of the kinase Hck, and even promoted Hsp90 binding (Scholz et al., 2000). Combined with the genetic studies in yeast, these results indicate that Cdc37 need not interact with Hsp90 directly to perform its function, and may have a role in promoting allosteric changes in kinase conformation that reveals the Hsp90 binding site.

Unlike Hsp90, Cdc37 is not induced in response to heat shock. However, it is up-regulated in cancer cells, where oncogenic kinases such as Cdk4, Akt, Raf, and others, are shown to have increased expression (Stepanova et al., 2000a). Besides, overexpression of Cdc37 leads to increased prostrate hyperplasia in mice

(Stepanova et al., 2000b) and enhanced proliferation of human prostrate epithelial cells (Schwarze et al., 2003).

The mechanism underlying Cdc37 function in protein-kinase folding has yet to be described. The binding of Cdc37 to the kinase N-domain suggests the need to stabilize this domain. Indeed, even in the crystal structure of unliganded protein kinase A, the N-domain represented the least stable part of the kinase (Akamine et al., 2003). In the presence of ATP and substrate, this domain is more stable (Taylor et al., 2004). It is possible, therefore, that Cdc37 can stabilize the unliganded kinase until nucleotide can bind. Recent studies of kinases and chaperones have shown that some kinases appear to have a constant chaperone requirement, distinguishing them from kinases that need chaperones for nascent chain folding only. The constant requirement is manifest in the speed with which some kinases are degraded in the presence of geldanamycin. For example, ErbB2 is degraded within a few hours of geldanamycin treatment whereas its paralog, epidermal growth factor receptor, has a half-life at least four times longer under the same conditions (Xu et al., 2001, 2005). Sequences in the loop between the α-C helix and β-4 strand mediate this effect, which is also a region implicated in Cdc37 binding to kinases. The constant requirement for chaperones may therefore represent the need to stabilize the kinase after each round of catalysis, when the ADP is released and before ATP rebinds. How the loop between the α-C helix and β-4 strand might mediate such an effect is unclear since a surface charge appears to mediate the difference in geldanamycin sensitivity rather than interactions in the interior of the protein that might stabilize kinase conformation.

In conclusion, Cdc37 is a chaperone that functions in association with Hsp90 to promote folding of protein kinases and some non-kinase clients. The kinase-chaperone interactions involve direct interactions with Hsp90, Cdc37, and the client although how each contributes to the overall reaction is unknown. Cdc37, while originally considered a targeting subunit of Hsp90, has now been shown to have a chaperone function of its own, and the kinase-binding domain is the essential domain of the protein.

References

Abbas-Terki, T., Briand, P. A., Donze, O., and Picard, D. (2002) The Hsp90 co-chaperones Cdc37 and Sti1 interact physically and genetically. *Biol Chem* 383:1335–42.

Akamine, P., Madhusudan, Wu, J., Xuong, N. H., Ten Eyck, L. F., and Taylor, S. S. (2003) Dynamic features of cAMP-dependent protein kinase revealed by apoenzyme crystal structure. *J Mol Biol* 327:159–71.

Aoyagi, Y., Fujita, N., and Tsuruo, T. (2005) Stabilization of integrin-linked kinase by binding to Hsp90. *Biochem Biophys Res Commun* 331:1061–8.

Arlander, S. J., Felts, S. J., Wagner, J. M., Stensgard, B., Toft, D. O., and Karnitz, L. M. (2006) Chaperoning checkpoint kinase 1 (Chk1), an Hsp90 client, with purified chaperones. *J Biol Chem* 281:2989–98.

Bandhakavi, S., McCann, R. O., Hanna, D. E., and Glover, C. V. (2003) A positive feedback loop between protein kinase CKII and Cdc37 promotes the activity of multiple protein kinases. *J Biol Chem* 278:2829–36.

Brugge, J. S. (1986) Interaction of the *Rous sarcoma* virus protein pp60src with the cellular proteins pp50 and pp90. *Curr Top Microbiol Immunol* 123:1–22.

Dai, K., Kobayashi, R., and Beach, D. (1996) Physical interaction of mammalian CDC37 with CDK4. *J Biol Chem* 271:22030–4.

Dey, B., Lightbody, J. J., and Boschelli, F. (1996) CDC37 is required for p60v-src activity in yeast. *Mol Biol Cell* 7:1405–17.

Fliss, A. E., Fang, Y., Boschelli, F., and Caplan, A. J. (1997) Differential *in vivo* regulation of steroid hormone receptor activation by Cdc37p. *Mol Biol Cell* 8:2501–9.

Frydman, J. (2001) Folding of newly translated proteins *in vivo*: The role of molecular chaperones. *Annu Rev Biochem* 70:603–47.

Gerber, M. R., Farrell, A., Deshaies, R. J, Herskowitz, I., and Morgan, D. O. (1995) Cdc37 is required for association of the protein kinase Cdc28 with G1 and mitotic cyclins. *Proc Natl Acad Sci U S A* 92:4651–5.

Grammatikakis, N., Lin, J. H., Grammatikakis, A., Tsichlis, P. N., and Cochran, B. H. (1999) p50(cdc37) acting in concert with Hsp90 is required for Raf-1 function. *Mol Cell Biol* 19:1661–72.

Hanks, S. K., and Hunter, T. (1995) Protein kinases 6. The eukaryotic protein kinase superfamily: Kinase (catalytic) domain structure and classification. *FASEB J* 9:576–96.

Harris, M. B., Bartoli, M., Sood, S. G., Matts, R. L., and Venema, R. C. (2006) Direct interaction of the cell division cycle 37 homolog inhibits endothelial nitric oxide synthase activity. *Circ Res* 98:335–41.

Harst, A., Lin, H., and Obermann, W. M. (2005) Aha1 competes with Hop, p50 and p23 for binding to the molecular chaperone Hsp90 and contributes to kinase and hormone receptor activation. *Biochem J* 387:789–96.

Hartson, S. D., Irwin, A. D., Shao, J., Scroggins, B. T., Volk, L., Huang, W., and Matts, R. L. (2000) p50(c)(dc37) Is a nonexclusive hsp90 cohort which participates intimately in hsp90-mediated folding of immature kinase mole. *Biochem* 39:7631–44.

Kimura, Y., Rutherford, S. L., Miyata, Y., Yahara, I., Freeman, B. C., Yue, L., Morimoto, R. I., and Lindquist, S. (1997) Cdc37 is a molecular chaperone with specific functions in signal transduction. *Genes Dev* 11:1775–85.

Lee, P., Rao, J., Fliss, A., Yang, E., Garrett, S., and Caplan, A. J. (2002) The Cdc37 protein kinase-binding domain is sufficient for protein kinase activity and cell viability. *J Cell Biol* 159:1051–9.

Lee, P., Shabbir, A., Cardozo, C., and Caplan, A. J. (2004) Sti1 and Cdc37 can stabilize Hsp90 in chaperone complexes with a protein kinase. *Mol Biol Cell*.

Lotz, G. P., Lin, H., Harst, A., and Obermann, W. M. (2003) Aha1 binds to the middle domain of hsp90, contributes to client protein activation, and stimulates the ATPase activity of the molecular chaperone. *J Biol Chem* 278:17228–35.

MacLean, M., and Picard, D. (2003) Cdc37 goes beyond Hsp90 and kinases. *Cell Stress Chap* 8:114–9.

Miyata, Y., and Nishida, E. (2004) CK2 controls multiple protein kinases by phosphorylating a kinase-targeting molecular chaperone, Cdc37. *Mol Cell Biol* 24:4065–74.

Panaretou, B., Siligardi, G., Meyer, P., Maloney, A., Sullivan, J. K., Singh, S., Millson, S. H., Clarke, P. A., Naaby-Hansen, S., Stein, R., Cramer, R., Mollapour, M., Workman, P., Piper, P. W., Pearl, L. H., and Prodromou, C. (2002) Activation of the ATPase activity of hsp90 by the stress-regulated cochaperone aha1. *Mol Cell* 10:1307–18.

Pearl, L. H. (2005) Hsp90 and Cdc37—a chaperone cancer conspiracy. *Curr Opin Genet Dev* 15:55–61.

Pratt, W. B., and Toft, D. O. (2003) Regulation of signaling protein function and trafficking by the hsp90/hsp70-based chaperone machinery. *Exp Biol Med (Maywood)* 228:111–33.

Prince, T., and Matts. R. L. (2004) Definition of protein kinase sequence motifs that trigger high affinity binding of Hsp90 and Cdc37. *J Biol Chem* 279:39975–81.

Prince, T., Sun, L., and Matts. R. L. (2005) Cdk2: A genuine protein kinase client of Hsp90 and Cdc37. *Biochemistry* 44:15287–95.

Prodromou, C., and Pearl, L. H. (2003) Structure and functional relationships of Hsp90. *Curr Cancer Drug Targets* 3:301–23.

Rao, J., Lee, P., Benzeno, S., Cardozo, C., Albertus, J., Robins, D. M., and Caplan, A. J. (2001) Functional interaction of human Cdc37 with the androgen receptor but not with the glucocorticoid receptor. *J Biol Chem* 276:5814–20.

Reed, S. I. (1980) The selection of *S. cerevisiae* mutants defective in the start event of cell division. *Genetics* 95:561–77.

Roe, S. M., Ali, M. M., Meyer, P., Vaughan, C. K., Panaretou, B., Piper, P. W., Prodromou, C., and Pearl, L. H. (2004) The mechanism of Hsp90 regulation by the protein kinase-specific cochaperone p50(cdc37). *Cell* 116:87–98.

Scholz, G., Hartson, S. D., Cartledge, K., Hall, N., Shao, J., Dunn, A. R., and Matts, R. L. (2000) p50(Cdc37) can buffer the temperature-sensitive properties of a mutant of Hck [In Process Citation]. *Mol Cell Biol* 20:6984–95.

Scholz, G. M., Cartledge, K., and Hall, N. E. (2001) Identification and characterization of Harc, a novel Hsp90-associating relative of Cdc37. *J Biol Chem* 276:30971–9.

Schwarze, S. R., Fu, V. X., and Jarrard. D. F. (2003) Cdc37 enhances proliferation and is necessary for normal human prostate epithelial cell survival. *Cancer Res* 63:4614–9.

Shaknovich, R., Shue, G., and Kohtz, D. S. (1992) Conformational activation of a basic helix–loop–helix protein (MyoD1) by the C-terminal region of murine HSP90 (HSP84). *Mol Cell Biol* 12:5059–68.

Shang, L., and Tomasi, T. B. (2006) The heat shock protein 90-CDC37 chaperone complex is required for signaling by types I and II interferons. *J Biol Chem* 281:1876–84.

Shao, J., Irwin, A., Hartson, S. D., and Matts, R. L. (2003a) Functional dissection of cdc37: Characterization of domain structure and amino acid residues critical for protein kinase binding. *Biochemistry* 42:12577–88.

Shao, J., Prince, T., Hartson, S. D., and Matts, R. L. (2003b) Phosphorylation of serine 13 is required for the proper function of the Hsp90 co-chaperone, Cdc37. *J Biol Chem* 278:38117–20.

Siligardi, G., Hu, B., Panaretou, B., Piper, P. W., Pearl, L. H, and Prodromou, C. (2004) Co-chaperone regulation of conformational switching in the Hsp90 ATPase cycle. *J Biol Chem* 279:51989–98.

Siligardi, G., Panaretou, B., Meyer, P., Singh, S., Woolfson, D. N., Piper, P. W., Pearl, L. H., and Prodromou, C. (2002) Regulation of Hsp90 ATPase activity by the co-chaperone Cdc37p/p50cdc37. *J Biol Chem* 277:20151–9.

Silverstein, A. M., Grammatikakis, N., Cochran, B. H., Chinkers, M., and Pratt, W. B. (1998) p50(cdc37) binds directly to the catalytic domain of Raf as well as to a site on hsp90 that is topologically adjacent to the tetratricopeptide repeat binding site. *J Biol Chem* 273:20090–5.

Stepanova, L., Finegold, M., DeMayo, F., Schmidt, E. V., and Harper, J. W. (2000a) The oncoprotein kinase chaperone CDC37 functions as an oncogene in mice and collaborates with both c-myc and cyclin d1 in transformation of multiple tissues. *Mol Cell Biol* 20:4462–73.

Stepanova, L., Leng, X., Parker, S. B., and Harper, J. W. (1996) Mammalian p50Cdc37 is a protein kinase-targeting subunit of Hsp90 that binds and stabilizes Cdk4. *Genes Dev* 10:1491–502.

Stepanova, L., Yang, G., DeMayo, F., Wheeler, T. M., Finegold, M., Thompson, T. C., and Harper, J. W. (2000b) Induction of human Cdc37 in prostate cancer correlates with the ability of targeted Cdc37 expression to promote prostatic hyperplasia [In Process Citation]. *Oncogene* 19:2186–93.

Taylor, S. S., Yang, J., Wu, J., Haste, N. M., Radzio-Andzelm, E., and Anand, G. (2004) PKA: A portrait of protein kinase dynamics. *Biochim Biophys Acta* 1697:259–69.

Terasawa, K., and Minami, Y. (2005) A client-binding site of Cdc37. *FEBS J* 272:4684–90.

Terasawa, K., Yoshimatsu, K., Iemura, S., Natsume, T., Tanaka, K., and Minami, Y. (2006) Cdc37 interacts with the glycine-rich loop of hsp90 client kinases. *Mol Cell Biol* 26:3378–89.

Turnbull, E. L., Martin, I. V., and Fantes, P. A. (2005) Cdc37 maintains cellular viability in *Schizosaccharomyces pombe* independently of interactions with heat-shock protein 90. *FEBS J* 272:4129–40.

Wang, X., Grammatikakis, N., and Hu, J. (2002) Role of p50/CDC37 in hepadnavirus assembly and replication. *J Biol Chem* 277:24361–7.

Whitelaw, M. L., Hutchison, K., and Perdew, G. H. (1991) A 50-kDa cytosolic protein complexed with the 90-kDa heat shock protein (hsp90) is the same protein complexed with pp60v-src hsp90 in cells transformed by the *Rous sarcoma* virus. *J Biol Chem* 266:16436–40.

Xu, W., Mimnaugh, E., Rosser, M. F., Nicchitta, C., Marcu, M., Yarden, Y., and Neckers, L. (2001) Sensitivity of mature Erbb2 to geldanamycin is conferred by its kinase domain and is mediated by the chaperone protein Hsp90. *J Biol Chem* 276:3702–8.

Xu, W., Yuan, X., Xiang, Z., Mimnaugh, E., Marcu, M., and Neckers, L. (2005) Surface charge and hydrophobicity determine ErbB2 binding to the Hsp90 chaperone complex. *Nat Struct Mol Biol*.

Yun, B. G., and Matts, R. L. (2005) Differential effects of Hsp90 inhibition on protein kinases regulating signal transduction pathways required for myoblast differentiation. *Exp Cell Res* 307:212–23.

Zhang, W., Hirshberg, M., McLaughlin, S. H., Lazar, G. A., Grossmann, J. G., Nielsen, P. R., Sobott, F., Robinson, C. V., Jackson, S. E., and Laue, E. D. (2004) Biochemical and structural studies of the interaction of Cdc37 with Hsp90. *J Mol Biol* 340:891–907.

Zhao, Q., Boschelli, F., Caplan, A. J., and Arndt, K. T. (2004) Identification of a conserved sequence motif that promotes Cdc37 and Cyclin D1 binding to Cdk4. *J Biol Chem*.

VI
Intracellular and Extracellular Stress Proteins in Human Disease

15
Targeting Hsp90 in Cancer and Neurodegenerative Disease

LEN NECKERS[1] AND PERCY IVY[2]

[1]*Urologic Oncology Branch, Center for Cancer Research, National Cancer Institute, Bethesda, MD;* [2]*Cancer Therapy Evolution Program, National Cancer Institute, Bethesda, MD*

1. Summary

Heat shock protein 90 (Hsp90) is a molecular chaperone required for the stability and function of a number of conditionally activated and/or expressed signaling proteins, as well as multiple mutated, chimeric, and/or overexpressed signaling proteins, that promote cancer cell growth and/or survival. Hsp90 inhibitors, by interacting specifically with a single molecular target, cause the inactivation, destabilization, and eventual degradation of Hsp90 client proteins, and they have shown promising anti-tumor activity in preclinical model systems. One Hsp90 inhibitor, 17-AAG, has completed phase I clinical trial, and more than 20 phase II trials of this agent are either planned or in progress. Phase I testing of a related Hsp90 inhibitor, 17-DMAG, is currently underway. Hsp90 inhibitors are unique in that, although they are directed toward a specific molecular target, they simultaneously inhibit multiple signaling pathways that frequently interact to promote cancer cell survival. Further, by inhibiting nodal points in multiple overlapping survival pathways utilized by cancer cells, combination of an Hsp90 inhibitor with standard chemotherapeutic agents may dramatically increase the in vivo efficacy of the standard agent. Hsp90 inhibitors may circumvent the characteristic genetic plasticity that has allowed cancer cells to eventually evade the toxic effects of most molecularly targeted agents. The mechanism-based use of Hsp90 inhibitors, both alone and in combination with other drugs, should be effective toward multiple forms of cancer. Further, because Hsp90 inhibitors also induce Hsf-1-dependent expression of Hsp70, and because certain mutated Hsp90 client proteins are neurotoxic, these drugs display ameliorative properties in several neurodegenerative disease models, suggesting a novel role for Hsp90 inhibitors in treating multiple pathologies involving neurodegeneration.

2. Introduction

Cancer is a disease of genetic instability. Although only a few specific alterations seem to be required for generation of the malignant phenotype, at least in colon carcinoma there are approximately 10,000 estimated mutations at time of diagnosis (Hahn and Weinberg, 2002; Stoler et al., 1999). This genetic plasticity of cancer cells allows them to frequently escape the precise molecular targeting of a single signaling node or pathway, making them ultimately non-responsive to molecularly targeted therapeutics. Even GleevecTM (Novartis Pharmaceuticals Corp.), a well-recognized clinically active Bcr-Abl tyrosine kinase inhibitor, can eventually lose its effectiveness under intense, drug-dependent selective pressure, due to either mutation of the drug interaction site or expansion of a previously existing resistant clone (La Rosee et al., 2002). Most solid tumors at the time of detection are already sufficiently genetically diverse to resist single agent molecularly targeted therapy (Kitano, 2003). Thus, a simultaneous attack on multiple nodes of a cancer cell's web of overlapping signaling pathways should be more likely to affect survival than would inhibition of one or even a few individual signaling nodes. Given the number of key nodal proteins that are Hsp90 clients (see the website maintained by D. Picard, http://www.picard.ch/DP/downloads/Hsp90interactors.pdf), inhibition of Hsp90 may serve the purpose of collapsing, or significantly weakening, a cancer cell's safety net. Indeed, following a hypothesis first proposed by Hanahan and Weinberg several years ago (Hanahan and Weinberg, 2000), genetic instability allows a cell eventually to acquire six capabilities that are characteristic of most if not all cancers. These are (1) self-sufficiency in growth signaling; (2) insensitivity to anti-growth signaling; (3) ability to evade apoptosis; (4) sustained angiogenesis; (5) tissue invasion and metastasis; and (6) limitless replicative potential. As is highlighted in Figure 1, Hsp90 plays a pivotal role in acquisition and maintenance of each of these capabilities. Several excellent reviews provide an in depth description of the many signaling nodes regulated by Hsp90 (Bagatell and Whitesell, 2004;

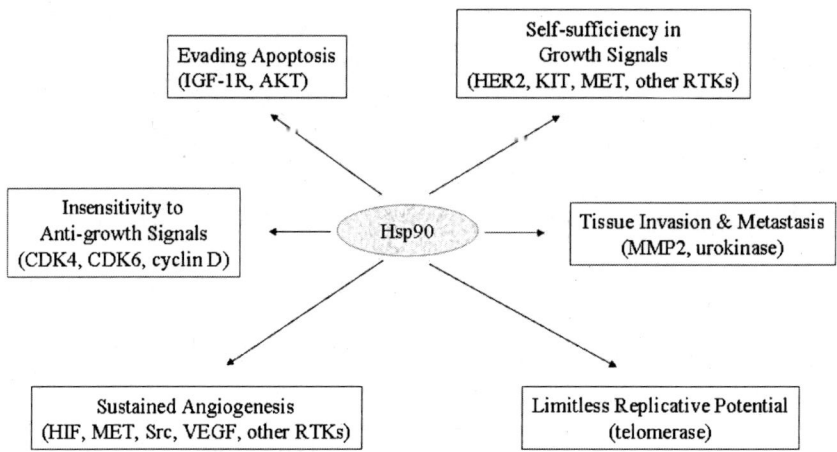

FIGURE 1. Hsp90 function is implicated in establishment of each of the hallmarks of cancer as first proposed by Hanahan and Weinberg (Hanahan and Weinberg, 2000). Importantly, Hsp90 function may also permit the genetic instability on which acquisition of the six hallmarks depends.

Chiosis et al., 2004; Goetz et al., 2003; Isaacs et al., 2003; Isaacs et al., 2005; Workman, 2004; Zhang and Burrows, 2004).

Cancer cells survive in the face of frequently extreme environmental stress, such as hypoxia and acidosis, as well as in the face of the exogenously applied environmental stresses of chemotherapy or radiation. These stresses tend to generate free radicals that can cause significant physical damage to cellular proteins. Given the combined protective role of molecular chaperones toward damaged proteins and the dependence of multiple signal transduction pathways on Hsp90, it is therefore not surprising that molecular chaperones in general, and Hsp90 in particular, are highly expressed in most tumor cells. However, Hsp90 may be elevated in tumor cells and may provide a unique molecular target therein for an additional reason. Using *Drosophila* and *Arabidopsis* as model systems, Lindquist and colleagues have shown that an ancient function of Hsp90 may be to permit accumulation at the protein level of inherent genetic mutations, and thus the chaperone may play a pivotal role in the evolutionary process itself (Queitsch et al., 2002; Rutherford and Lindquist, 1998). Extrapolating this hypothesis to genetically unstable cancer cells, it is not a great leap to think that Hsp90 may be critical to their ability to survive in the presence of an aberrantly high mutation rate.

The benzoquinoid ansamycin antibiotics, first isolated from the actinomycete, Streptomyces hygroscopicus var. geldanus var. nova (DeBoer et al., 1970) include geldanamycin (GA) and its semi-synthetic derivatives, 17-allylamino-17-demethoxygeldanamycin (17-AAG) and the more water-soluble 17-dimethylaminoethylamino-17-demethoxygeldanamycin (17-DMAG) (see Fig. 2). These small molecules inhibit the chaperone function of the heat shock protein Hsp90

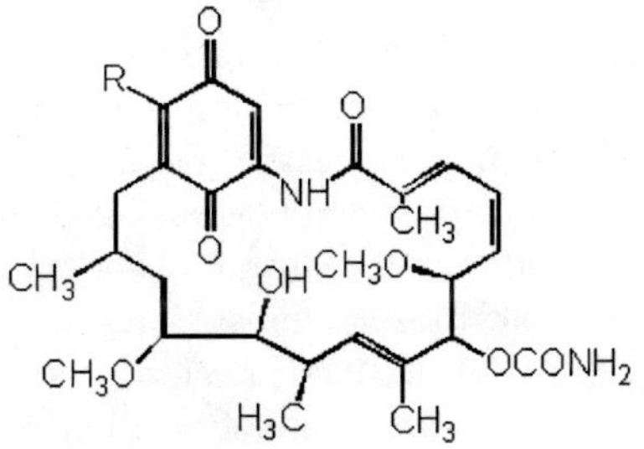

Compound	R Group
17-Allylaminogeldanamycin (17-AAG)	$CH_2=CHCH_2NH-$
17-Aminogeldanamycin (17-AG)	NH_2
17-Dimethylaminoethylamino-17-demethoxygeldanamycin (17-DMAG)	$(CH_3)_2NCH_2CH_2NH-$
Geldanamycin	CH_3O-

FIGURE 2. The chemical structures of geldanamycin, 17-AAG, its biologically active metabolite 17-AG, and 17-DMAG, highlighting the unique substitutions to the quinone moiety of the pharmacophore that characterize each molecule.

(Schulte and Neckers, 1998) and are currently being evaluated in phase 1 and 2 clinical trials. The parent compound, GA, is broadly cytotoxic in the NCI 60-cell line screen (Supko et al., 1995); its poor solubility and unacceptable liver toxicity in dogs precluded testing in humans. Because 17-AAG is less toxic than GA in rats and caused growth inhibition in breast (Paine-Murrieta et al., 1999), melanoma (Burger et al., 1998), and ovarian mouse xenograft models, the National Cancer Institute (NCI) initiated phase 1 trials in 1999.

3. Hsp90: A Chaperone of Oncogenes

Several recent, excellently detailed reviews of the mechanics of Hsp90 function are in the scientific literature (Bagatell and Whitesell, 2004; Chiosis, 2004; 2006; Prodromou and Pearl, 2003; Siligardi et al., 2004; Wegele et al., 2004; Zhang and Burrows, 2004). For the purposes of the current update on Hsp90-directed therapeutics, suffice it to say that Hsp90 is a conformationally flexible protein that associates with a distinct set of co-chaperones in dependence on nucleotide (ATP or ADP) occupancy of an amino-terminal binding pocket in Hsp90. Nucleotide exchange and ATP hydrolysis (by Hsp90 itself, with the assistance of co-chaperones) drive the so-called Hsp90 chaperone machine to bind, chaperone, and release client proteins. Indeed, identification of the GA binding site as a nucleotide pocket favoring purines led Chiosis and colleagues to design a series of highly potent purine scaffold Hsp90 inhibitors with markedly improved drug-like properties (Chiosis et al., 2002; 2003; He et al., 2006; Llauger et al., 2005). Workman and colleagues used a high-throughput screen based on inhibition of Hsp90 ATPase activity to identify 3,4-diarylpyrazoles as a novel class of Hsp90 inhibitors (Cheung et al., 2005; Dymock et al., 2005). By employing biochemical evaluation and crystallography, these investigators found that pyrazole inhibitors of Hsp90 provide a platform for extensive derivatization and provide an attractive starting point for hit to lead exploration.

The Hsp90 inhibitors currently in clinical trial (17-AAG and 17-DMAG), as well as those under development, all share the property of displacing nucleotide from the amino terminal pocket in Hsp90, and therefore short-circuiting the Hsp90 chaperone machine, much as one would stop the rotation of a bicycle wheel by inserting a stick between the spokes. Cycling of the chaperone machine is critical to its function. The Hsp90 inhibitors, by preventing nucleotide-dependent cycling, interfere with the chaperone activity of Hsp90, resulting in targeting of client proteins to the proteasome, the cell's garbage disposal, where they are degraded (Neckers, 2002). Even if the proteasome is inhibited, client proteins are not rescued from Hsp90 inhibition, but instead accumulate in a misfolded, inactive form in detergent-insoluble subcellular complexes (An et al., 2000).

3.1. Hsp90 Inhibitors Target Mutated and Chimeric Proteins Uniquely Expressed in Certain Cancers

Hsp90 characteristically chaperones a number of mutated or chimeric kinases that are key mediators of disease. Thus, anaplastic large cell lymphomas are characterized by expression of the chimeric protein NPM-ALK, which originates from a fusion of the nucleophosmin (NPM) and the membrane receptor anaplastic lymphoma kinase (ALK) genes. The chimeric kinase is constitutively active and capable of causing malignant transformation (Fujimoto et al., 1996). Bonvini and colleagues have shown that NPM-ALK kinase is an Hsp90 client protein, and that GA and 17-AAG destabilize the kinase and promote its proteasome-mediated degradation in several anaplastic large cell lymphoma cell lines (Bonvini et al., 2002).

FLT3 is a receptor tyrosine kinase that regulates proliferation, differentiation and survival of hematopoietic cells. FLT3 is frequently expressed in acute myeloid leukemia, and in 20 percent of patients with this cancer the tumor cells express a FLT3 protein harboring an internal tandem duplication in the juxtamembrane domain. This mutation is correlated with leukocytosis and a poor prognosis (Naoe et al., 2001). Minami and colleagues have reported that Hsp90 inhibitors cause selective apoptosis of leukemia cells expressing tandemly dublicated FLT3. Further, these investigators reported that mutated FLT3 was an Hsp90 client protein and that brief treatment with multiple Hsp90 inhibitors resulted in the rapid dissociation of Hsp90 from the kinase, accompanied by the rapid loss of kinase activity together with loss of activity of several downstream FLT3 targets including MAP kinase, Akt, and Stat5a (Minami et al., 2002). Minami et al. propose that Hsp90 inhibitors should be considered as promising compounds for the treatment of acute myeloid leukemia characterized by tandemly duplicated FLT3 expression.

BCR-ABL (p210Bcr-Abl) is an Hsp90 client protein that is also effectively inhibited by the novel tyrosine kinase inhibitor imatinib (An et al., 2000; Druker et al., 1996; Shiotsu et al., 2000). While imatinib has proven very effective in initial treatment of patients with chronic myelogenous leukemia, a majority of patients who are treated when their disease is in blast crisis stage (e.g., advanced) eventually relapse despite continued therapy (Sawyers et al., 2002). Relapse is correlated with loss of BCR-ABL inhibition by imatinib, due either to gene amplification or to specific point mutations in the kinase domain that preclude association of imatinib with the kinase (Shah et al., 2002). Gorre and colleagues have reported the very exciting finding that BCR-ABL protein that was resistant to imatinib remained dependent on Hsp90 chaperoning activity and thus retained sensitivity to Hsp90 inhibitors, including GA and 17-AAG. Both compounds induced the degradation of "wild-type" and mutant BCR-ABL, with a trend indicating more potent activity toward mutated imatinib-resistant forms of the kinase (Gorre et al., 2002). These findings were recently confirmed by other investigators (Nimmanapalli et al., 2002), thus providing a rationale for the use of 17-AAG in treatment of imatinib-resistant chronic myelogenous leukemia.

Mutations in the proto-oncogene c-kit cause constitutive kinase activity of its product, KIT protein, and are associated with human mastocytosis and gastrointestinal stromal tumors (GIST). Although currently available tyrosine kinase inhibitors are effective in the treatment of GIST, there has been limited success in the treatment of mastocytosis. Treatment with 17-AAG of the mast cell line HMC-1.2, harboring the Asp816Val and Val560Gly KIT mutations and the cell line HMC-1.1, harboring a single Val560Gly mutation, causes both the level and activity of KIT and downstream signaling molecules AKT and STAT3 to be down-regulated following drug exposure (Fumo et al., 2004). These data were validated using Cos-7 cells transfected with wild type and mutated KIT. 17-AAG promotes cell death of both HMC mast cell lines. In addition, neoplastic mast cells isolated from patients with mastocytosis and incubated with 17-AAG ex vivo are selectively sensitive to Hsp90 inhibition as compared to the mononuclear fraction as a whole. These data provide compelling evidence that 17-AAG may be effective in the treatment of c-kit related diseases including mastocytosis, GIST, mast cell leukemia, sub-types of acute myelogenous leukemia and testicular cancer.

More recently, several groups have reported that mutated B-Raf and mutated epidermal growth factor receptor (EGFR) develop strong dependence on Hsp90 and thus acquire marked sensitivity to Hsp90 inhibitors (da Rocha Dias et al., 2005; Grbovic et al., 2006; Shimamura et al., 2005). Since B-Raf is mutated in approximately 60% of melanomas and to a lesser degree in other cancers (Rajagopalan et al., 2002), and since cells expressing mutated B-Raf appear to be dependent on its activity for their survival, Hsp90 inhibitors may have wide applicability in melanoma. Indeed, results of a recent clinical trial support this hypothesis (Banerji et al., 2005). Similarly, the EGFR mutations described in a small percentage of non-small-cell lung cancer patients also confer Hsp90 dependence and sensitivity to Hsp90 inhibitors (Shimamura et al., 2005). While these patients initially respond to EGFR inhibitor therapy, they almost invariably become refractory with time. However, even tumors refractory to EGFR inhibitors remain very sensitive to Hsp90 inhibitors, suggesting that Hsp90 inhibitor therapy may be an efficacious second-line therapy in these patients (Shimamura et al., 2005).

3.2. Hsp90 Inhibitors Target the Androgen Receptor in Prostate Cancer

Androgen receptor continues to be expressed in the majority of hormone-independent prostate cancers, suggesting that it remains important for tumor growth and survival. Receptor overexpression, mutation, and/or post-translational modification may all be mechanisms by which androgen receptor can remain responsive either to low levels of circulating androgen or to anti-androgens. Vanaja et al. have shown that Hsp90 association is essential for the function and stability of the androgen receptor in prostate cancer cells (Vanaja et al., 2002). These investigators reported that androgen receptor levels in LNCaP cells were markedly reduced by GA, as was the ability of the receptor to become transcriptionally active in the presence of synthetic androgen. In addition, Georget et al. have shown that GA preferentially destabilized androgen receptor bound to anti-androgen, thus suggesting that the clinical efficacy of anti-androgens may be enhanced by combination with an Hsp90 inhibitor (Georget et al., 2002). These investigators also reported that GA prevented the nuclear translocation of ligand-bound androgen receptor, and inhibited the transcriptional activity of nuclear-targeted receptors, implicating Hsp90 in multiple facets of androgen receptor activity. Finally,

Solit and colleagues have reported that 17-AAG caused degradation of both wild-type and mutant androgen receptors and inhibited both androgen-dependent and androgen-independent prostate tumor growth in nude mice (Solit et al., 2002). Importantly, these investigators also demonstrated the loss of Her2 and Akt proteins, two Hsp90 clients that are upstream post-translational activators of the androgen receptor, in the tumor xenografts taken from 17-AAG-treated animals.

3.3. Hsp90 inhibitors exert anti-angiogenic activity by promoting oxygen- and VHL-independent inactivation and degradation of hypoxia inducible factor 1 α (HIF-1α) leading to inhibition of VEGF expression

Hypoxia-inducible factor-1α (HIF-1α) is a nuclear transcription factor involved in the transactivation of numerous target genes, many of which are implicated in the promotion of angiogenesis and adaptation to hypoxia (for a review, see Harris, 2002). Although these proteins are normally labile and expressed at low levels in normoxic cells, their stability and activation increase severalfold in hypoxia. The molecular basis for the instability of these proteins in normoxia depends upon VHL, the substrate recognition component of an E3 ubiquitin ligase complex that targets HIF-1α for proteasome-dependent degradation (Maxwell et al., 1999). Hypoxia normally impairs VHL function, thus allowing HIF to accumulate. HIF-1α expression has been documented in diverse epithelial cancers and most certainly supports survival in the oxygen-depleted environment inhabited by most solid tumors.

VHL can also be directly inactivated by mutation or hyper-methylation, resulting in constitutive overexpression of HIF in normoxic cells. In hereditary von Hippel-Lindau disease there is a genetic loss of VHL, and affected individuals are predisposed to an increased risk of developing highly vascular tumors in a number of organs. This is due, in large part, to deregulated HIF expression and the corresponding up-regulation of the HIF target gene vascular endothelial growth factor (VEGF). A common manifestation of VHL disease is the development of clear cell renal cell carcinoma (CC-RCC) (Seizinger et al., 1988). VHL inactivation also occurs in nonhereditary, sporadic CC-RCC.

HIF-1α interacts with Hsp90 (Gradin et al., 1996), and both GA and another Hsp90 inhibitor, radicicol, reduce HIF-dependent transcriptional activity (Hur et al., 2002; Isaacs et al., 2002). Hur et al. demonstrated that HIF protein from radicicol-treated cells was unable to bind DNA, suggesting that Hsp90 is necessary for mediating the proper conformation of HIF and/or recruiting additional cofactors. Likewise, Isaacs et al. reported GA-dependent, transcriptional inhibition of VEGF. Additionally, GA down-regulated HIF-1α protein expression by stimulating VHL-independent HIF-1α proteasomal degradation (Isaacs et al., 2002; Mabjeesh et al., 2002).

HIF-1α induction and VEGF expression has been associated with migration of glioblastoma cells in vitro and metastasis of glioblastoma in vivo. Zagzag et al., in agreement with the findings described above, have reported that GA blocks HIF-1α induction and VEGF expression in glioblastoma cell lines (Zagzag et al., 2003). Further, these investigators have shown that GA blocks glioblastoma cell migration, using an in vitro assay at non-toxic concentrations. This effect on tumor cell motility was independent of p53 and PTEN status, which makes Hsp90 inhibition an attractive modality in glioblastoma, where mutations in p53 and PTEN genes are common and where tumor invasiveness is a major therapeutic challenge.

Dias et al. have recently reported that VEGF promotes elevated Bcl2 protein levels and inhibits activity of the pro-apoptotic caspase-activating protein Apaf in normal endothelial cells and in leukemia cells bearing receptors for VEGF (Dias et al., 2002). Intriguingly, these investigators show that both phenomena require VEGF-stimulated Hsp90 association (e.g., with Bcl2 and Apaf), and that GA reverses both processes. Thus, GA blocked the pro-survival effects of VEGF by both preventing accumulation of anti-apoptotic Bcl2 and blocking the inhibition of pro-apoptotic Apaf.

3.4. Hsp90 Inhibitors Target MET and RET Receptor Tyrosine Kinases

The Met receptor tyrosine kinase is frequently overexpressed in cancer and is involved in angiogenesis, as well as in the survival and invasive ability of cancer cells. A recent report by Maulik et al. has demonstrated a role for Met in migration and survival of small cell lung cancer (Maulik et al., 2002). Met is an Hsp90 client protein, and these investigators went on to show that GA antagonized Met activity, reduced the Met protein level, and promoted apoptosis in several small cell lung cancer cell lines, even in the presence of excess Met ligand.

Hypoxia potentiates the invasive and metastatic potential of tumor cells. In an important recent study, Pennacchietti and colleagues reported that hypoxia (via two HIF-1α response elements) transcriptionally activated the Met gene, and synergized with Met ligand in promoting tumor invasion. Further, they showed that the pro-invasive effects of hypoxia were mimicked by Met overexpression, and that inhibition of Met expression prevented hypoxia-induced tumor invasion (Pennacchietti et al., 2003). Coupled with an earlier report describing induction of HIF-1 transcriptional activity by Met ligand (Tacchini et al., 2001), these data identify the HIF-VEGF-Met axis as a critical target for intervention using Hsp90 inhibitors, either alone or in conjunction with other inhibitors of angiogenesis. As Bottaro and Liotta recently pointed out (Bottaro and Liotta, 2003), the sole use of angiogenesis inhibitors to deprive tumors of oxygen might produce an unexpectedly aggressive phenotype in those cells that survived the treatment. These authors speculated that combination of Met inhibitors with anti-angiogenesis agents should therefore be beneficial. We would suggest that combination of an anti-angiogenesis drug with an Hsp90 inhibitor would not only potentiate the anti-tumor effects obtained by inhibiting angiogenesis, but would also break the HIF-Met axis by simultaneously targeting both Hsp90-dependent signaling proteins.

Mutation of a related receptor tyrosine kinase, RET, is associated with human cancer and several human neuroendocrine diseases. Point mutations of RET are responsible for multiple endocrine neoplasia type 2 (MEN2A, MEN2B, and familial medullary thyroid carcinoma [FMTC]). Somatic gene rearrangements juxtaposing the TK domain of RET to heterologous gene partners are found in papillary carcinomas of the thyroid (PTC) (Ichihara et al., 2004; Jhiang, 2000; Santoro et al., 2002).

Possible effects of 17-AAG on RET activity and cell growth of the TT MTC cell line have been examined (Cohen et al., 2002). Following treatment with 17-AAG, RET tyrosine kinase activity was inhibited by nearly 80%, as was the rate of cell growth. Thus, 17-AAG should be considered as an attractive pharmacologic agent for use as systemic therapy in patients with recurrent metastatic MTC for which non-surgical therapy has been ineffective.

3.5. Combined Inhibition of Hsp90 and the Proteasome Disrupt the Endoplasmic Reticulum and Demonstrate Enhanced Toxicity Toward Cancer Cells

Proteasome-mediated degradation is the common fate of Hsp90 client proteins in cells treated with Hsp90 inhibitors (Mimnaugh et al., 1996; Schneider et al., 1996). Proteasome inhibition does not protect Hsp90 clients in the face of chaperone inhibition—instead client proteins become insoluble (An et al., 2000; Basso et al., 2002). Since the deposition of insoluble proteins can be toxic to cells (French et al., 2001; Waelter et al., 2001), interest has arisen in combining proteasome inhibition with inhibition of Hsp90, the idea being that dual treatment will lead to enhanced accumulation of insoluble proteins and trigger apoptosis. This hypothesis is particularly appealing since a small molecule proteasome inhibitor has demonstrated efficacy in early clinical trials (Aghajanian et al., 2002; L'Allemain, 2002). Initial experimental support for such an hypothesis was provided by (Mitsiades et al., 2002) who reported that Hsp90 inhibitors enhanced multiple myeloma cell sensitivity to proteasome inhibition. Importantly, transformed cells are more sensitive to the cytotoxic effects of this drug combination than are non-transformed cells. Thus, 3T3 fibroblasts are fully resistant to combined administration of 17-AAG and Velcade TM at concentrations that prove cytotoxic to 3T3 cells transformed by HPV16 virus-encoding viral proteins E6 and E7 (Mimnaugh et al., 2004). In the same study, Mimnaugh et al. demonstrated that the endoplasmic reticulum is one of the main targets of this drug combination. In the presence of combined doses of both agents that show synergistic cytotoxicity, these investigators noted a nearly complete disruption of the architecture of the endoplasmic reticulum. Since all secreted and transmembrane proteins must pass through this organelle on their route to the extracellular space, it is not surprising that a highly secretory cancer such as multiple myeloma would be particularly sensitive to combined inhibition of Hsp90 and the proteasome. One might speculate that other highly secretory cancers, including hepatocellular carcinoma and pancreatic carcinoma, would also respond favorably to this drug combination.

3.6. Hsp90 Inhibitors Sensitize Cancer Cells to Radiation

Gius and colleagues have reported that 17-AAG potentiates both the in vitro and in vivo radiation response of cervical carcinoma cells (Bisht et al., 2003). An enhanced radiation response was noted when cells were exposed to radiation within 6 to 48 hours after drug treatment. Importantly, at 17-AAG concentrations that were themselves non-toxic, Hsp90 inhibition enhanced cell kill in response to an otherwise ineffective radiation exposure (2 Gy) by more than one log. Even at moderately effective levels of radiation exposure (4–6 Gy), addition of non-toxic amounts of 17-AAG enhanced cell kill by more than one log. Importantly, the sensitizing effects of 17-AAG observed in the cervical carcinoma cells were not seen in 3T3 cells, but were observed in HPV16-E6 and -E7 transformed 3T3 cells. The authors demonstrated convincingly that the effect of 17-AAG was multifactorial, since several pro-survival Hsp90 client proteins were rapidly down-regulated upon drug treatment. In vitro findings were confirmed by a murine xenograft study in which the anti-tumor activity of both single and fractionated radiation exposure was dramatically enhanced by treatment with 17-AAG, either 16 hours prior to single radiation exposure or on days 1 and 4 of a 6-day period during which the animals received fractionated radiation exposure. Machida and colleagues (2003) reported similar findings for lung carcinoma and colon adenocarcinoma cells in vitro. Thus, 17-AAG has been validated as a potential therapeutic agent that can be used at clinically relevant doses to enhance cancer cell sensitivity to radiation. It is reasonable to expect that other Hsp90 inhibitors will have a similar utility.

3.7. Targeting Hsp90 on the Cancer Cell Surface

Recently, Becker and colleagues reported that Hsp90 expression is dramatically up-regulated in malignant melanoma cells as compared to benign melanocytic lesions, and that Hsp90 is expressed on the surface of 7 out of 8 melanoma metastases (Becker et al., 2004). Eustace et al. have identified cell surface Hsp90 to be crucial for the invasiveness of HT-1080 fibrosarcoma cells in vitro (Eustace and Jay, 2004; Eustace et al., 2004). Taken together, these data implicate Hsp90 as an important determinant of tumor cell invasion and metastasis. Indeed, in the Eustace et al. study, the investigators demonstrated that GA covalently affixed to cell impermeable beads was able to significantly impair cell invasion across a Matrigel-coated membrane. These findings have been confirmed using a polar (and thus cell impermeable) derivative of 17-DMAG in place of GA-beads (Neckers et al., unpublished observations). Coincident with its inhibitory effects on cell invasiveness, cell impermeable GA also antagonized the maturation, via proteolytic self-processing, of the metalloproteinase MMP2, a cell-surface enzyme whose activity has been previously demonstrated as essential to cell invasion. Further, these investigators demonstrated that Hsp90 could be found in association with MMP2 in the culture medium bathing the HT-1080 cells. It is intriguing to speculate that association with Hsp90 on the cell surface is necessary for the self-proteolysis of MMP2. Thus, a possible chaperone function for cell surface Hsp90 may be directly implicated in tumor cell invasiveness and metastasis. As such, cell surface Hsp90 may represent a novel, perhaps cancer-specific target for cell-impermeant Hsp90 inhibitors.

4. Metabolism of 17-AAG and 17-DMAG In Vivo

In human or murine hepatic microsome assays, 17-aminogeldanamycin (17-AG), a diol, and an epoxide are the three major metabolites of 17-AAG (Egorin et al., 1998). The 17-AAG diol was the major metabolite in human hepatic microsomes, followed by 17-AG; in contrast, 17-AG was the most abundant metabolite in murine microsomes. Acrolein, a nephrotoxin, is a potential by-product of the 17-AG metabolite. Finally, the epoxide is probably formed by addition of oxygen across the double bond of the allylamino side chain. CYP3A4 enzymatic metabolism is responsible for 17-AG and epoxide formation. Microsomal epoxide hydrolase catalyzes the conversion of the diol to 17-AG, which does not undergo further microsomal metabolism. 17-AAG metabolites are active and may have clinical significance. The biologically active epoxides and acrolein may induce toxic effects in humans (Egorin et al., 1998). Pharmacodynamic studies show that the 17-AG metabolite (see Fig. 2) is as active as 17-AAG in decreasing cellular p185erbB2 in human breast cancer SKBr3 cells in culture (Schnur et al., 1995). 17-AG causes growth-inhibition in six human colon cancer lines and three ovarian cancer cell lines (Kelland et al., 1999).

In contrast to 17-AAG, 17-DMAG appears to be only minimally metabolized by CYP3A4 (Egorin et al., 2002). Therefore, intestinal CYP3A4 should not impede 17-DMAG's oral activity. 17-AG does not appear to be a metabolite of 17-DMAG based on the lack of conversion at the 17 position of the compound. The marked metabolic differences between 17-AAG and 17-DMAG suggest that they may have distinct toxicity profiles and therapeutic indices.

5. Why Are Tumor Cells Uniquely Sensitive to Hsp90 Inhibition?

It is apparent, from both preclinical and clinical observations, that Hsp90 inhibitors can be administered in vivo at doses and schedules that significantly impact tumor growth but with minimal target related toxicity to normal tissues. This is the case for several small molecule inhibitors, including 17-AAG and 17-DMAG, the synthetic purine mimetic PU24FCl, and it even applies to a novel peptidomimetic inhibitor of the N-terminal Hsp90 nucleotide binding site, shepherdin (Banerji et al., 2005; Eiseman et al., 2005; Plescia et al., 2005; Vilenchik et al., 2004; Xu et al., 2003). Since Hsp90 is highly expressed in most, if not all, normal tissues, these findings require an explanation. Indeed, when murine model systems are examined in vivo, Hsp90 inhibitors are found to concentrate in tumor tissue, while being rapidly cleared from normal tissue with a half-life similar to that of drug in plasama (Banerji et al., 2005; Eiseman et al., 2005; Vilenchik et al., 2004; Xu et al., 2003). The Hsp90 inhibitor 17-AAG also has been reported to actively concentrate in tumor cells in vitro (Chiosis et al., 2003).

Since preferential accumulation of these Hsp90 inhibitors in tumor vs. normal tissue may provide the observed therapeutic (or at least biologic) index, it is important to understand the reason for this phenomenon. A possible explanation put forth by Kamal and colleagues suggests that enhanced drug binding to tumor cell Hsp90 reflects the activity state of the Hsp90 chaperone machine in tumor vs. normal cells (Kamal et al., 2003). They proposed that enhanced the ATPase activity of the chaperone in tumor cells, which is dependent on preferential recruitment of Hsp90 to a multi-component chaperone complex, is responsible for the increased affinity of Hsp90 inhibitors in tumor cells.

Others have reported that expression of NAD(P)H:Quinone Oxidoreductase I (NQO1), also known as DT-diaphorase, dramatically enhances cellular sensitivity to 17-AAG (Banerji et al., 2005; Kelland et al., 1999). NQO1 generates the hydroquinone version of 17-AAG, which has recently been reported to bind more tightly to Hsp90 when compared to 17-AAG itself (Guo et al., 2005). Further, the presence of NQO1 in a cell seems also to lead to increased total accumulation of intracellular ansamycin molecules, presumably reflecting the increased water solubility of the 17-AAG dihydroquinone and its decreased propensity to cross membranes. Thus, by this model NQO1 serves to trap 17-AAG in cells while simultaneously enhancing its Hsp90 binding affinity. Intriguingly, these investigators and others have shown that the presence of NQO1 in tumor cells dramatically affects cellular sensitivity to 17-AAG (Banerji et al., 2005; Guo et al., 2005; Kelland et al., 1999). Since high levels of NQO1 have been observed in diverse tumor types (e.g., liver, lung, colon, breast) as compared to normal tissues of the same origin (Belinsky and Jaiswal, 1993), these data suggest an explanation for the disparate sensitivity of tumor and normal tissue to 17-AAG. However, the similar preference of other Hsp90 inhibitors, such as the synthetic purine analog PU24FCl and the peptidomimetic shepherdin, for tumor cells remains to be explained. Several groups are currently examining altered states of post-translational modification of Hsp90 in tumor vs. normal cells as a possible contributing factor to this phenomenon.

6. Clinical Trial Data

A phase 1 Institute of Cancer Research (UK) phase 1 trial of 17-AAG in malignant melanoma used a once weekly administration schedule. The starting dose was 10 mg/m^2/week administered IV once weekly in a cohort of three patients. Doses were doubled in each succeeding cohort (Banerji et al., 2001). Adverse events included grade 1/2 nausea and grade 1/2 fatigue in 3 and 9 of the first 15 patients, respectively. One patient experienced grade 3 vomiting at the 80-mg/m^2/week dose. Grade 3 nausea and vomiting occurred in two of six patients treated at the 320-mg/m^2/week dose, following which the dose was escalated by 40% to 450 mg/m^2/week (Banerji et al., 2002). A total of 28 patients have been treated to date on this trial. Among the six patients treated at the 320–450-mg/m^2/week dose range, two patients showed stable disease for 27 and 91 weeks, respectively.

Pharmacodynamic marker analysis of tumor biopsies done before and 24 hours after treatment in nine patients showed depletion of the Hsp90 client c-Raf in four of seven samples (where the marker was expressed), and cdk4 (Hsp90 client) depletion and Hsp70 induction in eight of the nine samples (Banerji et al., 2003). At the highest dose level, pharmacokinetic analysis indicated a t^1/$_2$ of 5.8 ± 1.9 h, Vdss of 274 ± 108 L, clearance of 35.5 ± 16.6 L/h, and Cmax of 16.2 ± 6.3 μM (Banerji et al., 2003), which is above the levels of 375 nM to10 μM reported to inhibit Hsp90 in vitro (Burger et al., 2000). Although a maximal tolerated dose was not established in this trial, the dose/schedule that will be taken forward to phase II evaluation is likely to be 450 mg/m^2/week, as there was evidence of tumor target inhibition at that dose level (Banerji et al., 2003). Updated results of this phase I trial have recently been published (Banerji et al., 2005).

Hsp90 inhibitors are a class of agents that affect a diverse group of client proteins involved in oncogenesis. Many of these clients are expressed in a disease-specific fashion. The development of these inhibitors as biomodulators is complex and not necessarily governed by standard approaches. The clinical approach taken with the Hsp90 inhibitors was to proceed simultaneously with single agent phase 2 studies as well as disease-specific combinations that would be used to evaluate the biomodulatory effects of 17-AAG and 17-DMAG. Results from the ongoing clinical trials outlined in the accompanying tables will be used to assess activity of the agents in a disease-specific fashion and to provide a response comparison for the phase 1 combinations to proceed into disease-specific phase 2 investigations. As these studies mature and reach completion, the role of Hsp90 inhibitors in the treatment of cancer should be better defined with regard to their activity and molecular targeted effects.

7. Hsp90 Inhibitors in Neurodegenerative Diseases

Unfolded or misfolded proteins have exposed hydrophobic segments that render them prone to aggregation. Protein aggregates are toxic to the cell (Taylor et al., 2002), and molecular chaperones, especially Hsp70, bind to hydrophobic surfaces of misfolded proteins to insure their continued solubility or to promote their degradation by the proteasome (Hershko and Ciechanover, 1998). Under pathologic conditions, the level of misfolded proteins may exceed the ability of the cell to either maintain them in a soluble form or to degrade them, allowing aggregation to proceed (Cohen, 1999; Zoghbi and Orr, 2000). Protein aggregates have been found in most chronic neurodegenerative diseases (Kakizuka, 1998; Taylor et al., 2002), as well as in global and focal ischemia and in hypoglycemic coma (Hu et al., 2000; Ouyang and Hu, 2001). Thus, pharmacologic induction of molecular chaperones in general, and Hsp70 in particular, may be ameliorative in these cases. The Hsp90 inhibitors currently in clinical trial, 17-AAG and 17-DMAG, have the property of inducing Hsp70 in normal cells and tissues, via disruption of Hsp90 sequestration of the heat shock transcription factor Hsf1 (Ali et al., 1998; Kim et al., 1999; Lu et al., 2002), and so they have become of interest in this regard.

Giffard et al. (2004) have reported that GA, an Hsp90 inhibitor structurally related to 17-AAG and 17-DMAG, via its ability to induce Hsp70, reduces protein aggregation in a rodent model of global ischemia and blocks apoptotic astrocyte death induced by glucose deprivation. These investigators showed that GA-treated astrocyte cultures were twice as viable as untreated cultures after 24 hours of glucose deprivation, and they make the point that, because Hsp70 can block both apoptotic and necrotic cell death, it is an intriguing target for anti-ischemic therapy.

The progressive loss of dopaminergic neurons in the substantia nigra is the defining pathogenic feature of Parkinson disease (PD). α-Synuclein is mutated in rare familial forms of PD and it is a major component of the pathologic protein aggregates characteristic of the disease (Kruger, 2004; Polymeropoulos et al., 1997; Tofaris and Spillantini, 2005). Expression of normal as well as mutant α-synuclein in *Drosophila melanogaster* causes selective loss of dopaminergic neurons (Feany and Bender, 2000), and this can be completely prevented by raising the level of Hsp70 by transgenic expression (Auluck et al., 2002). Thus, dopaminergic neurons may be sensitive to compromised chaperone levels. A recent study demonstrated that pharmacologic enhancement of Hsp70, via GA administration to adult *Drosophila* during a 3-week period, completely protected their dopaminergic neurons from a-synuclein-induced toxicity (Auluck and Bonini, 2002). Moreover, in contrast to the findings of a previous study, which treated developing flies in a similar fashion, prolonged exposure of adult flies to effective doses of GA caused no noticeable deleterious effects (Rutherford and Lindquist, 1998). Given the complete protection of dopaminergic neurons afforded by GA, the authors of this study propose that GA and its derivatives warrant a careful examination as cytoprotective agents for treating PD and other neurodegenerative diseases.

17-AAG has recently been shown to ameliorate polyglutamine-mediated motor neuron degeneration (Waza et al., 2005). Because mutated androgen receptor (AR) is a pathogenic gene product in spinal and bulbar muscular atrophy (SBMA), Waza and colleagues examined whether 17-AAG could potentiate degradation of the polyglutamine-expanded mutant AR. These investigators found that administration of 17-AAG markedly ameliorated motor impairments in the SBMA transgenic mouse model without detectable toxicity and reduced the amount of detectable monomeric and aggregated mutated AR protein. As expected, polyglutamine-expanded AR showed a higher affinity for Hsp90 as compared to wild-type AR, and it was preferentially degraded in the presence of 17-AAG in both cells and transgenic mice. 17-AAG also mildly induced Hsp70 in this model. These investigators suggest that 17-AAG thus provides a novel therapeutic approach to SBMA by promoting the degradation of a pathologic mutant Hsp90 client protein.

Finally, two groups have reported that, using both mouse and *Drosohpila* models of Huntington's disease (HD), pharmacologic induction of Hsp70 with Hsp90 inhibitors provides a useful therapeutic strategy. Hay et al. (2004) report that a progressive decrease in Hsp70 and other chaperones in brain tissue contributes to disease pathogenesis in the R6/2 mouse model of HD. Both radicicol and GA were able to maintain chaperone induction for at least three weeks and were able to improve the detergent-soluble properties of polyglutamine-containing aggregates over this time course. Meanwhile, Agrawal et al. (2005) have shown, using a *Drosophila* model of HD (flies transgenically express glutamine-expanded human huntingtin protein), that feeding affected flies GA alone or in combination with a histone deacetylase inhibitor (suberoylanilide hydroxamic acid, SAHA) strongly suppresses the degeneration of photoreceptor neurons while causing no overt toxicity in control flies. Intriguingly, we and others have recently shown that several classes of histone deacetylase inhibitors, including SAHA, have the unexpected property of inhibiting Hsp90 by promoting its hyperacetylation (Bali et al., 2005; Yu et al., 2002). Thus, in the *Drosophila* HD model the observed beneficial activity of each agent alone as well as the synergistic activity of their combination suggests that Hsp90 inhibition (and/or the resultant increase in Hsp70) is the primary mechanism of action of both drugs.

The apparent in vivo safety and efficacy of these benzoquinone ansamycin Hsp90 inhibitors in several models of neurodegeneration considerably extend the therapeutic application of these drugs (and perhaps other Hsp90 inhibitors) beyond oncology. Whether the primary mechanism is degradation of an Hsp90-dependent polyglutamine-expanded mutant protein, the pharmacologic induction of Hsp70, or a combination of the two processes, Hsp90 inhibitors have a promising future in the treatment of neurodegenerative pathologies.

References

Aghajanian, C., Soignet, S., Dizon, D. S., Pien, C. S., Adams, J., Elliott, P. J., Sabbatini, P., Miller, V., Hensley, M. L., Pezzulli, S., Canales, C., Daud, A. and Spriggs, D. R., (2002) A phase I trial of the novel proteasome inhibitor PS341 in advanced solid tumor malignancies. Clin Cancer Res 8:2505–11.

Agrawal, N., Pallos, J., Slepko, N., Apostol, B. L., Bodai, L., Chang, L.-W., Chiang, A.-S., Thompson, L. M. and Marsh, J. L. (2005) Identification of combinatorial drug regimens for treatment of Huntington's disease using Drosophila. Proc Natl Acad Sci USA 102:3777–81.

Ali, A., Bharadwaj, S., O'Carroll, R., and Ovsenek, N. (1998) HSP90 interacts with and regulates the activity of heat shock factor 1 in Xenopus oocytes. Mol Cell Biol 18:4949 60.

An, W. G., Schulte, T. W. and Neckers, L. M. (2000) The heat shock protein 90 antagonist geldanamycin alters chaperone association with p210bcr-abl and v-src proteins before their degradation by the proteasome. Cell Growth Differ 11:355–60.

Auluck, P. K., and Bonini, N. M. (2002) Pharmacological prevention of Parkinson disease in Drosophila. Nat Med 8:1185–6.

Auluck, P. K., Chan, H. Y., Trojanowski, J. Q., Lee, V. M.-Y., and Bonini, N. M. (2002) Chaperone suppression of alpha-synuclein toxicity in a Drosophila model for Parkinson's disease. Science 295:865–8.

Bagatell, R., and Whitesell, L. (2004) Altered Hsp90 function in cancer: A unique therapeutic opportunity. Mol Cancer Ther 3:1021–30.

Bali, P., Pranpat, M., Swaby, R., Fiskus, W., Yamaguchi, H., Balasis, M., Rocha, K., Wang, H. G., Richon, V. and Bhalla, K. (2005) Activity of suberoylanilide hydroxamic acid against human breast cancer cells with amplification of her-2. Clin Cancer Res 11: 6382–9.

Banerji, U., O'Donnell, A., Scurr, M., Benson, C., Hanwell, J., Clark, S., Raynaud, F., Turner, A., Walton, M., Workman, P. and Judson, I. (2001) Phase I trial of the heat shock protein 90 (HSP90) inhibitor 17-allylamino 17-demethoxygeldanamycin 17aag). Pharmacokinetic (PK) profile and pharmacodynamic (PD) endpoints. Proc Am Soc Clin Oncol 20:Abstract 326.

Banerji, U., O'Donnell, A., Scurr, M., Benson, C., Stapleton, S., Raynaud, F., Clarke, S., Turner, A., Workman, P. and Judson, I. (2003) A pharmacokinetically (PK)-pharmacodynamically (PD) guided phase I trial of the heat shock protein 90 (HSP90) inhibitor 17-allylamino,17-demethoxygeldanamycin (17AAG). Proc Am Soc Clin Oncol 22:Abstract 797.

Banerji, U., O'donnell, A., Scurr, M., Pacey, S., Stapleton, S., Asad, Y., Simmons, L., Maloney, A., Raynaud, F., Campbell, M., Walton, M., Lakhani, S., Kaye, S., Workman, P. and Judson, I. (2005) Phase I pharmacokinetic and pharmacodynamic study of 17-allylamino,17-demethoxygeldanamycin in patients with advanced malignancies. J Clin Oncol 23:4152–61.

Banerji, U., O'Donnell, A., Scurr, M., Benson, C., Brock, C., Hanwell, J., Stapleton, S., Raynaud, F., Simmons, L., Turner, A., Walton, M., Workman, P. and Judson, I. (2002) A pharmacokinetically (Pk)-pharmacodynamically (Pd) driven phase I trial of the Hsp90 molecular chaperone inhibitor 17-allyamino 17-demethoxygeldanamycin (17AAG). Proc 93rd Annu Meet Am Assoc Cancer Res 43:Abstract 1352.

Basso, A. D., Solit, D. B., Chiosis, G., Giri, B., Tsichlis, P. and Rosen, N. (2002) Akt forms an intracellular complex with heat shock protein 90 (Hsp90) and Cdc37 and is destabilized by inhibitors of Hsp90 function. J Biol Chem 277:39858–66.

Becker, B., Multhoff, G., Farkas, B., Wild, P. J., Landthaler, M., Stolz, W. and Vogt, T. (2004) Induction of Hsp90 protein expression in malignant melanomas and melanoma metastases. Exp Dermatol 13:27–32.

Belinsky, M., and Jaiswal, A. K. (1993) NAD(P)H:Quinone oxidoreductase1 (DT-diaphorase) expression in normal and tumor tissues. Cancer Metastasis Rev 12:103–17.

Bisht, K. S., Bradbury, C. M., Mattson, D., Kaushal, A., Sowers, A., Markovina, S., Ortiz, K. L., Sieck, L. K., Isaacs, J. S., Brechbiel, M. W., Mitchell, J. B., Neckers, L. M., and Gius, D. (2003) Geldanamycin and 17-allylamino-17-demethoxygeldanamycin potentiate the in vitro and in vivo radiation response of cervical tumor cells via the heat shock protein 90-mediated intracellular signaling and cytotoxicity. Cancer Res 63:8984–95.

Bonvini, P., Gastaldi, T., Falini, B. and Rosolen, A. (2002) Nucleophosmin-anaplastic lymphoma kinase (NPM-ALK), a novel Hsp90-client tyrosine kinase: down-regulation of NPM-ALK expression and tyrosine phosphorylation in ALK(+) CD30(+) lymphoma cells by the Hsp90 antagonist 17-allylamino,17-demethoxygeldanamycin. Cancer Res 62:1559–66.

Bottaro, D. P. and Liotta, L. A. (2003) Out of air is not out of action. Nature 423:593–5.

Burger, A. M., Fiebig, H. H., Newman, D. J., Camalier, R. F. and Sausville, E. A. (1998) Antitumor activity of 17-allylaminogeldanamycin (NSC 330507) in melanoma xenografts is associated with decline in Hsp90 protein expression. 10th NCI-EORTC Symposium on New Drugs in Cancer Therapy: Abstract 504.

Burger, A. M., Sausville, E. A., Carmalier, R. F., Newman, D. J. and Fiebig, H. H. (2000) Response of human melanomas to 17-AAG is associated with modulation of the molecular chaperone function of Hsp90. Proc Am Assoc Cancer Res 41: Abstract 2844.

Cheung, K. M., Matthews, T. P., James, K., Rowlands, M. G., Boxall, K. J., Sharp, S. Y., Maloney, A., Roe, S. M., Prodromou, C., Pearl, L. H., Aherne, G. W., McDonald, E. and Workman, P. (2005) The identification, synthesis, protein crystal structure and in vitro biochemical evaluation of a new 3,4-diarylpyrazole class of Hsp90 inhibitors. Bioorg Med Chem Lett 15:3338–43.

Chiosis, G. (2006) Targeting chaperones in transformed systems–a focus on Hsp90 and cancer. Expert Opin Ther Targets 10:37–50.

Chiosis, G., Huezo, H., Rosen, N., Mimnaugh, E., Whitesell, L. and Neckers, L. (2003) 17AAG: low target binding affinity and potent cell activity-finding an explanation. Mol Cancer Ther 2:123–9.

Chiosis, G., Lucas, B., Huezo, H., Solit, D., Basso, A. and Rosen, N. (2003) Development of purine-scaffold small molecule inhibitors of Hsp90. Curr Cancer Drug Targets 3:371–6.

Chiosis, G., Lucas, B., Shtil, A., Huezo, H. and Rosen, N. (2002) Development of a purine-scaffold novel class of Hsp90 binders that inhibit the proliferation of cancer cells and induce the degradation of Her2 tyrosine kinase. Bioorg Med Chem 10:3555–64.

Chiosis, G., Vilenchik, M., Kim, J. and Solit, D. (2004) Hsp90: The vulnerable chaperone. Drug Discov Today 9:881–8.

Cohen, F. E. (1999) Protein misfolding and prion diseases. J Mol Biol 293:313–20.

Cohen, M. S., Hussain, H. B., and Moley, J. F. (2002) Inhibition of medullary thyroid carcinoma cell proliferation and RET phosphorylation by tyrosine kinase inhibitors. Surgery 132:960-6;ee also discussion 966–7.

da Rocha Dias, S., Friedlos, F., Light, Y., Springer, C., Workman, P. and Marais, R., (2005) Activated B-RAF is an Hsp90 client protein that is targeted by the anticancer drug 17-allylamino-17-demethoxygeldanamycin. Cancer Res 65:10686–91.

DeBoer, C., Meulman, P. A., Wnuk, R. J., and Peterson, D. H. (1970) Geldanamycin, a new antibiotic. J Antibiot (Tokyo) 23:442–7.

Dias, S., Shmelkov, S. V., Lam, G., and Rafii, S. (2002) VEGF(165) promotes survival of leukemic cells by Hsp90-mediated induction of Bcl-2 expression and apoptosis inhibition. Blood 99:2532–40.

Druker, B. J., Tamura, S., Buchdunger, E., Ohno, S., Segal, G. M., Fanning, S., Zimmermann, J., and Lydon, N. B. (1996) Effects of a selective inhibitor of the Abl tyrosine kinase on the growth of Bcr–Abl positive cells. Nat Med 2:561–6.

Dymock, B. W., Barril, X., Brough, P. A., Cansfield, J. E., Massey, A., McDonald, E., Hubbard, R. E., Surgenor, A., Roughley, S. D., Webb, P., Workman, P., Wright, L., and Drysdale, M. J. (2005) Novel, potent small-molecule inhibitors of the molecular chaperone Hsp90 discovered through structure-based design. J Med Chem 48:4212 5.

Egorin, M. J., Lagattuta, T. F., Hamburger, D. R., Covey, J. M., White, K. D., Musser, S. M., and Eiseman, J. L. (2002) Pharmacokinetics, tissue distribution, and metabolism of 17-(dimethylaminoethylamino)-17-demethoxygeldanamycin (NSC 707545) in CD2F1 mice and Fischer 344 rats. Cancer Chemother Pharmacol 49:7–19.

Egorin, M. J., Rosen, D. M., Wolff, J. H., Callery, P. S., Musser, S. M., and Eiseman, J. L. (1998) Metabolism of 17-(allylamino)-17-demethoxygeldanamycin (NSC 330507) by murine and human hepatic preparations. Cancer Res 58:2385–96.

Eiseman, J. L., Lan, J., Lagattuta, T. F., Hamburger, D. R., Joseph, E., Covey, J. M., and Egorin, M. J. (2005) Pharmacokinetics and pharmacodynamics of 17-demethoxy 17-[[(2-dimethylamino)ethyl]amino]geldanamycin (17DMAG, NSC 707545) in C.B-17 SCID mice bearing MDA-MB-231 human breast cancer xenografts. Cancer Chemother Pharmacol 55:21–32.

Eustace, B. K., and Jay, D. G. (2004) Extracellular roles for the molecular chaperone, hsp90. Cell Cycle 3.

Eustace, B. K., Sakurai, T., Stewart, J. K., Yimlamai, D., Unger, C., Zehetmeier, C., Lain, B., Torella, C., Henning, S. W., Beste, G., Scroggins, B. T., Neckers, L., Ilag, L. L., and Jay, D. G. (2004) Functional proteomic screens reveal an essential extracellular role for hsp90 alpha in cancer cell invasiveness. Nat Cell Biol 6:507–14.

Feany, M. B., and Bender, W. W. (2000) A Drosophila model of Parkinson's disease. Nature 404:394–8.

French, B. A., van Leeuwen, F., Riley, N. E., Yuan, Q. X., Bardag-Gorce, F., Gaal, K., Lue, Y. H., Marceau, N. and French, S. W. (2001) Aggresome formation in liver cells in response to different toxic mechanisms: Role of the ubiquitin–proteasome pathway and the frameshift mutant of ubiquitin. Exp Mol Pathol 71:241–6.

Fujimoto, J., Shiota, M., Iwahara, T., Seki, N., Satoh, H., Mori, S., and Yamamoto, T. (1996) Characterization of the transforming activity of p80, a hyperphosphorylated protein in a Ki-1 lymphoma cell line with chromosomal translocation t(2;5). Proc Natl Acad Sci USA 93:4181–6.

Fumo, G., Akin, C., Metcalfe, D. D., and Neckers, L. (2004) 17-Allylamino-17-demethoxygeldanamycin (17-AAG) is effective in down-regulating mutated, constitutively activated KIT protein in human mast cells. Blood 103:1078–84.

Georget, V., Terouanne, B., Nicolas, J.-C., and Sultan, C. (2002) Mechanism of antiandrogen action: Key role of Hsp90 in conformational change and transcriptional activity of the androgen receptor. Biochemistry 41:11824–31.

Giffard, R. G., Xu, L., Zhao, H., Carrico, W., Ouyang, Y. B., Qiao, Y., Sapolsky, R., Steinberg, G., Hu, B. and Yenari, M. A. (2004) Chaperones, protein aggregation, and brain protection from hypoxic/ischemic injury. J Exp Biol 207:3213–20.

Goetz, M. P., Toft, D. O., Ames, M. M., and Erlichman, C. (2003) The Hsp90 chaperone complex as a novel target for cancer therapy. Ann Oncol 14:1169–76.

Gorre, M. E., Ellwood-Yen, K., Chiosis, G., Rosen, N. and Sawyers, C. L. (2002) BCR-ABL point mutants isolated from patients with STI571-resistant chronic myeloid leukemia remain sensitive to inhibitors of the BCR-ABL chaperone heat shock protein 90. Blood 100:3041–4.

Gradin, K., McGuire, J., Wenger, R. H., Kvietikova, I., fhitelaw, M. L., Toftgard, R., Tora, L., Gassmann, M., and Poellinger, L. (1996) Functional interference between hypoxia and dioxin signal transduction pathways: Competition for recruitment of the Arnt transcription factor. Mol Cell Biol 16:5221–31.

Grbovic, O. M., Basso, A., Sawai, A., Ye, Q., Friedlander, P., Solit, D. and Rosen, N. (2006) V600E B-Raf requires the Hsp90 chaperone for stability and is degraded in response to Hsp90 inhibitors. Proc Natl Acad Sci U S A 103:57–62.

Guo, W., Reigan, P., Siegel, D., Zirrolli, J., Gustafson, D., and Ross, D. (2005) Formation of 17-allylamino-demethoxygeldanamycin (17-AAG) hydroquinone by NAD(P)H:Quinone oxidoreductase 1: Role of 17-AAG hydroquinone in heat shock protein 90 inhibition. Cancer Res 65:10006–15.

Hahn, W. C., and Weinberg, R. A. (2002) Modelling the molecular circuitry of cancer. Nat Rev Cancer 2:331–41.

Hanahan, D., and Weinberg, R. A. (2000) The hallmarks of cancer. Cell 100:57–70.

Harris, A. L. (2002) Hypoxia—a key regulatory factor in tumor growth. Nat Rev Cancer 2:38–47.

Hay, D. G., Sathasivam, K., Tobaben, S., Stahl, B., Marber, M., Mestril, R., Mahal, A., Smith, D. L., Woodman, B., and Bates, G. P. (2004) Progressive decrease in chaperone protein levels in a mouse model of Huntington's disease and induction of stress proteins as a therapeutic approach. Hum Mol Genet 13:1389–405.

He, H., Zatorska, D., Kim, J., Aguirre, J., Llauger, L., She, Y., Wu, N., Immormino, R. M., Gewirth, D. T. and Chiosis, G. (2006) Identification of potent water soluble purine-scaffold inhibitors of the heat shock protein 90. J Med Chem 49:381–90.

Hershko, A., and Ciechanover, A. (1998) The ubiquitin system. Annu Rev Biochem 67:425–79.

Hu, B. R., Martone, M. E., Jones, Y. Z., and Liu, C. L. (2000) Protein aggregation after transient cerebral ischemia. J Neurosci 20:3191–9.

Hur, E., Kim, H. H., Choi, S. M., Kim, J. H., Yim, S., Kwon, H. J., Choi, Y., Kim, D. K., Lee, M. O., and Park, H. (2002) Reduction of hypoxia-induced transcription through the repression of hypoxia-inducible factor-1alpha/aryl hydrocarbon receptor nuclear translocator DNA binding by the 90-kDa heat-shock protein inhibitor radicicol. Mol Pharmacol 62:975–82.

Ichihara, M., Murakumo, Y., and Takahashi, M. (2004) RET and neuroendocrine tumors. Cancer Lett 204:197–211.

Isaacs, J. S. (2005) Heat-shock protein 90 inhibitors in antineoplastic therapy: Is it all wrapped up? Expert Opin Investig Drugs 14:569–89.

Isaacs, J. S., Jung, Y. J., Mimnaugh, E. G., Martinez, A., Cuttitta, F., and Neckers, L. M. (2002) Hsp90 regulates a von Hippel Lindau-independent hypoxia-inducible factor-1 alpha-degradative pathway. J Biol Chem 277:29936–44.

Isaacs, J. S., Xu, W., and Neckers, L. (2003) Heat shock protein 90 as a molecular target for cancer therapeutics. Cancer Cell 3:213–7.

Jhiang, S. M. (2000) The RET proto-oncogene in human cancers. Oncogene 19:5590–7.

Kakizuka, A. (1998) Protein precipitation: A common etiology in neurodegenerative disorders? Trends Genet 14:396–402.

Kamal, A., Thao, L., Sensintaffar, J., Zhang, L., Boehm, M. F., Fritz, L. C., and Burrows, F. J. (2003) A high-affinity conformation of Hsp90 confers tumour selectivity on Hsp90 inhibitors. Nature 425:407–10.

Kelland, L. R., Sharp, S. Y., Rogers, P. M., Myers, T. G., and Workman, P. (1999) DT-Diaphorase expression and tumor cell sensitivity to 17-allylamino, 17-demethoxygeldanamycin, an inhibitor of heat shock protein 90. J Natl Cancer Inst 91:1940–9.

Kim, H. R., Kang, H. S., and Kim, H. D. (1999) Geldanamycin induces heat shock protein expression through activation of HSF1 in K562 erythroleukemic cells. IUBMB Life 48:429–33.

Kitano, H. (2003) Cancer robustness: Tumour tactics. Nature 426:125.

Kruger, R. (2004) Genes in familial Parkinsonism and their role in sporadic Parkinson's disease. J Neurol 251(Suppl 6):VI/2-6.

L'Allemain, G. (2002) Update on. the proteasome inhibitor PS341. Bull Cancer 89:29–30.

La Rosee, P., O'Dwyer, M. E., and Druker, B. J. (2002) Insights from pre-clinical studies for new combination treatment regimens with the Bcr-Abl kinase inhibitor imatinib mesylate (Gleevec/Glivec) in chronic myelogenous leukemia: A translational perspective. Leukemia 16:1213–9.

Llauger, L., He, H., Kim, J., Aguirre, J., Rosen, N., Peters, U., Davies, P. and Chiosis, G. (2005) Evaluation of 8-arylsulfanyl, 8-arylsulfoxyl, and 8-arylsulfonyl adenine derivatives as inhibitors of the heat shock protein 90. J Med Chem 48:2892–905.

Lu, A., Ran, R., Parmentier-Batteur, S., Nee, A., and Sharp, F. R. (2002) Geldanamycin induces heat shock proteins in brain and protects against focal cerebral ischemia. J Neurochem 81:355–64.

Mabjeesh, N. J., Post, D. E., Willard, M. T., Kaur, B., Van Meir, E. G., Simons, J. W., and Zhong, H. (2002) Geldanamycin induces degradation of hypoxia-inducible factor 1α protein via the proteasome pathway in prostate cancer cells. Cancer Res 62:2478–82.

Machida, H., Matsumoto, Y., Shirai, M., and Kubota, N. (2003) Geldanamycin, an inhibitor of Hsp90, sensitizes human tumour cells to radiation. Int J Radiat Biol 79:973–80.

Maulik, G., Kijima, T., Ma, P. C., Ghosh, S. K., Lin, J., Shapiro, G. I., Schaefer, E., Tibaldi, E., Johnson, B. E., and Salgia, R. (2002) Modulation of the c-Met/hepatocyte growth factor pathway in small cell lung cancer. Clin Cancer Res 8:620–7.

Maxwell, P. H., Wiesener, M. S., Chang, G.-W., Clifford, S. C., Vaux, E. C., Cockman, M. E., Wykoff, C. C., Pugh, C. W., Maher, E. R., and Ratcliffe, P. J. (1999) The tumor suppressor protein VHL targets hypoxia-inducible factors for oxygen-dependent proteolysis. Nature 399:271–5.

Mimnaugh, E. G., Chavany, C., and Neckers, L. (1996) Polyubiquitination and proteasomal degradation of the p185c-erbB-2 receptor protein-tyrosine kinase induced by geldanamycin. J Biol Chem 271:22796–801.

Mimnaugh, E. G., Xu, W., Vos, M., Yuan, X., Isaacs, J. S., Bisht, K. S., Gius, D., and Neckers, L. (2004) Simultaneous inhibition of hsp 90 and the proteasome promotes protein ubiquitination, causes endoplasmic reticulum-derived cytosolic vacuolization, and enhances antitumor activity. Mol Cancer Ther 3:551–66.

Minami, Y., Kiyoi, H., Yamamoto, Y., Yamamoto, K., Ueda, R., Saito, H., and Naoe, T. (2002) Selective apoptosis of tandemly duplicated FLT3-transformed leukemia cells by Hsp90 inhibitors. Leukemia 16:1535–40.

Mitsiades, N., Mitsiades, C. S., Poulaki, V., Chauhan, D., Fanourakis, G., Gu, X., Bailey, C., Joseph, M., Libermann, T. A., Treon, S. P., Munshi, N. C., Richardson, P. G., Hideshima, T., and Anderson, K. C. (2002) Molecular sequelae of proteasome inhibition in human multiple myeloma cells. Proc Natl Acad Sci U S A 99:14374–9.

Naoe, T., Kiyoe, H., Yamamoto, Y., Minami, Y., Yamamoto, K., Ueda, R., and Saito, H. (2001) FLT3 tyrosine kinase as a target molecule for selective antileukemia therapy. Cancer Chemother Pharmacol 48:S27–30.

Neckers, L. (2002) Hsp90 inhibitors as novel cancer chemotherapeutic agents. Trends Mol Med 8:S55–61.

Nimmanapalli, R., O'Bryan, E., Huang, M., Bali, P., Burnette, P. K., Loughran, T., Tepperberg, J., Jove, R., and Bhalla, K. (2002) Molecular characterization and sensitivity of STI-571 (imatinib mesylate, Gleevec)-resistant, Bcr-Abl-positive, human acute leukemia cells to SRC kinase inhibitor PD180970 and 17-allylamino-17-demethoxygeldanamycin. Cancer Res 62:5761–9.

Ouyang, Y. B., and Hu, B. R. (2001) Protein ubiquitination in rat brain following hypoglycemic coma. Neurosci Lett 298:159–62.

Paine-Murrieta, G., Cook, P., Taylor, C. W., and Whitesell, L. (1999) The anti-tumor activity of 17-allylaminogeldanamycin is associated with modulation of target protien levels in vivo. Proc Am Assoc Cancer Res 40: Abstract 119.

Pennacchietti, S., Michieli, P., Galluzzo, M., Mazzone, M., Giordano, S., and Comoglio, P. M. (2003) Hypoxia promotes invasive growth by transcriptional activation of the met protooncogene. Cancer Cell 3:347–61.

Plescia, J., Salz, W., Xia, F., Pennati, M., Zaffaroni, N., Daidone, M. G., Meli, M., Dohi, T., Fortugno, P., Nefedova, Y., Gabrilovich, D. I., Colombo, G., and Altieri, D. C. (2005) Rational design of shepherdin, a novel anticancer agent. Cancer Cell 7:457–68.

Polymeropoulos, M. H., Lavedan, C., Leroy, E., Ide, S. E., Dehejia, A., Dutra, A., Pike, B., Root, H., Rubenstein, J., Boyer, R., Stenroos, E. S., Chandrasekharappa, S., Athanassiadou, A., Papapetropoulos, T., Johnson, W. G., Lazzarini, A. M., Duvoisin, R. C., Di Iorio, G., Golbe, L. I., and Nussbaum, R. L. (1997) Mutation in the alpha-synuclein gene identified in families with Parkinson's disease. Science 276: 2045–7.

Prodromou, C., and Pearl, L. H. (2003) Structure and functional relationships of Hsp90. Curr Cancer Drug Targets 3:301–23.

Queitsch, C., Sangster, T. A., and Lindquist, S. (2002) Hsp90 as a capacitor of phenotypic variation. Nature 417:618–24.

Rajagopalan, H., Bardelli, A., Lengauer, C., Kinzler, K. W., Vogelstein, B., and Velculescu, V. E. (2002) Tumorigenesis: RAF/RAS oncogenes and mismatch-repair status. Nature 418:934.

Rutherford, S. L., and Lindquist, S. (1998) Hsp90 as a capacitor for morphological evolution. Nature 396:336–42.

Santoro, M., Melillo, R. M., Carlomagno, F., Fusco, A., and Vecchio, G. (2002) Molecular mechanisms of RET activation in human cancer. Ann N Y Acad Sci 963:116–21.

Sawyers, C. L., Hochhaus, A., Feldman, E., Goldman, J. M., Miller, C. B., Ottmann, O. G., Schiffer, C. A., Talpaz, M., Guilhot, F., Deininger, M. W., Fischer, T., O'Brien, S. G., Stone, R. M., Gambacorti-Passerini, C. B., Russell, N. H., Reiffers, J. J., Shea, T. C., Chapuis, B., Coutre, S., Tura, S., Morra, E., Larson, R. A., Saven, A., Peschel, C., Gratwohl, A., Mandelli, F., Ben-Am, M., Gathmann, I., Capdeville, R., Paquette, R. L., and Druker, B. J. (2002) Imatinib induces hematologic and cytogenetic responses in patients with chronic myelogenous leukemia in myeloid blast crisis: Results of a phase II study. Blood 99:3530–9.

Schneider, C., Sepp-Lorenzino, L., Nimmesgern, E., Ouerfelli, O., Danishefsky, S., Rosen, N., and Hartl, F. U. (1996) Pharmacologic shifting of a balance between protein refolding and degradation mediated by Hsp90. Proc Natl Acad Sci U S A 93:14536–41.

Schnur, R. C., Corman, M. L., Gallaschun, R. J., Cooper, B. A., Dee, M. F., Doty, J. L., Muzzi, M. L., DiOrio, C. I., Barbacci, E. G., Miller, P. E., Pollack, V. A., Savage, D. M., Sloan, D. E., Pustilnik, L. R., Moyer, J. D., and Moyer, M. P. (1995) erbB-2 oncogene inhibition by geldanamycin derivatives: Synthesis, mechanism of action, and structure–activity relationships. J Med Chem 38:3813–20.

Schulte, T. W., and Neckers, L. M. (1998) The benzoquinone ansamycin 17-allylamino 17 demethoxygeldanamycin binds to HSP90 and shares important biologic activities with geldanamycin. Cancer Chemother Pharmacol 42:273–9.

Seizinger, B. R., Rouleau, G. A., Ozelius, L. J., Lane, A. H., Farmer, G. E., Lamiell, J. M., Haines, J., Yuen, J. W., Collins, D., Majoor-Krakauer, D., Bonner, T., Mathew, C., Rubenstein, A., Halperin, J., Mcconkie-Rosell, A., Green, J. S., Trofatter, J. A., Ponder, B. A., Eierman, L., Bowmer, M. I., Schimke, R., Oostra, B., Aronin, N., Smith, D. I., Drabkin, H., Waziri, M. H., Hobbs, W. J., Martuza, R. L., Conneally, P. M., Hsia, Y. E., and Gusella, J. F. (1988) Von Hippel-Lindau disease maps to the region of chromosome 3 associated with renal cell carcinoma. Nature 332:268–9.

Shah, N. P., Nicoll, J. M., Nagar, B., Gorre, M. E., Paquette, R. L., Kuriyan, J., and Sawyers, C. L. (2002) Multiple bcr-abl kinase domain mutations confer polyclonal resistance to the tyrosine kinase inhibitor imatinib (STI571) in chronic phase and blast crisis chronic myeloid leukemia. Cancer Cell 2:117–25.

Shimamura, T., Lowell, A. M., Engelman, J. A., and Shapiro, G. I. (2005) Epidermal growth factor receptors harboring kinase domain mutations associate with the heat shock protein 90 chaperone and are destabilized following exposure to geldanamycins. Cancer Res 65:6401–8.

Shiotsu, Y., Neckers, L. M., Wortman, I., An, W. G., Schulte, T. W., Soga, S., Murakata, C., Tamaoki, T., and Akinaga, S. (2000) Novel oxime derivatives of radicicol induce erythroid differentiation associated with preferential G(1) phase accumulation against chronic myelogenous leukemia cells through destabilization of Bcr-Abl with Hsp90 complex. Blood 96:2284–91.

Siligardi, G., Hu, B., Panaretou, B., Piper, P. W., Pearl, L. H., and Prodromou, C. (2004) Co-chaperone regulation of conformational switching in the Hsp90 ATPase cycle. *J Biol Chem* 279:51989–98.

Solit, D., Zheng, F., Drobnjak, M., Munster, P., Higgins, B., Verbel, D., Heller, G., Tong, W., Cordon-Cardo, C., Agus, D., Scher, H., and Rosen, N. (2002) 17-allylamino-17-demthoxygeldanamycin induces the degradation of androgen receptor and HER-2/neu and inhibits the growth of prostate cancer xenografts. Clin Cancer Res 986–93.

Stoler, D. L., Chen, N., Basik, M., Kahlenberg, M. S., Rodriguez-Bigas, M. A., Petrelli, N. J., and Anderson, G. R. (1999) The onset and extent of genomic instability in sporadic colorectal tumor progression. Proc Natl Acad Sci U S A 96:15121–6.

Supko, J. G., Hickman, R. L., Grever, M. R., and Malspeis, L. (1995) Preclinical pharmacologic evaluation of geldanamycin as an antitumor agent. Cancer Chemother Pharmacol 36:305–15.

Tacchini, L., Dansi, P., Matteucci, E., and Desiderio, M. A. (2001) Hepatocyte growth factor signalling stimulates hypoxia inducible factor-1 (HIF-1) activity in HepG2 hepatoma cells. Carcinogenesis 22:1363–71.

Taylor, J. P., Hardy, J., and Fischbeck, K. H. (2002) Toxic proteins in neurodegenerative disease. Science 296:1991–5.

Tofaris, G. K., and Spillantini, M. G. (2005) Alpha-synuclein dysfunction in Lewy body diseases. Mov Disord 20(Suppl 12):S37–44.

Vanaja, D. K., Mitchell, S. H., Toft, D. O., and Young, C. Y. F. (2002) Effect of geldanamycin on androgen receptor function and stability. Cell Stress Chap 7:55–64.

Vilenchik, M., Solit, D., Basso, A., Huezo, H., Lucas, B., He, H., Rosen, N., Spampinato, C., Modrich, P., and Chiosis, G. (2004) Targeting wide-range oncogenic transformation via PU24FCl, a specific inhibitor of tumor Hsp90. Chem Biol 11:787–97.

Waelter, S., Boeddrich, A., Lurz, R., Scherzinger, E., Lueder, G., Lehrach, H., and Wanker, E. E. (2001) Accumulation of mutant huntingtin fragments in aggresome-like inclusion bodies as a result of insufficient protein degradation. Mol Biol Cell 12:1393–407.

Waza, M., Adachi, H., Katsuno, M., Minamiyama, M., Sang, C., Tanaka, F., Inukai, A., Doyu, M., and Sobue, G. (2005) 17-AAG, an Hsp90 inhibitor, ameliorates polyglutamine-mediated motor neuron degeneration. Nat Med 11:1088–95.

Wegele, H., Muller, L., and Buchner, J. (2004) Hsp70 and Hsp90—a relay team for protein folding. Rev Physiol Biochem Pharmacol 151:1–44.

Workman, P. (2004) Combinatorial attack on multistep oncogenesis by inhibiting the Hsp90 molecular chaperone. Cancer Lett 206:149–57.

Xu, L., Eiseman, J. L., Egorin, M. J., and D'Argenio, D. Z. (2003) Physiologically-based pharmacokinetics and molecular pharmacodynamics of 17-(allylamino)-17-demethoxygeldanamycin and its active metabolite in tumor-bearing mice. J Pharmacokinet Pharmacodyn 30:185–219.

Yu, X., Guo, Z. S., Marcu, M. G., Neckers, L., Nguyen, D. M., Chen, G. A., and Schrump, D. S. (2002) Modulation of p53, ErbB1, ErbB2, and Raf-1 expression in lung cancer cells by depsipeptide FR901228. J Natl Cancer Inst 94:504–13.

Zagzag, D., Nomura, M., Friedlander, D. R., Blanco, C., Gagner, J. P., Nomura, N., and Newcomb, E. W. (2003) Geldanamycin inhibits migration of glioma cells in vitro: A potential role for hypoxia-inducible factor (HIF-1alpha) in glioma cell invasion. J Cell Physiol 196:394–402.

Zhang, H., and Burrows, F. (2004) Targeting multiple signal transduction pathways through inhibition of Hsp90. J Mol Med 82:488–99.

Zoghbi, H. Y., and Orr, H. T. (2000) Glutamine repeats and neurodegeneration. Annu Rev Neurosci 23:217–47.

16
gp96 and Tumor Immunity
A Simple Matter of Cross-Presentation Antigens?

CHRISTOPHER V. NICCHITTA
Department of Cell Biology, Duke University Medical Center, Durham, NC 27710

Overview

...the primary question is not what do we know, but how do we know it." —Aristotle

The focus of this commentary is gp96, the endoplasmic reticulum Hsp90 chaperone, its interactions with the innate and adaptive immune system, and how these interactions contribute to tumor immunity. The goal of this commentary, as embodied in the (paraphrased) quote from Aristotle, is to examine the "how do we know it." And to avoid any errors of assumption, "it" is the hypothesis that gp96 associates with the antigenic peptide repertoire of the cell, a hypothesis that, if correct, identifies gp96 as a universal cross-presentation antigen.

1. Introduction: The Guiding Principle—The Discovery of Tumor-Specific Transplantation Antigens

The field of heat shock protein tumor immunology can trace its beginnings to a series of highly influential experiments identifying the capacity of mice to acquire protective immunity to tumor challenge (Gross, 1943; Baldwin, 1955; Prehn and Main, 1957; Klein et al., 1960). The key observation reported in these early studies was that tumors expressed antigenic peptides that were effective targets for T-cell responses. Beyond this critical observation, which provided a compelling rationale for the development of immunological approaches to cancer therapy, these early studies also provided a somewhat unexpected finding; individual tumors displayed unique antigenic profiles; in other words, the capacity to elicit protective immunity was, generally speaking, tumor-restricted (Prehn and Main, 1957; Klein et al., 1960; Klein, 2001). From this observation comes the conclusion that tumors express tumor-specific transplantation antigens (TSTAs), and the hypothesis that such TSTAs would likely be of therapeutic benefit to human cancer. Considerable efforts have been focused on the molecular identification of TSTAs. In addition to the potential benefit of identifying target sequences for vaccine design, TSTA identification offers needed insights into the oncogenic events preceding tumor formation. It is through such efforts that numerous tumor antigens have been described and in their descriptions it has become clear that a broad diversity of oncogenic processes can contribute to malignancy. Included in this profile are mutations in oncogenes and tumor suppressors, which, intuitively, would be predicted. Other classes of tumor (type)-specific antigens have been identified and their antigenicity established. These include antigens, such as MAGE-3, which derive from normal germ cell proteins (Boon et al., 1997; Chaux et al., 1999); antigens derived from normal cellular proteins, such as tyrosinase, and which commonly represent differentiation antigens restricted to particular tissues (Kawakami et al., 1993; Boon et al., 1994); antigens derived from normal cellular proteins yet which are grossly overexpressed, with HER-2 neu representing a prominent example (Disis and Cheever, 1998); normal proteins that undergo inappropriate or insufficient post-translational modifications (e.g., mucin) (Vlad et al., 2004); and oncoviral proteins, such as those arising from human papilloma virus (Sadovnikova and Stauss, 1994; Klein and Klein, 2005). The consistent theme in these studies is that most cancer antigens derive from normal, non-mutated cellular proteins, many of which are differentiation markers and which are common to particular tissue malignancies, such as melanoma, breast, ovary and cervix (Robbins and Kawakami, 1996; Kawakami et al., 2004). Given the extensive literature on the expression and origins of these normal, tissue-derived antigens, the question of the antigenic "uniqueness" of individual tumors can be asked. Practically speaking, the methodologies used to identify tumor antigens and to evaluate their expression patterns are by their very nature far more successful at identifying prominent shared antigens than individual, unique antigens. And in this context, the available data make clear that a rather limited number of tumor-derived peptides have the clinically relevant property of undergoing recognition by CD8+ T cells. Given this intrinsic bias, it is difficult to formally conclude that individual tumors are, or are not, antigenically unique. However, the more significant question may be whether the absolute magnitude of antigenic diversity of a given tumor has a significant therapeutic correlate, and thus whether such supposed antigenic diversity could be exploited for clinical application. In one view, those tumor-derived peptides whose structural identities have been confirmed and which undergo recognition by CD8+ T cells in the patient population are of highest clinical interest (Rosenberg, 2001; Laheru et al., 2005). Yet, and largely because of the complex interplay between the tumor cells, the tumor microenvironment and the immune system, the induction of tumor antigen-directed cytotoxic T lymphocytes is rarely of prognostic significance (Pardoll, 2003; Rosenberg, 2004). These findings point prominently to the need for therapeutic approaches to the phenomenon of "immunological escape" commonly displayed by tumors, if immunotherapy is to become a prominent treatment modality (Pardoll, 2003; Rosenberg, 2004). In a different view, the presumed antigenic diversity of individual tumors is exploited in the form of highly multivalent peptide vaccines. In one expression of this approach, pioneered by the Srivastava group, the highly multivalent peptide vaccine takes the form of tumor-derived heat shock protein-peptide complexes (Srivastava and Maki, 1991; Srivastava, 2000). This intriguing development extends from the discovery that gp96 (GRP94), the endoplasmic reticulum (ER) heat shock protein 90 (Hsp90) served as the TSTA of MethA sarcomas (Srivastava et al., 1986). Importantly, the ability of gp96 to elicit tumor-directed immune responses was reported to be tumor-restricted, thereby identifying gp96 as both a general (capable of eliciting anti-tumor immune responses regardless of tumor identity) and specific (only effective against its tumor of origin) tumor vaccine (Srivastava et al., 1986).

2. Heat Shock Proteins: The TSTAs of MethA Sarcoma

To answer how gp96, an abundant, highly conserved chaperone expressed in essentially all mammalian cells, could elicit tumor-specific immune responses, regardless of its tumor of origin, Pramod Srivastava and colleagues proposed that gp96 associates with the entire antigenic peptide repertoire of the cell (Srivastava and Heike, 1991; Srivastava and Maki, 1991). In this property, gp96 appears extraordinary. Because its capacity to stably associate with antigenic peptides is apparently unbiased, the need for a priori knowledge of tumor antigen identity is minimized or eliminated; the induction of an anti-tumor immune response being considered proof positive evidence that tumor (peptide) antigens stably associate with gp96. In the years following publication of this proposal, a multitude of studies have been published whose results mirror, in part, the original observation—that tumor-derived gp96 elicits anti-tumor immune responses (reviewed in: (Srivastava and Udono, 1994; Srivastava et al., 1998; Srivastava, 2002)). Only in very few instances however, has it been demonstrated that gp96 activity is tumor-restricted (Srivastava et al., 1986; Tamura et al., 1997). To the contrary, and where it has been examined, significant "tumor cross-reactivity" and/or partial to substantive anti-tumor immune responses to administration of normal tissue derived gp96 has been reported (Tamura et al., 1997; Baker-LePain et al., 2002). On some reflection, this is not surprising and points to an obvious conundrum: studies devoted to identifying tumor antigens in spontaneous cancer settings (i.e., non-chemical carcinogenesis models) have demonstrated a predominance of antigens derived from normal cellular proteins that are inappropriately expressed in the transformed state and frequently encountered in cancers derived from different tissues. Carcinoembryonic antigen (CEA) and human telomerase reverse transcriptase (hTER) are representative examples, and can be found in cancers of the colon, rectum, pancreas, stomach, breast, lung, or in the case of hTER, nearly universally (Huang and Kaufman, 2002; Shay and Wright, 2002; Vonderheide, 2002; Marshall, 2003). And so the conundrum: If gp96 binds the entire antigenic peptide repertoire of the cell and if many cancers share tumor antigens, how then can gp96 function as a tumor-specific antigen, as originally proposed? There are possible solutions to the conundrum, but most, if not all, fall in the category of hypothetical tumor antigens, that although significantly immunogenic, have resisted identification and are highly restricted to individual tumor isolates.

3. The Peptide Antigens of gp96

As we approach the twentieth anniversary of the discovery of gp96 as the TSTA of MethA sarcoma, and the fifteenth anniversary of the gp96-peptide chaperone hypothesis, one is struck by the wealth of studies demonstrating the phenomenon of gp96-elicited tumor immunity, and the poverty of knowledge regarding the structural characteristics of the bound peptide pool and the mechanism of peptide binding and release. The lack of insight into the gp96-bound peptide pool is problematic; without these needed data, the validity of the primary hypothesis remains in question. In that vein, however, it is important to note that complexes of gp96 and synthetic peptides, prepared in vitro, are recognized and internalized by professional antigen-presenting cells (APC), and their bound peptides represented on APCMHCclass I molecules, to yield CD8+ T cell activation (Suto and Srivastava, 1995; Blachere et al., 1997; Berwin et al., 2002). These data provide critical proof-of-principle evidence that gp96 can function as a peptide chaperone and direct access of bound peptides into the MHC class I re-presentation pathways of APC. However, caution must be exercised in extending these findings to the in vivo scenario. In particular, in very similar assays, where paired studies of in vitro-assembled gp96-synthetic peptide complexes and tumor cell-derived gp96-peptide complexes were performed, little to no activity was seen in the tumor cell-derived gp96 (Fleischer et al., 2004). As has been argued by this author in the past, the biological validity of this proof of principle observation is entirely dependent on whether the peptide binding interaction used for in vitro assembly of gp96-peptide complexes recapitulates the (presumed) peptide-binding process occurring in vivo (Nicchitta, 2003; Nicchitta et al., 2004). This concern is amplified by the current lack of understanding regarding the molecular regulation of gp96-(poly)peptide interactions. To any real degree, the molecular basis for (poly)peptide and peptide recognition by gp96 is unknown and it is also uncertain how the lifetime of the gp96 (poly)peptide complex is determined and more specifically, what role ATP binding and hydrolysis play in defining the lifetimes of the chaperone-substrate interaction.

By strict analogy to Hsp90, it can be presumed that these parameters are under the regulatory guidance of ATP binding and hydrolysis but recent crystallographic studies of the apo and ATP-bound forms of the gp96 N-terminal domain may prove this to be an unfounded presumption. Most importantly, the conformational changes occurring on ATP binding to gp96 are opposite to those proposed by the current DNA gyrase B and MutL-influenced models, which are central to the current model of the Hsp90 chaperone cycle (Prodromou et al., 1997; Young et al., 2001; Prodromou and Pearl, 2003). Also significant are the crystallographic data demonstrating that the ADP- and ATP-bound forms of the Hsp90 N-terminal domain display identical conformations (Prodromou et al., 1997; Dehner et al., 2003; Immormino et al., 2004). Returning to the theme of this commentary, we do know that gp96 can indeed interact with peptides and under suitable experimental conditions form very stable (SDS-resistant) complexes (Blachere et al., 1997; Vogen et al., 2002; Gidalevitz et al., 2004). Evaluating the ultimate physiological significance of this finding requires a discussion of the conditions necessary to obtain significant molar stoichiometries of peptide binding. The experimental conditions are unusual, in particular with respect to either the markedly elevated temperatures (i.e., 50°C) necessary for significant peptide binding and/or the extended incubation times required for significant binding at more physiological temperatures (ca. 24 hours at 37°C) (Blachere et al., 1997; Vogen et al., 2002; Gidalevitz et al., 2004). Biophysical analyses of gp96 structure demonstrate that exposure to such conditions elicits irreversible alterations in tertiary conformation that closely parallel the appearance of peptide binding activity (Wearsch et al., 1998). Also, analyses of the binding of environment sensitive fluorophores, such as Nile Red and 8-aminonapthalenesulfonate indicate that the observed temperature-induced, irreversible tertiary conformational changes are accompanied by exposure of hydrophobic domains (Wearsch et al., 1998). Could it be that peptide binding is occurring to domains of the protein that are excluded from solvent accessibility in the native state? This question harkens back to one of the more puzzling characteristics of the in vitro peptide-binding reaction, the observation that once bound, gp96-peptide complexes are stable to SDS (Blachere et al., 1997; Vogen et al., 2002; Gidalevitz et al., 2004). This observation alone suggests that peptide binding is occurring via an adsorptive process requiring the above-mentioned irreversible conformational change (Nicchitta et al., 2004). In fact, very recent studies of the real-time interactions of gp96 with the peptide substrate VSV8 have also demonstrated that peptide binding is both weak and likely to occur via a non-specific adsorptive process (Ying and Flatmark, 2006). Formally, that gp96 might bind peptide via a non-selective, irreversible, adsorptive process does not refute the hypothesis that gp96 functions as a peptide binding protein in vivo. In fact, it can be argued to the contrary that such a binding function might best serve its function as an antigen chaperone: the non-selective nature of the interaction would provide the needed promiscuous basis for substrate recognition and the irreversibility would imbue the complex with the physical stability needed to retain activity following biochemical isolation and immunization. In this regard, the recent studies from Argon and colleagues are valuable in their identification of the N terminal domain as a prominent site for peptide association as well as the experimental conditions necessary to achieve high stoichiometry binding to this site (Gidalevitz et al., 2004).

Accepting this view leads to the previously proposed hypothesis that gp96 associates with the entire antigenic peptide population of the cell, but neither supports nor refutes the hypothesis that the proposed peptide binding activity represents an in vivo function of gp96. The obvious solution to such an arduous and circuitous line of argument is to do the direct test of the prediction. That is, to isolate gp96 from different sources and demonstrate the diversity and compositional identity of the bound peptide pool by the well established and validated mass spectrometric methods used to identify and characterize the bound peptide composition of, for example, MHC Class I molecules. The data on this point are intriguing. For example, the van Bleek laboratory has reported the isolation of VSV8 in the gp96 fraction isolated from VSV8-infected cells (Nieland et al., 1996). As noted above, the interaction of VSV8 with gp96 has been reported to be very weak (apparent Kd of ca. 800 µM), though ultimately irreversible (Ying and Flatmark, 2006), (Vogen et al., 2002). More significantly, this study identified VSV8 in cells lacking the cognate MHC class I binding partner (Nieland et al., 1996). Prior global analyses of cellular peptide composition have concluded, however, that in the absence of a cognate MHC class I binding partner, a given peptide is highly unstable and unlikely to be recovered at any significant level (Falk et al., 1990). In a limited number of other studies, individual peptides have been recovered from purified gp96 and identified by mass spectrometric analysis (Nieland et al., 1996; Ishii et al., 1999; Meng et al., 2001; Liu et al., 2004; Demine and Walden, 2005). Curiously, and with the notable exception of the study of (Demine and Walden, 2005), the peptide of interest was readily identified, yet structural descriptions of the bound peptide pool composition were not included. This important gap has now been addressed in a study from Demine and Walden (Demine and Walden, 2005). In this study, the authors biochemically isolated gp96 and MHC class I from MyLa cells, a T-cell lymphoma, and analyzed the composition of the trifluoracetic acid-soluble peptide pool by reverse phase HPLC/MALDI-TOF mass spectrometry. In contrast to the MHC class I-derived fraction, a structurally diverse and low-abundance peptide pool was recovered in the gp96-derived extract with estimates of peptide binding stoichiometries revealing that peptide loading was markedly sub-stoichiometric, with values of 0.1 to 0.4% occupancy suggested. More significantly, the identified peptides bore only a weak resemblance to MHC class I peptide ligands, in particular at their critical C-terminal residues. On the basis of the low yield of gp96-bound peptides and their lack of structural similarity to known MHC class I binding peptides, these authors concluded that gp96 is unlikely to serve as a significant source of peptides for MHC class I loading (Demine and Walden, 2005). In similar experiments, gp96 was purified from porcine pancreas, the putative bound peptide extracted in guanidinium chloride/trifouoracetic acid and the composition analyzed by combined LC/MALDI-TOF (Nicchitta and Burlingame, unpublished observations). As in the Demine and Walden report, relatively few peptides could be unequivocally identified, though milligram quantities (ca. 50 nmol) of gp96 were extracted. Most notably, the total ion chromatograms of interest were unexpectedly "quiet," which is to say there were relatively few species detected in the 800–1,600 molecular weight window (Nicchitta and Burlingame, unpublished observations). In summarizing the available data on the mass spectrometry-based analyses of the gp96-bound peptide pool, the data are consistent; all studies reported the identification of peptides in acid extracts of gp96, though in the four available published studies, the predicted diverse and abundant peptide population (i.e., the antigenic peptide repertoire) could not be identified. To this author it is puzzling why unbiased screens for gp96-bound MHC class I peptide ligands reveal few to no precise matches to canonical class I peptide ligands, yet such peptides are readily identified in scenarios where the sequence information is know in advance (i.e., VSV8). In one attempt to resolve this dilemma, recombinant vaccinia virus (rVV) constructs encoding the dominant Kb epitope, SIINFEKL, in frame with an N-terminal signal sequence, were used to over-express peptides in the ER compartment. In these experiments, the appearance of SIINFEKL in the ER was confirmed by the subsequent and near complete conversion of surface MHC class I molecules to Class I-SIINFEKL complexes. Even under these circumstances, little to no SIINFEKL peptide could be identified in the gp96 fraction obtained from rVV-SIINFEKL-infected cells (Reed and Nicchitta, unpublished observations). These results, albeit negative, do highlight an absolutely pressing concern in the field. And that is, direct chemical evidence of a diverse and cell/tumor specific population of bound peptides needs to be obtained before the postulated TSTA function of gp96 can be validated. Furthermore, it needs to be determined whether the peptides that can be identified represent precursor forms of MHC class I ligands or simply "dead-end" proteolysis products of uncertain biological significance. To further emphasize this concern, recent studies have demonstrated that gp96 preparations purified by current biochemical methods contain numerous contaminants, any of which could be formally considered to serve as the (a) source of the peptides in mass spectrometric studies and/or sources of antigenic (poly)peptides in animal studies of tumor rejection (Srivastava, 1994; Reed et al., 2002; Nicchitta, 2003; Nicchitta et al., 2004).

4. Cross-Presentation Antigens—The Cellular Basis for gp96 Function?

In its original embodiment, the hypothesis that gp96 binds the antigenic repertoire of the cell provided an explanation for the decades old question regarding the molecular identity of the tumor-specific antigens of chemically induced sarcomas. This hypothesis has subsequently evolved to identify gp96 as a fundamental, if not necessary and sufficient, cross-presentation antigen (Srivastava et al., 1998; Binder and Srivastava, 2005). In this view, gp96-peptide complexes (as well as other heat shock/chaperone protein-peptide species) provide the answer to yet another decades old question—how does the adaptive immune system generate MHC class I responses to intracellular pathogens that attack parenchymal tissue, yet spare professional antigen presenting cells (APCs)? To rephrase this view, knowing that the immune system uses MHC class I and Class II systems to communicate information, in the form of peptides, to CD8+ and CD4+ T cells, there must exist a mechanism to provide APCs that have not been infected with virus a means for obtaining viral (poly) peptides, so that an adaptive immune response can be initiated. In this scenario, an abundant ER chaperone, such as gp96, thought to bind the antigenic repertoire of the cell, could serve as the ideal cross-presentation antigen. In addition, APCs have been demonstrated to display cell surface receptors for gp96 (and other chaperones/heat shock proteins) and these receptors have also been demonstrated to access the MHC class I antigen processing pathway(s) of APC (Binder et al., 2000; Berwin and Nicchitta, 2001; Berwin et al., 2002; 2003). Here, too, a pivotal "proof-of-principle" experiment clearly demonstrates that gp96 complexed with synthetic peptides can be recognized in the extracellular space, internalized, and the bound peptide re-present on APC class I molecules, to yield T-cell activation (Suto and Srivastava, 1995; Berwin et al., 2002a; b). What remains to be determined (an experimentally difficult question) is the source(s) of cross-presentation antigen in vivo. What is clear, is that cross-presentation antigens derive from a different source than do "endogenous" antigens (self-generated peptides presented on self-MHC class I molecules). With regard to endogenous antigens, the pioneering work of Yewdell, Bennink, and colleagues has demonstrated that the generation of direct antigens is intimately coupled to on-going protein synthesis through an as yet uncharacterized, highly unstable (poly)peptide form (s) referred to as DriPs (defective ribosomal products) (Yewdell et al., 1996). Consistent with the DRiP hypothesis, inhibition of cellular protein synthesis results in a rapid loss of MHC class I-peptide complex generation and in a particularly interesting experiment, specific suppression of antigen-encoding gene expression correlated with the loss of the cognate MHC class I-peptide complex, even though the translation product was highly stable (Norbury et al., 2004; Wolkers et al., 2004). In this particular scenario, it can be concluded that the fully folded and stable translation product is not a significant source of the MHC class I peptide ligand.

In contrast to direct presentation, cross-presentation utilizes stable forms of the antigen (or encoded antigen) and strongly favors particulate rather than soluble antigens (Rock and Shen, 2005). This latter characteristic highlights the role of phagocytic uptake and phagosome-derived trafficking itineraries in the successful processing of cross-presentation antigens (Groothuis and Neefjes, 2005; Rock and Shen, 2005). By these criteria, it would not be immediately presumed that gp96, or other heat shock/chaperone proteins, would be successful as cross-presentation antigens—which is not to say that they are not, in principle, capable. This conclusion emerges from the understanding that gp96 would be present in the extracellular space in a soluble form and would be internalized by APC via receptor-mediate endocytosis (Srivastava et al., 1994). Three recently published studies substantiate the conclusion that gp96-peptide complexes are not dominant cross-priming antigens (Norbury et al., 2004; Shen and Rock, 2004; Wolkers et al., 2004). In one study, the consequence of suppressed proteasome activity, and/or expression of antigen in precursor or processed forms, on the efficiency of cross-priming was evaluated. Using the well-established ovalbumin model, Norbury et al., demonstrated that proteasome-precursor forms of antigens, rather than processed forms, were substantially more efficient as cross-priming antigens (Norbury et al., 2004). Notably, targeting of the ovalbumin Kb ligand SIINFEKl to the ER, a scenario that would be expected to greatly favor the formation of gp96-SIINFEKL complexes, resulted in the loss, rather than the enhancement, of cross-priming activity (Norbury et al., 2004). In another study, Wolkers et al., prepared chimeric genes containing an MHC class I epitope in either an unstable (signal sequence) or stable (C-terminus of mature protein) context and evaluated the efficiency of cross-presentation in vivo (Wolkers et al., 2004). In this study, MHC class I epitopes originating from a stable protein were far more effective as cross-priming antigens than epitopes present in short-lived (signal sequence) domains of the protein. In a third study, also utilizing ovalbumin as the model antigen, stable cell lines expressing wildtype ovalbumin (secretory form), a cytosolic form of ovalbumin (sans signal sequence), or a plasma membrane form (fusion with transferring receptor) were studied (Shen and Rock, 2004). The primary and consistent conclusion was that cross-priming antigen activity correlated in all formats with the expression and location of the full-length intact protein, a finding confirmed by direction immunodepletion experiments using antibodies against native ovalbumin (Shen and Rock, 2004). Contrasting with these studies, Binder and Srivastava report that heat shock protein-peptide complexes are "necessary and sufficient" cross-priming antigens (Binder and Srivastava, 2005). In their studies, crude homogenates, or detergent extracts, of β-galactosidase or ovalbumin-expressing cells were evaluated for their ability to provide cross-priming activity. As would be expected, both forms of cell extract exhibited robust cross-priming activity, which could be subsequently eliminated by combined immunoaffinity depletion of the Hsp90, Hsp70, gp96, and calreticulin component of the cell extracts. Importantly, supplementation of the depleted extracts with either Hsp70 or gp96 from the original extracts restored immunogenicity. Surprisingly, these authors also depleted intact ovalbumin from their lysates and reported that loss of intact protein was, in their dose-administration protocol, without effect on cross-priming activity (Binder and Srivastava, 2005). At present, there is no generally accepted model that reconciles these utterly juxtaposed conclusions regarding gp96-peptide function in cross-presentation. Presumably, the strengths and weaknesses of each approach would be revealed in direct, "side-by-side" analyses of gp96 cross-priming antigen function. In this case, the engine of science, with its emphasis on independent verification, will bring clarity to what now is a confusing and seemingly contradictory body of data.

5. Summary

Over the past two decades, the immunogenic functions of heat shock/chaperone proteins have grown from relatively obscure beginnings, as the TSTA of MethA sarcoma, to prominence, where gp96 functions biologically as a universal cross-priming antigen, with application in the immunotherapeutic treatment of human cancer and viral infection. The strength of this ascendance owes a debt to the many experiments demonstrating heat shock/chaperone protein-elicited immune responses. Missing, though, is the rigorous, detailed analysis of how, when and if gp96 intersects with the antigen processing pathways of the cell and how such interactions yield the postulated association with the antigenic peptide repertoire of the cell. In the limited number of studies in which these questions have been asked, the results have not yielded the expected answers. If indeed the immunogenic activity of gp96 resides in a pool of bound peptides, then the direct study of such peptides, their origins, and their capacity to enter either the direct and/or cross-presentation pathways of APCs via their association with gp96 is where this attractive hypothesis will be validated, or refuted.

References

Baker-LePain, J. C., Sarzotti, M., Fields, T. A., Li, C. Y., and Nicchitta, C. V. (2002) GRP94 (gp96) and GRP94 N-terminal geldanamycin binding domain elicit tissue nonrestricted tumor suppression. *J Exp Med* 196(11):1447–59.

Baldwin, R. W. (1955) Immunity to methylcholanthrene-induced tumours in inbred rats following atrophy and regression of the implanted tumours. *Br J Cancer* 9(4): 652–7.

Berwin, B., and Nicchitta, C. V. (2001) To find the road traveled to tumor immunity: The trafficking itineraries of molecular chaperones in antigen-presenting cells. *Traffic* 2(10):690–7.

Berwin, B., Hart, J. P., Pizzo, S. V., and Nicchitta, C. V. (2002a) Cutting edge: CD91-independent cross-presentation of GRP94(gp96)-associated peptides. *J Immunol* 168(9):4282–6.

Berwin, B., Rosser, M. F., Brinker, K. G., and Nicchitta, C. V. (2002b) Transfer of GRP94(Gp96)-associated peptides onto endosomal MHC class I molecules. *Traffic* 3(5):358–66.

Berwin, B., Hart, J. P., Rice, S., Gass, C., Pizzo, S. V., Post, S. R., and Nicchitta, C. V. (2003) Scavenger receptor-A mediates gp96/GRP94 and calreticulin internalization by antigen-presenting cells. *EMBO J* 22(22):6127–36.

Binder, R. J., Han, D. K., and Srivastava, P. K. (2000) CD91: A receptor for heat shock protein gp96. *Nat Immunol* 1(2):151–5.

Binder, R. J., and Srivastava, P. K. (2005) Peptides chaperoned by heat-shock proteins are a necessary and sufficient source of antigen in the cross-priming of CD8+ T cells. *Nat Immunol* 6(6):593–9.

Blachere, N. E., Li, Z., Chandawarkar, R. Y., Suto, R., Jaikaria, N. S., Basu, S., Udono, H., and Srivastava, P. K. (1997) Heat shock protein–peptide complexes, reconstituted *in vitro*, elicit peptide-specific cytotoxic T lymphocyte response and tumor immunity. *J Exp Med* 186(8):1315–22.

Boon, T., Cerottini, J. C., Van den Eynde, B., van der Bruggen, P., and Van Pel, A. (1994) Tumor antigens recognized by T lymphocytes. *Annu Rev Immunol* 12:337–65.

Boon, T., Coulie, P. G., and Van den Eynde, B. (1997) Tumor antigens recognized by T cells. *Immunol Today* 18(6):267–8.

Chaux, P., Vantomme, V., Stroobant, V., Thielemans, K., Corthals, J., Luiten, R., Eggermont, A. M., Boon, T., and van der Bruggen, P., (1999) Identification of MAGE-3 epitopes presented by HLA-DR molecules to CD4(+) T lymphocytes. *J Exp Med* 189(5):767–78.

Dehner, A., Furrer, J., Richter, K., Schuster, I., Buchner, J., and Kessler, H. (2003) NMR chemical shift perturbation study of the N-terminal domain of Hsp90 upon binding of ADP, AMP-PNP, geldanamycin, and radicicol. *Chem biochem* 4(9):870–7.

Demine, R., and Walden, P. (2005) Testing the role of gp96 as peptide chaperone in antigen processing. *J Biol Chem* 280(18):17573–8.

Disis, M. L., and Cheever, M. A. (1998) HER-2/neu oncogenic protein: Issues in vaccine development. *Crit Rev Immunol* 18(1–2):37–45.

Falk, K., Rotzschke, O., and Rammensee, H. G. (1990) Cellular peptide composition governed by major histocompatibility complex class I molecules. *Nature* 348(6298):248–51.

Fleischer, K., Schmidt, B., Kastenmuller, W., Busch, D. H., Drexler, I., Sutter, G., Heike, M., Peschel, C., and Bernhard, H. (2004) Melanoma-reactive class I-restricted cytotoxic T cell clones are stimulated by dendritic cells loaded with synthetic peptides, but fail to respond to dendritic cells pulsed with melanoma-derived heat shock proteins *in vitro*. *J Immunol* 172(1):162–9.

Gidalevitz, T., Biswas, C., Ding, H., Schneidman-Duhovny, D., Wolfson, H. J., Stevens, F., Radford, S., and Argon, Y. (2004) Identification of the N-terminal peptide binding site of glucose-regulated protein 94. *J Biol Chem* 279(16):16543–52.

Groothuis, T. A., and Neefjes, J. (2005) The many roads to cross-presentation. *J Exp Med* 202(10):1313–8.

Gross, L. (1943) Intradermal immunization of C3H mice against a sarcoma that originated in an animal of the same line. *Cancer Res* 3:323–6.

Huang, E. H., and Kaufman, H. L. (2002) CEA-based vaccines. *Expert Rev Vaccines* 1(1):49–63.

Immormino, R. M., Dollins, D. E., Shaffer, P. L., Soldano, K. L., Walker, M. A., and Gewirth, D. T. (2004) Ligand-induced conformational shift in the N-terminal domain of GRP94, an Hsp90 chaperone. *J Biol Chem* 279(44):46162–71.

Ishii, T., Udono, H., Yamano, T., Ohta, H., Uenaka, A., Ono, T., Hizuta, A., Tanaka, N., Srivastava, P. K., and Nakayama, E. (1999) Isolation of MHC class I-restricted tumor antigen peptide and its precursors associated with heat shock proteins hsp70, hsp90, and gp96. *J Immunol* 162(3):1303–9.

Kawakami, Y., Fujita, T., Matsuzaki, Y., Sakurai, T., Tsukamoto, M., Toda, M., and Sumimoto, H. (2004) Identification of human tumor antigens and its implications for diagnosis and treatment of cancer. *Cancer Sci* 95(10):784–91.

Kawakami, Y., Nishimura, M. I., Restifo, N. P., Topalian, S. L., O'Neil, B. H., Shilyansky, J., Yannelli, J. R., and Rosenberg, S. A. (1993) T-cell recognition of human melanoma antigens. *J Immunother* 14(2):88–93.

Klein, G. (2001) The strange road to the tumor-specific transplantation antigens (TSTAs). *Cancer Immun* 1:6.

Klein, G., and Klein, E. (2005) Surveillance against tumors—is it mainly immunological? *Immunol Lett* 100(1):29–33.

Klein, G., Sjogren, H. O., Klein, E., and Hellstrom, K. E. (1960) Demonstration of resistance against methylcholanthrene-induced sarcomas in the primary autochthonous host. *Cancer Res* 20:1561–72.

Laheru, D. A., Pardoll, D. M., and Jaffee, E. M. (2005) Genes to vaccines for immunotherapy: How the molecular biology revolution has influenced cancer immunology. *Mol Cancer Ther* 4(11):1645–52.

Liu, C., Ewing, N., and DeFilippo, M. (2004) Analytical challenges and strategies for the characterization of gp96-associated peptides. *Methods* 32(1):32–7.

Marshall, J. (2003) Carcinoembryonic antigen-based vaccines. *Semin Oncol* 30(3 Suppl. 8):30–6.

Meng, S. D., Gao, T., Gao, G. F., and Tien, P. (2001) HBV-specific peptide associated with heat-shock protein gp96. *Lancet* 357(9255):528–9.

Nicchitta, C. V. (2003) Re-evaluating the role of heat-shock protein–peptide interactions in tumour immunity. *Nat Rev Immunol* 3(5):427–32.

Nicchitta, C. V., Carrick, D. M., and Baker-Lepain, J. C. (2004) The messenger and the message: gp96 (GRP94)–peptide interactions in cellular immunity. *Cell Stress Chap* 9(4):325–31.

Nieland, T. J., Tan, M. C., Monne-van Muijen, M., Koning, F., Kruisbeek, A. M., and van Bleek, G. M. (1996) Isolation of an immunodominant viral peptide that is endogenously bound to the stress protein GP96/GRP94. *Proc Natl Acad Sci USA* 93(12):6135–9.

Norbury, C. C., Basta, S., Donohue, K. B., Tscharke, D. C., Princiotta, M. F., Berglund, P., Gibbs, J., Bennink, J. R., and Yewdell, J. W. (2004) CD8+ T cell cross-priming via transfer of proteasome substrates. *Science* 304(5675):1318–21.

Pardoll, D. (2003) Does the immune system see tumors as foreign or self? *Annu Rev Immunol* 21:807–39.

Prehn, R. T., and Main, J. M. (1957) Immunity to methylcholanthrene-induced sarcomas. *J Natl Cancer Inst* 18(6):769–78.

Prodromou, C., and Pearl, L. H. (2003) Structure and functional relationships of Hsp90. *Curr Cancer Drug Targets* 3(5):301–23.

Prodromou, C., Roe, S. M., O'Brien, R., Ladbury, J. E., Piper, P. W., and Pearl, L. H. (1997) Identification and structural characterization of the ATP/ADP-binding site in the Hsp90 molecular chaperone. *Cell* 90(1):65–75.

Prodromou, C., Roe, S. M., Piper, P. W., and Pearl, L. H. (1997) A molecular clamp in the crystal structure of the N-terminal domain of the yeast Hsp90 chaperone. *Nat Struct Biol* 4(6):477–82.

Reed, R. C., Zheng, T., and Nicchitta, C. V. (2002) GRP94-associated enzymatic activities. Resolution by chromatographic fractionation. *J Biol Chem* 277(28):25082–9.

Robbins, P. F., and Kawakami, Y. (1996) Human tumor antigens recognized by T cells. *Curr Opin Immunol* 8(5):628–36.

Rock, K. L., and Shen, L. (2005) Cross-presentation: Underlying mechanisms and role in immune surveillance. *Immunol Rev* 207:166–83.

Rosenberg, S. A. (2001) Progress in human tumour immunology and immunotherapy. *Nature* 411(6835):380–4.

Rosenberg, S. A. (2004) Shedding light on immunotherapy for cancer. *N Engl J Med* 350(14):1461–3.

Sadovnikova, E., and Stauss, H. J. (1994) T cell epitopes in human papilloma virus proteins. *Behring Inst Mitt* (94):87–93.

Shay, J. W., and Wright, W. E. (2002) Telomerase: A target for cancer therapeutics. *Cancer Cell* 2(4):257–65.

Shen, L., and Rock, K. L. (2004) Cellular protein is the source of cross-priming antigen *in vivo*. *Proc Natl Acad Sci USA* 101(9):3035–40.

Srivastava, P. (2002) Roles of heat-shock proteins in innate and adaptive immunity. *Nat Rev Immunol* 2(3):185–94.

Srivastava, P. K. (1994) Endo-beta-D-glucuronidase (heparanase) activity of heat-shock protein/tumour rejection antigen gp96. *Biochem J* 301(Pt 3):919.

Srivastava, P. K. (2000) Heat shock protein-based novel immunotherapies. *Drug News Perspect* 13(9):517–22.

Srivastava, P. K., DeLeo, A. B., and Old, L. J. (1986) Tumor rejection antigens of chemically induced sarcomas of inbred mice. *Proc Natl Acad Sci USA* 83(10):3407–11.

Srivastava, P. K., and Heike, M. (1991) Tumor-specific immunogenicity of stress-induced proteins: Convergence of two evolutionary pathways of antigen presentation? *Semin Immunol* 3(1):57–64.

Srivastava, P. K., and Maki, R. G. (1991) Stress-induced proteins in immune response to cancer. *Curr Top Microbiol Immunol* 167:109–23.

Srivastava, P. K., Menoret, A., Basu, S., Binder, R. J., and McQuade, K. L. (1998) Heat shock proteins come of age: Primitive functions acquire new roles in an adaptive world. *Immunity* 8(6):657–65.

Srivastava, P. K., and Udono, H. (1994) Heat shock protein–peptide complexes in cancer immunotherapy. *Curr Opin Immunol* 6(5):728–32.

Srivastava, P. K., Udono, H., Blachere, N. E., and Li, Z. (1994) Heat shock proteins transfer peptides during antigen processing and CTL priming. *Immunogenetics* 39(2):93–8.

Suto, R., and Srivastava, P. K. (1995) A mechanism for the specific immunogenicity of heat shock protein-chaperoned peptides. *Science* 269(5230):1585–8.

Tamura, Y., Peng, P., Liu, K., Daou, M., and Srivastava, P. K. (1997) Immunotherapy of tumors with autologous tumor-derived heat shock protein preparations. *Science* 278(5335):117–20.

Vlad, A. M., Kettel, J. C., Alajez, N. M., Carlos, C. A., and Finn, O. J. (2004) MUC1 immunobiology: From discovery to clinical applications. *Adv Immunol* 82:249–93.

Vogen, S., Gidalevitz, T., Biswas, C., Simen, B. B., Stein, E., Gulmen, F., and Argon, Y. (2002) Radicicol-sensitive peptide binding to the N-terminal portion of GRP94 *J Biol Chem* 277(43):40742–50.

Vonderheide, R. H. (2002) Telomerase as a universal tumor-associated antigen for cancer immunotherapy. *Oncogene* 21(4):674–9.

Wearsch, P. A., Voglino, L., and Nicchittam, C. V. (1998) Structural transitions accompanying the activation of peptide binding to the endoplasmic reticulum Hsp90 chaperone GRP94. *Biochemistry* 37(16):5709–19.

Wolkers, M. C., Brouwenstijn, N., Bakker, A. H., Toebes, M., and Schumacher, T. N. (2004) Antigen bias in T cell cross-priming. *Science* 304(5675):1314–7.

Yewdell, J. W., Anton, L. C., and Bennink, J. R. (1996) Defective ribosomal products (DRiPs): A major source of antigenic peptides for MHC class I molecules? *J Immunol* 157(5):1823–6.

Young, J. C., Moarefi, I., and Hartl, F. U. (2001) Hsp90: A specialized but essential protein-folding tool. *J Cell Biol* 154(2):267–73.

17
Immunoregulatory Activities of Extracellular Stress Proteins

A. GRAHAM POCKLEY AND MUNITTA MUTHANA

Immunobiology Research Unit, School of Medicine and Biomedical Sciences, L Floor, Royal Hallamshire Hospital, Glossop Road, Sheffield, S10 2JF, United Kingdom

1. Introduction

Although the generally accepted concept is that exogenous stress proteins act as inflammatory mediators and "danger" signals to the immune system (Matzinger, 1994), there is now much evidence to indicate that stress proteins such as heat shock protein (Hsp)10 (Cavanagh, 1996; Rolfe et al., 1984), Hsp27 (De et al., 2000; Laudanski et al., 2005, Hsp60 (Anderton et al., 1995; Anderton and van Eden, 1996; Birk et al., 1999; de Graeff-Meeder et al., 1995; Macht et al., 2000; Paul et al., 2000; Prakken et al., 2003; Quintana et al., 2003; Quintana and Cohen, 2005; Thompson et al., 1990; van den Broek et al., 1989; van Eden et al., 2005; van Roon et al., 1996; Zanin-Zhorov et al., 2005), Hsp70 (Kingston et al., 1996; Quintana et al., 2004; Tanaka et al., 1999; Wendling et al., 2000), glucose-regulated protein (grp)78 (Brownlie et al., 2003; Corrigall et al., 2001, 2004; Corrigall and Panayi, 2005; Panayi and Corrigall, 2006; Sattar et al., 2003), and gp96 (Chandawarkar et al., 1999, 2004; Kovalchin et al., 2006) can, under certain circumstances, exhibit anti-inflammatory activities.

For many years the dogma has been that stress proteins are exclusively intracellular molecules which are only released into the extracellular environment as a consequence of cellular damage. However, it is now known that these molecules can be released from a variety of viable (non-necrotic) cell types (Bassan et al., 1998; Child et al., 1995; Hightower and Guidon 1989; Hunter-Lavin et al., 2004; Liao et al., 2000). Moreover, we and a number of others have reported Hsp60 and Hsp70 to be present in the peripheral circulation of normal individuals, in some instances at levels that have been shown to elicit inflammatory responses in vitro (>1,000 ng/ml; (Lewthwaite et al., 2002; Njemini et al., 2003; Pockley et al., 1998, 1999, 2000, 2002, 2003; Rea et al., 2001; Xiao et al., 2005; Xu et al., 2000)). These observations require that current thinking on the pro-inflammatory activities of extracellular stress proteins be reconsidered.

2. Current Dogma

Stress proteins are immunodominant molecules and a significant element of the immune response to pathogenic micro-organisms is directed toward stress-protein-derived peptides (Kaufmann, 1990; Young, 1990). Stress proteins of microbial and mammalian origin also exhibit a high degree of phylogenetic similarity (~50%), and it has long been thought that these proteins might act as potentially harmful autoantigens (Kaufmann, 1990), and that immune recognition of cross-reactive stress protein epitopes provides a link between infection and autoimmunity (Lamb et al., 1989). Another dogma of stress protein biology is that the high degree of sequence homology between equivalent stress protein family members derived from prokaryotes and eukaryotes is reflected in a high degree of functional conservation. However, this concept has been questioned by a number of studies which have demonstrated that the biological activities of highly homologous stress proteins, even those from the same organism, can differ considerably (reviewed in Coates and Tormay, 2005; Pockley, 2003). It is therefore essential that studies clearly identify the precise nature and origin of the stress proteins to which functional properties are being assigned.

If extracellular stress proteins are indeed inflammatory in nature, then the identification of stress proteins in the peripheral circulation (Lewthwaite et al., 2002; Njemini et al., 2003; Pockley et al., 1998, 1999, 2000, 2002, 2003; Rea et al., 2001; Xiao et al., 2005; Xu et al., 2000) and their physiological release from a number of different cell types in the absence of pathological events (Bassan et al., 1998; Child et al., 1995; Hightower and Guidon, 1989; Hunter-Lavin et al., 2004; Liao et al., 2000) indicate that the recognition of, and responsiveness to these endogenous, "self" proteins must be tightly controlled unless a chronic inflammatory state is to ensue. An alternative proposition is that the qualitative nature of adaptive immune responses to endogenous (self, "en") stress proteins and exogenous (prokaryotic, "ex") stress proteins differ, and in the case of the stress proteins Hsp60 and Hsp70 this differential reactivity has been elegantly demonstrated in a number of studies.

3. Differential Recognition of Endogenous and Exogenous HSP60

The immune system has the potential to recognize and respond to endogenous stress proteins, as the normal T-cell repertoire includes low-affinity T cells that are reactive against such proteins (Cohen, 1996; Kaufmann, 1990; Macht et al., 2000; Munk et al., 1989; Ramage et al., 1999). Although it could be suggested that stress proteins are normally shielded from self-reactive T cells due to their intracellular location, this is not the case, as stress proteins are released from cells (Hightower and Guidon, 1989; Child et al., 1995; Bassan et al., 1998; Liao et al., 2000; Hunter-Lavin et al., 2004) and are present in the peripheral circulation (Lewthwaite et al., 2002; Njemini et al., 2003; Pockley et al., 1998, 1999, 2000,

2002, 2003; Rea et al., 2001; Xiao et al., 2005; Xu et al., 2000). The consequences of immune recognition with respect to the type of response induced reflect the qualitative nature of that response, and it is therefore important to distinguish between immune recognition and functional consequence.

Peripheral T cells are capable of distinguishing between enHsp60 and exHsp60, as the phenotype of T cells responding to eukaryotic and prokaryotic Hsp60 differs, as does the profile of cytokines secreted by T cells responding to endogenous and exogenous stress proteins. Whereas human Hsp60 activates $CD45RA^+RO^-$ (naïve) human peripheral blood T cells and bacterial-specific peptides activate $CD45RA^-RO^+$ (memory) T cells, bacterial Hsp60 (which contains both conserved (human) and non-conserved (bacterial) sequences) activates $CD45RA^+RO^-$ and $CD45RA^-RO^+$ T cells (Ramage et al., 1999). The cytokine secretion profile and regulatory activity of the $CD45RA^+RO^-$ and $CD45RA^-RO^+$ T cells that differentially recognize enHsp60 and exHsp60 is currently unknown. However, many findings suggest that the recognition of endogenous stress proteins results in an adaptive immune response which displays an anti-inflammatory phenotype and regulates inflammatory disease processes, whereas the recognition of exogenous stress proteins results in an adaptive immune response which promotes and/or enhances inflammatory disease.

4. Anti-Inflammatory Properties of HSP60 and HSP70

Although many studies have implicated immunity to stress proteins, particularly Hsp60 and Hsp70, in inflammatory conditions such as arthritis (de Graeff-Meeder et al., 1991; Gaston et al., 1990; Res et al., 1988), multiple sclerosis (Georgopoulos and McFarland, 1993; Stinissen et al., 1995; Wucherpfennig et al., 1992), diabetes (Child et al., 1993; Elias et al., 1990; Tun et al., 1994), and cardiovascular disease (Wick et al., 1995, 1996, 2001, 2004), others have demonstrated that immunity to stress proteins and its induction can attenuate inflammatory disease.

T cells isolated from the synovial fluid of patients with rheumatoid arthritis respond to enHsp60 by predominantly producing regulatory Th2-type cytokine responses, whereas the response to exHsp60 produces higher levels of IFN-γ, which is consistent with a pro-inflammatory Th1-type response (van Roon et al., 1997). In addition, T-cell lines generated from the synovial fluid of patients with rheumatoid arthritis in response to enHsp60 suppress the production of the pro-inflammatory cytokine TNF-α by peripheral blood mononuclear cells, whereas cells generated using mycobacterial Hsp65 have no such regulatory effect (van Roon et al., 1997).

The apparent anti-inflammatory nature of endogenous stress proteins has been confirmed by studies that have reported an inverse association between the severity of disease and the production of regulatory cytokines such as IL-4 and IL-10 by T cells stimulated with enHsp60 in patients with rheumatoid arthritis (de Graeff-Meeder et al., 1995; Macht et al., 2000; van Eden et al., 2005; van Roon et al., 1996). As a specific example, in patients with juvenile chronic arthritis, in whom the disease follows a relapsing-remitting rather than progressive course, circulating T cells responsive to human Hsp60 are of the regulatory Th2 phenotype and their presence is beneficial. In contrast, T cells reactive with the 65 kDa mycobacterial antigen Hsp65 display the inflammatory Th1 phenotype and their presence correlates with disease severity (van Roon et al., 1996). More recently, it has been reported that the spontaneous remission of juvenile idiopathic arthritis is associated with the presence of enHsp60-reactive CD30$^+$ T cells that produce the anti-inflammatory cytokine IL-10 (de Kleer et al., 2003). Although CD30 can be expressed by pro-inflammatory Th1 clones, its expression has been shown to be a feature of cultured immunoregulatory Th2 cells (Del Prete et al., 1995; Falini et al., 1995) and to define a subset of human-activated CD45$^+$RO$^+$ T cells (Ellis et al., 1993)—the population of peripheral blood T cells that responds to enHsp60 (Ramage et al., 1999).

The induction of T-cell reactivity to enHsp60 and enHsp70 has also been shown to down-regulate disease in a number of experimental arthritis models, by a mechanism that also appears to involve the development of Th2-type CD4$^+$ T cells producing the regulatory cytokines IL-4 and IL-10 (Anderton et al., 1995; Anderton and van Eden, 1996; Kingston et al., 1996; Paul et al., 2000; Tanaka et al., 1999; Thompson et al., 1990; van den Broek et al., 1989; Wendling et al., 2000). DNA vaccines encoding for Hsp60 and Hsp70 can also inhibit experimental adjuvant arthritis and diabetes in non-obese diabetic mice (Elias et al., 1991; Quintana et al., 2002, 2003, 2004).

An interesting feature of the immunoregulatory properties of Hsp60 is that the anti-inflammatory capacity of enHsp60 reactivity appears to dominate, as the administration of whole mycobacterial Hsp65, which contains the epitope that induces T-cell activation and can induce arthritis in rats when administered alone, does not induce the disease. It therefore appears that the concomitant presence of conserved (self)-epitopes can dominantly down-regulate the arthritogenic capacity of the non-conserved (non-self) epitopes (Anderton et al., 1995).

Much less studied has been the regulatory role of stress proteins in transplant rejection (Pockley, 2001). Studies have shown that immunizing recipient mice with enHsp60, or Hsp60 peptides that have the capacity to shift Hsp60 reactivity from a pro-inflammatory Th1 phenotype towards a regulatory Th2 phenotype delays murine skin allograft rejection (Birk et al., 1999). In the clinical situation, it also appears that the development of immune responses to enHsp60 can influence allograft rejection, as IL-10 production in response to enHsp60 peptides is increased in the late post-transplantation period (longer than 1 year) (Caldas et al., 2004). The recognition of peptides from the intermediate and C-terminal regions of the protein dominates at this time (Caldas et al., 2004).

5. Anti-Inflammatory Properties of HSP10, BiP, and GP96

Other stress proteins have been shown to exhibit anti-inflammatory activity. Human Hsp10 (formerly known as early pregnancy factor (Cavanagh, 1996; Rolfe et al.,

1984)) can inhibit inflammation in experimental allergic encephalomyelitis (the animal model of multiple sclerosis), delayed-type hypersensitivity responses and allograft rejection (Athanasas-Platsis et al., 2003; Morton et al., 2000; Zhang et al., 2003).

The protein BiP (grp78), a member of the 70-kDa family of stress protein molecules is an autoantigen in rheumatoid arthritis which stimulates specific responses in populations of synovial T cells from patients with rheumatoid arthritis (Corrigall et al., 2001). However, this response is not accompanied by the secretion of IFN-γ, which suggests that its phenotype is non-inflammatory (Corrigall et al., 2001). These anti-inflammatory properties have been confirmed by in vivo studies in which the intravenous or subcutaneous administration of BiP effectively prevents the induction of collagen-induced arthritis in mice and treats ongoing disease. The mechanism appears to involve a switch towards an immunoregulatory phenotype involving a heightened secretion of IL-4 and IL-10 (Brownlie et al., 2003; Corrigall et al., 2001; Panayi and Corrigall, 2006).

Gp96, also known as glucose-regulated protein (grp) 94 (Lee, 1981), endoplasmin (Koch et al., 1986) and 99-kDa endoplasmic reticulum protein Erp99 (Lewis et al., 1985) is a 94–96-kDa member of the 90-kDa family of molecular chaperones/stress proteins which resides within the lumen of the endoplasmic reticulum. Research in the mid-1980s demonstrated that the administration of a 96-kDa protein fractionated from a tumor cell lysate (subsequently shown to be gp96) to mice induced resistance to the same tumor from which the protein had been originally isolated (Srivastava et al., 1986). As intracellular chaperones, stress proteins bind a large number of peptides derived from the cells from which they are isolated (Gething and Sambrook, 1992; Young et al., 1993), the so-called "antigenic fingerprint" or repertoire of that cell (Srivastava and Udono, 1994). It has been proposed that the administration of gp96 isolated from murine tumor cells induces tumor-specific cytolytic T cells and anti-tumor immunity in mice, both of which are specific for tumor-derived peptides that are associated with the gp96 rather than the gp96 itself (Chandawarkar et al., 1999; Udono and Srivastava, 1994). Although it is widely thought that it is via this mechanism that gp96 induces tumor-specific immunity, this has been questioned (Demine and Walden, 2005) and an alternative mechanism involving the primary involvement of innate (non-antigen-specific) immune activation has been proposed (Baker-LePain et al., 2003; Nicchitta, 2003). Further details on the ability of gp96 to induce tumor immunity and the mechanisms involved can be found elsewhere in this volume.

Although gp96 has been reported to induce peptide-specific protective immunity, at high doses anti-tumor immunity is not apparent and it can attenuate autoimmune diseases. This observation was originally made in murine studies demonstrating that the induction of immunity to methylcholanthrene-induced (Meth A) fibrosarcoma by the administration of gp96-purified from Meth A fibrosarcoma cells displays a consistent dose restriction, in that two intra-dermal injections of 1 μg protect, whereas two injections of 10 μg do not (Chandawarkar et al., 1999). It was proposed that the suppressive activity was source-specific, i.e., that for the suppression of tumor immunity to occur, the gp96 must be purified from that tumor (Chandawarkar et al., 1999). However, subsequent work from the same investigators has shown that the administration of high-dose gp96 (2 × 100 μg subcutaneously) purified from normal liver tissue is equally able to suppress tumor immunity in mice, the onset of diabetes in non-obese diabetic mice and the induction of autoimmune encephalomyelitis by myelin basic protein and proteolipid protein in SJL mice (Chandawarkar et al., 2004).

High-dose gp96 administration has also been shown to improve the survival of murine skin transplants exhibiting minor and major antigenic disparity (Kovalchin et al., 2006). In this study, the capacity of gp96 to prolong graft survival was not dependent on the source of the tissue from which it was isolated (Kovalchin et al., 2006), thereby confirming that, unlike the influence of low-dose gp96 on tumor immunity is source-specific, the immunoregulatory activities of high-dose gp96 does not appear to be (Chandawarkar et al., 1999, 2004). The timing of the high-dose gp96 administration is important, as the immunoregulatory activity is only apparent if the gp96 is administered at the time of, or shortly after encounter with the antigenic stimulus (Chandawarkar et al., 2004). This suggests that the primary targets of the suppressive activity are recently activated rather than memory T cells (Chandawarkar et al., 2004). We have recently demonstrated that high-dose gp96 can prolong the survival of rat cardiac allografts; however, in our studies it appears that the immunoregulatory effect is specific, as the prolongation of donor organ survival requires the administration of donor tissue-derived gp96 (manuscript submitted for publication). The basis to these apparently discrepant results has yet to be defined.

6. Mechanistic Basis to the Anti-Inflammatory Properties of Stress Proteins?

There are a number of different stages in the development of an immune response at which stress proteins could influence the qualitative nature of that response. Firstly, stress proteins might influence the phenotype (antigen expression, cytokine secretion profile) of antigen-presenting cells (APCs), particularly dendritic cells (DCs), and their capacity to activate and influence the functional phenotype of responding T-cell populations. Secondly, stress proteins might directly interact with, and influence the functionality of different T-cell populations.

6.1. At the Level of the Antigen-Presenting Cell

Many studies have reported the effects of the stress proteins Hsp60, Hsp70, and gp96 on monocytes, macrophages, and DCs. Human Hsp60 induces the secretion of IL-6 from macrophages, as does chlamydial Hsp60 (Kol et al., 2000). With kinetics similar to those induced by lipopolysaccharide (LPS), enHsp60 has also been shown to induce a rapid release of TNF-α and nitric oxide from macrophages, as well as the expression of IL-12 and IL-15 (Chen et al., 1999). Human Hsp60 has also been shown to up-regulate CD86 and CD40 expression on a murine macrophage cell line and to enhance the maturation of immature DCs and the antigen-presenting capacity of Hsp60-stimulated cells (Flohé et al., 2003).

Endogenous Hsp70 has been shown to activate monocytes, the consequences of which are intracellular Ca^{2+} fluxes and the induction of pro-inflammatory cytokine (IL-1β, IL-6, TNF-α) secretion (Asea et al., 2000, 2002). Although endogenous gp96 has been reported by many groups to induce the maturation of, and cytokine secretion from DCs and other APCs, and to increase their capacity to stimulate the proliferation of T cells (Baker-LePain et al., 2002; Basu et al., 2000; Reed et al., 2003; Singh-Jasuja et al., 2000; Zheng et al., 2001), we (Mirza et al., in press) and others (Bethke et al., 2002) have found gp96 to have no effect on rat bone marrow-derived DCs or human peripheral blood monocyte-derived DCs, respectively.

Although the issue has not been specifically raised in the context of explaining their anti-inflammatory properties, it has been suggested that the pro-inflammatory effects of stress proteins such as Hsp60 and Hsp70 result from in vitro artifacts. The observations that stress proteins Hsp60, Hsp70 and gp96 bind to receptors that are also used by LPS (CD14, CD40, TLR) has prompted the proposition that some, if not all of the pro-inflammatory activities of stress proteins might result from the effects of LPS or other proteins, present as either a contaminant of the preparation or chaperoned by the stress protein under investigation (Bausinger et al., 2002; Gao and Tsan, 2003a,b; Gaston, 2002; Reed et al., 2003; Tsan and Gao, 2004; Wallin et al., 2002). This is especially so given that many studies use E. coli-derived recombinant protein preparations. Were this indeed to be the case, then the primary function of stress proteins could be more easily considered to be as antagonists rather than protagonists of inflammatory disease. However, much evidence argues against the inflammatory effects of stress proteins being a result of LPS contamination (discussed in Coates and Tormay, 2005; Vabulas and Wagner, 2005):

- Stress protein activity can be detected when LPS is undetectable, and the removal of traces of LPS with polymyxin B does not have any effect on activity (Asea et al., 2000; Bethke et al., 2002; Lewthwaite et al., 2001; Wang et al., 2001).
- The activity of stress proteins, but not that of LPS is lost following protease- or heat-treatment (Asea et al., 2000; MacAry et al., 2004; Panjwani et al., 2002).
- LPS is known to signal from the cell surface and does not need endocytosis for this activity (Ahmad-Nejad et al., 2002; Latz et al., 2002), whereas Hsp60, Hsp70, and gp96 binding is accompanied by endocytosis, and this is required to initiate the signaling and for biological activity (Basu et al., 2001; Becker et al., 2002; Delneste et al., 2002; Vabulas et al., 2001, 2002a).

An additional key piece of evidence which argues against the "LPS contamination theory" for the activity of Hsp70 is the observation that, unlike LPS, Hsp70 induces a transient cytoplasmic Ca^{2+} wave in monocytes which is clearly distinguishable from that induced by LPS (Asea et al., 2000).

On the basis of the above evidence, it is difficult to rationalize the apparent pro-inflammatory effects of endogenous stress proteins at the level of the innate immune system with their apparent in vivo anti-inflammatory effects, particularly with respect to Hsp60, Hsp70, and gp96. It is somewhat easier in the case of other stress protein, namely Hsp10, Hsp27, and BiP, all of which have been shown to confer an anti-inflammatory phenotype on APCs.

Recombinant human Hsp10 substantially free of LPS has been shown to inhibit LPS-induced activation of NF-κB, to reduce LPS-induced secretion of TNF-α, RANTES, and IL-6 and to enhance IL-10 secretion (Johnson et al., 2005). Although the precise mechanism underlying the anti-inflammatory activities of Hsp10 remain unclear, it has been suggested that this molecule might inhibit Toll-like receptor (TLR4) signaling, possibly via an interaction with extracellular Hsp60 (Johnson et al., 2005). The anti-inflammatory properties of BiP can also be attributed, in part at least, to its direct effects of APCs.

BiP induces the secretion of large amounts of IL-10, but a minimal and transient secretion of TNF-α from peripheral blood mononuclear cells. It also influences the capacity of APCs to activate T cells by down-modulating the expression of the essential co-stimulatory molecules CD80 and CD86 (Corrigall et al., 2004). On the basis of cytokine gene expression patterns, it appears that BiP "alternatively activates" monocytes (Gordon, 2003) and results in an anti-inflammatory cell which produces IL-10 (Corrigall and Panayi, 2005). BiP also inhibits the differentiation of monocytes into DCs and results in a reduced ability to induce allogeneic T-cell proliferation (Vittecoq et al., 2003). Although its clinical significance has yet to be established, another stress protein which has been shown, on the basis of in vitro experiments, to exhibit anti-inflammatory effects is Hsp27 (reviewed in (Laudanski et al., 2005)).

Human IL-27 enhances the secretion of IL-10 from human monocytes without concomitantly increasing the secretion of TNF-α (De et al., 2000). Its effects on IL-10 secretion results in an inhibition of IL-1β, TNF-α, IL-6 and IL-12 secretion, a decrease the expression of co-stimulatory molecules on DCs and prevents the development of APCs. Hsp27 also induces prostaglandin E_2 production by monocytes, which has a number of anti-inflammatory consequences (Hwang, 2000), and their production of macrophage-colony stimulating factor (M-CSF). M-CSF suppresses immunity in pregnancy and depletes APC precursors from the peripheral blood monocyte pool (De et al., 2003; Hashimoto et al., 1997). Hsp27 markedly reduces the expression of TLR4, slightly increases the expression of TLR2 and has no effect on the expression of TLR1 on monocytes, the consequences of the former being to reduce the pro-inflammatory responsiveness of these cells to LPS activation (Laudanski et al., 2005). The differentiation of monocytes into immature DCs is also inhibited by Hsp27 and this might profoundly influence the induction and qualitative nature of adaptive immunity (Laudanski et al., 2005). In contrast to its effects on monocyte differentiation, Hsp27 potently induces the maturation of DCs (Laudanski et al., 2005).

6.2. At the Level of the Adaptive Immune System

There are many stages in the induction and progression of adaptive immune responses at which endogenous stress proteins might influence the phenotype and immunoregulatory nature of those responses. It is possible that stress protein-reactive T cells recognizing endogenous stress protein epitopes via low-affinity interactions leads to the generation of Th2 (IL-4-producing), Th3 (TGFβ-producing), or Tr1 (IL-10-producing) regulatory T-cell responses. It might also be that T-cell responses to microbial stress proteins in the tolerizing environment of the gut leads to a regulatory response, or that T cells recognize endogenous stress protein epitopes as altered peptide ligands which do not fully activate T cells, and that this leads to the generation of regulatory cytokines (van Eden et al., 1998). Indeed, a lack of prior microbial exposure has been shown to predispose animals to autoimmunity (Rook and Stanford, 1998). There is much evidence to indicate that the anti-inflammatory activities of Hsp60, Hsp70, and BiP involve the induction of T-cell populations with a preponderance to secrete the immunoregulatory cytokines IL-4 and IL-10 (detailed above). However, the precise identity of these cells and the mechanism via which they become activated, or are regulated in vivo remains to be defined.

With regard to the immunoregulatory activity of gp96, high-dose gp96 administration appears to induce and/or activate immunoregulatory $CD4^+$ T-cell populations, as the inhibition of tumor immunity and protection from diabetes and EAE can be achieved by the adoptive transfer of $CD4^+$ T cells from animals that have been treated with high-dose gp96 (Chandawarkar et al., 1999; 2004). Given that the anti-inflammatory activity of $CD4^+$ T cells has been shown in a number of well-characterized model systems to partition, in part at least, into a naturally occurring $CD25^+$ subset (Gavin and Rudensky, 2003; Lee et al., 2004; Shevach, 2002; Waldmann et al., 2004; Wood and Sakaguchi, 2003), one possibility is that gp96 specifically influences the presence and/or functional activities of such cells. This is especially so given that gp96 has been reported to be a ligand for TLRs (Binder et al., 2004; Vabulas et al., 2002a,b) which are expressed on such regulatory T cells (Caramalho et al., 2003). Although the triggering of TLRs on $CD4^+CD25^+$ T cells by LPS has been shown to induce a tenfold increase in their suppressive activity (Caramalho et al., 2003), the suppressive effect of high-dose gp96 has not been shown to partition with the $CD4^+CD25^+$ or $CD4^+CD25^-$ phenotypes (Chandawarkar et al., 2004).

To date there are no definitive reports on the effects of gp96 on the induction or activities of specific regulatory T cell populations, nor is the mechanism by which high-dose gp96 can down-regulate inflammatory events known. However, it is possible that gp96 has direct effects on T cells as we (Mirza et al., in press) and others (Banerjee et al., 2002) have demonstrated that it can act as a co-stimulatory molecule for T cells and promote the secretion of an immunoregulatory Th2-like (IL-4, IL-10) cytokine profile.

7. Concluding Statement

Although this chapter has focused on the anti-inflammatory properties of stress proteins, there is also a wealth of literature indicating the stress proteins such as

Hsp60, Hsp70 and gp96 can be potent pro-inflammatory molecules (Henderson and Pockley, 2005; Pockley, 2003). The apparent dichotomous immune functionality of stress proteins has yet to be fully explained. The report that the treatment of human monocytes with enHsp60 suppresses their production of TNF-α following re-stimulation with enHsp60 or treatment with LPS, yet enhances their production of IL-1β, and that it down-regulates the expression of HLA-DR, CD86 and Toll-like receptor 4 (Kilmartin and Reen, 2004) highlights the complexity of stress protein-mediated immunoregulation and the difficulties that will be encountered in attempting to understand the balance between the ability of these proteins to control inflammatory and anti-inflammatory responses.

It is interesting to note that the anti-inflammatory capacity of enHsp60 re-activity appears to dominate, in that the administration of whole mycobacterial Hsp65, which contains the epitope which induces T-cell activation and can induce arthritis in rats when administered alone, does not induce the disease (Anderton et al., 1995). It therefore appears that the concomitant presence of conserved (self) epitopes can dominantly down-regulate the pro-inflammatory capacity of the non-conserved (non-self) epitopes (Anderton et al., 1995). Rather than promoting pro-inflammatory immune responses, endogenous stress proteins might be inherently anti-inflammatory and constitute a complex network and physiological mechanism for regulating and/or resolving, inflammatory disease processes (Panayi et al., 2004; van Eden et al., 1998; 2005).

Acknowledgments. The authors are grateful to the National Heart, Lung and Blood Institute, the Association for International Cancer Research, the British Heart Foundation, the Food Standards Agency, and Yorkshire Cancer Research for financial support.

References

Ahmad-Nejad, P., Hacker, H., Rutz, M., Bauer, S., Vabulas, R. M., and Wagner, H. (2002) Bacterial CpG-DNA and lipopolysaccharides activate toll-like receptors at distinct cellular compartments. *Eur J Immunol* 32:1958–68.

Anderton, S. M., van der Zee, R., Prakken, B., Noordzij, A., and van Eden, W. (1995) Activation of T cells recognizing self 60-kD heat shock protein can protect against experimental arthritis. *J Exp Med* 181:943–52.

Anderton, S. M., and van Eden, W. (1996) T lymphocyte recognition of hsp60 in experimental arthritis. In van Eden, W., and Young, D. (eds.): *Stress Proteins in Medicine.* Marcel Dekker, New York, pp. 73–91.

Asea, A., Kraeft, S.-K., Kurt-Jones, E. A., Stevenson, M. A., Chen, L. B., Finberg, R. W., Koo, G. C., and Calderwood, S. K. (2000) Hsp70 stimulates cytokine production through a CD14-dependent pathway, demonstrating its dual role as a chaperone and cytokine. *Nature Med* 6:435–42.

Asea, A., Rehli, M., Kabingu, E., Boch, J. A., Baré, O., Auron, P. E., Stevenson, M. A., and Calderwood, S. K. (2002) Novel signal transduction pathway utilized by extracellular HSP70. Role of Toll-like receptor (TLR) 2 and TLR4. *J Biol Chem* 277:15028–34.

Athanasas-Platsis, S., Zhang, B., Hillyard, N. C., Cavanagh, A. C., Csurhes, P. A., Morton, H., and McCombe, P. A. (2003) Early pregnancy factor suppresses the infiltration of lymphocytes and macrophages in the spinal cord of rats during experimental autoimmune encephalomyelitis but has no effect on apoptosis. *J Neurol Sci* 214:27–36.

Baker-LePain, J. C., Reed, R. C., and Nicchitta, C. V. (2003) ISO: A critical evaluation of the role of peptides in heat shock/chaperone protein-mediated tumor rejection. *Cur Opin Immunol* 15:89–94.

Baker-LePain, J. C., Sarzotti, M., Fields, T. A., Li, C. Y., and Nicchitta, C. V. (2002) GRP94 (gp96) and GRP94 N-terminal geldanamycin binding domain elicit tissue nonrestricted tumor suppression. *J Exp Med* 196:1447–59.

Banerjee, P. P., Vinay, D. S., Mathew, A., Raje, M., Parekh, V., Prasad, D. V., Kumar, A., Mitra, D., and Mishra, G. C. (2002) Evidence that glycoprotein 96 (B2), a stress protein, functions as a Th2-specific costimulatory molecule. *J Immunol* 169:3507–18.

Bassan, M., Zamostiano, R., Giladi, E., Davidson, A., Wollman, Y., Pitman, J., Hauser, J., Brenneman, D. E., and Gozes, I. (1998) The identification of secreted heat shock 60-like protein from rat glial cells and a human neuroblastoma cell line. *Neurosci Lett* 250:37–40.

Basu, S., Binder, R. J., Ramalingam, T., and Srivastava, P. K. (2001) CD91 is a common receptor for heat shock proteins gp96, hsp90, hsp70 and calreticulin. *Immunity* 14:303–13.

Basu, S., Binder, R. J., Suto, R., Anderson, K. M., and Srivastava, P. K. (2000) Necrotic but not apoptotic cell death releases heat shock proteins, which deliver a partial maturation signal to dendritic cells and activates the NF-κB pathway. *Int Immunol* 12:1539–46.

Bausinger, H., Lipsker, D., Ziylan, U., Manie, S., Briand, J. P., Cazenave, J. P., Muller, S., Haeuw, J. F., Ravanat, C., de la Salle, H., and Hanau, D. (2002) Endotoxin-free heat-shock protein 70 fails to induce APC activation. *Eur J Immunol* 32:3708–13.

Becker, T., Hartl, F. U., and Wieland, F. (2002) CD40, an extracellular receptor for binding and uptake of Hsp70-peptide complexes. *J Cell Biol* 158:1277–85.

Bethke, K., Staib, F., Distler, M., Schmitt, U., Jonuleit, H., Enk, A. H., Galle, P. R., and Heike, M. (2002) Different efficiency of heat shock proteins to activate human monocytes and dendritic cells: Superiority of HSP60. *J Immunol* 169:6141–8.

Binder, R. J., Vatner, R., and Srivastava, P. (2004) The heat-shock protein receptors: Some answers and more questions. *Tissue Antigens* 64:442–51.

Birk, O. S., Gur, S. L., Elias, D., Margalit, R., Mor, F., Carmi, P., Bockova, J., Altmann, D. M., and Cohen, I. R. (1999) The 60-kDa heat shock protein modulates allograft rejection. *Proc Nat Acad Sci USA* 96:5159–63.

Brownlie, R., Sattar, Z., Corrigall, V. M., Bodman-Smith, M. D., Panayi, G. S., and Thompson, S. (2003) Immunotherapy of collagen induced arthritis with BiP. *Rheumatology (Oxford)* 42(suppl):13.

Caldas, C., Spadafora-Ferreira, M., Fonseca, J. A., Luna, E., Iwai, L. K., Kalil, J., and Coelho, V. (2004) T-cell response to self HSP60 peptides in renal transplant recipients: A regulatory role? *Transplant Proc* 36:833–5.

Caramalho, I., Lopes-Carvalho, T., Ostler, D., Zelenay, S., Haury, M., and Demengeot, J. (2003) Regulatory T cells selectively express toll-like receptors and are activated by lipopolysaccharide. *J Exp Med* 197:403–11.

Cavanagh, A. C. (1996) Identification of early pregnancy factor as chaperonin 10: Implications for understanding its role. *Rev Reprod* 1:28–32.

Chandawarkar, R. Y., Wagh, M. S., Kovalchin, J. T., and Srivastava, P. (2004) Immune modulation with high-dose heat-shock protein gp96: Therapy of murine autoimmune diabetes and encephalomyelitis. *Int Immunol* 16:615–24.

Chandawarkar, R. Y., Wagh, M. S., and Srivastava, P. K. (1999) The dual nature of specific immunological activity of tumor-derived gp96 preparations. *J Exp Med* 189:1437–42.

Chen, W., Syldath, U., Bellmann, K., Burkart, V., and Kold, H. (1999) Human 60-kDa heat-shock protein: A danger signal to the innate immune system. *J Immunol* 162:3212–9.

Child, D., Smith, C., and Williams, C. (1993) Heat shock protein and the double insult theory for the development of insulin-dependent diabetes. *J Royal Soc Med (Eng)* 86:217–9.

Child, D. F., Williams, C. P., Jones, R. P., Hudson, P. R., Jones, M., and Smith, C. J. (1995) Heat shock protein studies in Type 1 and Type 2 diabetes and human islet cell culture. *Diabetic Med* 12:595–9.

Coates, A. R. M., and Tormay, P. (2005) Cell–cell signalling properties of chaperonins. In Henderson, B., and Pockley, A. G. (eds.): *Molecular Chaperones and Cell Signalling*. Cambridge University Press, New York, pp. 99–112.

Cohen, I. R. (1996) Heat shock protein 60 and the regulation of autoimmunity. In van Eden, W., and Young, D. B. (eds.): *Stress Proteins in Medicine*. Marcel Dekker, Inc, New York, pp. 93–102.

Corrigall, V. M., Bodman-Smith, M. D., Brunst, M., Cornell, H., and Panayi, G. S. (2004) Inhibition of antigen-presenting cell function and stimulation of human peripheral blood mononuclear cells to express an antiinflammatory cytokine profile by the stress protein BiP: relevance to the treatment of inflammatory arthritis. *Arthritis Rheum* 50:1164–71.

Corrigall, V. M., Bodman-Smith, M. D., Fife, M. S., Canas, B., Myers, L. K., Wooley, P., Soh, C., Staines, N. A., Pappin, D. J., Berlo, S. E., van Eden, W., van der Zee, R., Lanchbury, J. S., and Panayi, G. S. (2001) The human endoplasmic reticulum molecular chaperone BiP is an autoantigen for rheumatoid arthritis and prevents the induction of experimental arthritis. *J Immunol* 166:1492–8.

Corrigall, V. M., and Panayi, G. S. (2005) BiP, a negative regulator involved in rheumatoid arthritis. In Henderson, B., and Pockley, A. G. (eds.): *Molecular Chaperones and Cell Signalling*. Cambridge University Press, New York, pp. 234–48.

De, A. K., Kodys, K. M., Yeh, B. S., and Miller-Graziano, C. (2000) Exaggerated human monocyte IL-10 concomitant to minimal TNF-α induction by heat-shock protein 27 (Hsp27) suggests Hsp27 is primarily an anti-inflammatory stimulus. *J Immunol* 165:3951–8.

De, A. K., Laudanski, K., and Miller-Graziano, C. L. (2003) Failure of monocytes of trauma patients to convert to immature dendritic cells is related to preferential macrophage-colony-stimulating factor-driven macrophage differentiation. *J Immunol* 170:6355–62.

de Graeff-Meeder, E. R., van der Zee, R., Rijkers, G. T., Schuurman, H. J., Kuis, W., Bijlsma, J. W. J., Zegers, B. J. M., and van Eden, W. (1991) Recognition of human 60 kD heat shock protein by mononuclear cells from patients with juvenile chronic arthritis. *Lancet* 337:1368–72.

de Graeff-Meeder, E. R., van Eden, W., Rijkers, G. T., Prakken, B. J., Kuis, W., Voorhorst Ogink, M. M., van der Zee, R., Schuurman, H. J., Helders, P. J., and Zegers, B. J. (1995) Juvenile chronic arthritis: T cell reactivity to human HSP60 in patients with a favorable course of arthritis. *J Clin Invest* 95:934–40.

de Kleer, I. M., Kamphuis, S. M., Rijkers, G. T., Scholtens, L., Gordon, G., de Jager, W., Hafner, R., van de Zee, R., van Eden, W., Kuis, W., and Prakken, B. J. (2003) The spontaneous remission of juvenile idiopathic arthritis is characterized by $CD30^+$ T cells directed to human heat-shock protein 60 capable of producing the regulatory cytokine interleukin-10. *Arthritis Rheum* 48:2001–10.

Del Prete, G., De Carli, M., D'Elios, M. M., Daniel, K. C., Almerigogna, F., Alderson, M., Smith, C. A., Thomas, E., and Romagnani, S. (1995) CD30-mediated signaling promotes the development of human T helper type 2-like T cells. *J Exp Med* 182:1655–61.

Delneste, Y., Magistrelli, G., Gauchat, J., Haeuw, J., Aubry, J., Nakamura, K., Kawakami-Honda, N., Goetsch, L., Sawamura, T., Bonnefoy, J., and Jeannin, P. (2002) Involvement of LOX-1 in dendritic cell-mediated antigen cross-presentation. *Immunity* 17:353–62.

Demine, R., and Walden, P. (2005) Testing the role of gp96 as peptide chaperone in antigen processing. *J Biol Chem* 280:17573–8.

Elias, D., Markovits, D., Reshef, T., van der Zee, R., and Cohen, I. R. (1990) Induction and therapy of autoimmune diabetes in the non-obese diabetic mouse by a 65-kDa heat shock protein. *Proc Natl Acad Sci USA* 87:1576–80.

Elias, D., Reshef, T., Birk, O. S., van der Zee, R., Walker, M. D., and Cohen, I. R. (1991) Vaccination against autoimmune mouse diabetes with a T cell epitope of the human 65-kDa heat shock protein. *Proc Natl Acad Sci USA* 88:3088–91.

Ellis, T. M., Simms, P. E., Slivnick, D. J., Jack, H. M., and Fisher, R. I. (1993) CD30 is a signal-transducing molecule that defines a subset of human activated CD45RO[+] T cells. *J Immunol* 151:2380–9.

Falini, B., Pileri, S., Pizzolo, G., Durkop, H., Flenghi, L., Stirpe, F., Martelli, M. F., and Stein, H. (1995) CD30 (Ki-1) molecule: A new cytokine receptor of the tumor necrosis factor receptor superfamily as a tool for diagnosis and immunotherapy. *Blood* 85:1–14.

Flohé, S. B., Bruggemann, J., Lendemans, S., Nikulina, M., Meierhoff, G., Flohé, S., and Kolb, H. (2003) Human heat shock protein 60 induces maturation of dendritic cells versus a Th1-promoting phenotype. *J Immunol* 170:2340–8.

Gao, B., and Tsan, M. F. (2003a) Endotoxin contamination in recombinant human Hsp70 preparation is responsible for the induction of TNFα release by murine macrophages. *J Biol Chem* 278:174–9.

Gao, B., and Tsan, M. F. (2003b) Recombinant human heat shock protein 60 does not induce the release of tumor necrosis factor alpha from murine macrophages. *J Biol Chem* 278:22523–9.

Gaston, J. S. H. (2002) Heat shock proteins and innate immunity. *Clin Exp Immunol* 127:1–3.

Gaston, J. S. H., Life, P. F., Jenner, P. J., Colston, M. J., and Bacon, P. A. (1990) Recognition of a mycobacteria-specific epitope in the 65kD heat shock protein by synovial fluid derived T cell clones. *J Exp Med* 171:831–41.

Gavin, M., and Rudensky, A. (2003) Control of immune homeostasis by naturally arising regulatory CD4[+] T cells. *Cur Opin Immunol* 15:690–6.

Georgopoulos, C., and McFarland, H. (1993) Heat shock proteins in multiple sclerosis and other autoimmune diseases. *Immunology Today* 14:373–5.

Gething, M. J., and Sambrook, J. (1992) Protein folding in the cell. *Nature* 355:33–45.

Gordon, S. (2003) Alternative activation of macrophages. *Nat Rev Immunol* 3:23–35.

Hashimoto, S., Yamada, M., Motoyoshi, K., and Akagawa, K. S. (1997) Enhancement of macrophage colony-stimulating factor-induced growth and differentiation of human monocytes by interleukin-10. *Blood* 89:315–21.

Henderson, B., and Pockley, A. G. (eds.) (2005) *Molecular Chaperones and Cell Signalling*. Cambridge University Press, New York.

Hightower, L. E., and Guidon, P. T. (1989) Selective release from cultured mammalian cells of heat-shock (stress) proteins that resemble glia-axon transfer proteins. *J Cell Physiol* 138:257–66.

Hunter-Lavin, C., Davies, E. L., Bacelar, M. M., Marshall, M. J., Andrew, S. M., and Williams, J. H. (2004) Hsp70 release from peripheral blood mononuclear cells. *Biochem Biophys Res Commun* 324:511–7.

Hwang, D. (2000) Fatty acids and immune responses—a new perspective in searching for clues to mechanism. *Ann Rev Nutr* 20:431–56.

Johnson, B. J., Le, T. T., Dobbin, C. A., Banovic, T., Howard, C. B., Flores F de, M., Vanags, D., Naylor, D. J., Hill, G. R., and Suhrbier, A. (2005) Heat shock protein 10 inhibits lipopolysaccharide-induced inflammatory mediator production. *J Biol Chem* 280:4037–47.

Kaufmann, S. H. E. (1990) Heat shock proteins and the immune response. *Immunol Today* 11:129–36.

Kilmartin, B., and Reen, D. J. (2004) HSP60 induces self-tolerance to repeated HSP60 stimulation and cross-tolerance to other pro-inflammatory stimuli. *Eur J Immunol* 34:2041–51.

Kingston, A. E., Hicks, C. A., Colston, M. J., and Billingham, M. E. J. (1996). A 71-kD heat shock protein (hsp) from *Mycobacterium tuberculosis* has modulatory effects on experimental rat arthritis. *Clin Exp Immunol* 103:77–82.

Koch, G., Smith, M., Macer, D., Webster, P., and Mortara, R. (1986) Endoplasmic reticulum contains a common, abundant calcium-binding glycoprotein, endoplasmin. *J Cell Sci* 86:217–22.

Kol, A., Lichtman, A. H., Finberg, R. W., Libby, P., and Kurt-Jones, E. A. (2000) Heat shock protein (HSP) 60 activates the innate immune response: CD14 is an essential receptor for HSP60 activation of mononuclear cells. *J Immunol* 164:13–17.

Kovalchin, J. T., Mendonca, C., Wagh, M. S., Wang, R., and Chandawarkar, R. Y. (2006) *In vivo* treatment of mice with heat shock protein, gp96, improves survival of skin grafts with minor and major antigenic disparity. *Transpl Immunol* 15:179–85.

Lamb, J. R., Bal, V., Mendez-Samperio, A., Mehlert, A., So, J., Rothbard, J. B., Jindal, S., Young, R. A., and Young, D. B. (1989) Stress proteins may provide a link between the immune response to infection and autoimmunity. *Int Immunol* 1:191–6.

Latz, E., Visintin, A., Lien, E., Fitzgerald, K. A., Monks, B. G., Kurt-Jones, E. A., Golenbock, D. T., and Espevik, T. (2002) Lipopolysaccharide rapidly traffics to and from the Golgi apparatus with the toll-like receptor 4-MD-2-CD14 complex in a process that is distinct from the initiation of signal transduction. *J Biol Chem* 277:47834–43.

Laudanski, K., De, A. K., and Miller-Graziano, C. L. (2005) Hsp27 as an anti-inflammatory protein. In Henderson, B., and Pockley, A. G. (eds.): *Molecular Chaperones and Cell Signalling*. Cambridge University Press, New York, pp. 220–33.

Lee, A. S. (1981) The accumulation of three specific proteins related to glucose-regulated proteins in a temperature-sensitive hamster mutant cell line K12. *J Cell Physiol* 106:119–25.

Lee, M. K., Moore, D. J., Jarrett, B. P., Lian, M. M., Deng, S., Huang, X., Markmann, J. W., Chiaccio, M., Barker, C. F., Caton, A. J., and Markmann, J. F. (2004) Promotion of allograft survival by $CD4^+CD25^+$ regulatory T cells: Evidence for *in vivo* inhibition of effector cell proliferation. *J Immunol* 172:6539–44.

Lewis, M. J., Mazzarella, R. A., and Green, M. (1985) Structure and assembly of the endoplasmic reticulum. The synthesis of three major endoplasmic reticulum proteins during lipopolysaccharide-induced differentiation of murine lymphocytes. *J Biol Chem* 260:3050–7.

Lewthwaite, J., Owen, N., Coates, A., Henderson, B., and Steptoe, A. (2002) Circulating human heat shock protein 60 in the plasma of British civil servants. *Circulation* 106:196–201.

Lewthwaite, J. C., Coates, A. R. M., Tormay, P., Singh, M., Mascagni, P., Poole, S., Roberts, M., Sharp, L., and Henderson, B. (2001) Mycobacterium tuberculosis chaperonin 60.1 is a more potent cytokine stimulator than chaperonin 60.2 (hsp 65) and contains a CD14-binding domain. *Infect Immun* 69:7349–55.

Liao, D.-F., Jin, Z.-G., Baas, A. S., Daum, G., Gygi, S. P., Aebersold, R., and Berk, B. C. (2000) Purification and identification of secreted oxidative stress-induced factors from vascular smooth muscle cells. *J Biol Chem* 275:189–96.

MacAry, P. A., Javid, B., Floto, R. A., Smith, K. G. C., Singh, M., and Lehner, P. J. (2004) HSP70 peptide binding mutants separate antigen delivery from dendritic cell stimulation. *Immunity* 20:95–106.

Macht, L. M., Elson, C. J., Kirwan, J. R., Gaston, J. S. H., Lamont, A. G., Thompson, J. M., and Thompson, S. J. (2000) Relationship between disease severity and responses by blood mononuclear cells from patients with rheumatoid arthritis to human heat-shock protein 60. *Immunology* 99:208–14.

Matzinger, P. (1994) Tolerance, danger, and the extended family. *Ann Rev Immunol* 12:991–1045.

Mirza, S., Muthana, M., Fairburn, B., Slack, L. K., Hopkinson, K., Pockley, A. G. The stress protein gp96 is not an activator of resting rat bone marrow-derived dendritic cells, but is a co-stimulator and activator of CD3+ T cells. Cell Stress & Chaperones, in press.

Morton, H., McKay, D. A., Murphy, R. M., Somodevilla-Torres, M. J., Swanson, C. E., Cassady, A. I., Summers, K. M., and Cavanagh, A. C. (2000) Production of a recombinant form of early pregnancy factor that can prolong allogeneic skin graft survival time in rats. *Immunol Cell Biol* 78:603–7.

Munk, M. E., Schoel, B., Modrow, S., Karr, R. W., Young, R. A., and Kaufmann, S. H. E. (1989) T lymphocytes from healthy individuals with specificity to self-epitopes shared by the mycobacterial and human 65-kilodalton heat shock protein. *J Immunol* 143:2844–9.

Nicchitta, C. V. (2003) Re-evaluating the role of heat-shock protein-peptide interactions in tumour immunity. *Nat Rev Immunol* 3:427–32.

Njemini, R., Lambert, M., Demanet, C., and Mets, T. (2003) Elevated serum heat-shock protein 70 levels in patients with acute infection: use of an optimized enzyme-linked immunosorbent assay. *Scand J Immunol* 58:664–9.

Panayi, G. S., and Corrigall, V. M. (2006) BiP regulates autoimmune inflammation and tissue damage. *Autoimmun Rev* 5:140–42.

Panayi, G. S., Corrigall, V. M., and Henderson, B. (2004) Stress cytokines: Pivotal proteins in immune regulatory networks;opinion. *Cur Opin Immunol* 16:531–4.

Panjwani, N. N., Popova, L., and Srivastava, P. K. (2002) Heat shock proteins gp96 and hsp70 activate the release of nitric oxide by APCs. *J Immunol* 168:2997–3003.

Paul, A. G. A., van Kooten, P. J. S., van Eden, W., and van der Zee, R. (2000) Highly autoproliferative T cells specific for 60-kDa heat shock protein produce IL-4/IL-10 and IFN-γ and are protective in adjuvant arthritis. *J Immunol* 165:7270–7.

Pockley, A. G. (2001) Heat shock proteins, heat shock protein reactivity and allograft rejection. *Transplantation* 71:1503–7.

Pockley, A. G. (2003) Heat shock proteins and their role as regulators of the immune response. *Lancet* 362:469–76.

Pockley, A. G., Bulmer, J., Hanks, B. M., and Wright, B. H. (1999) Identification of human heat shock protein 60 (Hsp60) and anti-Hsp60 antibodies in the peripheral circulation of normal individuals. *Cell Stress Chap* 4:29–35.

Pockley, A. G., de Faire, U., Kiessling, R., Lemne, C., Thulin, T., and Frostegård, J. (2002) Circulating heat shock protein and heat shock protein antibody levels in established hypertension. *J Hypertension* 20:1815–20.

Pockley, A. G., Georgiades, A., Thulin, T., de Faire, U., and Frostegård, J. (2003) Serum heat shock protein 70 levels predict the development of atherosclerosis in subjects with established hypertension. *Hypertension* 42:235–8.

Pockley, A. G., Shepherd, J., and Corton, J. (1998) Detection of heat shock protein 70 (Hsp70) and anti-Hsp70 antibodies in the serum of normal individuals. *Immunol Invest* 27:367–77.

Pockley, A. G., Wu, R., Lemne, C., Kiessling, R., de Faire, U., and Frostegård, J. (2000) Circulating heat shock protein 60 is associated with early cardiovascular disease. *Hypertension* 36:303–7.

Prakken, B. J., Roord, S., Ronaghy, A., Wauben, M., Albani, S., and van Eden, W. (2003) Heat shock protein 60 and adjuvant arthritis: A model for T cell regulation in human arthritis. *Springer Semin Immunopathol* 25:47–63.

Quintana, F. J., Carmi, P., Mor, F., and Cohen, I. R. (2002) Inhibition of adjuvant arthritis by a DNA vaccine encoding human heat shock protein 60. *J Immunol* 169:3422–8.

Quintana, F. J., Carmi, P., Mor, F., and Cohen, I. R. (2003) DNA fragments of the human 60-kDa heat shock protein (HSP60) vaccinate against adjuvant arthritis: Identification of a regulatory HSP60 peptide. *J Immunol* 3533–41.

Quintana, F. J., Carmi, P., Mor, F., and Cohen, I. R. (2004) Inhibition of adjuvant-induced arthritis by DNA vaccination with the 70-kd or the 90-kd human heat-shock protein: Immune cross-regulation with the 60-kd heat-shock protein. *Arthritis Rheum* 50:3712–20.

Quintana, F. J., and Cohen, I. R. (2005) Heat shock proteins regulate inflammation by both molecular and network cross-reactivity. In Henderson, B. and Pockley, A. G. (eds.): *Molecular Chaperones and Cell Signalling*. Cambridge University Press, New York, pp. 263–87.

Ramage, J. M., Young, J. L., Goodall, J. C., and Hill Gaston, J. S. (1999) T cell responses to heat shock protein 60: Differential responses by CD4$^+$ T cell subsets according to their expression of CD45 isotypes. *J Immunol* 162:704–10

Rea, I. M., McNerlan, S., and Pockley, A. G. (2001) Serum heat shock protein and anti-heat shock protein antibody levels in aging. *Exp Gerontol* 36:341–52.

Reed, R. C., Berwin, B., Baker, J. P., and Nicchitta, C. V. (2003) GRP94/gp96 elicits ERK activation in murine macrophages. A role for endotoxin contamination in NF-κB activation and nitric oxide production. *J Biol Chem* 278:31853–60.

Res, P. C., Schaar, C. G., Breedveld, F. C., van Eden, W., van Embden, J. D. S., Cohen, I. R., and De Vries, R. R. P. (1988) Synovial fluid T cell reactivity against 65 kDa heat shock protein of mycobacteria in early chronic arthritis. *Lancet* ii:478–80.

Rolfe, B., Cavanagh, A., Forde, C., Bastin, F., Chen, C., and Morton, H. (1984) Modified rosette inhibition test with mouse lymphocytes for detection of early pregnancy factor in human pregnancy serum. *J Immunol Methods* 70:1–11.

Rook, G. A., and Stanford, J. L. (1998) Give us this day our daily germs. *Immunol Today* 19:113–6.

Sattar, Z., Brownlie, R., Corrigall, V. M., Bodman-Smith, M. D., Staines, N. A., Panayi, G. S., et al. (2003) CD4+ T cells specific for the stress protein BiP modulate the development of collagen induced arthritis. *Rheumatology (Oxford)* 42 suppl:124.

Shevach, E. M. (2002) CD4+CD25+ suppressor T cells: More questions than answers. *Nat Rev Immunol* 2:389–400.

Singh-Jasuja, H., Scherer, H. U., Hilf, N., Arnold-Schild, D., Rammensee, H.-G., Toes, R. E. M., and Schild, H. (2000) The heat shock protein gp96 induces maturation of dendritic cells and down-regulation of its receptor. *Eur J Immunol* 30:2211–5.

Srivastava, P. K., DeLeo, A. B., and Old, L. J. (1986) Tumor rejection antigens of chemically induced sarcomas of inbred mice. *Proc Natl Acad Sci USA* 83:3407–11.

Srivastava, P. K., and Udono, H. (1994) Heat shock protein–peptide complexes in cancer immunotherapy. *Cur Opin Immunol* 6:728–32.

Stinissen, P., Vandevyver, C., Medaer, R., Vandegaar, L., Nies, J., Tuyls, L., Hafler, D. A., Raus, J., and Zhang, J. (1995) Increased frequency of $\gamma\delta$ T cells in cerebrospinal fluid and peripheral blood of patients with multiple sclerosis: Reactivity, cytotoxicity, and T cell receptor V gene rearrangements. *J Immunol* 154:4883–94.

Tanaka, S., Kimura, Y., Mitani, A., Yamamoto, G., Nishimura, H., Spallek, R., Singh, M., Noguchi, T., and Yoshikai, Y. (1999) Activation of T cells recognizing an epitope of heat-shock protein 70 can protect against rat adjuvant arthritis. *J Immunol* 163: 5560–5.

Thompson, S. J., Rook, G. A. W., Brealey, R. J., van der Zee, R., and Elson, C. J. (1990) Autoimmune reactions to heat shock proteins in pristane induced arthritis. *Eur J Immunol* 20:2479–84.

Tsan, M. F., and Gao, B. (2004) Heat shock protein and innate immunity. *Cell Mol Immunol.*:274–9.

Tun, R. Y. M., Smith, M. D., Lo, S. S. M., Rook, G. A. W., Lydyard, P., and Leslie, R. D. G. (1994) Antibodies to heat shock protein 65 kD in Type 1 diabetes mellitus. *Diabetic Med* 11:66–70.

Udono, H., and Srivastava, P. K. (1994) Comparison of tumor-specific immunogenicities of stress-induced proteins gp96, hsp90 and hsp70. *J Immunol* 152:5398–403.

Vabulas, R. M., Ahmad-Nejad, P., da Costa, C., Miethke, T., Kirschning, C. J., Hacker, H., and Wagner, H. (2001) Endocytosed HSP60s use toll-like receptor 2 (TLR2) and TLR4 to activate the toll/interleukin-1 receptor signaling pathway in innate immune cells. *J Biol Chem* 276:31332–9.

Vabulas, R. M., Braedel, S., Hilf, N., Singh-Jasuja, H., Herter, S., Ahmad-Nejad, P., Kirschning, C. J., Da Costa, C., Rammensee, H. G., Wagner, H., and Schild, H. (2002a) The endoplasmic reticulum-resident heat shock protein Gp96 activates dendritic cells via the Toll-like receptor 2/4 pathway. *J Biol Chem* 277:20847–53.

Vabulas, R. M., and Wagner, H. (2005) Toll-like receptor-dependent activation of antigen presenting cells by Hsp60, Hsp70 and gp96. In Henderson, B., and Pockley, A. G., (eds.): *Molecular Chaperones and Cell Signalling*. Cambridge University Press, New York, pp. 113–32.

Vabulas, R. M., Wagner, H., and Schild, H. (2002b) Heat shock proteins as ligands of toll-like receptors. *Cur Top Microbiol Immunol* 270:169–84.

van den Broek, M. F., Hogervorst, E. J. M., van Bruggen, M. C. J., van Eden, W., van der Zee, R., and van den Berg, W. (1989) Protection against streptococcal cell wall induced arthritis by pretreatment with the 65kD heat shock protein. *J Exp Med* 170:449–66.

van Eden, W., van der Zee, R., Paul, A. G. A., Prakken, B. J., Wendling, U., Anderton, S. M., and Wauben, M. H. M. (1998) Do heat shock proteins control the balance of T-cell regulation in inflammatory diseases? *Immunol Today* 19:303–7.

van Eden, W., van der Zee, R., and Prakken, B. (2005) Heat shock proteins induce T-cell regulation of chronic inflammation. *Nat Immunol* 5:318–30.

van Roon, J., van Eden, W., Gmelig-Meylig, E., Lafeber, F., and Bijlsma, J. (1996) Reactivity of T cells from patients with rheumatoid arthritis towards human and mycobacterial hsp60. *FASEB J* 10:A1312.

van Roon, J. A. G., van Eden, W., van Roy, J. L. A. M., Lafeber, F. J. P. G., and Bijlsma, J. W. J. (1997) Stimulation of suppressive T cell responses by human but not bacterial 60-kD heat shock protein in synovial fluid of patients with rheumatoid arthritis. *J Clin Invest* 100:459–63.

Vittecoq, O., Corrigall, V. M., Bodman-Smith, M. D., and Panayi, G. S. (2003) The molecular chaperone BiP (GRP78) inhibits the differentiation of normal human monocytes into immature dendritic cells. *Rheumatology (Oxford)* 42 (suppl):43.

Waldmann, H., Graca, L., Cobbold, S., Adams, E., Tone, M., and Tone, Y. (2004) Regulatory T cells and organ transplantation. *Semin Immunol* 16:119–26.

Wallin, R. P. A., Lundqvist, A., Moré, S. H., von Bonin, A., Kiessling, R., and Ljunggren, H.-G. (2002) Heat-shock proteins as activators of the innate immune system. *Trends Immunol* 23:130–5.

Wang, Y., Kelly, C. G., Karttunen, T., Whittall, T., Lehner, P. J., Duncan, L., MacAry, P., Younson, J. S., Singh, M., Oehlmann, W., Cheng, G., Bergmeier, L., and Lehner, T. (2001) CD40 is a cellular receptor mediating mycobacterial heat shock protein 70 stimulation of CC-chemokines. *Immunity* 15:971–83.

Wendling, U., Paul, L., van der Zee, R., Prakken, B., Singh, M., and van Eden, W. (2000) A conserved mycobacterial heat shock protein (hsp) 70 sequence prevents adjuvant arthritis upon nasal administration and induces IL-10-producing T cells that cross-react with the mammalian self-hsp70 homologue. *J Immunol* 164:2711–7.

Wick, G., Kleindienst, R., Amberger, A., Schett, G., Seitz, C., Dietrich, H., and Xu, Q. (1996) The role of heat shock protein 65/60 in the initial immune-mediated stages of atherosclerosis. In Hansson, G., and Libby, P. (eds.): *Immune Functions of the Vessel Wall*. Harwood Academic Publishers GmbH, Amsterdam, pp. 173–83.

Wick, G., Kleindienst, R., Schett, G., Amberger, A., and Xu, Q. (1995) Role of heat shock protein 65/60 in the pathogenesis of atherosclerosis. *Int Arch Allergy Immunol* 107:130–1.

Wick, G., Knoflach, M., and Xu, Q. (2004) Autoimmune and inflammatory mechanisms in atherosclerosis. *Ann Rev Immunol* 22:361–403.

Wick, G., Perschinka, H., and Millonig, G. (2001) Atherosclerosis as an autoimmune disease: An update. *Trends Immunol* 22:665–9.

Wood, K. J., and Sakaguchi, S. (2003) Regulatory T cells in transplantation tolerance. *Nat Rev Immunol* 3:199–210.

Wucherpfennig, K., Newcombe, J., Li, H., Keddy, C., and Cuzner, M. L. (1992) $\gamma\delta$T cell receptor repertoire in acute multiple sclerosis lesions. *Proc Natl Acad Sci USA* 89:4588–92.

Xiao, Q., Mandal, K., Schett, G., Mayr, M., Wick, G., Oberhollenzer, F., Willeit, J., Kiechl, S., and Xu, Q. (2005) Association of serum-soluble heat shock protein 60 with carotid atherosclerosis: Clinical significance determined in a follow-up study. *Stroke* 36:2571–6.

Xu, Q., Schett, G., Perschinka, H., Mayr, M., Egger, G., Oberhollenzer, F., Willeit, J., Kiechl, S., and Wick, G. (2000) Serum soluble heat shock protein 60 is elevated in subjects with atherosclerosis in a general population. *Circulation* 102:14–20.

Young, D., Romain, E., Moreno, C., O'Brien, R., and Born, W. (1993) Molecular chaperones and the immune system response. *Phil Trans Royal Soc Lond* 339:363–7.

Young, R. A. (1990) Stress proteins and immunology. *Ann Rev Immunol* 8:401–20.

Zanin-Zhorov, A., Bruck, R., Tal, G., Oren, S., Aeed, H., Hershkoviz, R., Cohen, I. R., and Lider, O. (2005) Heat shock protein 60 inhibits Th1-mediated hepatitis model via innate regulation of Th1/Th2 transcription factors and cytokines. *J Immunol* 174:3227–36.

Zhang, B., Walsh, M. D., Nguyen, K. B., Hillyard, N. C., Cavanagh, A. C., McCombe, P. A., and Morton, H. (2003) Early pregnancy factor treatment suppresses the inflammatory response and adhesion molecule expression in the spinal cord of SJL/J mice with experimental autoimmune encephalomyelitis and the delayed-type hypersensitivity reaction to trinitrochlorobenzene in normal BALB/c mice. *J Neurol Sci* 212:37–46.

Zheng, H., Dai, J., Stoilova, D., and Li, Z. (2001) Cell surface targeting of heat shock protein gp96 induces dendritic cell maturation and anti-tumor immunity. *J Immunol* 167:6734–5.

18
Heat Shock Proteins and Neurodegenerative Diseases

IAN R. BROWN

Centre for the Neurobiology of Stress, Department of Life Sciences, University of Toronto at Scarborough, Toronto, Ontario, Canada M1C 1A4

1. Introduction

The prevalence in the aging human population of neurodegenerative diseases such as Alzheimer's disease, Parkinson's disease, amyotrophic lateral sclerosis, and Huntington's disease is a burden on present society in terms of cost to health care systems and emotional stress on family members. What is the current status of research on these neurodegenerative disorders? Are these diseases entirely separate in character or are there commonalities? Are attempts to investigate the molecular basis of neurodegenerative diseases revealing clues that might lead to the development of rational therapeutics? Increasing evidence supports the view that the above-mentioned neurodegenerative diseases have common molecular mechanisms associated with protein misfolding and aggregation. These diseases have been termed "protein-misfolding disorders" or "protein-conformational disorders" that are characterized by the accumulation of intracellular and extracellular protein aggregates. This review explores the possibility that manipulation of the cellular stress response offers opportunities to counter conformational changes in proteins that trigger pathogenic cascades that result in neurodegenerative diseases.

2. Characteristics of Neurodegenerative Diseases

Alzheimer's disease is a late-onset dementing illness that results in a progressive loss of memory. Neurons in the brain degenerate and protein aggregates form. The extracellular aggregates are termed neuritic plaques and they contain a peptide derived from the proteolytic cleaving of the amyloid precursor protein (APP). The intracellular aggregates are neurofibrillary tangles of the microtubule associated protein tau (Martin, 1999; Selkoe, 2004). Parkinson's disease is associated with resting tremor, rigidity, and slow movement. Dopaminergic neurons in the substantia nigra of the midbrain degenerate. Lewy bodies with high levels of alpha-synuclein protein form inside these neurons (Schapira and Olanow, 2004; Selkoe, 2004). Amyotrophic lateral sclerosis (ALS) is a fatal disease resulting from the degeneration of motor neurons in the spinal cord and cerebral cortex. The familial form of ALS appears to be associated with mutations in the superoxide dismutase (SOD1) gene. Aggregates of the SOD1 protein are observed in transgenic mice overexpressing mutant SOD1 (Bruijn et al., 2004). Huntington's disease is neuronal degenerative disease caused by an expansion of a CAG repeat coding for polyglutamine in the N terminus of the huntingtin protein. When expansion of the repeat exceeds a threshold, the disease is triggered and protein aggregates are observed in neurons (Ross and Poirier, 2004).

3. Heat Shock Proteins and Quality Control of Protein Folding

Heat shock proteins play important roles in the modulation of protein folding in vivo. In non-stressed cells, constitutively expressed heat shock proteins transiently bind to newly synthesized proteins to promote correct conformational folding and prevent inappropriate protein interactions and aggregation. Stressful stimuli can trigger the misfolding of cellular proteins. Stress-induced heat shock proteins facilitate the correct refolding of proteins (Morimoto et al., 1990; 1994). In addition to this refolding mechanism, cells can also degrade misfolded proteins via the ubiquitin-proteasome pathway and lysosome-mediated autophagy, processes that involve heat shock proteins Hsc70 and Hsp70. The refolding and degradative mechanisms function to prevent the buildup of misfolded cellular proteins. In neurodegenerative diseases these surveillance systems that monitor protein folding quality control appear to be overwhelmed and misfolded proteins accumulate in aggregates exhibiting a fibrillar structure (Sherman and Goldberg, 2001; Soto, 2003; Barral et al., 2004; Ross and Poirier, 2004; Meriin and Sherman, 2005; Muchowski and Wacker, 2005). Controversy exits as to whether these aggregates are toxic or inert or represent a protective "sequestering" mechanism for sidetracking potentially damaging misfolded proteins. Toxic soluble, oligomers, rather than mature fibrils, may be the misfolded proteins than actually trigger cascades that results in neuronal cell death (Conway et al., 2000; Kayed et al., 2003; Agorogiannis et al., 2004; Schaffar et al., 2004).

Dynamic imaging on living cells has demonstrated that Hsp70 exhibits a rapid cycle of association and dissociation with huntingtin inclusion bodies, perhaps reflecting attempts to refold, solubilize or degrade aggregated proteins or sequester toxic soluble oligomers (Kim et al., 2002). In contrast, the proteosome appears to be sequestered irreversibly within aggregates of overexpressed N terminal mutant huntingtin fragment or polyglutamine expansion proteins (Holmberg et al., 2004) and within ALS-associated mutant SOD1 aggregates (Matsumoto et al., 2005). This impairs the functional ability of the proteosome system to degrade and clear misfolded proteins as the aggregated proteins are kinetically trapped within the proteosome. Live cell imaging reveals distinct aggregate structures for specific disease-associated proteins. Mutant huntingtin forms aggregates composed of a dense core, inaccessible to other cellular proteins, surrounded by a surface that sequesters other proteins and interacts transiently with heat shock proteins (Matsumoto et al., 2006). Mutant SOD1 forms aggregates that exhibit a porous structure through which other cellular proteins can diffuse. The huntingtin aggregate contains a single exterior surface surrounding the core to which other proteins can bind. The mutant SOD1 aggregates are porous and exhibit binding sites for other cellular proteins throughout the aggregate. The porous structure of the SOD1 aggregates could generate a higher level of aberrant interaction and sequestering of cellular protein compared to the single exterior surface on the huntingtin aggregates (Matsumoto et al., 2006).

4. Neural Heat Shock Response

4.1. Stress-Inducible and Constitutively Expressed Heat Shock Proteins

In response to a wide range of stressful stimuli, cells induce the highly conserved heat shock response that results in the rapid induction of genes encoding heat shock proteins that play important roles in cellular repair and protective mechanisms (Morimoto et al., 1990; 1994). This superfamily is organized by molecular size and function into the Hsp100, Hsp90, Hsp70, Hsp60, Hsp40, and small heat shock protein families. Some members of these families are constitutively expressed in unstressed cells. Many heat shock proteins act as molecular chaperones involved in the folding, intracellular disposition and degradation of proteins (Hartl, 1996; Bukau and Horwich, 1998). Induction of the heat shock response protects cells from subsequent stress that would normally be lethal. This mechanism, termed "induced thermotolerance," was discovered in tissue culture experiments, but the ability of heat shock proteins to protect against cellular damage has been demonstrated in vivo in the brain after a range of stressful stimuli (Brown, 1994; Brown and Sharp, 1999).

In response to a fever-like increase in body temperature, glial cells in the mammalian brain show a robust induction the Hsp70, Hsp32, and Hsp27 (Sprang and Brown, 1987; Manzerra and Brown, 1992a,b; Manzerra et al., 1993; Foster et al., 1995; Foster and Brown, 1997; Manzerra et al., 1997; Bechtold and Brown, 2000, 2003). Neuronal cell populations, which exhibit high levels of constitutively expressed Hsc70 and Hsp90, demonstrate comparatively little induction of Hsp70 following hyperthermia (Manzerra and Brown, 1992b; Manzerra et al., 1993; Foster et al., 1995; Quraishi and Brown, 1995). More severe stress, such as ischemia, results in the induction of Hsp70 in neurons (Franklin et al., 2005). The high levels of constitutively expressed Hsc70 in neuronal cells may buffer or preprotect neurons against stress (Manzerra and Brown, 1996; Manzerra et al., 1997). Motor and sensory neurons in the brain stem and spinal demonstrate particularly high levels of constitutively expressed Hsp27 and a high threshold for induction of Hsp70 (Batulan et al., 2003; Franklin et al., 2005). The constitutive expression of heat shock proteins is developmentally regulated in the nervous system (D'Souza and Brown, 1998).

4.2. Cell-to-Cell Transfer of Stress Proteins

Heat shock proteins may be transferred between cell types in the nervous system. The first report was the observation that thermal stress induces the synthesis of Hsp70 in glial cells that are situated in the sheath that surrounds the squid giant axon and that this glial Hsp70 is rapidly transported into the adjacent axon (Tytell et al., 1986). The "glial to neuron" transfer system provides a mechanism for fast delivery of neuroprotective Hsp70 to cellular processes that are distant for the neuronal cell body. In the mammalian nervous system, hyperthermic stress triggers the appearance of Hsp27 and Hsp32 in supportive "perisynaptic" glial processes that surround synaptic termini (Bechtold and Brown, 2000). These glial stress proteins are transferred to the postsynaptic element, particularly the postsynaptic density that provides a structural framework for signal transduction complexes involved in neurotransmission. Tissue culture experiments indicate the glial cells can release stress-induced Hsp70 and that neurons exposed to exogenous Hsp70 exhibit an enhancement of tolerance to neuronal stress (Guzhova et al., 2001). In vivo administration of exogenous Hsp70 has been shown to inhibit motor and sensory nerve degeneration (Tidwell et al., 2004; Robinson et al., 2005; Tytell, 2005).

4.3. Protection of Synapses by Heat Shock Proteins

Induction of the heat shock response protects the nervous system at a functional level. Neurons communicate via neurotransmission events at synaptic connections. During stressful conditions, neurotransmission must be preserved in order to prevent breakdown of intercellular communication in the nervous system. Prior heat shock protects synaptic neurotransmission and synapses are able to function at temperature conditions that would normally be disruptive (Karunanithi et al., 1999). Selective overexpression of Hsp70 further enhances the level of synaptic protection (Karunanithi et al., 2002). Addition of recombinant Hsp70 to the medium of a brain slice culture system preserved normal synaptic transmission after hyperthermic stress and was as effective as induction of heat shock proteins by hyperthermic pre-conditioning (Kelty et al., 2002). Hsp70, Hsp32, and Hsp27 have been localized to synaptic termini (Bechtold and Brown, 2000; Bechtold et al., 2000). Thus induction of the heat shock response protects the nervous system at a functional level to permit neurotransmission events to proceed at synaptic connections under stressful conditions.

4.4. Association of Stress Proteins with Lipid Raft Organizational Platforms

Lipid rafts are specialized membrane microdomains that serve as major assembly and sorting platforms for signal transduction complexes. The brain is enriched in lipid rafts as more than 1% of total brain protein is recovered in a lipid raft fraction whereas less of 0.1% of total protein is associated with lipid rafts isolated from non-neural tissue (Maekawa et al., 2003). A range of neurotransmitters receptors and constitutively expressed heat shock proteins are present in neural lipid rafts (Chen et al., 2005). Following hyperthermia, Hsp70 is detected in lipid rafts suggesting that it may play roles in maintaining the stability of raft-associated signal transduction complexes following neural stress (Chen et al., 2005).

Heat shock proteins accumulated in extracellular plaques in the brains of Alzheimer's patients (Muchowski and Wacker, 2005); however, these stress proteins do not contain a secretory signal sequence. It has been suggested that the interaction of Hsp70 with lipid rafts might play a role in the extracellular accumulation of this stress protein (Broquet et al., 2003; Hunter-Lavin et al., 2004). Others have suggested that exosomes, small membrane vesicles that are released by numerous cell types, may be involved in the secretory pathway for stress proteins (Lancaster and Febbraio, 2005).

4.5. Release of Heat Shock Proteins into the Bloodstream

Increasing evidence suggests that heat shock proteins are released into the bloodstream after stressful stimuli and that this may represent a previously unrecognized feature of the stress response (Fleshner and Johnson, 2005). Exercise stress has been reported to induce the release of Hsp70 from the human brain into the blood stream in vivo (Lancaster et al., 2004). The biological significance of the neural release of Hsp70 remains to be determined. Research has focused on the role of intracellular Hsp70 and very little is known about the potential function of extracellular Hsp70 or its possible relevance to neurodegenerative diseases.

4.6. Neural Cell Death and the Heat Shock Response

Stressful stimuli can elicit two reactive cellular responses, the heat shock response and the activation of cell death pathways (Samali and Orrenius, 1998). Most studies on the effects of hyperthermia and the mammalian nervous system and other body tissues have focused on the heat shock response (Brown and Sharp, 1999). A fever-like increase in body temperature triggers a differential response in tissues of the adult rat. Apoptosis is triggered in dividing cell populations of the testis and thymus but not in mature, postmitotic cells of the adult cerebellum (Khan et al., 2002). At early stages of neural development when neural cells were engaged in cell division, proliferative neural regions of the cerebellum and neocortex were highly susceptible to hyperthermia-induced apoptosis. These results suggest the actively dividing cell populations are more prone to cell death induced by hyperthermia than fully differentiated postmitotic neural cells.

Analysis at the cellular level has revealed that cell types that were not pushed into cell death by a fever-like temperature either induction Hsp70 (Belay and Brown, 2003) or demonstrated high levels of constitutively expressed Hsc70 (Belay and Brown, 2006). Neurons are postmitotic cells that must engage in continual, highly complex neurotransmission events at synaptic connections throughout life.

Protective mechanisms are likely highly evolved in neurons. There is a premium on mounting efforts to protect and preserve these critical postmitotic cells and prevent their loss by inhibiting the induction of cell death. Why do protein aggregates and inclusion bodies accumulation in neuronal cells during several neurodegenerative diseases? Is this reflective of cellular attempts to mount a protective response and keep neurons functional as long as possible even in the face of major protein misfolding problems? Could enhancement of neural heat shock response heighten neural protective mechanisms to keep more neurons viable during the progression of neurodegenerative diseases?

4.7. Ectopic Expression of Cell-Cycle-Related Proteins as Early Markers of Neuronal Distress

A close association has been observed between the pattern of neuronal cell death in Alzheimer's disease and cellular processes that normally occur only during the mitotic cell cycle. Neurons that are destined to die in Alzheimer's disease express cell-cycle-related proteins at an early stage, well before the first amyloid deposits are observed (Arendt, 2000; Yang et al., 2001; Herrup and Arendt, 2002; Yang et al., 2003). This ectopic expression of cell-cycle-related proteins is a useful marker and an early sign of neuronal distress. The same anatomically pattern of ectopic expression has been observed in both the human disease and in mouse models of the disease (Yang et al., 2006). It has been proposed that Alzheimer's disease is a manifestation of loss of differentiation control in a subset of neurons that express immature features in the adult brain (Arendt, 2000).

Ectopic expression of cell-cycle-related proteins leads to a cascade of events that results in neuronal cell death (Herrup and Arendt, 2002). Why do normally postmitotic neuronal cells attempt a process that strongly resembles mitotic cell division? Could this be a protective response by a neuron under stress? This phenomenon, in which neuronal cell death is associated with re-entrance into a lethal cell cycle, has been observed in the retinoblastoma-deficient mouse (Lee et al., 1992), cerebellar target-related cell death (Herrup and Busser, 1995), oncogene expression in maturing neurons (Feddersen et al., 1992), the oxidative stress model of the harlequin mouse (Klein et al., 2002), stroke models (Katchanov et al., 2001), a mouse model of ALS (Ranganathan et al., 2001; Ranganathan and Browser, 2003), a mouse model of ataxia-telangiectasia (Yang and Herrup, 2005), and in tissue culture systems (Farinelli and Greene, 1996; Giovanni et al., 1999; Greene et al., 2004). Initiation of cell cycling in neurons appears to be a causative factor in the death of neurons on these models. However, the death of neurons in Alzheimer's disease occurs months after expression of cycle-cell-related proteins (Busser et al., 1998). Cell-cycle events may be a necessary first step, but they are not the immediate cause of neuronal cell death in mouse models, and a "second hit" is likely needed (Yang et al., 2006). It has been suggested that tau isoforms or expression patterns could be a second trigger (Andorfer et al., 2005).

In a variety of neurodegenerative conditions, ectopic expression of cell-cycle-related proteins is an early marker of neuronal distress. The pattern of ectopic expression of these cell cycle proteins in human Alzheimer's disease is mirrored in mouse Alzheimer's models (Yang et al., 2006). This indicates that overexpression of a single human disease gene in mouse models reproduces the complex anatomical pattern of neuronal distress that is observed in the human neurodegenerative disease.

5. Protective Roles of Heat Shock Proteins in Neurodegenerative Diseases

Heat shock proteins appear to be protective in animal models of neurodegenerative diseases. In transgenic mouse models of Alzheimer's disease, intracellular beta-amyloid peptide triggers neuronal dysfunction before it accumulates in extracellular plaques (Chui et al., 1999; Hsia et al., 1999; Moechars et al., 1999; Kumar-Singh et al., 2000). Overexpression of Hsp70 rescues neurons from intracellular beta-amyoid-mediated toxicity (Magrane et al., 2004; Smith et al., 2005). In a *Drosophila* model of Parkinson's disease, expression of wild-type or mutant alpha-synuclein in dopaminergic neurons results in formation of inclusion bodies and a loss of neurons. Co-expression of human Hsp70 prevents alpha-synuclein-mediated toxicity (Auluck et al., 2002). A recent study indicates that Hsp70 inhibits alpha-synuclein fibril formation via preferential binding to prefibrillar species and that cellular toxicity arises from these prefibrillar forms (Dedmon et al., 2005). In an ALS model, co-injection of expression vectors for HSP70 and mutant SOD1 into primary culture motor neurons reduces the toxicity of mutant SOD1, decreases SOD1 aggregation and enhances motor neuron survival compared with the injection of the SOD1 vector alone (Bruening et al., 1999).

Huntington's disease is caused by expansion of a CAG repeat coding for polyglutamine. Studies in yeast, *C. elegans*, *Drosophila* and mice models of polyglutamine expansion diseases have demonstrated that heat shock proteins such as Hsp70 and Hsp40 suppresses polyQ toxicity (Warrick et al., 1998; 1999; Chai et al., 1999; Jana et al., 2000; Muchowski et al., 2000; Cummings et al., 2001; Kobayashi and Sobue 2001; Zhou et al., 2001; Meriin et al., 2002; Adachi et al., 2003; Hsu et al., 2003; Schaffar et al., 2004). Genetic screens in yeast, *Drosophila* and *C. elegans* models of polyQ aggregation support the role of heat shock proteins such as Hsp70 and Hsp40 as suppressors of polyQ-mediated toxicity (Fernandez-Funez et al., 2000; Kazemi-Esfarjani and Benzer, 2000; Willingham et al., 2003; Nollen et al., 2004). In vitro and in vivo studies have also demonstrated neuroprotective effects of the small heat shock protein Hsp27 (Latchman, 2005).

Heat shock proteins likely promote protection against toxic proteins that arise in neurodegenerative diseases at several levels that include prevention of inappropriate protein interactions, promotion of protein degradation, sequestering of toxic protein forms, and blocking signal events that lead to neuronal dysfunction and cell death (Muchowski and Wacker, 2005).

6. Regulation of the Heat Shock Response

6.1. HSF1, *the Master Stress-Inducible Regulator*

In mammalian cells the heat shock response is controlled at the transcriptional level by heat shock transcription factor 1 (HSF1) (Morimoto, 1998). This factor is constitutively expressed in cells and present as an inert monomer. Following stress, HSF1 is de-repressed, trimerizes, and the trimers bind with high affinity to heat shock elements (HSEs) that are present in the promoter regions of target genes. Transcriptional activation of this stress-inducible gene set requires the post-translational phosphorylation of HSF1 (Guettouche et al., 2005; Holmberg et al., 2001, 2002; Hong et al., 2001; Kim et al., 2005). Mathematical modeling suggests that the balance of kinase and phosphatase action on HSF1 regulates the heat shock response (Rieger et al, 2005). Little is known about the signaling cascades that regulate these kinases and phosphatases but such information could identify targets for modulating levels of induced heat shock proteins. HSF1 is also post-translationally modified by sumoylation (Holmberg et al., 2002; Hietakangas et al., 2003).

When heat shock proteins are induced, they are able to autorepress their induction through interactions with HSF1 (Morimoto, 1998). This interaction also modulates the peak of the response and mathematical modeling suggests that disruption of the interaction may be a target for enhancing the levels of heat shock proteins (Rieger et al, 2005). However, chronic high levels of expression of these stress proteins could results in negative effects as heat shock proteins interact with multiple components of signaling pathways that regulate growth and development (Nollen and Morimoto, 2002). A balance is required between increased cellular protection against misfolded proteins and negative effects on cellular function and viability.

The molecular biology of the heat shock response has been extensively analyzed in tissue culture systems with less investigation of the intact nervous system. In mammalian tissue culture systems, supra-physiological temperature elevations are required to elicit a robust activation of HSF1 to DNA-binding form and subsequent triggering of the induction of heat shock genes (Jurivich et al., 1994). Such temperatures increases are lethal to a mammal. How do intact, thermoregulating mammals react to lower temperature elevations of 2–3°C like those attained during fever and inflammation?

In contrast to mammalian tissue culture experiments, a fever-like increase in body temperature is sufficient to activate HFS1 to a DNA-binding form and induce Hsp70 in the rabbit brain (Brown and Rush, 1996). This study suggests that additional factors are present in the brain in vivo that modulate the threshold temperature for activation of HSF1 and facilitate the induction of the heat shock response in the nervous system under conditions that are physiological relevant. These factors have been lost in tissue culture systems and supra-physiological temperature levels are required to induce synthesis of heat shock proteins in vitro.

Modulation of the threshold for activation of HSF1 in tissue culture systems has been reported. Low concentrations of arachidonate, a central mediator of the inflammatory response, that alone do not activate HSF1 to a DNA-binding form, can reduce the temperature threshold for HSF1 activation in tissue culture cells to levels that are physiologically relevant (Jurivich et al., 1994). A beneficial approach for the therapeutic treatment of neurodegenerative disorders could be the development of drugs that reduce of the threshold for in vivo activation of HSF1 in the nervous system.

Immunocytochemical studies indicate that HSF1 is prepositioned at abundant levels in the nucleus of both neuronal and glial cells in the unstressed mammalian brain (Brown and Rush, 1999). Following a fever-like temperature, glial cells rapidly induce Hsp70 whereas populations of neurons do not. The lack of Hsp70 induction in neurons does not appear to be due to deficiencies in levels of nuclear HSF1 but rather to deficiencies in its activation to a DNA-binding form (Batulan et al., 2003). During postnatal development of the cerebellum, levels of HSF1 increase progressively from birth to day 30 (Brown and Rush, 1999; Morrison et al., 2000).

6.2. Aging, Control of Life Span, and Expression of Heat Shock Proteins

During aging, a progressive decrease in HSF1 activity and heat shock protein expression has been reported (Walker et al., 2003; Shamovsky and Gershon, 2004; Tonkiss and Calderwood, 2005). Does this phenomenon place neural cells at risk for the development of neurodegenerative diseases? Neuroprotective mechanisms in which heat shock proteins suppress misfolded, aggregation-prone proteins may be operational in human brain cells for decades until the aging process diminishes cellular capacity to deal with toxic misfolded proteins due to a reduced presence of heat shock proteins.

Mutational or pharmacologically impairment of Hsp90 suggests that this heat shock protein plays a key role in facilitating the correct folding of proteins that have acquired misfolding errors (Rutherford and Lindquist, 1998; Queitsch et al., 2002; Sangster et al., 2004). If this Hsp90 buffering capacity is compromised during aging, misfolded, aggregation-prone proteins that were phenotypically silent earlier in life might be exposed. The process of protein aggregation in the presence of heat shock proteins has been studied using mathematical modeling (Rieger et al., 2006). This investigation indicates that slight variations in the concentration of heat shock proteins have important consequences for the threshold of onset of protein aggregation both in vitro and vivo.

Elevation of HSF1 increases life span in *C. elegans* and *Drosophila* and inhibition of HSF1 expression or function decreases longevity (Garigan et al., 2002; Hsu et al., 2003). The increase in life span appears to be due to an increased transcription of genes encoding small heat shock proteins (Kurapati et al., 2000; Hsu et al., 2003). Inactivation of the Hsp22 gene in *Drosophila* greatly decreases life span (Morrow et al., 2004a,b). Expression of this small heat shock protein in motor neurons appears to play a key role in longevity.

Strategies to extend life span appear to delay the onset of aging-related diseases characterized by the appearance of misfolded and aggregation-prone proteins. For example, age-dependent aggregation of polyglutamine proteins is suppressed in C elegans long-lived age-1 mutants suggesting that longevity is associated with protection against misfolded proteins (Morley et al., 2002). HSF1 and heat shock proteins have been shown to be regulators of longevity in *C. elegans*, hence the cellular stress response and pathways controlling life span are linked. (Morley and Morimoto, 2004).

7. Therapeutic Approaches to Neurodegenerative Diseases

7.1. Enhancement of the Heat Shock Response

Overexpression studies on Hsp70, Hsp40, and Hsp27 have demonstrated protective effects of heat shock proteins in animal models of neurodegenerative diseases (Muchowski and Waker, 2005). These studies usually involve the overexpression of a single heat shock protein. However, a set of stress proteins is induced after stress and perhaps induction of a set could be more neuroprotective than overexpression of one type of heat shock protein. In a mouse model of Huntington's disease, a progressive decrease in a set of heat shock proteins has been observed (Hay et al., 2004). Can knowledge that has been attained on the neuroprotective effects of heat shock proteins in animal model systems be translated into approaches to combat human neurodegenerative diseases? Animal models use genetic manipulation approaches that are not easy to transfer to human application. An approach that could be employed in humans is the development of drugs that could trigger or enhance the induction of a set of heat shock proteins in the brain. Such drugs would need to have the additional characteristics of having the ability, at non-toxic levels, to cross the blood brain and induce appropriate levels of heat shock proteins in particular neural cell types that can confer protection against the progression of neurodegenerative protein-misfolding disorders.

In neurodegenerative diseases, distortion of protein folding quality control mechanisms results in the accumulation of protein species that form oligomers, aggregates and inclusion bodies. A common feature of these diseases is the formation of off-pathway folding intermediates that are unstable, self-associated and lead to a chronic imbalance in protein homeostatis and resultant impairment of cell function. Enhancing components of the quality control machinery, particularly levels of heat shock proteins, could be a general approach to the treatment of neurodegenerative misfolding diseases. This has stimulated interest in the regulation of heat shock proteins by HSF1, the master stress-inducible regulator, and the investigation of pharmacologically active small molecules that regulate HSF1, as a therapeutic strategy for neurodegenerative diseases that have alterations in protein conformation that imbalances homeostatis (Westerheide and Morimoto, 2005).

7.2. Hsp90 Inhibitors as Tools to Activate HSF1

The activity of HSF1 is negatively regulated by a complex that includes Hsp90 (Zou et al., 1998). The drug geldanamycin binds to the ATP site on Hsp90 and blocks its interaction with HSF1, promoting HSF1 activation and the induction of heat shock proteins (Prodromou et al., 1997; Zou et al., 1998; Kim et al., 1999; Bagatell et al., 2000). The antibiotic radicicol exhibits a similar functional mechanism but has a 50-fold greater affinity for Hsp90 (Roe et al., 1999). Geldanamycin activates the heat shock response and inhibits the formation of huntingtin aggregates in a cell culture model of Huntington's disease (Sittler et al., 2001). Radicicol and geldanamycin induce stress proteins in a hippocampal slice culture system derived from a mouse model of Huntington's disease and alter the detergent soluble properties of polyQ aggregates (Hay et al., 2004). Geldanamycin induces Hsp70 and protects against dopaminergic neurotoxicity in *Drosophila* and mouse models of Parkinson's diseases (Auluck et al. 2005; Shen et al., 2005). Another Hsp90 inhibitor, 17-AAG, reduces polyglutamine-mediated motor neurons degeneration (Waza et al., 2005).

Radicicol and geldanamycin were initially identified in screens for anti-cancer drugs and subsequently developed for their potential as cancer therapies (Whitesell et al., 2003). 17-AAG has completed a multi-institutional phase I clinical trial and a phase II trial is underway (Bagatell and Whitesell, 2004). Application of these drugs to human neurodegenerative diseases will require analysis of their abilities to cross the blood brain barrier and investigation of neurotoxicity issues. It has been suggested the Hsp90 plays an important role in facilitating the correct folding of proteins that have acquired misfolding errors (Rutherford and Lindquist, 1998; Queitsch et al., 2002; Sangster et al., 2004). Hence the application of inhibitors that compromise the function of Hsp90 could have unanticipated negative outcomes in exposing misfolded, aggregation-prone proteins that are phenotypically silent (Sangster et al., 2004; Whitesell and Lindquist, 2005).

7.3. Co-Inducers of the Heat Shock Response—Arimoclomol Delays Disease Progression in ALS Mice

The hydroxylamine derivate bimoclomol, and its potent analogue BRX-220, are co-inducers of the heat shock response. Bimoclomol elevates the expression of heat shock proteins in response to stress and has been shown to be beneficial in both in vitro and in vivo models of ischemia and diabetic complications (Biro et al., 1997, 1998; Jednakovits et al., 2000). BRX-220 up-regulated heat shock proteins and rescued motor neurons from axotomy-induced cell death in neonatal rats (Kalmar et al., 2002). It has also been shown to promoted functional properties in a sensory nerve system following peripheral nerve injury (Kalmar et al, 2003).

Arimoclomol, another analog of bimoclomol that acts as a co-inducer of heat shock proteins, shows particular promise as a potential therapeutic agent for the treatment of neurodegenerative disorders. This drug has been shown to delay disease progression in a mouse model of amylotrophic lateral sclerosis (ALS) (Kieran et al., 2004; reviewed by Benn and Brown, 2004; Kalmar et al., 2005). ALS is a fatal neurodegenerative condition in humans involving progressive paralysis resulting from the death of motor neurons in the spinal cord and motor cortex (Bruijn et al., 2004). There is no cure and death occurs 1 to 5 years after diagnosis. Twenty percentage of the familial case of ALS carry mutations in the gene encoding Cu/Zn superoxide dismutase-1 (SOD1) (Rosen et al., 1993).

An animal model of ALS has been created using transgenic mice that overexpress human mutant SOD1 and exhibit a phenotype and pathology similar to that observed in human ALS patients (Gurney et al., 1994; Wong et al., 1995). Treatment of these human mutant SOD1 transgenic mice with arimoclomol, a co-inducer of heat shock proteins, delayed disease progression in that a marked improvement in hind limb muscle function and a decrease in motor neuron cell death were observed that resulted in a 22% increase in life span (Kieran et al., 2004). A significant delay in ALS disease progression was observed even when the arimoclomol is administered after the visible onset of disease symptoms. Observations on the mutant SOD1 mice indicted that the arimoclomol prolonged the activation of HSF1, resulting in an increased expression of Hsp70 and Hsp90 in motor neurons in the spinal cord. Interestingly, cultured motor neurons fail to express Hsp70 after heat shock or after exposure to excitotoxic glutamate or expression of mutant SOD1 (Batulan et al., 2003), hence these neurons are hampered in their ability to react to cellular stress. Treatment of SOD1 mice with the co-inducer arimoclomol overcomes the intrinsic inability of motor neurons to induce the Hsp70 and rescues these neurons in the mouse model of ALS (Kieran et al., 2004).

The arimoclomol report demonstrates that pharmacological activation of the heat shock response is a valid therapeutic approach to treating ALS that could be applied to other neurodegenerative diseases (Kieran et al., 2004). Heat shock proteins are known to confer protection to cells under acute stress situations. However, in chronic diseases the heat shock response does not appear to be able to deal with prolonged exposure to a stressful environment. Increased levels of heat shock proteins have been observed in brain tissue from human ALS patients and in mutant SOD1 mice (Garofalo et al., 1991; Vleminckx, et al., 2002), but motor neurons are still pushed into cell death. Pharmacological amplification of the heat shock response, by an agent such as arimoclomol, alters disease progression, reducing the death rate of motor neurons, improving hind limb muscle function and prolonging life span (Kieran et al., 2004).

A multi-center phase II clinical trial on arimoclomol, sponsored by CytRx, started in October 2005 in the USA (www.clinicaltrials.gov/ct/show/NCT00244 244). The primary purpose of this multi-center trial is to evaluate the safety and tolerability of arimocolomol in ALS patients following 90 days of dosing. In addition, the amount of arimoclomol in blood and cerebrospinal fluid is being measured.

While the arimoclomol results in the animal ALS model are impressive, it is necessary to realize that many of the mouse models of neurodegenerative diseases facilitate the study of inherited forms of the disease not the sporadic forms that are much more common. For example, 90% of all ALS cases in the human population are the sporadic form (Soto, 2003) for which there is no universally accepted animal model (Lipton, 2004). Transgenic mice overexpressing mutant human SOD1 are useful for studying hereditary ALS. In the case of Alzheimer's disease, 95% of the cases are sporadic and 5% inherited and Parkinson's disease is mostly sporadic and rarely inherited (Soto, 2003).

Circumstantial evidence has pointed to glutamate, the major excitatory transmitter of the brain, as a possible candidate in sporadic ALS (Lipton and Rosenberg, 1994; Rothstein, 1995). Recent studies suggest that abnormal post-transcriptional modification of mRNA encoding the glutamate receptor subunit GluR2 may promote sporadic ALS (Kawahara et al., 2004). This abnormal RNA editing may lead to enhanced glutamate receptor activity and calcium influx and resultant excitotoxicity that triggers cell death. Abnormal editing of GluR2 has been observed in spinal cord motor neurons from ALS patients (Kawahara et al., 2004). These findings suggest a route to the creation of mouse models of sporadic ALS. There is a transgenic mouse line with calcium permeable GluR2 subunits that develops a motor neuron disease in late life (Feldmeyer et al., 1999). A refinement would be the development of a mouse line that specifically expresses unedited GluR2 subunits in spinal cord motor neurons. The development of such models would facilitate the testing of whether therapeutic drugs such as arimoclomol, that amplify the heat shock response, can protect motor neurons from cell death in sporadic forms of ALS or whether these types of drugs are only effective for hereditary forms of the disease.

8. Neurodegenerative Drug Screening Consortium

8.1. New Uses for Old Drugs

In a quest to speed up the search for new treatments of neurodegenerative diseases, investigators from 26 academic laboratories joined forces and tested 1,040 existing drugs, three-quarters of which have been approved by the US Food and Drug Authority (FDA) (Abbott, 2002; Heemskerk et al., 2002). The design strategy being that since the toxicological profiles of the bulk of the drugs was known, beneficial compounds would experience a shorter path to clinical trials that novel drugs. The testing program was sponsored by the National Institute of Neurological Disorders and Stroke, the Huntington's Disease Society of America, the Amyotrophic Lateral Sclerosis Association, and the Hereditary Disease Foundation. The objective was to find new uses for known drugs employing a battery of experimental assays for neurodegeneration. The investigators each carried out blind tests on the same set of compounds which were selected in association with Microsource Discovery Systems (Gaylordsville, CT). By design, the selected compounds were enriched in compounds known to cross the blood–brain barrier and the tests employed concentrations that are achievable in humans. The collection included Federal Drug Administration (FDA)-approved drugs, controlled substances, and natural products.

A total of 29 assays, mostly cell culture based, were employed to simulate various aspects of neurodegeneration. Different investigators were assigned different sets of test assays. Some assays targeted protein aggregation and others tested for drugs that blocked the toxicity of disease causing proteins (Heemskerk et al., 2002). The proteins tested included mutant SOD, an enzyme altered in familial ALS; several polyglutamine-containing proteins thought to be involved in Huntington's disease; and alpha-synuclein, which has been implicated in Parkinson's disease. The 29 screens generated 294 compounds from the list of 1,040 that showed activity in one or more assays. Of these positives, 17 compounds were active on three or more assays. Of that subgroup, eight compounds showed overlap in two or more of the polyglutamine toxicity assays and seven overlaps were found in the ALS-related assays (Heemskerk et al, 2002).

8.2. Celastrol as a Novel Inducer of the Heat Shock Response

One of the laboratories involved in the screening consortium was that of Richard Morimoto at Northwestern University, Evanston, Illinois. The positive compounds that showed one or more hits in the primary screen were subsequently screened in his laboratory in human cells with an hsp70.1 promoter-luciferase reporter to determine if any of these compounds increased expression of the inducible hsp70 promoter (Westerheide et al., 2004). Of the 7 positive compounds identified in this assay, one compound, celastrol, a quinone methide triterpene, was of great interest as it was independently identified in five other consortium laboratories using 6 different assays for huntingtin aggregation and toxicity. Celastrol is a natural product derived from the Celastraceae family of plants (Ngassapa, et al., 1994; Chen et al., 1999). Extracts of these plants are used in traditional Chinese herbal medicine for the treatment of fever, joint pain, inflammation, edema, rheumatoid arthritis, and bacterial infection (Li and Shun, 1989; Gunatilaka et al., 1996).

The Morimoto laboratory then proceeded to characterize celastrol as a novel inducer of the heat shock response in both human Hela and neuroblastoma cells grown in tissue culture (Westerheide et al., 2004). A structure/function study revealed that the celastrol structure is very specific and activated HSF1 with kinetics similar to that observed following heat stress as determined by induction of HSF1 DNA binding, hyperphosphorylation of HSF1, and expression of Hsp70, Hsp40, and Hsp27 genes. In addition, celastrol exhibited cytoprotective properties permitting cells to survive a normally lethal heat stress and avoid apoptotic cell death as assayed by a decrease in cytosolic nucleosomes. These studies have identified celastrol as a new class of pharmacologically active regulators of the heat shock response. At the optimal concentration of 3 μM, the observed levels of induction in cultured human cells are comparable or even greater than that obtained with standard 42°C heat treatment of culture cells. Higher concentrations (10 μM) have been reported to exhibit toxic effects in tissue culture cells (Piccioni et al., 2004). Celastrol has been independently identified in five other laboratories in the Neurodegenerative Drug Screening Consortium as being positive in six different cell-based screens for reducing huntingtin aggregation and toxicity.

Activators of the heat shock response often generate unfolded proteins or inactivate molecular chaperones that participate in feedback inhibition of the HSF1 transcriptional response. Tests in the Morimoto laboratory reveal that this is not the case for celastrol (Westerheide et al., 2004). Celastrol has several advantages compared to other small molecular modulators of the heat shock response. Its kinetics of activation of the heat shock response is rapid and the magnitude of induction is comparable or greater than obtained with heat shock. The optimal induction concentration of celastrol is low at 3 μM compared to millimolar concentrations required by other compounds. Extracts containing celastrol have been given to Chinese patients for many years without reports of carcinogenic or other side effects (Li and Shun, 1989; Gunatilaka et al., 1996). In rat models of Alzheimer's disease, celastrol at 7 μg/kg improved memory, learning, and psychomotor activity (Allison et al., 2001). Given the pleiotropic functions of HSF1 and heat shock proteins, a challenge for the future is the development of modulators that alter the expression of heat shock regulated genes with some degree of tissue and cell specificity (Westerheide and Morimoto, 2005).

8.3. Reversal of a Full-Length Mutant Huntingtin Neuronal Cell Phenotype by Celastrol

Another group who participated in the Neurodegenerative Drug Screening Consortium is Marcy MacDonald and James Gusella at Massachusetts General Hospital in Boston. They screened the collection of 1040 FDA approved drugs and natural compounds for their ability to prevent in vitro aggregation of an amino terminal fragment of huntingtin that contains the expanded polyglutamine tracts (Wang et al., 2005). Ten positive compounds were identified that included celastrol. This was a useful screening method, however, mouse models of Huntington's disease (HD) that carry the amino terminal fragment of huntingtin become ill very fast and do not entirely model the development of HD in humans. The problem is that such HD mice have only the amino-terminal of the HD protein, while in humans the fragmentation of the HD protein and the aggregation of the fragments are now understood to be downstream in the HD disease process. Other mouse models have been developed that have the genetic coding for the entire mutant huntingtin protein. These mice are slower to develop HD symptoms in a time course than parallel the human progression of HD.

The MacDonald/Gusella group then subjected the ten compounds that were positive in the first screen to a second screen using a striatal cell line derived from mice that express the full-length mutant huntingtin protein (Wang et al., 2005). The goal was to determine what compounds reversed the abnormal localization of the full-length protein. Using this design the researchers hoped to identify if any of the ten compounds were effective earlier in the disease process. They found that celastrol, an inhibitor of polyglutamine-mediated aggregation, reversed the full-length mutant huntingtin neuronal cell phenotype. Thus identification and testing of compounds that alter in vitro aggregation is a viable approach for defining potential therapeutic compounds that may counter the negative conformational property of full-length mutant huntingtin.

8.4. Neuroprotective Effects of Celastrol in Animal Models of Neurodegenerative Diseases

A third line of investigation suggesting that celastrol is a potential therapeutic tool for neurodegenerative disease, is in vivo investigations from the laboratory of Flint Beal at the Weill Medical College on Cornell University, New York Presbyterian Hospital (Cleren et al., 2005). Mice were injected intraperitoneally with celastrol before and after injections of MPTP, a dopaminergic neurotoxin which produces a model of Parkinson's disease. A 48% loss of dopaminergic neurons induced by MPTP in the substantia nigra was significantly reduced by celastrol treatment. In addition, depletion of dopamine concentration induced by MPTP was reduced by celastrol treatment.

Administration of 3-nitropropionic acid, a neurotoxin, induces lesions in the striatum and is a rat model for Huntington's disease. Intraperitoneal injection of celastrol significantly decreased the striatal lesions induced by this neurotoxin (Cleren et al., 2005). Celastrol was observed to induce Hsp70 in dopaminergic neurons and reduce astrogliosis. Further experiments are required to investigate whether the neuroprotective properties of celastrol stem from its induction of heat shock proteins in the nervous system.

9. Conclusion

In summary, current research suggests that manipulation of the cellular heat shock response is a promising avenue for the development of therapeutic tools for the treatment of neurodegenerative diseases that are associated with protein conformational disorders. For example, arimoclomol, a co-inducer of the heat shock response, commenced a phase II clinical trial in October 2005 to evaluate safety and tolerability in ALS patients. Celastrol, a natural product derived from the Celastraceae family of plants, has been identified as a new class of pharmacologically active regulators of the heat shock response. Recent work indicates that it facilitates reversal of a full-length mutant huntingtin neuronal cell phenotype and confers protection in vivo against neurotoxicity in animal models of Parkinsons's disease and Huntington's disease.

Acknowledgments. The author holds a Canada Research Chair (Tier 1) in Neuroscience and is Director of the Centre for the Neurobiology of Stress at the University of Toronto. This work was supported by grants for NSERC Canada and the Canada Research Chair program.

References

Abbott, A. (2002) Neurologists strike gold in drug screen effort. *Nature* 417:109.
Adachi, H., Katsuno, M., Minamiyama, M., Sang, C., Pagoulatos, G., Angelidis, C., Kusakabe, M., Yoshiki, A., Kobayashi Y., Doyu, M., and Sobue, G. (2003) Heat shock protein 70 chaperone over-expression ameliorates phenotypes of the spinal and bulbar

muscular atrophy transgenic mouse model by reducing nuclear-localized mutant androgen receptor protein. *J Neurosci* 23:2203–11.

Agorogiannis, E. I., Agrorogiannis, G. I., Papadimitriou, A., and Hadjigeorgiou, G. M. (2004) Protein misfolding in neurodegenerative diseases. *Neuropath Appl Neurobiol* 30:215–24.

Allison, A. C., Cacabelos, R., Lombardi, V. R., Alvarez, X. A., and Vigo, C. (2001) Celastrol, a potent anti-oxidant and anti-inflammatory drug, as a possible treatment for Alzheimer's disease. *Prog Neuropsycholopharmacol Biol Psychiatry* 25:1341–57.

Andorfer, C., Acker, C. M., Kress, Y., Hof, P. R., Duff, K., and Davies, P. (2005) Cell-cycle re-entry and cell death in transgenic mice expressing nonmutant human tau isoforms. *J Neurosci* 25:5446–54.

Arendt, T. (2000) Alzheimer's disease as a loss of differentiation control in a subset of neurons that retain immature feature in the adult brain. *Neurobiol Aging* 21:783–96.

Auluck, P. K., Chan, H. Y., Trojanowski, J. Q., Lee, V. M., and Bonini, N. M. (2002) Chaperone suppression of alpha-synuclein toxicity in a *Drosophila* model of Parkinson's disease. *Science* 295:865–8.

Auluck, P. K., Meulener, M. C., and Bonini, N. M (2005) Mechanisms of suppression of alpha-synuclein neurotoxicity by geldanamycin in *Drosophila*. *J Biol Chem* 280:2873–8.

Bagatell, R., Paine-Murrieta, G. D., Taylor, C. W., Pulcini, E. J., Akinga, S., Benjamin, I. J., and Whitesell, L. (2000) Induction of heat shock factor 1-dependent stress response alter the cytotoxic activity of hsp90-binding agent. *Clin Cancer Res* 6:3312–8.

Bagatell, R., and Whitesell, L. (2004) Altered Hsp90 function in cancer: A unique therapeutic opportunity. *Mol Cancer Ther* 3:1021–30.

Barral, J. M., Broadley, G., and Hartl, F. U. (2004) Roles of molecular chaperones in protein misfolding diseases. *Semin Cell Dev Biol* 15:17–29.

Batulan, Z., Shinder, G. A., Minotti, S., He, B. P., Doroudchi, M. M., Nalbantoglu, J., Strong, M. J., and Durham, H. D. (2003) High threshold for induction of the stress response in motor neurons is associated with failure to activate HSF1. *J Neurosci* 23:5789–98.

Bechtold, D. A., and Brown, I. R. (2000) Heat shock proteins Hsp27 and Hsp32 localize to synaptic sites in the rat cerebellum following hyperthermia. *Mol. Brain Res* 75:309–20.

Bechtold, D. A., and Brown, I. R. (2003) Induction of Hsp27 and Hsp32 stress proteins and vimentin in glial cells of the rat hippocampus following hyperthermia. *Neurochem Res* 28:1163–74.

Bechtold, D. A., Rush, S. J., and Brown, I. R. (2000) Localization of the heat shock protein Hsp70 to the synapse following hyperthermic stress in the brain. *J Neurochem* 74:641–6.

Belay, H. T., and Brown, I. R. (2003) Spatial analysis of cell death and stress protein Hsp70 induction in brain, thymus and bone marrow of the hyperthermic rat. *Cell Stress Chap* 8:395–404.

Belay, H. T., and Brown, I. R. (2006) Cell death and expression of heat shock (stress) protein Hsc70 in the hyperthermic rat brain. *J Neurochem 97, Suppl* 1:116–119.

Benn, S. C., and Brown, R. H. (2004) Putting the heat on ALS. *Nat Med* 10:345–7.

Biro, K., Jednakovits, A., Kukorelli, T., Hegedus, E., and Koranyi, L. (1997) Bimoclomol (BRLP-42) ameliorates peripheral neuropathy in streptozotocin-induced diabetic rats. *Brain Res Bull* 44:259–63.

Biro, K., Palhalmi, J., Toth, A. J., Kukorelli, T., and Juhasz, G. (1998) Bimoclomol improves early electrophysiological signs of retinopathy in diabetic rats. *Neuroreport* 9:2029–33.

Broquet, A. H., Thomas, G., Masliah, J., Trugnan, G., and Bachelet, M. (2003) Expression of the molecular chaperone Hsp70 in detergent-resistant microdomains correlates with its membrane delivery and release. *J Biol Chem* 278:21601–6.

Brown, I. R. (1994) Induction of heat shock genes in the mammalian brain by hyperthermia and tissue injury. In Mayer, J., and Brown, I. R. (eds.): *Heat Shock Proteins in the Nervous System*. Academic Press, London, pp. 31–53.

Brown, I. R., and Rush, S. J. (1996) *In vivo* activation of neural heat shock transcription factor HSF1 by a physiologically relevant increase in body temperature. *J Neurosci Res* 44:52–7.

Brown, I. R., and Rush, S. J. (1999) Cellular localization of the heat shock transcription factors HSF1 and HSF2 in the rat brain during postnatal development and following hyperthermia. *Brain Res* 821:333–40.

Brown, I. R., and Sharp, F. R. (1999) The cellular stress gene response in brain. In Latchmann, D. S. (ed.): *Stress Proteins, Handbook of Experimental Pharmacology*. Vol. 136. Springer-Verlag, Heidelberg, pp. 243–63.

Bruening, W., Roy, J., Giasson, B., Figlewisz, D. A., Mushynski, W. E., and Durham, H. D. (1999) Up-regulation of protein chaperones preserves viability of cells expressing toxic Cu/Zn-superoxide-dismutase mutants associated with amyotropic lateral sclerosis. *J Neurochem* 72:693–9.

Bruijn, L. I., Miller, T. M., and Cleveland, D. W. (2004) Unraveling the mechanisms involved in motor neuron degeneration in ALS. *Annu Rev Neurosci* 27:723–49.

Bukau, B., and Horwich, A. L. (1998) The Hsp70 and Hsp60 chaperone machines. *Cell* 92:351–66.

Busser, J., Geldmacher, D. S., and Herrup, D. S. (1998) Ectopic cell cycle proteins predict the sites of neuronal cell death in Alzheimer's disease brain. *J Neurosci* 18:2801–7.

Chai, Y., Koppenhafer, S. L., Bonini, N. M., and Paulson, H. L. (1999) Analysis of the role of heat shock protein (Hsp) molecular chaperones in polyglutamine diseases. *J Neurosci* 19:10338–47.

Chen, B., Duan, H., and Takaishi, Y. (1999) Triterpene caffeoyl esters and diterpenes for *Celastrus stephanotifolius*. *Phytochemistry* 51:683–97.

Chen, S., Bawa, D., Besshoh, S., Gurd, J. W., and Brown, I. R. (2005) Association of heat shock proteins and neuronal membrane components with lipid rafts from rat brain. *J Neurosci Res* 81:522–9.

Chui, D. H. Tanahashi, H., Ozawa, K., Ikeda, S., Checler, F., Ueda, O., Suzuki, H., Arak, W., Inoue, H., Shirotani, K., Takahashi, K., Gallyas, F., and Tabir, T. (1999) Transgenic mice with Alzheimer presenilin 1 mutations show accelerated neurodegeneration without amyloid plaque formation. *Nat Med* 5:560–4.

Cleren, C., Calingasan, N. Y., Chen, J., and Beal, M. F. (2005) Celastrol protects against MPTP- and 3-nitropropionic acid-induced neurotoxicity. *J Neurochem* 94:995–1004.

Conway, K. A., Lee, S., Rochet, I., Ding, T. T., Wiliamson, R. E, and Lansbury, P. T. (2000) Acceleration of oligomerization, not fibrillization, is a shared property of both alpha-synuclein mutations linked to early-onset Parkinson's disease: Implications for pathogenesis and therapy. *PNAS* 97:571–6.

Cummings, C. J., Sun, Y., Opal, P., Antalffy, B., Mestril, R., Orr, H. T., Dillman, W. H., and Zoghbi, H. Y. (2001) Over-expression of inducible Hsp70 chaperone suppresses neuropathology and improves motor function in SCA1 mice. *Hum Mol Genet* 10:1511–8.

Dedmon, M. M., Christodoulou, J., Wilson, M. R., and Dobson, C. M. (2005) Heat shock protein 70 inhibits alpha-synuclein fibril formation via preferential binding to prefibrillar species. *J Biol Chem* 280;14733–40.

D'Souza, S. M., and Brown, I. R. (1998) Constitutive expression of heat shock proteins hsp90, hsc70, hsp70 and hsp60 in neural and non-neural tissues of the rat during postnatal development. *Cell Stress Chap* 3:188–99.

Farinelli, S., and Greene, L. (1996) Cell cycle blockers mimosine, ciclopirox, and deferoxamine prevent the death of PC12 cells and postmitotic sympathetic neurons after removal of trophic support. *J Neurosci* 16:1150–62.

Feddersen, R. M., Ehlenfeldt, R., Yunis, W. S., Clark, H. B., and Orr, H. T. (1992) Disrupted cerebellar cortical development and progressive degeneration of Purkinje cells in SV40 T antigen transgenic mice. *Neuron* 9:955–66.

Feldmeyer, D., Kask, K., Brusa, R., Koman, H. C., Kolhekar, R., Rozov, A., Burnashev, N., Jensen, V., Hvalby, O., Sprengel, R., and Seeburg, P. H. (1999) Neurological dysfunctions in mice expressing different levels of the Q/R site-unedited AMPAR submit GluR-B. *Nat Neurosci* 2:57–64.

Fernandez-Funez, P., Nino-Rosales, M. L., de Gouyon, B., She, W. C., Luchak, J. M., Martinez, P., Turiegano, E., Benito, J., Capovilla, M., Skinner, P. J., McCall. A., Canal, I., Orr, H. T., and Zoghbi, H. Y. (2000) Identification of genes that modify ataxin-1-induced neurodegeneration. *Nature* 408:101–6.

Fleshner, M., and Johnson, J. D. (2005) Endogeneous extra-cellular heat shock protein 72: Releasing signal(s) and function. *Int J Hyperthermia* 21:457–71.

Foster, J. A., and Brown, I. R. (1997) Differential induction of heat shock mRNA in oligodendrocytes, microglia, and astrocytes following hyperthermia. *Mol Brain Res* 45:207–18.

Foster, J. A., Rush, S. J., and Brown, I. R. (1995) Localization of constitutive and hyperthermia-inducible heat shock mRNAs (hsc70 and hsp70) in the rabbit cerebellum and brainstem by non-radioactive in situ hybridization. *J Neurosci Res* 41:603–12.

Franklin, T. B., Krueger-Naug, A. M., Clarke, D. B., Arrigo, A. P., and Currie, R. W. (2005) The role of heat shock proteins Hsp70 and Hsp27 in cellular protection of the central nervous system. *Int J Hyperthermia* 21:379–92.

Garigan, D., Hsu, A. L., Fraser, A. G., Kamath, R. S., Ahringer, J., and Kenyon, C. (2002) Genetic analysis of tissue aging in *C. elegans*: A role for heat shock factor. *Genetics* 161:1101–12.

Garofalo, O., Kennedy, P. G., Swash, M., Martin, J. E., Luthert, P., Anderton, B. H., and Leigh, P. N. (1991) Ubiquitin and heat shock protein expression in amyotrophic lateral sclerosis. *Neuropath Appl Neurobiol* 17:39–45.

Giovanni, A., Wirtz-Brugger, F., Keramaris, E., Slack, R., and Park, D. S. (1999) Involvement of cell cycle elements, cyclin-dependent kinases, pRb, and E2FxDP, in B-amyloid-induced neuronal death. *J Biol Chem* 274:19011–6.

Greene, L. A., Biswas, S. C., and Liu, D. X. (2004) Cell cycle molecules and vertebrate neuron death: E2F at the hub. *Cell Death Differ* 11:49–60.

Guettouche, T., Boellmann, F., Lane, W. S., and Voellmy, R. (2005) Analysis of phosphorylation of human heat shock factor 1 in cells experiencing a stress. *BMC Biochem* 6:4.

Gunatilaka, A. A., Herz, W., Kirby, G. W., Moore, R. E., Steglich, W., and Tamm, C. (1996) *Triterpenoid Quinomethides and Related Compounds (Celastroids)*. Springer-Verlag, Vienna.

Gurney, M. E., Pu, H., Chiu, A. Y., Dal Canto, M. C., Polchow, C. Y., Alexander, D. D., Caliendo, J., Hentati, A., Kwon, Y. W., and Deng, H. X. (1994) Motor neuron degeneration in mice that express a human Cu, Zn superoxide dismutase mutation. *Science* 264: 1772–5.

Guzhova, I., Kislyakova, K., Moskoliova, O., Fridlanskaya, I., Tytell, M., Cheetham, M., and Margulis, B. (2001) *In vitro* studies show that Hsp70 can be released by glia and that exogenous Hsp70 can enhance neuronal stress tolerance. *Brain Res* 914:66–73.

Hartl, F. U. (1996) Molecular chaperones in cellular protein folding. *Nature* 381:571–9.

Hay, D. G., Sathasivam, K., Tobaben, S., Stahl, B., Marber, M., Mestril, R., Mahal, A., Smith, D. L., Woodman, B., and Bates, G. P. (2004) Progressive decrease in chaperone protein levels in a mouse model of Huntington's disease and induction of stress proteins as a therapeutic approach. *Hum Mol Genet* 13:1389–405.

Heemskerk, J., Tobin, A. J., and Bain, L. J. (2002) Teaching old drugs new tricks. *Trends Neurosci* 25:494–6.

Herrup, K., and Arendt, T. (2002) Re-expression of cell cycle proteins induces neuronal cell death during Alzheimer's disease. *J Alzheimer Dis* 4:243–7.

Herrup, K., and Busser, J. C. (1995) The induction of multiple cell cycle events precedes target-related neuronal death. *Development* 121:2385–95.

Hietakangas, V., Ahskog, J. K., Jakobsson, A. M., Hellesuo, M., Sahlberg, N. M., Holmberg, C. I., Mikhailov, A., Palvimo, J. J., Pirkkala, L., and Sistonen, L. (2003) Phosphorylation of serine 303 is a prerequisite for the stress-inducible SUMO modification of heat shock factor 1. *Mol Cell Biol* 23:2953–68.

Holmberg, C. I., Hietakangas, V., Mikhailov, A., Rantanen, J. O., Kallio, M., Meinander, A., Hellman, J., Morrice, N., MacKintosh, C., Morimoto, R. I., Eriksson, J. E., and Sistonen, L. (2001) Phosphorylation of serine 230 promotes inducible transcriptional activity of heat shock factor 1. *EMBO J* 20:3800–10.

Holmberg, C. I., Tran, S. E., Eriksson, J. E., and Sistonen, L. (2002) Multi-site phosphorylation provides a sophisticated regulation of transcription factors. *Trends Biochem Sci* 27:619–27.

Holmberg, C. I., Staniszewski, K., Mensah, K. N., Matouschek, A., and Morimoto, R. I. (2004) Inefficient degradation of truncated polyglutamine proteins by the proteosome. *EMBO J* 23:4307–18.

Hong, Y., Rogers, R., Matunis, M. J., Mayhew, C. N., Goodson, M. L., Park-Sarge, O. K., and Sarge, K. D. (2001) Regulation of heat shock transcription factor 1 by stress-induced SUMO-1 modification. *J Biol Chem* 276:40263–7.

Hsia, A. Y., Masliah, E., McConlogue, L., Yu, G. Q., Tatsuno, G., Hu, K., Kholodenko, D., Malenka, R. C., Nicoll, R. A., and Mucke, L. (1999) Plaque-independent disruption of neural circuits in Alzheimer's disease mouse models. *Proc Natl Acad Sci USA* 96:3228–33.

Hsu, A. L., Murphy, C. T., and Kenyon, C. (2003) Regulation of aging and age-related diseases by DAF-16 and heat shock factor. *Science* 300:1142–5.

Hunter-Lavin, C., Davies, E. L., Bacelar, M. M., Marshall, M. J., Andrew, S. A., and Williams, J. H. (2004) Hsp70 release from peripheral bood mononuclear cells. *Biochem Biophys Res Commun* 324:511–7.

Jana, N. R., Tanaka, M., Wang, G., and Nukina, N. (2000) Polyglutamine length-dependent interaction of Hsp40 and Hsp70 family chaperones with truncated N-terminal huntingtin: Their role in suppression of aggregation and cellular toxicity. *Hum Mol Genet* 9:2009–18.

Jednakovits, A., Kurucz, I., and Nanasi, P. (2000) Effect of sub-chronic bimoclomol treatment on vascular responsiveness and heat shock protein production in spontaneously hypertensive rats. *Life Sci* 67:1791–7.

Jurivich, D. A., Sistonen, L., Sarge, K. D., and Morimoto, R. I. (1994) Arachidonate is a potent modulator of human heat shock gene transcription. *Proc Natl Acad Sci USA* 91:2280–4.

Kalmar, B., Burnstock, G., Vrbova, G., Urbanics, R., Csermely, P., and Greensmith, L. (2002) Upregulation of heat shock proteins rescues motoneurones from axotomy-induced cell death in neonatal rats. *Exp Neurology* 176:87–97.

Kalmar, B., Greensmith, L., Malcangio, M., McMahon, S. B., Csermely, P., and Burnstock, G. (2003) The effect of treatment with BRX-220, a co-inducer of heat shock proteins, on sensory fibers of the rat following peripheral nerve injury. *Exp Neurol* 184:636–47.

Kalmar, B., Kieran, D., and Greensmith, L. (2005) Molecular chaperones as therapeutic targets in amyotrophic lateral sclerosis. *Biochem Soc Trans* 33:551–2.

Karunanithi, S., Barclay, J. W., Robertson, R. M., Brown, I. R., and Atwood, H. L. (1999) Neuroprotection at *Drosophila* synapses conferred by prior heat shock. *J Neurosci* 19:4360–9.

Karunanithi, S., Barclay, J. W., Brown, I. R., Robertson, R. M., and Atwood, H. L. (2002) Enhancement of presynaptic performance in transgenic *Drosophila* overexpressing heat shock protein Hsp70. *Synapse* 44:8–14.

Katchanov, J., Harms, C., Gertz, K., Hauck, L., Waeber, C., Hirt, L., Priller, J., von Harsdorf, R., Bruck, W., Hortnagl, H., Dirnagl, U., Bhide, P. G., and Endres, M. (2001) Mild cerebral ischemia induces loss of cyclin-dependent kinase inhibitors and activation of cell cycle machinery before delayed neuronal cell death. *J Neurosci* 21:5045–53.

Kawahara, Y., Ito, K., Sun, H., Aizawa, H., Kanazawa, I., and Kwak, S. (2004) Glutamate receptors: RNA editing and death of motor neurons. *Nature* 427:801.

Kayed, R., Head, E., Thompson, J. L., McIntire, T. M., Milton, S. C., Cotman, C. W., and Glabe, C. G. (2003) Common structures of soluble amyloid oligomers implies common mechanism of pathogenesis. *Science* 300:486–9.

Kazemi-Esfarjani, P., and Benzer, S. (2000) Genetic suppression of polyglutamine toxicity in *Drosophila*. *Science* 287:1837–40.

Kelty, J. D., Noseworthy, P. A., Feder, M. E., Robertson, R. M., and Ramirez, J. M. (2002) Thermal pre-conditioning and heat shock protein 72 preserves synaptic transmission during thermal stress. *J Neurosci* 22:RC193.

Khan, V. R., and Brown, I. R. Brown. (2002) The effect of hyperthermia on the induction of cell death in brain, testis, and thymus of the adult and developing rat *Cell Stress Chap* 7:73–90.

Kieran, D., Kalmar, B., Dick, J. R., Riddoch-Contreras, J., Burnstock, G., and Greensmith, L. (2004) Treatment with arimoclomol, a co-inducer of heat shock proteins, delays disease progression in ALS mice. *Nat Med* 10:402–5.

Kim, H. R., Kang, H. S., and Kim, H. D. (1999) Geldanamycin induces heat shock protein expression through activation of HSF1 in K562 erythroleukemic cells. *IUBMB Life* 48:429–33.

Kim, S., Nollen, E. A., Kitagawa, K., Bindokast, V. P., and Morimoto, R. I. (2002) Polyglutamine protein aggregates are dynamic. *Nat Cell Biol* 4:826–31.

Kim, S. A., Yoon, J. H., Lee, S. H., and Ahn, S. G. (2005) Polo-like kinase 1 phosphorylates heat shock transcription factor 1 and mediates its nuclear translocation during heat stress. *J Biol Chem* 280:12653–7.

Klein, J. A., Longo-Guess, C. M., Rossmann, M. P., Seburn, K. L., Hurd, R. E., Frankel, W. N., Bronson, R. T., and Ackerman, S. L. (2002) The harlequin mouse mutation downregulatates apoptosis-inducing factor. *Nature* 419:367–74.

Kobayashi, Y., and Sobue, G. (2001) Protective effect of chaperones on polyglutamine diseases. *Brain Res Bull* 56:165–8.

Kumar-Singh, S., Dewachter, I., Moechars, D., Lubke, U., De Jonghe, C., Ceuterick, C., Checler, F., Naidu, A., Cordell, B., Cras, P.,Van Broeckhoven, C., and Van Leuven, F. (2000) Behavioural disturbances without amyloid deposits in mice overexpressing human amyloid precursor protein with Flemish (A692G) or Dutch (E693Q) mutation. *Neurobiol Dis* 7:9–22.

Kurapati, R., Passananti, H. B., Rose, M. R., and Tower, J. (2000) Increased hsp22 RNA levels in *Drosophila* lines genetically selected for increased longevity. *J Gerontol A Biol Sci Med Sci* 55:B552–9.

Lancaster, G. I., and Febbraio, M. A. (2005) Exosome-dependent trafficking of HSP70. *J Biol Chem* 280:23349–55.

Lancaster, G. I., Moller, K., Nielsen, B., Secher, N. H., Febbraio, M. A., and Nybo, L. (2004) Exercise induces the release of heat shock protein 72 from the human brain *in vivo*. *Cell Stress Chap* 9:276–80.

Latchman, D. S. (2005) Hsp27 and cell survival in neurons. *Int J Hyperthermia* 21:393–402.

Lee, E. Y., Chang, C. Y., Hu, N., Wang, Y., Lai, C., Herrup, K., Lee, W., and Bradley, A. (1992) Mice deficient for Rb are nonviable and show effects in neurogenesis and haematopoiesis. *Nature* 359:288–94.

Li, R. L., and Shun, D. F. (1989) *Investigations and Clinical Applications Trysterygium Wilfordii Hook F*. China Science and Technology Press, Beijing.

Lipton, S. A., and Rosenberg, P. A. (1994) Excitatory amino acids as a final common pathway for neurologic disorders. *N Engl J Med* 330:613–22.

Lipton, R. A. (2004) Sporadic ALS: Blame it on the editor. *Nat Med* 10:347.

Maekawa, S., Iino, S., and Miyata, S. (2003) Molecular characterization of detergent-insoluble cholesterol-rich membrane microdomain (raft) of central nervous system. *Biochim Biophys Acta* 1610:216–70.

Magrane, J., Smith, R. C., Walsh, K., and Querfurth, H. W. (2004) Heat shock protein 70 participates in the neuroprotective response to intracellularly expressed beta-amyloid in neurons. *J Neurosci* 24:1700–6.

Manzerra, P., and Brown, I. R. (1992a) Distribution of constitutive and hyperthermia-inducible heat shock mRNA species (hsp70) in the Purkinje layer of the rabbit cerebellum. *Neurochem Res* 17:559–64.

Manzerra, P., and Brown, I. R. (1992b) Expression of heat shock genes (hsp70) in the rabbit spinal cord: Localization of constitutive and hyperthermia-inducible mRNA species. *J Neurosci Res* 31:606–15.

Manzerra, P., and Brown, I. R. (1996) The neuronal stress response: Nuclear translocation of heat shock proteins as an indicator of hyperthermic stress. *Exp Cell Res* 229:35–47.

Manzerra, P., Rush, S. J., and Brown, I. R. (1993) Temporal and spatial distribution of heat shock mRNA and protein (hsp70) in the rabbit cerebellum in response to hyperthermia. *J Neurosci Res* 36:480–90.

Manzerra, P., Rush, S. J., and Brown, I. R. (1997) Tissue-specific differences in heat shock protein hsc70 and hsp70 in the control and hyperthermic rabbit. *J Cell Physiol* 170:130–7.

Martin, J. B. (1999) Molecular basis of neurodegenerative disorders. *N Engl J Med* 340:1970–80.

Matsumoto, G., Stojanovic, A., Holmberg, C. I., Kim, S., and Morimoto, R. I. (2005) Structural properties and neuronal toxicity of amyotrophic lateral sclerosis-associated Cu/Zn superoxidase dismutase 1 aggregates. *J Cell Biol* 171:75–85.

Matsumoto, G., Kim, S., and Morimoto, R. I. (2006) Hungtingtin and mutant SOD1 form aggregate structures with distinct molecular properties in human cells. *J Biol Chem* (in press).

Meriin, A. B., Zhang, X., He, X., Newnam, G. P., Chernoff, Y. O., and Sherman, M. Y. (2002) Huntington toxicity in yeast model depends on polyglutamine aggregation mediated by a prion-like protein Rnq1. *J Cell Biol* 157:997–1004.

Meriin, A. B., and Sherman, M. Y. (2005) Role of molecular chaperones in neurodegenerative disorders. *Int J Hyperthermia* 21:403–19.

Moechars, D., Dewachter, I., Lorent, K., Reverse, D., Baekelandt, V., Naidu, A., Tesseur, I., Spittaels, K., Haute, C. V., Checler, F., Godaux, E., Cordell, B., and Van Leuven, F. (1999) Early phenotypic changes in transgenic mice overexpress different mutants of amyloid precursor protein in brain. *J Biol Chem* 274:6483–92.

Morimoto, R. I. (1998) Regulation of the heat shock transcriptional response: Cross talk between a family of heat shock factors, molecular chaperones, and negative regulators. *Genes Dev* 12:3788–96.

Morimoto, R. I., Tissieres, A., and Georgopoulos, C. (1990) *Stress Proteins in Biology and Medicine*. Cold Spring Harbor Laboratory Press, Cold Spring Harbor, New York.

Morimoto, R. I., Tissieres, A., and Georgopoulos, C. (1994) *The Biology of Heat Shock Proteins and Molecular Chaperones*. Cold Spring Harbor Laboratory Press, Cold Spring Harbor, New York.

Morley, J. F., and Morimoto, R. I. (2004) Regulation of longevity in *Caenorhabditis elegans* by heat shock factor and molecular chaperones. *Mol Biol Cell* 15:657–64.

Morley, J. F., Brignull, H. R., Weyers, J. J., and Morimoto, R. I. (2002) The threshold for polyglutamine-expansion protein aggregation and cellular toxicity is dynamic and influenced by aging in *Caenorhabditis elegans*. *PNAS* 99:10417–22.

Morrison, A. J., Rush, S. J., and Brown, I. R. (2000) Heat shock transcription factors and the hsp70 induction response in brain and kidney of the hyperthermic rat during postnatal development. *J Neurochem* 75:363–72.

Morrow, G., Battistini, S., Zhang, P., and Tanguay, R. M. (2004a) Decreased lifespan in absence of expression of mitochondrial small heat shock protein Hsp22 in *Drosophila*. *J Biol Chem* 279:43382–5.

Morrow, G., Samson, M., Michaud, S., and Tanguay, R. M. (2004b) Over-expression of the small mitochondrial Hsp22 extends *Drosophila* life span and increases resistance to oxidative stress. *FASEB J* 18:598–9.

Muchowski, P. J., Schaffar, G., Sittler, A., Wanker, E. E., Hayer-Hartl, M. K., Hartl, F. U. (2000) Hsp70 and hsp40 chaperones can inhibit self-assembly of polyglutamine proteins into amyloid-like fibrils. *Proc Natl Acad Sci USA* 97:7841–6.

Muchowski, P. J., and Wacker, J. L. (2005) Modulation of neurodegeneration by molecular chaperones. *Nat Rev Neurosci* 6:11–22.

Ngassapa, O., Soejart, D. D., Pezzuto, J. M., and Farnsworth, N. R. (1994) Quinone-methide triterpenes and salaspermic acid from *Kokoona ochracea*. *J Nat Prod* 57:1–8.

Nollen, E. A., and Morimoto, R. I. (2002) Chaperoning signaling pathways: Molecular chaperones as stress-sensing 'heat shock' proteins. *J Cell Sci* 115:2809–16.

Nollen, N. A., Garcia, S. M., van Haaften, G., Kim, S., Chavez, A., Morimoto, R. I., and Plasterk, R. H. A. (2004) Genome-wide RNA interference screen identifies previously undescribed regulators of polyglutamine aggregation. *Proc Natl Acad Sci USA* 101:6403–8.

Piccioni, F., Roman, B. R., Fishbeck, K. H., and Taylor, J. P. (2004) A screen for drugs that protect against the cytotoxicity of polyglutamine-expanded androgen receptor. *Human Mol Gen* 3:437–46.

Prodromou, Roe, S. M., O'Brien, R., Ladbury, J. E., Piper, P. W., and Pearl, L. H. (1997) Identification and structural characterization of the ATP/ADP-binding site in the Hsp90 molecular chaperone. *Cell* 90:65–75.

Queitsch, C., Sangster, T. A., and Lindquist, S. (2002) Hsp90 as a capacitor of phenotypic variation. *Nature* 417:618–24.

Quraishi, H., and Brown, I. R. (1995) Expression of heat shock protein 90 (hsp90) in neural and nonneural tissues of the control and hyperthermic rabbit. *Exp Cell Res* 219:358–63.

Ranganathan, S., and Bowser, R. (2003) Alterations in G(1) to S phase cell-cycle regulators during amyotrophic lateral sclerosis. *Am J Pathol* 162:823–35.

Ranganathan, S., Scudiere, S., and Bowser, R. (2001) Hyperphosphorylation of the retinoblastoma gene product and altered subcellular distribution of E2F-1 during Alzheimer's disease and amyotrophic lateral sclerosis. *J Alzheimer's Dis* 3:377–85.

Rieger, T. R., Morimoto, R. I., and Hatzimanikatis, V. (2005) Mathematical modeling of the eukaryotic heat shock response: Dynamics of the hsp70 promoter. *Biophys J* 88:1646–8.

Rieger, T. R., Morimoto, R. I., and Hatzimanikatis, V. (2006) Bistability explains threshold phenomena in protein aggregation both *in vitro* and *in vivo*. *Biophys J* 90:886–95.

Robinson, M. C., Tidwell, J. L., Gould, T., Taylor, A. R., Newbern, J. M., Graves, J., Tytell, M., and Milligan, C. E. (2005) Extracellular heat shock protein 70: A critical component for motorneuron survival. *J Neurosci* 25:9735–45.

Roe, S. M., Prodromou, C., O'Brien, R., Ladbury, J. E., Piper, P. W., and Pearl, L. H. (1999) Structural basis for inhibition of the Hsp90 molecular chaperone by the antitumor antibiotics radicicol and geldanamycin. *J Med Chem* 42:260–6.

Rosen, D. R., Siddique, T., Patterson, D., Figlewicz, D. A., Sapp, P., Hentati, A., Donaldson, D., Goto, J., O'Regan, J. P., and Deng, H. X. (1993) Mutations in Cu/Zn superoxide dismutase are associated with familial amyotrophic lateral sclerosis. *Nature* 362:59–62.

Ross, C. A., and Poirier, M. A. (2004) Protein aggregation and neurodegenerative diseases. *Nat Med* 10:S10–17.

Rothstein, J. D. (1995) Excitotoxic mechanisms in the pathogenesis of amyotrophic lateral sclerosis. *Adv Neurol* 68:7–20.

Rutherford, S. L., and Lindquist, S. (1998) Hsp90 as a capacitor for morphological evolution. *Nature* 396:336–42.

Samali, A., and Orrenius, S. (1998) Heat shock proteins: Regulators of stress response and apoptosis. *Cell Stress Chap* 3:228–36.

Sangster, T. A., Lindquist, S., and Queitsch, C. (2004) Under cover: Causes, effects and implications of Hsp90-mediated genetic capacitance. *Bioessays* 26:348–62.

Schaffar, G., Breuer, P., Boteva, R., Behrends, C., Tzvetkov, N., Strippel, N., Sakahira, H., Siegers, K., Hayer-Hartl, M., and Hartl, F. U. (2004) Cellular toxicity of polyglutamine expansion proteins: Mechanism of transcription factor deactivation. *Mol Cell* 15:95–105.

Schapira, A. H. V., and Olanow, C. W. (2004) Neuroprotection in Parkinson disease: Mysteries, myths, and misconceptions. *JAMA* 291:358–64.

Selkoe, D. J. (2004) Cell biology of protein misfolding: The examples of Alzheimer's and Parkinson's diseases. *Nat Cell Biol* 6:1054–61.

Shamovsky, I., Gershon, D. (2004) Novel regulatory factors of HSF-1 activation: Facts and perspectives regarding their involvement in the age-associated attenuation of the heat shock response. *Mechanics of Ageing Development* 125:767–75.

Shen, H., He, J., Wang, Y., Huang, Q., and Chen, J. (2005) Geldanamycin induces heat shock protein 70 and protects against MPTP-induced dopaminergic neurotoxicity in mice. *J Biol Chem* 280:39962–9.

Sherman, M. Y., and Goldberg, A. L. (2001) Cellular defenses against unfolded proteins: A cell biologist thinks about neurodegenerative diseases. *Neuron* 29:15–32.

Sittler, A., Lurz, R., Lueder, G., Priller, J., Lehrach, H., Hayer-Hartl, M. K., Hartl, F. U., and Wanker, E. E. (2001) Geldanamycin activates a heat shock response and inhibits huntingtin aggregation in a cell culture model of Hungtington's disease. *Hum Mol Genet* 10:1307–15.

Smith, R. C., Rosen, K. M., Pola, R., and Magrane, J. (2005) Stress proteins in Alzheimer's disease. *Int J Hyperthermia* 21:421–31.

Soto, C. (2003) Unfolding the role of protein misfolding in neurodegenerative diseases. *Nat Rev Neurosci* 4:49–60.

Sprang, G. K., and Brown, I. R. (1987) Selective induction of a heat shock gene in fibre tracts and cerebellar neurons of the rabbit brain detected by in situ hybridization. *Mol Brain Res* 3:89–93.

Tidwell, J. L., Houenou, L. J., and Tytell, M. (2004) Administration of Hsp70 *in vivo* inhibits motor and sensory neuron degeneration. *Cell Stress Chap* 9:88–98.

Tonkiss, J., and Calderwood, S. K. (2005) Regulation of heat shock gene transcription in neuronal cells. *Int J Hyperthermia* 21:433–44.

Tytell, M. (2005) Release of heat shock proteins (Hsps) and the effects of extracellular hsps on neural cells and tissues. *Int J Hyperthermia* 21:445–55.

Tytell, M., Greenberg, S. G., and Lasek, R. J. (1986) Heat shock-like protein is transferred from glia to axon. *Brain Res* 363:161–4.

Vleminckx, V., Van Damme, P., Goffin, K., Delye, H., Van Den Bosch. L., and Robberecht, W. (2002) Upregulation of HSP27 in a transgenic model of ALS. *J Neuropath Exp Neurol* 61:968–74.

Walker, G. A., Thompson, F. J., Brawley, A., Scanlon, T., and Devaney, E. (2003) Heat shock factor functions at the convergence of the stress response and developmental pathways in *C. elegans*. *FASEB J* 17:1960–2.

Wang, J., Gines, S., MacDonald, M. E., and Gusella, J. F. (2005) Reversal of a full-length mutant huntingtin neuronal phenotype by chemical inhibitors of polyglutamine-mediated aggregation. *BMC Neurosci* 6:1.

Warrick, J. M., Paulson, H. L., Gray-Board, G. L., Bui, Q. T., Fischeck, K. H., Pittman, R. N., and Bonini, N. M. (1998) Expanded polyglutamine protein forms nuclear inclusion inclusions and causes neural degeneration in *Drosophila*. *Cell* 93:939–49.

Warrick, J. M., Chan, H. Y., Gray-Board, G. I., Chai, Y., Paulson, H. I., and Bonini, N. M. (1999) Suppression of polyglutamine-mediated neurodegneration in *Drosophila* by the molecular chaperone Hsp70. *Nat Genet* 23:425–8.

Waza, M., Adachi, H., Katsuno, M., Minamiyama, M., Sang, C., Tanaka, F., Inukai, A., Doyu, M., and Sobue, G. (2005) 17-AAG, an Hsp90 inhibitor, ameliorates polyglutamine-mediated motor neuron degeneration. *Nat Med* 11:1088–95.

Westerheide, S. D., Bosman, J. D., Mbadugha, B. N., Kawahara, T. L., Matsumoto, G., Kim, S., Gu, W., Devlin, J. P., Silverman, R. B., and Morimoto, R. I. (2004) Celastrols as inducers of the heat shock response and cytoprotection. *J Biol Chem* 279:56053–60.

Westerheide, S. D., and Morimoto, R. I. (2005) Heat shock response modulators as therapeutic tools for diseases of protein conformation. *J Biol Chem* 280:33097–100.

Whitesell, L., Bagatell, R., and Falsey, R. (2003) The stress response: Implications for the clinical development of hsp90 inhibitors. *Curr Cancer Drug Targets* 3:349–58.

Whitesell, L., and Lindquist, L. (2005) HSP90 and the chaperoning of cancer. *Nat Rev Cancer* 5:761–72.

Willingham, S., Outeiro, T. F., DeVit, M. J., Lindquist, S. L., and Muchowski, P. J. (2003) Yeast genes that enhance the toxicity of a mutant huntingtin fragment of alpha-synuclein. *Science* 302:1769–72.

Wong, P. C., Pardo, C. A., Borchelt, D. R., Lee, M. K., Copeland, N. G., Jenkins, N. A., Sisodia, S. S., Cleveland, D. W., and Price, D. L. (1995) An adverse property of a familial ALS-linked SOD1 mutation causes motor neuron disease characterized by vacuolar degeneration of mitochondria. *Neuron* 14:1105–16.

Yang, Y., and Herrup, K. (2005) Loss of neuronal cell cycle control in ataxia-telangiectasia: A unified disease mechanism. *J Neurosci* 2522–9.

Yang, Y., Geldmacher, D. S., and Herrup, K. (2001) DNA replication precedes neuronal cell death in Alzheimer's disease. *J Neurosci* 21:2661–266.

Yang, Y., Mufson, E. J., and Herrup, K. (2003) Neuronal cell death is preceded by cell cycle events at all stages of Alzheimer's disease. *J Neurosci* 23:2557–63.

Yang, Y., Varvel, N. H., Lamb, B. T., and Herrup, K. (2006) Ectopic cell cycle events link human Alzheimer's disease and amyloid precursor protein transgenic mouse models. *J Neurosci* 26:775–84.

Zhou, H., Li, S. H., and Li, X. J. (2001) Chaperone suppression of cellular toxicity of huntingtin is independent of poyglutamine aggregation. *J Biol Chem* 276:48417–24.

Zou, J., Guo, Y., Guettouche, T., Smith, D. F., and Voellmy, R. (1998) Repression of heat shock transcription factor HSF1 activation by HSP90 (HSP90 complex) that forms a stress–sensitive complex with HSF1. *Cell* 94:471–80.

19
Heat Shock Proteins in the Progression of Cancer

STUART K. CALDERWOOD,[1,2] ABDUL KHALIQUE,[1] AND DANIEL R. CIOCCA[3]

[1]*Division of Molecular and Cellular Radiation Oncology, Beth Israel Deaconess Medical Center, Harvard Medical School, Boston, MA 02115;*
[2]*Department of Medicine, Boston University School of Medicine, Boston, MA 02118;*
[3]*Oncology Laboratory, Institute of Experimental Medicine and Biology of Cuyo (CRICYT-CONICET), and Argentine Foundation for Cancer Research (FAIC), Mendoza, C. C. 855, 5500 Mendoza, Argentina*

1. Summary

The cohort of heat shock proteins (HSP) induced by cell stress becomes expressed at high levels in a wide range of tumors, and elevated levels of HSP are closely associated with a poor prognosis and treatment resistance. Increased HSP transcription in tumor cells is due both to loss of p53 function and elevated expression of proto-oncogenes such as HER2 and c-Myc and plays an essential role in tumorigenesis. The HSP family members overexpressed in cancer play overlapping, essential roles in tumor growth both by promoting autonomous cell proliferation and by inhibiting multiple death pathways. The HSP have thus become important and novel targets for rational anti-cancer drug design and HSP 90 inhibitors such as geldanomycin and 17-AAG are currently showing much promise in clinical trial while elevated HSP in tumors form the basis for chaperone-based immunotherapy.

2. Introduction

2.1. Heat Shock Proteins, Molecular Chaperones, and Cancer

Expression of the heat shock proteins (HSP) is increased in a large number of cancers of diverse morphologies, and these abundant concentrations the HSP play significant roles in the emergence of cancer cells (Nylandsted et al., 2000; van't Veer et al., 2002; Ciocca and Calderwood, 2005; Rohde et al., 2005). However, the mechanisms involved are only recently being elucidated. At first sight, HSP, with their ancient roles as stress proteins handed down almost unchanged over many millions of years from the most primitive of cellular organisms, might seem unlikely candidates for mediators of tumor formation (Lindquist and Craig, 1988). Clearly, however, high concentrations of HSP play an essential and almost ubiquitous role in the evolution of the disease and the evasion of strategies for cancer therapy.

The HSP are in fact the products of a number of distinct gene families that mediate the folding of proteins into functional tertiary structures (Lindquist and Craig, 1988; Bukau and Horwich, 1998). The major mammalian *HSP* families are named for their approximate Mr as HSP 10, 27, 40, 60, 70, 90, and 110 (Tang et al., 2005). These proteins manipulate the structures of cellular polypeptides within the context of "chaperone machines," high Mr complexes formed from the multimeric self-association as in HSP27 and HSP60, or within multi-protein complexes as with HSP70 and HSP90 (Netzer and Hartl, 1998; Pratt et al., 2004). Most of the HSP can then use the energy of ATP hydrolysis to release from such cellular substrates which using intrinsic ATPase domains (Bukau and Horwich, 1998; Georgopolis and Welch, 1993). Early in the course of evolution, the *HSP* genes developed the capacity to be induced massively by environmental insults, and in all species, exposure to protein stress coordinately induces the *HSP* gene families and inhibits the pathways of programmed cell death (PCD) (Georgopolis and Welch, 1993). Cytoprotection involves multiple mechanisms including the inhibition by HSP of lethal protein aggregate formation and direct blockage of the PCD pathways (Li and Werb, 1982; Beere, 2001; Cashikar et al., 2005). The eukaryotic and prokaryotic heat shock responses although remarkably matched in terms of effector proteins diverge in the mechanisms employed to regulate their expression (Wu, 1995; Guisbert et al., 2004). The mammalian stress response is regulated by the heat shock transcription factors (HSF), of which HSF1 is the dominant factor (Wu, 1995). HSF1 contributes to expression of each family of HSP through interaction with the heat shock elements (HSE) in their promoters. In addition other factors contribute to expression with STAT1 leading to HSP70 and HSP90 expression, E-box factors such as c-Myc regulating HSP60 and HSP90 and the Brn-3a factor regulating the expression of small HSP such as HSP27 (Stephanou and Latchman, 1999; Farooqui-Kabir et al., 2004). Moreover, activation of estrogen receptors by ligand binding stimulates transcription of HSP27 in a range of tissues (Ciocca et al., 1993).

2.2. Elevated HSP Expression in Human Cancer

Under normal conditions cells may express constitutive levels of HSP (known as constitutive or cognate heat shock proteins), while during the process of carcinogenesis the transformed cells begin to express an elevated levels of HSP and this induction of HSP continues, in certain cases, during tumor progression. One clear example occurs in the uterine cervix, in which the subcolumnar cells of the endocervix show strong expression of Hsp27 during the process of transformation into foci of metaplasia (Ciocca et al., 1986). In the majority of cases cervical intraepithelial neoplasia arises from a metaplastic epithelium, and in human cervical samples with intraepithelial neoplasia as well as in invasive carcinomas Hsp27 is persistently expressed at high levels in an elevated number of cases (Puy et al., 1989). Moreover, Hsp27 expression in the uterine cervix is affected by HPV

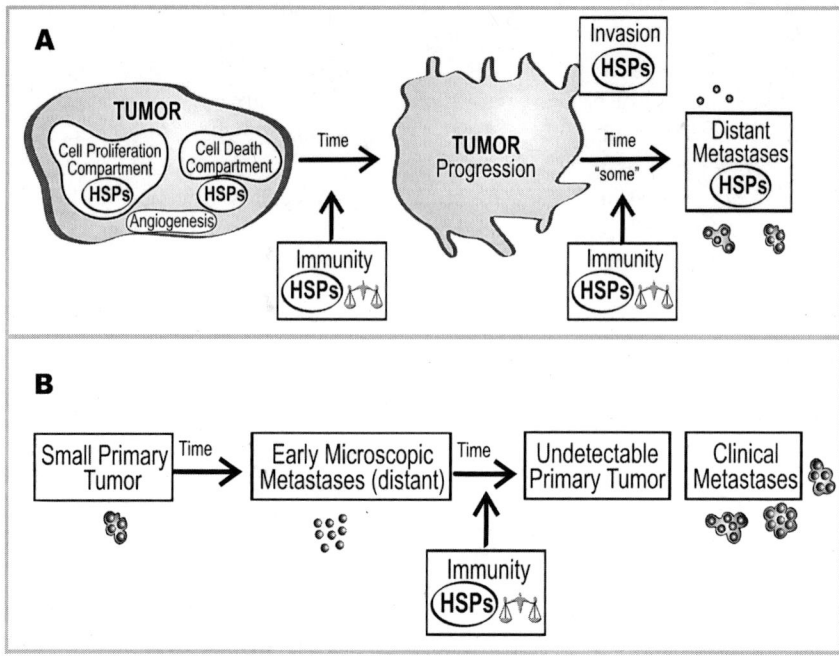

FIGURE 1. Schema of the most common form of tumor progression of a solid tumor. **A**: The primary tumor progresses due to the imbalance between the cell proliferation compartment (which is maintained by angiogenesis) and the cell death compartment, the heat shock proteins (HSPs) participate in these processes. With time, and thanks to the lack of adequate immunity against the cancer cells, the tumor invades the blood and lymphatic vessels and distant metastases might appear (again the HSPs participate in these processes). **B**: In a low percentage of patients the primary tumor might disappear or remain undetectable but the tumor is present/detectable at clinical level in the form of metastases.

Infection, a well known agent of malignant transformation (Ciocca et al., 1992). Other examples of elevated HSP expression during the process of carcinogenesis of human tissues have been documented in oral cancer (Kaur et al., 1998), in oesophageal cancer (Lambot et al., 2000), in colorectal carcinoma (Cappello et al., 2003a), and in prostate cancer (Cappello et al., 2003b). During the process of tumor progression, each tumor follows a unique pattern of molecular changes, and these changes in several occasions involve the elevated expression of certain HSP (Fig. 1, Table 1). The biological and clinical significances of the elevated expression of the different HSP in human cancers has recently been reviewed (Ciocca and Calderwood, 2005; Calderwood, 2006 #1059).

2.3. Essential Steps in Tumor Progression

Most tumors are formed by stepwise progression of cells from a minimally altered state able to grow and form nodules or polyps (in the case of solid tumors)

TABLE 1. Examples of heat shock proteins found at elevated levels in human tumors

Tumor	Heat shock proteins	References
Breast cancer	Hsp27, Hsp70, Hsp90, Grp78	Ciocca et al., 1983, 1993; Jameel et al., 1992; Fernandez et al., 2000
Endometrial cancer	Hsp27, Hsp70, Hsp90	Ciocca et al., 1985, 1989; Nambu et al., 1996
Ovarian cancer	Hsp27, Hsp72, Hsp90	Mileo et al., 1990; Langdon et al., 1995; Athanassiadou et al., 1998
Uterine cervix	Hsp10, Hsp27, Hsp60, Hsp70	Ciocca et al., 1989; Kaur and Rathan, 1995; Cappello et al., 2002, 2003
Oral cancer	Hsp27, Hsp60, Hsp70, Hsp90	Ito et al., 1998; Kaur et al., 1998; Kumamoto et al., 2002
Gastrointestinal, liver, pancreas	Hsp10, Hsp27, Hsp47, Hsp60, Hsp70, Hsp90, Hsp110, gp96	Liu et al., 1999; Delhaye et al., 1992; Chuma et al., 2004; Cappello et al., 2003a; Hwang et al., 2003; Heike et al., 2000; Maitra et al., 2002; Ogata et al., 2000
Lung cancer	Ubiquitin, Hsp27, Hsp63, Hsp70, Hsp90, grp94	Bonay et al., 1994; Michils et al., 2001; Malusecka et al., 2001; Wang et al., 2002
Urinary system	Hsp27, Hsp60, Hsp70, Hsp90	Storm et al., 1993; Kamishima et al., 1997; Takashi et al., 1998; Cardillo et al., 2000
Prostate cancer	HSF1, Hsp10, Hsp27, Hsp60, Hsp70	Storm et al., 1993; Hoang et al., 2000; Cornford et al., 2000; Cappello et al., 2002, 2003b
Leukemia, lymphoma	Hsp27, Hsp60, Hsp70, Hsp90	Chant et al., 1995; Xiao et al., 1996; Hsu and Hsu, 1998
Brain tumors	Hsp27, Hsp70, Hsp90	Kato et al., 1992; Gandour-Edwards et al., 1995; Hitotsumatsu et al., 1996; Assimakopoulou et al., 1997, 2000
Skin cancer, melanoma	Hsp27, Hsp70, Hsp90 gp96	Kanitakis et al., 1989; Bayerl et al., 1999; Lazaris et al., 1995; Missotten et al., 2003.
Sarcomas	Hsp27, Hsp47, Hsp60, Hsp72, Hsp90	Tetu et al., 1992; Uozaki et al., 2000; Trieb et al., 2000.

to multiply deviated cells able of unlimited growth, manipulation of their local environment, invasion of surrounding tissues, and escape into the circulation to found new colonies of secondary tumors or metastases (Vogelstein and Kinzler, 1993). Such a progression involves a vast array of molecular and morphological changes. In their landmark review, Hanahan and Weinberg have suggested organizing these traits into six essential alterations in cell physiology (Hanahan and Weinberg, 2000): (1) self-sufficiency in growth signals; (2) insensitivity to growth inhibition; (3) evasion of PCD; (4) limitless replicative potential; (5) sustained angiogenesis, and (6) tissue invasion and metastasis. The occurrence of this series of radical mostly genetic changes in physiology is envisaged to be at least partially dependent on the evolving instability of the tumor genome due to a breakdown in the DNA repair pathways (Fishel et al., 1993; Fishel and Kolodner, 1995). In addition to progressing in terms of acquiring enhanced malignant capabilities most human tumors undergo a further selections through exposure to various forms of cytotoxic therapy leading, in surviving cells, various phenotypes resistant to the cytotoxic therapies used in cancer treatment. Increases in heat shock protein levels appear to be involved in both tumor progression and in the acquisition of treatment resistance.

Tumor progression can be defined by the ability of the malignant cells to growth and spread beyond the initial micro-location where they arose. During tumor progression the degree of differentiation of the cells usually (but not always) changes from more differentiated to more undifferentiated states. Tumor progression is easier to characterize at morphological level in epithelial tumors due to the existence of sequential pathways leading from in situ carcinomas to invasive carcinomas and then to metastatic carcinomas. In contrast, tumor progression is more difficult to characterize at histological level in sarcomas, lymphomas and other tumor types (for instance, high-grade anaplastic gliomas kill the host by local progression not by metastatic dissemination). However, it is clear that during tumor progression there is a very complex array of molecular alterations that change the malignant phenotype and mediate tumor progression (Fig. 1).

2.4. Molecular Mechanisms for HSP Expression in Cancer

Elevated HSP gene transcription in cancer is coupled to some of the basic oncogenic pathways. Hsp promoters contain a number of response elements, including HSE, which binds to activated HSF1 trimers, and a CCAAT element, which binds to a complex between the trimeric factor NF-Y and HSP-CBF (Hunt and Calderwood, 1990; Wu, 1995; Imbriano et al., 2001) (Fig. 2). A primary mechanism for HSP regulation in normal cells involves the tumor repressor p53 and the related protein p63. These proteins repress transcription of HSPs through binding sites for the transcription factor NF-Y present within their promoters (Taira et al., 1999; Chae et al., 2005; Wu et al., 2005). During transformation, p53 mutation, a genetic change associated with over 45% of all cancers at many organ sites, reverses this effect and leads to enhanced HSP70 transcription through loss of HSP70 promoter repression (Agoff et al., 1993; Tsutsumi-Ishii et al., 1995; Madden et al., 1997; Ghioni et al., 2002). Alterations in p63 are closely associated with HSP70 expression in cancer and expression of the isoform $\Delta Np63\alpha$ which acts as a dominant negative inhibitor of wild-type p63 upregulates HSP70 and HSP40 levels in head and neck cancer (Wu et al., 2005). In addition to reversal of repression by p53 family proteins, induction of tumor HSP also involves positive effects on transcription through the signaling circuitry of the heat shock response pathway (Khaleque et al., 2005). During the heat shock response massive levels of HSP gene expression occur through interaction of HSF1 with the HSE present in all *HSP* promoters (Lindquist and Craig, 1988; Georgopolis and Welch, 1993; Wu, 1995). It was shown recently shown that the tumorigenic factor heregulin-β1 binds to the cell surface of breast cancer cells and leads to increased HSP expression, enhanced survival, and transformation through induced stabilization of HSF1 (Khaleque

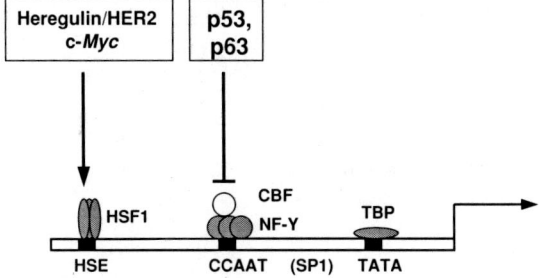

FIGURE 2. Structure of the proximal promoter of *hsp70.1* and potential regulatory input in cancer. HSP70 and HSP40 genes are regulated through sequences in the proximal promoter, including the heat shock element (HSE) and a CCAAT box element. These genes are normally repressed by interaction of p53 and its homologue p63 with the CBF-HSP/NF-Y complexes that bind to the CCAAT box element. Transforming mutations that inactivate p53 and p63 can lead to elevated HSP expression through relief of such p53/p63 mediated repression. In addition, HSP genes can be activated in cancer through the HSE elements. Heregulin, HER2 and c-myc can activate HSP gene transcription through HSE.

et al., 2005). Heregulin activates HSF1 through a signaling pathway involving activation of HER2 and PI-3 kinase at the cell surface and leads to expression of *HSP* genes through the HSE in their promoters (Khaleque et al., 2005). As PI-3 kinase is a key enzyme in malignant progression, particularly through its activation by PTEN mutation and induction of c-Myc, this may be an important mechanism for HSP elevation in cancer (Bader et al., 2005). In addition, the proto-oncogene c-Myc, which is activated by heregulin and HER2 also positively regulates HSP transcription through activation of HSE (Taira et al., 1999). Indeed, c-Myc activates the HSP90A promoter and inhibition of this activation decreases the transforming effects of c-Myc expression (Teng et al., 2004). HSF1 also plays a number of additional roles in the malignant phenotype, including the override of cell-cycle checkpoints leading to tumor aneuploidy and enhanced metastasis which may involve non-HSP dependent effects of HSF1 (Hoang et al., 2000; Wang et al., 2004). In addition, HSP90 is the intrinsic repressor of HSF1 under non-stress conditions, and one can envisage a mechanism for HSP elevation that includes increased sequestration of HSP90 by unstable mutated tumor proteins and derepression of HSF1, resulting in expression of HSP (Zou et al., 1998). In addition, HSP27 is activated by factors in addition to HSF1, including the POU domain protein Brn3a (Farooqui-Kabir et al., 2004; Arrigo, 2005; Lee et al., 2005). Overall, therefore, elevated expression of HSP occurs through relief of repression by p53, which is inactivated in many cancers, and through positive regulation by oncogenic signaling pathways which lead to activation of *HSP* promoters.

It also seems possible that HSP expression could be induced by the microenvironmental stress imposed by the tumor milieu (Folkman, 2002). However, little information is available to encourage this suggestion, and the available data indicate that another transcriptional stress response, the unfolded protein response rather than HSP expression, is activated by the tumor milieu and growth of tumor cells as xenografts leads to the inhibition rather than enhanced expression of HSP in cells growing in tumors (Tang et al., 2005). HSP90 could, however, be involved in the evolution of new phenotypes in nutritionally deprived tumor cells as discussed above.

3. HSP, HSF, and the Biology of Cancer

3.1. HSP90 and Self-Sufficiency in Growth Signals

The effects of the HSP on the anabolic pathways leading to the first of the essential alterations in the Hanahan and Weinberg scheme leading to cancer—self-sufficiency in growth signals—are mediated largely through HSP90. This molecular chaperone plays an essential role in stabilizing the fragile structures of many of the receptors, protein kinases and transcription factors that lie along the pathways of normal cellular growth (Neckers and Ivy, 2003). HSP90 is required to maintain signaling proteins in active conformation that can be rapidly triggered by growth signals (Neckers and Ivy, 2003; Pratt and Toft, 2003). HSP90 may thus be viewed as a facilitator of the rapid and fluid responses to extracellular signals required particularly in development and cell renewal (Neckers and Ivy, 2003; Pratt et al., 2004). Transformation involves the overexpression or mutation of many of these HSP90-dependent signaling molecules and HSP90 is increasingly required to maintain such proteins in active conformation. The degree to which HSP90 clients are essential in signal transduction that ultimately become subverted in cancer is illustrated in Figure 3, which depicts signaling pathways emanating from the proto-oncogene HER2. HSP90 is essential for the stability and activity of HER2 itself and downstream proteins including the protein kinases Akt, c-*src* and Raf-1 that play key roles in cell growth and survival (Neckers and Ivy, 2003). Over 100 similar HSP90 clients exist, each of which shows a similar dependency on HSP90 binding (Neckers and Ivy, 2003). HSP90 performs these molecular chaperone functions as the dominant component of a high Mr chaperone machine, a large complex incorporating five core proteins found in all complexes: HSP90 itself, the scaffold protein Hop, the p23 protein which mediates substrate choice and a HSP70 / HSP40 complex that mediates formation of the HSP90-substrate complexes (Pratt et al., 2004). HSP90 complexes are however heterogeneous and steroid hormone receptors for instance contain, in addition to the core proteins the immunophilins FKBP51, FKBP52 and CyP40 necessary for receptor function (Pratt et al., 2004). Pharmacological targeting of HSP90 using specific chemical inhibitors leads to the degradation of the client proteins and inhibition of tumor growth through G_1 arrest, morphological and functional differentiation and activation of apoptosis (Neckers and Ivy, 2003). This strongly implicates HSP90 as a key component required for "self-sufficiency in growth signals." However, overexpression of HSP90 in tumor

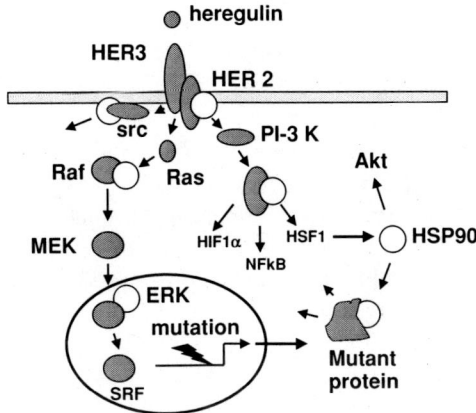

FIGURE 3. Role of HSP90 in intrinsic growth signaling and the accumulation of protein mutations in cancer. HSP90 (O) is associated with many of the key signaling molecules that mediate autonomous growth in cancer. We show here, for an example, the pathways of growth and survival signaling initiated by heregulin association with HER3/HER2 heterodimers and (1) induction of the growth promoting ERK pathway (yellow circles) leading to activation of SRF and transcription of immediate early genes and (2) the pro-survival PI-3 kinase pathway (blue symbols). HSP90 binding is required for the stability of many proteins in this pathway, including HER2, c-*src*, c-*Raf*, ERK1, and Akt. The figure also depicts one of the mechanisms for HSP induction in cancer and shows heregulin induced HSF1 activation through PI-3 kinase and Akt leading to HSP expression. In addition, HSP90 is involved in a second related process, the stabilization of mutant proteins. Mutant proteins are depicted binding to HSP90 and arrows suggest the multiple influences on the tumor cell phenotype of such stabilized proteins.

cells compared to normal tissues has been observed only in sporadic cases (Teng et al., 2004; Ciocca and Calderwood, 2005). More subtle alterations are also involved. For example, the splice variant HSP90N, which lacks an ATP binding site, is observed in some cancers and mediates transformation (Grammatikakis et al., 2002; Zhou et al., 2004). In addition, the co-chaperone cdc37, which is essential for the function of a subset of growth-related HSP90 binding protein kinases in normal cells is an oncogene in itself when overexpressed in prostate carcinoma (Pearl, 2005). Cdc37 binding to protein kinases and cyclophilin binding to nuclear receptors are mutually exclusive interactions, pointing to the existence of unique classes of individual HSP90 complexes that could be targeted in therapy. These changes in the abundance and composition of HSP90 complexes in cancer increase the chaperoning efficiency for oncogenic proteins (Scheibel and Buchner, 1998; Neckers, 2002; Kamal et al., 2003; Neckers and Lee, 2003). In addition, the increased susceptibility of HSP90 in tumors to ansamycin family drugs (which target their ATPase domains) reflects the concentration of tumor HSP90 within the chaperone machine complexes in which form it has a high affinity for the drugs as opposed to "free HSP90" in unbound form which predominates in normal cells and has low affinity for drugs (Kamal et al., 2003).

In addition to permitting autonomous growth in cancer, HSP90 plays a second, related role in enabling the emergence of polymorphisms and mutations that support the evolution of resistant clones. HSP90 stabilizes the conformations of mutant proteins which arise during transformation such as v-src, Bcr-Abl and p53 (Nimmanapalli et al., 2001; Neckers, 2002). Lindquist and co-workers have suggested that HSP90 stabilization of mutated, quasi-unstable protein conformations plays a generalized function in mediating the evolution of variant organisms and potentially new species. By permitting the accumulation of mutant proteins, new phenotypes emerge when free HSP90 levels are compromised by environmental stress (Rutherford and Lindquist, 1998; Queitsch et al., 2002). In this context HSP90 has been described as a "capacitor for evolution" (Rutherford and Lindquist, 1998).

HSP90 plays such a role in the evolution of growing tumor cell populations, which undergo a process similar to natural selection in overcoming the tumor suppression mechanisms of normal cells. The attainment of a "mutator phenotype" by tumor cells through the inactivation of DNA mismatch repair protein hMSH2, is believed to play an enabling role in cancer. HSP90, by buffering the conformations of resultant mutated proteins, may help to drive tumor progression (Fishel et al., 1993; Fishel and Kolodner, 1995). Indeed a number of DNA damage-response pathways, including DNA double-strand repair and cell cycle checkpoint control, become inactivated during tumor progression and these changes may contribute to the pool of mutated proteins buffered by HSP90 (van Gent et al., 2001). The current data indicates a process involving the production of proteins with mutations in individual amino acids that tend to be malfolded at normal temperatures and thus targeted for degradation. The intervention of the HSP90 chaperone machinery results in accumulation of a repertoire of these novel proteins more or less fit for the generation of new tumor cell phenotypes. HSP90 is capable not only of chaperoning point-mutated proteins, but also of stabilizing molecules with gross alteration in structure as in the case of fusion oncogenes formed by chromosomal translocation such as Bcr-Abl (Rahmani et al., 2005). HSP90 may thus play a dual role in allowing a cell to attain many of the steps required for tumor progression, permitting the autonomous function of unstable growth signaling molecules and the origin of new transforming phenotypes by buffering the growing pool of mutant proteins.

HSP90 functions at the interface between phenotype and environment due to (1) its role of licensing the function of unstable regulatory proteins and (2) its responsiveness to stress (Rutherford and Lindquist, 1998; Queitsch et al., 2002). In addition to its modulation by heat shock, HSP90 is highly sensitive to energy deprivation and requires a relatively high intracellular concentration of ATP and low ADP levels for function (Peng et al., 2005). Thus microenvironmental deprivation characterized by the hypoxic and glucose deprived cores of tumors is a condition under which HSP90 function is inhibited and new phenotypes might be expected to emerge in cells that survive this environment. Indeed, hypoxia has been shown to lead to a large increase in genomic instability which is accompanied by altered phenotypes such as enhanced invasive and metastatic potential (Folkman, 2002). It has been shown previously in a number of species that, in addition to the transient exposure of new phenotypes by HSP90 inactivation, these phenotypes become heritable due to enrichment by selection (Rutherford and Lindquist, 1998; Queitsch et al., 2002). Microenvironmental increases in mutation may thus be propagated by selection in the metabolically hostile milieu of the tumor (Mihaylova et al., 2003; Bindra and Glazer, 2005).

3.2. Role of HSP as Inhibitors of Cell Death Pathways

Unlike HSP90, HSP70 and HSP27 are not direct mediators of proliferation. They are however elevated in a wide spectrum of human cancers and mediate tumorigenesis through alternative strategies, involving inhibition of the programmed cell death (PCD) and senescence, two other essential traits in cancer. The HSP are powerful inhibitors of stress-mediated cell killing in all cellular organisms and mediate a profoundly resilient state (Gerner and Schneider, 1975). They play an essential role in the survival of a large proportion of human cancer cells and inactivation or knockdown, particularly of HSP70 or HSP27 leads to spontaneous activation of PCD not observed in normal tissues of origin (Beere, 2001) (Fig. 4). It has been shown that multiple PCD pathways must be inhibited to permit tumor progression (Tenniswood et al., 1992; Nylandsted et al., 2000; Paul et al., 2002). The enhanced activity of a number of oncogenes, most notably c-Myc or Ras, induce PCD pathways including apoptosis and autophagy and therefore a number of systems involved in PCD regulation, including the p53 and Bcl-2 family mediated networks, are subverted and inactivation of these network are important steps in the emergence of cancer cells (Hanahan and Weinberg, 2000; Nelson and White, 2004). The relationship of HSP27 and HSP70 to the p53 and Bcl-2 pathways is currently not clear although each can function independently in countering death signals (Beere, 2001; Jaattela, 2004). Convincing evidence for the inhibition of caspase-dependent apoptosis has been shown when the expression of either HSP70 or HSP27 is elevated (Beere, 2001). Notable molecular targets for HSP27 or HSP70 within the caspase dependent apoptosis pathway include c jun kinase, apaf-1, and caspase 8 (Beere, 2001) (Fig. 4). HSP70 is additionally involved in other pathways of PCD in addition to caspase-dependent apoptosis, and has been shown to inhibit a death pathway involving cell digestion by lysosomally derived cathepsins (Nylandsted et al., 2004). HSP70 was shown to be enriched in the lysosomal membranes of tumor cells and depletion of such HSP70 leads to spontaneous death related to the leakage of cathepsins into the bulk cytoplasm of a wide range of cancer cells (Nylandsted et al., 2004). It has been shown that the blockage of cell death through apoptosis and autophagy can lead to a proportion of such cells dying through default by necrosis (Proskuryakov et al., 2003). This form of death may be less efficient than other death pathways, permitting enhanced growth (Nelson and White, 2004). In addition necrosis, arising from inhibition of PCD and to ischemia in the poorly perfused tumor core is not opposed by HSP70 overexpression

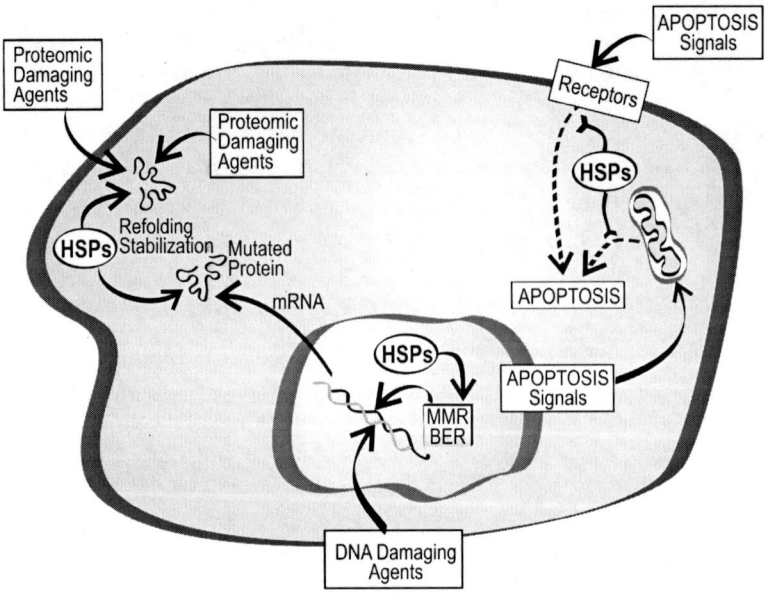

FIGURE 4. Schematic representation of a tumor cell showing the main molecular pathways where the heat shock proteins (HSPs) are involved with mechanisms of tumor cell protection. HSP are shown (a) maintaining the integrity of altered proteins damaged by external or internal damaging agents, (b) maintaining/protecting mutated proteins, (c) blocking/interfering with mechanism of apoptosis awaked by internal or external apoptotic signals, and (d) protecting the integrity of the DNA, here the HSPs interact with the mismatch repair (MMR) system and with the base excision repair (BER) system.

and leads to the release of cell contents into the tumor milieu and initiation of an inflammatory environment in the vicinity of the tumor that favors angiogenesis, tumor cell invasion and metastasis (Proskuryakov et al., 2003; Nelson and White, 2004; Viatour et al., 2005).

A key factor during tumor progression is the lack of balance between cell proliferation and cell death; in other words, a very large number of cells are in proliferation while cell death is relatively reduced (Fig. 4). This reduced level of cell death occurs even when there are endogenous and exogenous causes that force cell death. Among the endogenous factors are the insufficient blood supply which causes changes in the available nutrients, in oxygenation and in pH. Among the exogenous factors are the different treatments aimed to kill the tumor cells (e.g., radiotherapy, chemotherapy). One of the first pieces of evidence connecting HSP with resistance to tumor cell death was reported when in in vitro experiments tumor cells were exposed to nonlethal elevated temperatures (heat shock) previously to the exposure to chemotherapy agents (Donaldson et al., 1978). In contrast, more effective cell killing was reported when hyperthermia was combined with chemotherapy (Hahn, 1983). These contrasting effects can now be explained by the induction of a

HSP response due to the heat shock; then, the induced HSP protect the tumor cells against damaging agents. Experiments performed with human breast cancer cells showed that specific modifications (increased mRNA and protein expression and protein phosphorylation) in HSP were associated with resistance to certain anticancer drugs (Ciocca et al., 1992). This study showed that Hsp70 and Hsp27 were involved in resistance to doxorubicin but not to other cytotoxic drugs, and that this mechanism of drug resistance was not associated with that conferred by the P170 (*mdr*-1 gene) multidrug resistance protein, in fact the level of protection conferred by the HSP response was of lower level than that produced by the P170. In line with this, elevated levels of HSP27 have been found in human colon carcinoma cell lines expressing differential anthracycline sensitivity and similar levels of P170 glycoprotein (Boquete et al., 2001). However, there are potential connections between P170 and HSP, such as the presence of heat shock responsive elements in the promoter region of the *mdr-1* gene, this and another considerations have been reviewed elsewhere (Vargas-Roig et al., 1998). The association between HSP27 and drug resistance have been confirmed by transfection experiments in human breast, colon and testis cancer cells (Garrido et al., 1996; Richards et al., 1996). In other experimental studies Hsp70 has also been implicated in protection against the cytotoxic effects of hyperthermia and chemotherapy (Barnes et al., 2001; Ciocca et al., 2003). At the clinical level, high HSP70 levels in recurrent breast tumors correlated with low response to radiation and hyperthermia treatments (Liu and Hill, 1996), while the disease free survival of patients concomitantly overexpressing HSP27 and HSP70 was significantly shorter in breast cancer patients treated with neoadjuvant chemotherapies (Vargas-Roig et al., 1998). In a recent review we have presented the clinical data relating the HSP with the response to different anticancer treatments (Ciocca and Calderwood, 2005). Figure 4 shows the main pathways implicating HSP in resistance to cell death. One of the primary properties of the HSP is to catalyze the correct folding of nonnative or misfolded proteins and prevent their aggregation (Hartl and Hayer-Hartl, 2002), and thus if a tumor has mutated p53 this is protected by HSP which enhances tumor growth tumor cell but bad for the patient (Fig. 3).

The HSP have been involved with resistance to apoptosis through a variety of molecular mechanisms. HSP70 seems to be specifically required for tumor cell survival, depletion of HSP70 can activate a tumor-specific death program that is independent of caspases and bypasses Bcl-2 (Nylandsted et al., 2000). In addition, HSP27 has been involved in resistance to apoptosis preferentially blocking the mitochondrial cytochrome c release, whereas HSP72 interferes with apoptosomal caspase activation (Samali et al., 2001). HSP27 can also sequester procaspase-3 and cytochrome c inhibiting caspase-3 activation (Assimakopoulou et al., 1997). ASK1 (apoptosis signal-regulating kinase 1) elicits an apoptotic response under conditions of cellular stress (including oxidative stress, tumor necrosis factor 1 and Fas ligation) and the activity of this mitogen-activated protein kinase (MAP-KKK) is regulated in part by HSP70. The C-terminus of Hsp70-interacting protein (CHIP) interacts as a cochaperone with HSP70 and facilitates ASK1 ubiquitylation and degradation which in turn inhibits ASK1-dependent apoptosis (Hwang et al., 2005). In another recent study, it has been found that apoptosis mediated by CD95 (a death membrane receptor, implicated with Fas, Apo-1 and procaspase-8 in the formation of the death-inducing signalling complex) results in increased levels and stability of HSP70; and although this increase cannot be enough to completely block caspase activation, it can have a role in malignancy increasing the resistance to CD95-mediated apoptosis (Concannon et al., 2005). Tumor cells suffer oxidative stress, and it has been found that heat shock pre-treatment can inhibit the release of Smac/DIABLO from mitochondria (inhibiting the activation of caspase-9 and caspase-3) protecting the cells against H_2O_2-induced apoptosis (Jiang et al., 2005). In summary, elevated levels of HSP can inhibit the two major pathways of apoptosis involving either mitochondria or cell surface death receptors and may thus contribute to tumor growth and progression.

Apoptosis can be massive in tumors treated with anticancer agents, and there is evidence indicating that in these cases HSP27 and HSP70 mediate resistance to such massive cell death, with the surviving HSP27/70-positive tumor cells located around the blood vessels (Ciocca et al., 2003). It is highly likely that these surviving cells are responsible for tumor progression. In a clinical study it has been reported that HSP70 overexpression in breast cancer patients appeared associated with a worse prognosis/tumor progression (Ciocca et al., 1993). In fact several clinical studies have implicated a number of HSP with poor cancer prognosis (Ciocca and Calderwood, 2005). On the other hand, there is evidence that HSP70 when located at the cell membrane of the tumor cells might influence the NK (natural killer)-mediated cell death, low HSP70 membrane levels have correlated with a lower sensitivity to NK lysis and altered growth morphology/differentiation (Gehrmann et al., 2005).

Another mechanism of HSP protection in cancer cells occurs at the level of the cell nucleus. In biopsy samples from patients treated with chemotherapy, several of the surviving cells showed increased nuclear translocation of HSP27 and mainly HSP70 (Vargas-Roig et al., 1998). In order to know how this protein translocation could influence drug resistance, the authors performed in vitro studies with human peripheral blood mononuclear cells exposed to heat shock and doxorubicin, and they have shown that the heat shock response influences the DNA damage-repair capacity of the cells, suggesting that HSP70 participates in the DNA repair process mediated by the MMR (mismatch repair) system (Nadin et al., 2003). Immunoprecipitation and co-localization studies have confirmed that HSP27 and HSP70 contribute to the DNA repair function of hMLH1 and hMSH2 in peripheral blood lymphocytes from healthy subjects and cancer patients (Nadin et al., unpublished observations). In addition, base excision repair (BER) is another important mechanism of DNA repair, and HSP70 has been involved in this mechanism (Bases, 2005). This author has found that pretreatment of human leukemia cells with HSP70 protected them from the toxic effect of a topoisomerase II inhibitor and enhanced the repair of sublethal radiation damage. HSP70 can enhance 10–100-fold the BER enzyme activities (Mendez et al., 2003). Therefore, the HSP are involved with the protection of cells from many different lethal stressors, both endogenous and exogenous, and this protection is of crucial importance to cancer cell progression. The understanding of the involved molecular mechanisms is of great importance because then we can use this knowledge to block specific pathways to increase/facilitate cell death (see L. Neckers, this volume).

3.3. HSP and Replicative Senescence

Activation of HSF1 and elevated expression of HSP has a remarkably potent effect in increasing longevity in a number of species, suggesting a role in resistance to senescence (Tatar et al., 1997; Hsu et al., 2003). All somatic cells possess replicative checkpoints which place limits on the number of permitted cell divisions and, on arriving at such checkpoints, cells enter the pathways of senescence (Nelson and White, 2004; Campisi, 2005). To escape senescence and undergo unlimited growth, tumor cells must bypass "crisis" at which point the telomeres on chromosomes have shortened sufficiently to prevent successful future cell divisions (Campisi, 2005). p53-Sensitive expression of the enzyme telomerase in tumor cells is sufficient to bypass crisis and permit unlimited growth in some cells (Campisi, 2005). HSP90 has been shown to be essential for telomerase stability further indicating the importance of HSP90 in transformation (Workman, 2004). Recent studies also indicate that HSP70.2, a non-stress-inducible HSP70 whose normal expression is restricted to spermatogenesis, is expressed to high level in breast cancer and inhibits the onset of senescence (Rohde et al., 2005). HSP70.2 antagonizes the engagement of the senescence pathways by decreasing both the p53- (and p21)-dependent and -independent mechanisms of senescence (Rohde et al., 2005). Another member of the HSP70 family, the mitochondrial protein HSP75, or *mortalin*, plays an analogous role in countering replicative senescence and increasing the number of cell divisions in transformed cells and senescence can be induced by mortalin knockdown (Wadhwa et al., 2002). Mortalin performs this function at least in part by inhibiting p53 activity (Wadhwa et al., 2002; Proskuryakov et al., 2003). HSP70 and HSP27 are thus highly versatile cytoprotective proteins that may play essential roles in the emerging malignant cell and permit it to evade both fast (PCD) and slow (senescence) pathways of cell inactivation. Consistent with this, elevated expression of both HSP27 and HSP70 is correlated with chemotherapy resistance (Vargas-Roig et al., 1998). It may be that evasion of cell death through HSP expression originally evolved to permit cell populations to survive the toxic effects of environmental stress until protein repair occurred, and that this mechanism has subsequently been co-opted by tumor cells to permit override of the barriers to transformation of the intrinsic PCD and senescence pathways.

3.4. HSP Promote Angiogenesis

Angiogenesis is crucial for tumor progression; the generation of new blood vessels formed by proliferating endothelial cells can support the expanding tumor cell population, avoiding anoxia or decreasing hypoxia and bringing the nutrients to the tumor cells. Numerous compounds released from tumors stimulate angiogenesis and a number of pharmacological agents and strategies are being developed and tested as angiogenesis inhibitors (Vargas-Roig et al., 1998; Malusecka et al., 2001). One of the major problems with successful anti-angiogenic therapy is that with tumor progression there is an increasing number of angiogenic agents produced by the tumor cells, some are even produced in response to anti-angiogenic drugs (Kerbel, 2005) A major stimulus to induce angiogenesis is hypoxia, when a tumor is growing the tumor cells located more distant ($\cong$200 μm) around a capillary vessel suffer hypoxia and the tumor cells and the endothelial cells produce growth factors and survival factors (Camphausen et al., 2001).

In an experimental animal model, sarcoma cells surviving cytotoxic treatments were observed located around the blood vessels and where the tumor cells had the highest levels of HSP25 and HSP70 (Ciocca et al., 2003). Moreover, endothelial cells are relatively very tolerant to hypoxia and under hypoxia stress they produce hypoxia-associated proteins (different to HSP) (Graven and Farber, 1998), in certain tumors the endothelial cells may express HSP, like HSP25 (Ciocca et al., 2003). HSP70 can also be produced by endothelial cells (Zhang et al., 2003); however, so far the most-studied HSP in endothelial cells and angiogenesis is HSP90. For example, in a rabbit model of chronic ischemia it has been reported that HSP90 is capable of inducing angiogenesis via nitric oxide; liposomal HSP90 cDNA induced neovascularization preceding the growth of larger conductance vessels (Pfosser et al., 2005). This is consistent with the finding of Sun and Liao (2004) who found that HSP90 is involved in Akt and eNOS (endothelial nitric oxide synthase) phosphorylation and in eNOS gene transcription (these molecules are critical for angiogenesis). Specific inhibition of HSP90-mediated events by 17-AAG decreased Akt and eNOS expression in a concentration- and time-dependent manner, HSP90 and Akt modulate angiopoietin-1 induced angiogenesis (Sun and Liao, 2004; Chen et al., 2004). Another HSP90 modulator, 17-DMAG (a water-soluble benzoquinone ansamycin), has shown anti-angiogenic properties affecting endothelial cell functions (Kaur et al., 2004). Moreover, it has been reported that an angiogenesis inhibitor (apigenin, a plant-derived flavone) also interferes with the function of HSP90 activating the degradation of HIF-1alpha (hypoxia-inducible factor 1) protein, which in terms caused an inhibition of VEGF expression in human umbilical artery endothelial cells (Osada et al., 2004). VEGF can increase the association of HSP90 with VEGFR2, HSP90 is important for the phosphorylation of FAK (focal adhesion kinase) and the recruitment of vinculin to VEGFR2, all of these are necessary angiogenesis pathways (Masson-Gadais et al., 2003). The implications of HSP90 with angiogenesis is of special interest for cancer therapy since this protein is a molecular target of the anti-HSP90 compounds, thus the combined therapy of HSP90 inhibitors/modulators with anti-angiogenic drugs is an attractive therapeutic approach. In an experimental breast cancer model, it has been shown that tumorigenesis can occur in mice with defective angiogenesis but that tumors developing in such an environment seems to be highly sensitive to inhibitors of HER2/neu and HSP90-inhibitors (de Candia et al., 2003).

It has also been shown that, when tumors are exposed to antivascular and anti-angiogenic drugs, tumor cells begin to express the endoplasmic reticulum localized stress protein grp78 as a form of acquired drug resistance in response to glucose deprivation, anoxia and acidosis (Dong et al., 2005). However, grp78 may also be expressed on the surface of proliferating endothelial cells, and can be a target for Kringle 5 of human plasminogen a induceing apoptosis of endothelial cells and tumor cells (Davidson et al., 2005). Grp78 may thus be a "double-edged sword" promoting cell survival under some circumstances but targeting cells for apoptosis under others.

3.5. HSPs, HSF1, Invasion, and Metastasis

Invasion and metastasis are hallmarks of advanced cancer. Tumor cells that overexpress HSF1 and heat shock proteins show increased tendency to invade their microenvironment and spread to distant organs, although the mechanisms involved are not clear (Hoang et al., 2000). In addition, it was shown recently that heregulin stimulates cells to an anchorage-independent state necessary for metastasis by mechanisms that require the *hsf1* gene (Khaleque et al., 2005). In addition, a large number of clinical studies indicate positive correlation between increased levels of the downstream products of HSF1, including HSP27 and HSP70 and invasive / metastatic capacity of tumors (Ciocca and Calderwood, 2005). Recent studies have indicated an important extracellular role for HSP90 in the invasion step of metastasis through its binding to matrix metalloprotein 2 a key protein in invasion (Eustace and Jay, 2004). Other mechanisms for elevated HSP in invasion and metastasis include increased ability of tumor cells to survive in the bloodstream due to the death-inhibitory properties of HSP27 and HSP70, the emergence of genetic changes favoring these processes due to the effects of HSP90 on stabilizing mutant proteins, and alterations in the inflammatory nature of the tumor microenvironment at least partially due to HSP70 release from necrotic cells (Ciocca and Calderwood, 2005). Assessment of the role of HSP90 in this process is complicated, recent studies indicating that HSP90 inhibitory drugs enhance bone metastasis in breast cancer (Price et al., 2005). The mechanisms involved are not clear although HSP90 inhibitors are potent activators of HSF1, a known pro-metastatic protein (Neckers and Ivy, 2003).

In addition, HSP70 has been shown to be released from cells undergoing necrosis, enter the extracellular space and eventually circulate in the bloodstream (Asea et al., 2000; Calderwood et al., 2005). Such extracellular HSP70 exerts a profoundly pro-inflammatory effect due to interaction with receptors on inflammatory cells and the secondary release from monocytes and macrophages of inflammatory cytokines and NO (Asea et al., 2000). Extracellular HSP70 exerts both positive and negative effects on tumor growth. At a moderate rate of necrosis, these inflammatory effects of HSP70 may enhance tumor progression through the activation NFkB which has been shown to regulate the expression of a large number of proteins in both stromal and tumor cells that enhance tumor growth and spread (Viatour et al., 2005). In addition, many tumor cells express HSP70 receptors, which mediate internalization of HSP70 and enhance cell growth due to the pro-survival properties of the HSP (Calderwood et al., 2005). However, under conditions of widespread tumor cell necrosis such as occur after exposure to cytotoxic drugs, the massive release of HSP may lead to a specific CD8 T cell mediated anti-tumor immune response which can mediate tumor regression (Daniels et al., 2004). This appears to be due both to the pro- inflammatory effects of the HSP which stimulate the innate immune response and the ability of the HSP to carry tumor-associated antigenic peptides as cargo and stimulate a specific antitumor immune response (Asea et al., 2000; Calderwood et al., 2005).

4. Targeting Heat Shock Proteins in Cancer Therapy

On initial consideration, the HSPs, as highly abundant molecules that work in a stoichiometric rather than catalytic manner with substrates would seem unlikely targets for cancer drug discovery. However, the ATPase domain of HSP90 has recently been effectively targeted and a very active and unique family of anti-cancer drugs has been produced (Workman, 2004) (Fig. 5). The ability to specifically target HSP90 depends on the unique structure of its ATPase domain that can be selectively inhibited by the ansamycin family of drugs. While one might anticipate that the HSP90-targeted drugs, attacking proteins expressed in both normal and malignant cells, would lack specificity and inflict normal tissue damage, this has not proven to be the case and tumor cells are selectively sensitive to the anti-HSP90 drug 17AAG (Kamal et al., 2003; Neckers and Lee, 2003). Interestingly, these drugs are protective under some circumstances in normal cells, probably due to the potent induction of other, cytoprotective heat shock proteins including HSP70 (Peng et al., 2005). In tumor cells, however, the principle effects of HSP90 inhibition are degradation of proteins required for autonomous growth and for cytoprotection. In addition, the selective effect of HSP90 targeting drugs also involves the increased concentration of HSP90 substrates found in tumor cells, including overexpressed oncogenes and mutant proteins generated through the "mutator phenotype" of advanced cancers. Anti-HSP90 drugs are therefore highly promising through the targeting of a novel protein with potential to block a wide spectrum of the major pathways of autonomous tumor growth (Scheibel and Buchner, 1998; Neckers, 2002; Kamal et al., 2003; Neckers and Ivy, 2003; Neckers and Lee, 2003; Workman, 2004) (Fig. 3). One possible side effect of such drugs is resistance to cancer therapy through selection of resistant traits due to the unmasking (as discussed above) of cryptic phenotypes normally buffered by HSP90 (Rutherford and Lindquist, 1998). Nonetheless, these drugs target a wide spectrum of proteins required for malignant cell growth and are showing considerable promise in phase I and II clinical trials (Neckers and Ivy, 2003; Workman, 2004; Banerji et al., 2005).

The other HSPs have not yet been effectively targeted although efforts to do this are proceeding. One would predict high promise for such drugs due to the widespread role of HSP27 and HSP70 in blocking PCD and senescence during tumor progression, although the high concentrations of these proteins may place some constraints on the feasibility of this approach (Fig. 5). However, the free, unbound levels of these proteins are low due to the proliferation of quasi-stable client proteins in tumor cells (Craig and Gross, 1991). Drugs that target HSP70 family

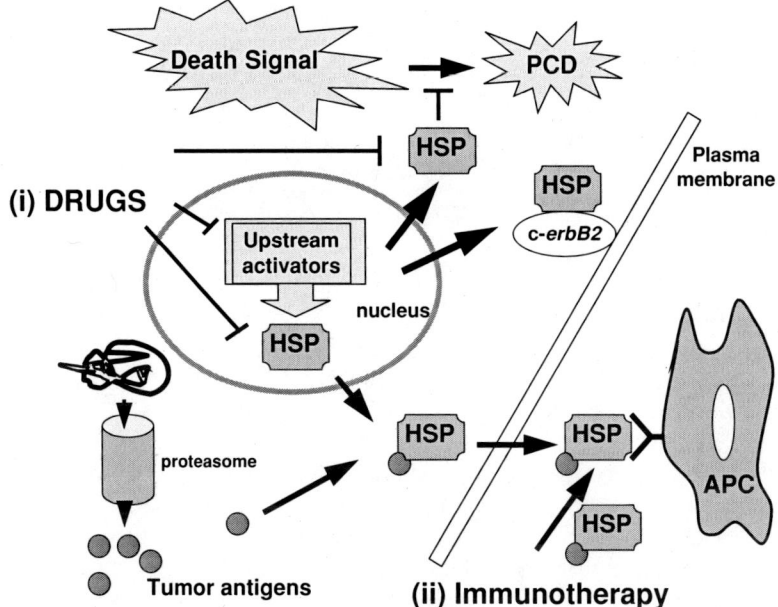

FIGURE 5. The double jeopardy of elevated HSP expression: Targeting HSP with novel pharmaceuticals and immunotherapy. HSP levels are essential for the growth of many tumor types. They are thus susceptible to (i) drugs that inhibit HSP expression by targeting upstream activators and drugs that block HSP function in chaperoning client proteins. In addition (ii) HSP70 and HSP90 can chaperone antigenic peptides produced by digestion of intracellular antigenic tumor proteins through the proteasome. Such HSP70 peptide complexes are secreted from viable cells or shed from necrotic cells and thus encounter antigen-presenting cells (APC) in the extracellular milieu and bind to specific HSP receptors. HSP-peptide complexes are unique immunogens in stimulating adaptive immunity through the cross presentation of antigenic peptides and inducing innate immunity through receptors on APC. HSP70 peptide complexes are also shown (ii) being introduced as anti-cancer vaccines in immunotherapy approaches.

proteins in general would likely be toxic through inhibition of the basic molecular chaperone function needed in folding most intracellular proteins. However, discrete *HSP70* family members, many of which are overexpressed in cancer, have distinct mechanisms for increasing cell survival and can be targeted individually (Rohde et al., 2005). In addition, additive toxicity can be achieved by combining treatments that target individual HSP70 family members (Rohde et al., 2005).

The elevated HSP levels in tumors also represent an opportunity for cancer immunotherapy due to the innate ability of many HSP to function as biological adjuvants and to chaperone tumor antigens (Calderwood et al., 2005) (Fig. 5). HSP70 and HSP110 associate with tumor antigens in vivo and can be easily extracted, purified, and used as autologous vaccines (Srivastava, 2002; Manjili et al., 2003). In addition, when HSP are elevated in cancer cells and such cells are destroyed through a necrotic mechanism, cells at distant sites are destroyed by a specific anti-tumor immune response which is strictly dependent on elevated HSP release from the primary cells (Calderwood, 2005). Maximal utilization of molecular chaperone-based immunotherapy will probably await full characterization of mechanisms of tumor antigen uptake by HSP and the nature of HSP receptors on antigen presenting cells and their role in antigen presentation (Calderwood et al., 2005).

5. Conclusions

In conclusion, therefore, heat shock proteins are expressed at elevated levels in many cancers due to the de-repression of *hsp* genes during malignant progression. At these elevated levels, HSP family members play an essential facilitating role in cancer, permitting autonomous growth through the accumulation of overexpressed and mutated oncogenes and inhibiting the programmed cell death of tumor cells. However, these abundant HSP levels also offer a tempting target for oncologists to design treatments that can inhibit broad areas of the malignant phenotype and promising approaches to cancer treatment based on inhibiting the properties of the HSP are currently emerging.

References

Agoff, S. N., Hou, J., Linzer, D. I., and Wu, B. (1993) Regulation of the human hsp70 promoter by p53. *Science* 259:84–7.

Arrigo, A. P. (2005) Heat shock proteins as molecular chaperones. *Med Sci (Paris)* 21:619–25.

Asea, A., Kraeft, S. K., Kurt-Jones, E. A., Stevenson, M. A., Chen, L. B., Finberg, R. W., Koo, G. C., and Calderwood, S. K. (2000) HSP70 stimulates cytokine production through a CD14-dependant pathway, demonstrating its dual role as a chaperone and cytokine. *Nat Med* 6:435–42.

Assimakopoulou, M. (2000) Human meningiomas: Immunohistochemical localization of progesterone receptor and heat shock protein 27 and absence of estrogen receptor and PS2. *Cancer Detect Prev* 24:163–8.

Assimakopoulou, M., Sotiropoulou-Bonikou, G., Maraziotis, T., and Varakis, I. (1997) Prognostic significance of Hsp-27 in astrocytic brain tumors: An immunohistochemical study. *Anticancer Res* 17:2677–82.

Athanassiadou, P., Petrakakou, E., Sakelariou, V., Zerva, C., Liossi, A., Michalas, S., and Athanassiades, P. (1998) Expression of p53, bcl-2 and heat shock protein (hsp72) in malignant and benign ovarian tumours. *Eur J Cancer Prev* 7:225–31.

Bader, A. G., Kang, S., Zhao, L., and Vogt, P. K. (2005) Oncogenic PI3K deregulates transcription and translation. *Nat Rev Cancer* 5:921–9.

Banerji, U., Walton, M., Raynaud, F., Grimshaw, R., Kelland, L., Valenti, M., Judson, I., and Workman, P. (2005) Pharmacokinetic-pharmacodynamic relationships for the heat shock protein 90 molecular chaperone inhibitor 17-allylamino, 17-demethoxygeldanamycin in human ovarian cancer xenograft models. *Clin Cancer Res* 11:7023–32.

Barnes, J. A., Dix, D. J., Collins, B. W., Luft, C., and Allen, J. W. (2001) Expression of inducible Hsp70 enhances the proliferation of MCF-7 breast cancer cells and protects against the cytotoxic effects of hyperthermia. *Cell Stress Chap* 6:316–25.

Bases, R. (2005) Clonogenicity of human leukemic cells protected from cell-lethal agents by heat shock protein 70. *Cell Stress Chap* 10:37–45.

Bayerl, C., Dorfner, B., Rzany, B., Fuhrmann, E., Coelho, C. C., and Jung, E. G. (1999) Heat shock protein HSP 27 is expressed in all types of basal cell carcinoma in low and high risk UV exposure groups. *Eur J Dermatol* 9:281–4.

Beere, H. M. (2001) Stressed to death: Regulation of apoptotic signaling pathways by the heat shock proteins. *Sci STKE* 2001:RE1.

Bindra, R. S., and Glazer, P. M. (2005) Genetic instability and the tumor microenvironment: Towards the concept of microenvironment-induced mutagenesis. *Mutat Res* 569:75–85.

Bonay, M., Soler, P., Riquet, M., Battesti, J. P., Hance, A. J., and Tazi, A. (1994) Expression of heat shock proteins in human lung and lung cancers. *Am J Respir Cell Mol Biol* 10:453–61.

Boquete, A. L., Vargas Roig, L., Lopez, G. A., Gude, R., Binda, M. M., Gonzalez, A. D., Ciocca, D. R., and Bonfil, R. D. (2001) Differential anthracycline sensitivity in two related human colon carcinoma cell lines expressing similar levels of P-glycoprotein. *Cancer Lett* 165:111–6.

Bukau, B., and Horwich, A. L. (1998) The Hsp70 and Hsp60 chaperone machines. *Cell* 92:351–66.

Calderwood, S. K. (2005) Chaperones and slow death—a recipe for tumor immunotherapy. *Trends Biotechnol* 23:57–9.

Calderwood, S. K., Theriault, J. R., and Gong, J. (2005) Message In A Bottle: Role of the 70 kilodalton heat shock protein family in anti-tumor immunity. *Eur J immunol* in press.

Camphausen, K., Moses, M. A., Beecken, W. D., Khan, M. K., Folkman, J., and O'Reilly, M. S. (2001) Radiation therapy to a primary tumor accelerates metastatic growth in mice. *Cancer Res* 61:2207–11.

Campisi, J. (2005) Senescent cells, tumor suppression, and organismal aging: good citizens, bad neighbors. *Cell* 120:513–22.

Cappello, F., Bellafiore, M., David, S., Anzalone, R., and Zummo, G. (2003a) Ten kilodalton heat shock protein (HSP10) is overexpressed during carcinogenesis of large bowel and uterine exocervix. *Cancer Lett* 196:35–41.

Cappello, F., Bellafiore, M., Palma, A., Marciano, V., Martorana, G., Belfiore, P., Martorana, A., Farina, F., Zummo, G., and Bucchieri, F. (2002) Expression of 60-kD heat shock protein increases during carcinogenesis in the uterine exocervix. *Pathobiology* 70: 83–8.

Cappello, F., Rappa, F., David, S., Anzalone, R., and Zummo, G. (2003b) Immunohistochemical evaluation of PCNA, p53, HSP60, HSP10 and MUC-2 presence and expression in prostate carcinogenesis. *Anticancer Res* 23:1325–31.

Cardillo, M. R., Sale, P., and Di Silverio, F. (2000) Heat shock protein-90, IL-6 and IL-10 in bladder cancer. *Anticancer Res* 20:4579–83.

Cashikar, A. G., Duennwald, M., and Lindquist, S. L. (2005) A chaperone pathway in protein disaggregation: HSP26 alters the nature of protein aggregates to facilitate reactivation by hsp104. *J Biol Chem* 280:23869–75.

Chae, H. D., Yun, J., and Shi, D. Y. (2005) Transcription repression of a CCAAT-binding transcription factor CBF/HSP70 by p53. *Exp Mol Med* 37:488–91.

Chant, I. D., Rose, P. E., and Morris, A. G. (1995) Analysis of heat-shock protein expression in myeloid leukaemia cells by flow cytometry. *Br J Haematol* 90:163–8.

Chuma, M., Saeki, N., Yamamoto, Y., Ohta, T., Asaka, M., Hirohashi, S., and Sakamoto, M. (2004) Expression profiling in hepatocellular carcinoma with intrahepatic metastasis: identification of high-mobility group I(Y) protein as a molecular marker of hepatocellular carcinoma metastasis. *Keio J Med* 53:90–7.

Ciocca, D. R., and Calderwood, S. K. (2005) Heat shock proteins in cancer: Diagnostic, prognostic, predictive, and treatment applications. *Cell Stress Chap* 10:86–103.

Ciocca, D. R., Clark, G. M., Tandon, A. K., Fuqua, S. A., Welch, W. J., and McGuire, W. L. (1993) Heat shock protein hsp70 in patients with axillary lymph node-negative breast cancer: Prognostic implications. *J Natl Cancer Inst* 85:570–4.

Ciocca, D. R., Fuqua, S. A., Lock-Lim, S., Toft, D. O., Welch, W. J., and McGuire, W. L. (1992) Response of human breast cancer cells to heat shock and chemotherapeutic drugs. *Cancer Res* 52:3648–54.

Ciocca, D. R., Lo Castro, G., Alonio, L. V., Cobo, M. F., Lotfi, H., and Teyssie, A. (1992) Effect of human papillomavirus infection on estrogen receptor and heat shock protein hsp27 phenotype in human cervix and vagina. *Int J Gynecol Pathol* 11:113–21.

Ciocca, D. R., Oesterreich, S., Chamness, G. C., McGuire, W. L., and Fuqua, S. A. (1993) Biological and clinical implications of heat shock protein 27,000 (Hsp27): A review. *J Natl Cancer Inst* 85:1558–70.

Ciocca, D. R., Puy, L. A., Edwards, D. P., Adams, D. J., and McGuire, W. L. (1985) The presence of an estrogen-regulated protein detected by monoclonal antibody in abnormal human endometrium. *J Clin Endocrinol Metab* 60:137–43.

Ciocca, D. R., Puy, L. A., and Fasoli, L. C. (1989) Study of estrogen receptor, progesterone receptor, and the estrogen-regulated Mr 24,000 protein in patients with carcinomas of the endometrium and cervix. *Cancer Res* 49:4298–304.

Ciocca, D. R., Puy, L. A., and Lo Castro, G. (1986) Localization of an estrogen-responsive protein in the human cervix during menstrual cycle, pregnancy, and menopause and in abnormal cervical epithelia without atypia. *Am J Obstet Gynecol* 155:1090–6.

Ciocca, D. R., Rozados, V. R., Cuello Carrion, F. D., Gervasoni, S. I., Matar, P., and Scharovsky, O. G. (2003) Hsp25 and Hsp70 in rodent tumors treated with doxorubicin and lovastatin. *Cell Stress Chap* 8:26–36.

Concannon, C. G., FitzGerald, U., Holmberg, C. I., Szegezdi, E., Sistonen, L., and Samali, A. (2005) CD95-mediated alteration in Hsp70 levels is dependent on protein stabilization. *Cell Stress Chap* 10:59–65.

Cornford, P. A., Dodson, A. R., Parsons, K. F., Desmond, A. D., Woolfenden, A., Fordham, M., Neoptolemos, J. P., Ke, Y., and Foster, C. S. (2000) Heat shock protein expression independently predicts clinical outcome in prostate cancer. *Cancer Res* 60: 7099–105.

Craig, E. A., and Gross, C. A. (1991) Is HSP70 the molecular thermometer. *Trends Biochem Sci* 16:135–40.

Daniels, G. A., Sanchez-Perez, L., Diaz, R. M., Kottke, T., Thompson, J., Lai, M., Gough, M., Karim, M., Bushell, A., Chong, H., Melcher, A., Harrington, K., and Vile, R. G. (2004) A simple method to cure established tumors by inflammatory killing of normal cells. *Nat Biotechnol* 22:1125–32.

Davidson, D. J., Haskell, C., Majest, S., Kherzai, A., Egan, D. A., Walter, K. A., Schneider, A., Gubbins, E. F., Solomon, L., Chen, Z., Lesniewski, R., and Henkin, J. (2005) Kringle 5 of human plasminogen induces apoptosis of endothelial and tumor cells through surface-expressed glucose-regulated protein 78. *Cancer Res* 65:4663–72.

de Candia, P., Solit, D. B., Giri, D., Brogi, E., Siegel, P. M., Olshen, A. B., Muller, W. J., Rosen, N., and Benezra, R. (2003) Angiogenesis impairment in Id-deficient mice cooperates with an Hsp90 inhibitor to completely suppress HER2/neu-dependent breast tumors. *Proc Natl Acad Sci U S A* 100:12337–42.

Delhaye, M., Gulbis, B., Galand, P., and Mairesse, N. (1992) Expression of 27-kD heat-shock protein isoforms in human neoplastic and nonneoplastic liver tissues. *Hepatology* 16:382–9.

Donaldson, S. S., Gordon, L. F., and Hahn, G. M. (1978) Protective effect of hyperthermia against the cytotoxicity of actinomycin D on Chinese hamster cells. *Cancer Treat Rep* 62:1489–95.

Dong, D., Ko, B., Baumeister, P., Swenson, S., Costa, F., Markland, F., Stiles, C., Patterson, J. B., Bates, S. E., and Lee, A. S. (2005) Vascular targeting and antiangiogenesis agents induce drug resistance effector GRP78 within the tumor microenvironment. *Cancer Res* 65:5785–91.

Eustace, B. K., and Jay, D. G. (2004) Extracellular roles for the molecular chaperone, hsp90. *Cell Cycle* 3:1098–100.

Farooqui-Kabir, S. R., Budhram-Mahadeo, V., Lewis, H., Latchman, D. S., Marber, M. S., and Heads, R. J. (2004) Regulation of Hsp27 expression and cell survival by the POU transcription factor Brn3a. *Cell Death Differ* 11:1242–4.

Fernandez, P. M., Tabbara, S. O., Jacobs, L. K., Manning, F. C., Tsangaris, T. N., Schwartz, A. M., Kennedy, K. A., and Patierno, S. R. (2000) Overexpression of the glucose-regulated stress gene GRP78 in malignant but not benign human breast lesions. *Breast Cancer Res Treat* 59:15–26.

Fishel, R., and Kolodner, R. D. (1995) Identification of mismatch repair genes and their role in the development of cancer. *Curr Opin Genet Dev* 5:382–95.

Fishel, R., Lescoe, M. K., Rao, M. R., Copeland, N. G., Jenkins, N. A., Garber, J., Kane, M., and Kolodner, R. (1993) The human mutator gene homolog MSH2 and its association with hereditary nonpolyposis colon cancer. *Cell* 75:1027–38.

Folkman, J. (2002) Role of angiogenesis in tumor growth and metastasis. *Semin Oncol* 29:15–8.

Garrido, C., Mehlen, P., Fromentin, A., Hammann, A., Assem, M., Arrigo, A. P., and Chauffert, B. (1996) Inconstant association between 27-kDa heat-shock protein (Hsp27) content and doxorubicin resistance in human colon cancer cells. The doxorubicin-protecting effect of Hsp27. *Eur J Biochem* 237:653–9.

Gehrmann, M., Schonberger, J., Zilch, T., Rossbacher, L., Thonigs, G., Eilles, C., and Multhoff, G. (2005) Retinoid- and sodium-butyrate-induced decrease in heat shock protein 70 membrane-positive tumor cells is associated with reduced sensitivity to natural killer cell lysis, growth delay, and altered growth morphology. *Cell Stress Chap* 10. 136–46.

Georgopolis, C., and Welch, W. J. (1993) Role of the major heat shock proteins as molecular chaperones. *Ann Rev Cell Biol* 9:601–34.

Gerner, E. W., and Schneider, M. J. (1975) Induced thermal resistance in HeLa cells. *Nature* 256:500–2.

Ghioni, P., Bolognese, F., Duijf, P. H., Van Bokhoven, H., Mantovani, R., and Guerrini, L. (2002) Complex transcriptional effects of p63 isoforms: identification of novel activation and repression domains. *Mol Cell Biol* 22:8659–68.

Grammatikakis, N., Vultur, A., Ramana, C. V., Siganou, A., Schweinfest, C. W., Watson, D. K., and Raptis, L. (2002) The role of Hsp90N, a new member of the Hsp90 family, in signal transduction and neoplastic transformation. *J Biol Chem* 277:8312–20.

Graven, K. K., and Farber, H. W. (1998) Endothelial cell hypoxic stress proteins. *J Lab Clin Med* 132:456–63.

Guisbert, E., Herman, C., Lu, C. Z., and Gross, C. A. (2004) A chaperone network controls the heat shock response in *E. coli*. *Genes Dev* 18:2812–21.

Hahn, G. M. (1983) *Hyperthermia to Enhance Drug Delivery*. Plenum Press, New York.

Hanahan, D., and Weinberg, R. A. (2000) The hallmarks of cancer. *Cell* 100:57–70.

Hartl, F. U., and Hayer-Hartl, M. (2002) Molecular chaperones in the cytosol: From nascent chain to folded protein. *Science* 295:1852–8.

Heike, M., Frenzel, C., Meier, D., and Galle, P. R. (2000) Expression of stress protein gp96, a tumor rejection antigen, in human colorectal cancer. *Int J Cancer* 86:489–93.

Hitotsumatsu, T., Iwaki, T., Fukui, M., and Tateishi, J. (1996) Distinctive immunohistochemical profiles of small heat shock proteins (heat shock protein 27 and alpha B-crystallin) in human brain tumors. *Cancer* 77:352–61.

Hoang, A. T., Huang, J., Rudra-Ganguly, N., Zheng, J., Powell, W. C., Rabindran, S. K., Wu, C., and Roy-Burman, P. (2000) A novel association between the human heat shock transcription factor 1 (HSF1) and prostate adenocarcinoma. *Am J Pathol* 156:857–64.

Hsu, A. L., Murphy, C. T., and Kenyon, C. (2003) Regulation of aging and age-related disease by DAF-16 and heat-shock factor. *Science* 300:1142–5.

Hsu, P. L., and Hsu, S. M. (1998) Abundance of heat shock proteins (hsp89, hsp60, and hsp27) in malignant cells of Hodgkin's disease. *Cancer Res* 58:5507–13.

Hunt, C., and Calderwood, S. K. (1990) Characterization and sequence of a mouse HSP70 gene and its expression in mouse cell lines. *Gene* 87:199–204.

Hwang, J. R., Zhang, C., and Patterson, C. (2005) C-terminus of heat shock protein 70— interacting protein facilitates degradation of apoptosis signal-regulating kinase 1 and inhibits apoptosis signal-regulating kinase 1—dependent apoptosis. *Cell Stress Chap* 10:147–56.

Hwang, T. S., Han, H. S., Choi, H. K., Lee, Y. J., Kim, Y. J., Han, M. Y., and Park, Y. M. (2003) Differential, stage-dependent expression of Hsp70, Hsp110 and Bcl-2 in colorectal cancer. *J Gastroenterol Hepatol* 18:690–700.

Imbriano, C., Bolognese, F., Gurtner, A., Piaggio, G., and Mantovani, R. (2001) HSP-CBF is an NF-Y-dependent coactivator of the heat shock promoters CCAAT boxes. *J Biol Chem* 276:26332–9.

Ito, T., Kawabe, R., Kurasono, Y., Hara, M., Kitamura, H., Fujita, K., and Kanisawa, M. (1998) Expression of heat shock proteins in squamous cell carcinoma of the tongue: An immunohistochemical study. *J Oral Pathol Med* 27:18–22.

Jaattela, M. (2004) Multiple cell death pathways as regulators of tumour initiation and progression. *Oncogene* 23:2746–56.

Jameel, A., Skilton, R. A., Campbell, T. A., Chander, S. K., Coombes, R. C., and Luqmani, Y. A. (1992) Clinical and biological significance of HSP89 alpha in human breast cancer. *Int J Cancer* 50:409–15.

Jiang, B., Xiao, W., Shi, Y., Liu, M., and Xiao, X. (2005) Heat shock pretreatment inhibited the release of Smac/DIABLO from mitochondria and apoptosis induced by hydrogen peroxide in cardiomyocytes and C2C12 myogenic cells. *Cell Stress Chap* 10:252–62.

Kamal, A., Thao, L., Sensintaffar, J., Zhang, L., Boehm, M. F., Fritz, L. C., and Burrows, F. J. (2003) A high-affinity conformation of Hsp90 confers tumour selectivity on Hsp90 inhibitors. *Nature* 425:407–10.

Kamishima, T., Fukuda, T., Usuda, H., Takato, H., Iwamoto, H., and Kaneko, H. (1997) Carcinosarcoma of the urinary bladder: expression of epithelial markers and different expression of heat shock proteins between epithelial and sarcomatous elements. *Pathol Int* 47:166–73.

Kanitakis, J., Zambruno, G., Viac, J., Tommaselli, L., and Thivolet, J. (1989) Expression of an estrogen receptor-associated protein (p29) in epithelial tumors of the skin. *J Cutan Pathol* 16:272–6.

Kato, S., Kato, M., Hirano, A., Takikawa, M., and Ohama, E. (2001) The immunohistochemical expression of stress-response protein (srp) 60 in human brain tumours: Relationship of srp 60 to the other five srps, proliferating cell nuclear antigen and p53 protein. *Histol Histopathol* 16:809–20.

Kaur, G., Belotti, D., Burger, A. M., Fisher-Nielson, K., Borsotti, P., Riccardi, E., Thillainathan, J., Hollingshead, M., Sausville, E. A., and Giavazzi, R. (2004) Antiangiogenic properties of 17-(dimethylaminoethylamino)-17-demethoxygeldanamycin: An orally bioavailable heat shock protein 90 modulator. *Clin Cancer Res* 10:4813–21.

Kaur, J., Das, S. N., Srivastava, A., and Ralhan, R. (1998) Cell surface expression of 70 kDa heat shock protein in human oral dysplasia and squamous cell carcinoma: Correlation with clinicopathological features. *Oral Oncol* 34:93–8.

Kaur, J., and Ralhan, R. (1995) Differential expression of 70-kDa heat shock-protein in human oral tumorigenesis. *Int J Cancer* 63:774–9.

Kerbel, R. S. (2005) Therapeutic implications of intrinsic or induced angiogenic growth factor redundancy in tumors revealed. *Cancer Cell* 8:269–71.

Khaleque, M. A., Bharti, A., Sawyer, D., Gong, J., Benjamin, I. J., Stevenson, M. A., and Calderwood, S. K. (2005) Induction of heat shock proteins by heregulin beta1 leads to protection from apoptosis and anchorage-independent growth. *Oncogene* 24:6564–73.

Kumamoto, H., Suzuki, T., and Ooya, K. (2002) Immunohistochemical analysis of inducible nitric oxide synthase (iNOS) and heat shock proteins (HSPs) in ameloblastomas. *J Oral Pathol Med* 31:605–11.

Lambot, M. A., Peny, M. O., Fayt, I., Haot, J., and Noel, J. C. (2000) Overexpression of 27-kDa heat shock protein relates to poor histological differentiation in human oesophageal squamous cell carcinoma. *Histopathology* 36:326–30.

Langdon, S. P., Rabiasz, G. J., Hirst, G. L., King, R. J., Hawkins, R. A., Smyth, J. F., and Miller, W. R. (1995) Expression of the heat shock protein HSP27 in human ovarian cancer. *Clin Cancer Res* 1:1603–9.

Lazaris, A. C., Theodoropoulos, G. E., Aroni, K., Saetta, A., and Davaris, P. S. (1995) Immunohistochemical expression of C-myc oncogene, heat shock protein 70 and HLA-DR molecules in malignant cutaneous melanoma. *Virchows Arch* 426:461–7.

Lee, S. A., Ndisang, D., Patel, C., Dennis, J. H., Faulkes, D. J., D'Arrigo, C., Samady, L., Farooqui-Kabir, S., Heads, R. J., Latchman, D. S., and Budhram-Mahadeo, V. S. (2005) Expression of the Brn-3b transcription factor correlates with expression of HSP-27 in breast cancer biopsies and is required for maximal activation of the HSP-27 promoter. *Cancer Res* 65:3072–80.

Li, G. C., and Werb, Z. (1982) Correlation between the synthesis of heat shock proteins and the development of thermotolerance in CHO fibroblasts. *Proc Natl Acad Sci U S A* 79:3218–22.

Lindquist, S., and Craig, E. A. (1988) The heat shock proteins. *Ann Rev Genet* 22:631–7.

Liu, F. F., and Hill, R. P. (1996) Potential role of HSP70 as an indicator of response to radiation and hyperthermia treatments for recurrent breast cancer. *Int J Hyperthermia* 12:301–2.

Liu, X., Ye, L., Wang, J., and Fan, D. (1999) Expression of heat shock protein 90 beta in human gastric cancer tissue and SGC7901/VCR of MDR-type gastric cancer cell line. *Chin Med J (Engl)* 112:1133–7.

Madden, S. L., Galella, E. A., Zhu, J., Bertelsen, A. H., and Beaudry, G. A. (1997) SAGE transcript profiles for p53-dependent growth regulation. *Oncogene* 15:1079–85.

Maitra, A., Iacobuzio-Donahue, C., Rahman, A., Sohn, T. A., Argani, P., Meyer, R., Yeo, C. J., Cameron, J. L., Goggins, M., Kern, S. E., Ashfaq, R., Hruban, R. H., and Wilentz, R. E. (2002) Immunohistochemical validation of a novel epithelial and a novel stromal marker of pancreatic ductal adenocarcinoma identified by global expression microarrays: Sea urchin fascin homolog and heat shock protein 47. *Am J Clin Pathol* 118: 52–9.

Malusecka, E., Zborek, A., Krzyzowska-Gruca, S., and Krawczyk, Z. (2001) Expression of heat shock proteins HSP70 and HSP27 in primary non-small cell lung carcinomas. An immunohistochemical study. *Anticancer Res* 21:1015–21.

Manjili, M. H., Wang, X. Y., Chen, X., Martin, T., Repasky, E. A., Henderson, R., and Subjeck, J. R. (2003) HSP110-HER2/neu chaperone complex vaccine induces protective immunity against spontaneous mammary tumors in HER-2/neu transgenic mice. *J Immunol* 171:4054–61.

Masson-Gadais, B., Houle, F., Laferriere, J., and Huot, J. (2003) Integrin alphavbeta3, requirement for VEGFR2-mediated activation of SAPK2/p38 and for Hsp90-dependent phosphorylation of focal adhesion kinase in endothelial cells activated by VEGF. *Cell Stress Chap* 8:37–52.

Mendez, F., Kozin, E., and Bases, R. (2003) Heat shock protein 70 stimulation of the deoxyribonucleic acid base excision repair enzyme polymerase beta. *Cell Stress Chap* 8:153–61.

Michils, A., Redivo, M., Zegers de Beyl, V., de Maertelaer, V., Jacobovitz, D., Rocmans, P., and Duchateau, J. (2001) Increased expression of high but not low molecular weight heat shock proteins in resectable lung carcinoma. *Lung Cancer* 33:59–67.

Mihaylova, V. T., Bindra, R. S., Yuan, J., Campisi, D., Narayanan, L., Jensen, R., Giordano, F., Johnson, R. S., Rockwell, S., and Glazer, P. M. (2003) Decreased expression of the DNA mismatch repair gene Mlh1 under hypoxic stress in mammalian cells. *Mol Cell Biol* 23:3265–73.

Mileo, A. M., Fanuele, M., Battaglia, F., Scambia, G., Benedetti-Panici, P., Mancuso, S., and Ferrini, U. (1990) Selective over-expression of mRNA coding for 90 KDa stress-protein in human ovarian cancer. *Anticancer Res* 10:903–6.

Missotten, G. S., Journee-de Korver, J. G., de Wolff-Rouendaal, D., Keunen, J. E., Schlingemann, R. O., and Jager, M. J. (2003) Heat shock protein expression in the eye and in uveal melanoma. *Invest Ophthalmol Vis Sci* 44:3059–65.

Nadin, S. B., Vargas-Roig, L. M., Cuello-Carrion, F. D., and Ciocca, D. R. (2003) Deoxyribonucleic acid damage induced by doxorubicin in peripheral blood mononuclear cells: Possible roles for the stress response and the deoxyribonucleic acid repair process. *Cell Stress Chap* 8:361–72.

Nanbu, K., Konishi, I., Mandai, M., Kuroda, H., Hamid, A. A., Komatsu, T., and Mori, T. (1998) Prognostic significance of heat shock proteins HSP70 and HSP90 in endometrial carcinomas. *Cancer Detect Prev* 22:549–55.

Neckers, L. (2002) Hsp90 inhibitors as novel cancer chemotherapeutic agents. *Trends Mol Med* 8:S55–61.

Neckers, L., and Ivy, S. P. (2003) Heat shock protein 90. *Curr Opin Oncol* 15:419–24.

Neckers, L., and Lee, Y. S. (2003) Cancer: The rules of attraction. *Nature* 425:357–9.

Nelson, D. A., and White, E. (2004) Exploiting different ways to die. *Genes Dev* 18:1223–6.

Netzer, W. F., and Hartl, F. U. (1998) Protein folding in the cytosol: Chaperonin-dependent and-independent mecanisms. *TIBS* 23:68–74.

Nimmanapalli, R., O'Bryan, E., and Bhalla, K. (2001) Geldanamycin and its analogue 17-allylamino-17-demethoxygeldanamycin lowers Bcr-Abl levels and induces apoptosis and differentiation of Bcr-Abl-positive human leukemic blasts. *Cancer Res* 61:1799–804.

Nylandsted, J., Brand, K., and Jaattela, M. (2000) Heat shock protein 70 is required for the survival of cancer cells. *Ann N Y Acad Sci* 926:122–5.

Nylandsted, J., Gyrd-Hansen, M., Danielewicz, A., Fehrenbacher, N., Lademann, U., Hoyer-Hansen, M., Weber, E., Multhoff, G., Rohde, M., and Jaattela, M. (2004) Heat shock protein 70 promotes cell survival by inhibiting lysosomal membrane permeabilization. *J Exp Med* 200:425–35.

Ogata, M., Naito, Z., Tanaka, S., Moriyama, Y., and Asano, G. (2000) Overexpression and localization of heat shock proteins mRNA in pancreatic carcinoma. *J Nippon Med Sch* 67:177–85.

Osada, M., Imaoka, S., and Funae, Y. (2004) Apigenin suppresses the expression of VEGF, an important factor for angiogenesis, in endothelial cells via degradation of HIF-1alpha protein. *FEBS Lett* 575:59–63.

Paul, C., Manero, F., Gonin, S., Kretz-Remy, C., Virot, S., and Arrigo, A. P. (2002) Hsp27 as a negative regulator of cytochrome C release. *Mol Cell Biol* 22:816–34.

Pearl, L. H. (2005) Hsp90 and Cdc37—a chaperone cancer conspiracy. *Curr Opin Genet Dev* 15:55–61.

Peng, X., Guo, X., Borkan, S. C., Bharti, A., Kuramochi, Y., Calderwood, S., and Sawyer, D. B. (2005) Heat shock protein 90 stabilization of ErbB2 expression is disrupted by ATP depletion in myocytes. *J Biol Chem* 280:13148–52.

Peng, X., Guo, X., Borkan, S. C., Bharti, A., Kuramochi, Y., Calderwood, S., and Sawyer, D. B. (2005) Heat shock protein 90 stabilization of erbB2 expression is disrupted by ATP depletion in myocytes. *J Biol Chem* 280:13148–52.

Pfosser, A., Thalgott, M., Buttner, K., Brouet, A., Feron, O., Boekstegers, P., and Kupatt, C. (2005) Liposomal Hsp90 cDNA induces neovascularization via nitric oxide in chronic ischemia. *Cardiovasc Res* 65:728–36.

Pratt, W. B., Galigniana, M. D., Harrell, J. M., and DeFranco, D. B. (2004) Role of hsp90 and the hsp90-binding immunophilins in signalling protein movement. *Cell Signal* 16:857–72.

Pratt, W. B., and Toft, D. O. (2003) Regulation of signaling protein function and trafficking by the hsp90/hsp70-based chaperone machinery. *Exp Biol Med (Maywood)* 228:111–33.

Price, J. T., Quinn, J. M., Sims, N. A., Vieusseux, J., Waldeck, K., Docherty, S. E., Myers, D., Nakamura, A., Waltham, M. C., Gillespie, M. T., and Thompson, E. W. (2005) The heat shock protein 90 inhibitor, 17-allylamino-17-demethoxygeldanamycin, enhances osteoclast formation and potentiates bone metastasis of a human breast cancer cell line. *Cancer Res* 65:4929–38.

Proskuryakov, S. Y., Konoplyannikov, A. G., and Gabai, V. L. (2003) Necrosis: A specific form of programmed cell death? *Exp Cell Res* 283:1–16.

Puy, L. A., Lo Castro, G., Olcese, J. E., Lotfi, H. O., Brandi, H. R., and Ciocca, D. R. (1989) Analysis of a 24-kilodalton (KD) protein in the human uterine cervix during abnormal growth. *Cancer* 64:1067–73.

Queitsch, C., Sangster, T. A., and Lindquist, S. (2002) Hsp90 as a capacitor of phenotypic variation. *Nature* 417:618–24.

Rahmani, M., Reese, E., Dai, Y., Bauer, C., Kramer, L. B., Huang, M., Jove, R., Dent, P., and Grant, S. (2005) Cotreatment with suberanoylanilide hydroxamic acid and 17-allylamino 17-demethoxygeldanamycin synergistically induces apoptosis in Bcr-Abl+ cells sensitive and resistant to STI571 (imatinib mesylate) in association with down-regulation of Bcr-Abl, abrogation of signal transducer and activator of transcription 5 activity, and Bax conformational change. *Mol Pharmacol* 67:1166–76.

Richards, E. H., Hickey, E., Weber, L., and Master, J. R. (1996) Effect of overexpression of the small heat shock protein HSP27 on the heat and drug sensitivities of human testis tumor cells. *Cancer Res* 56:2446–51.

Rohde, M., Daugaard, M., Jensen, M. H., Helin, K., Nylandsted, J., and Jaattela, M. (2005) Members of the heat-shock protein 70 family promote cancer cell growth by distinct mechanisms. *Genes Dev* 19:570–82.

Rutherford, S. L., and Lindquist, S. (1998) Hsp90 as a capacitor for morphological evolution. *Nature* 396:336–42.

Samali, A., Robertson, J. D., Peterson, E., Manero, F., van Zeijl, L., Paul, C., Cotgreave, I. A., Arrigo, A. P., and Orrenius, S. (2001) Hsp27 protects mitochondria of thermotolerant cells against apoptotic stimuli. *Cell Stress Chap* 6:49–58.

Scheibel, T., and Buchner, J. (1998) The Hsp90 complex—a super-chaperone machine as a novel drug target. *Biochem Pharmacol* 56:675–82.

Srivastava, P. (2002) Interaction of heat shock proteins with peptides and antigen presenting cells: Chaperoning of the innate and adaptive immune responses. *Annu Rev Immunol* 20:395–425.

Stephanou, A., and Latchman, D. S. (1999) Transcriptional regulation of the heat shock protein genes by STAT family transcription factors. *Gene Expr* 7:311–9.

Storm, F. K., Mahvi, D. M., and Gilchrist, K. W. (1993) Hsp-27 has no diagnostic or prognostic significance in prostate or bladder cancers. *Urology* 42:379–82.

Taira, T., Sawai, M., Ikeda, M., Tamai, K., Iguchi-Ariga, S. M., and Ariga, H. (1999) Cell cycle-dependent switch of up-and down-regulation of human hsp70 gene expression by interaction between c-Myc and CBF/NF-Y. *J Biol Chem* 274:24270–9.

Takashi, M., Katsuno, S., Sakata, T., Ohshima, S., and Kato, K. (1998) Different concentrations of two small stress proteins, alphaB crystallin and HSP27 in human urological tumor tissues. *Urol Res* 26:395–9.

Tang, D., Khaleque, A. A., Jones, E. R., Theriault, J. R., Li, C., Wong, W. H., Stevenson, M. A., and Calderwood, S. K. (2005) Expression of heat shock proteins and HSP messenger ribonucleic acid in human prostate carcinoma *in vitro* and in tumors *in vivo*. *Cell Stress Chap* 10:46–59.

Tatar, M., Khazaeli, A. A., and Curtsinger, J. W. (1997) Chaperoning extended life. *Nature* 390–30.

Teng, S. C., Chen, Y. Y., Su, Y. N., Chou, P. C., Chiang, Y. C., Tseng, S. F., and Wu, K. J. (2004) Direct activation of HSP90A transcription by c-Myc contributes to c-Myc-induced transformation. *J Biol Chem* 279:14649–55.

Tenniswood, M. P., Guenette, R. S., Lakins, J., Mooibroek, M., Wong, P., and Welsh, J. E. (1992) Active cell death in hormone-dependent tissues. *Cancer Metastasis Rev* 11:197–220.

Tetu, B., Lacasse, B., Bouchard, H. L., Lagace, R., Huot, J., and Landry, J. (1992) Prognostic influence of HSP-27 expression in malignant fibrous histiocytoma: A clinicopathological and immunohistochemical study. *Cancer Res* 52:2325–8.

Trieb, K., Kohlbeck, R., Lang, S., Klinger, H., Blahovec, H., and Kotz, R. (2000) Heat shock protein 72 expression in chondrosarcoma correlates with differentiation. *J Cancer Res Clin Oncol* 126:667–70.

Tsutsumi-Ishii, Y., Tadokoro, K., Hanaoka, F., and Tsuchida, N. (1995) Response of heat shock element within the human HSP70 promoter to mutated p53 genes. *Cell Growth Differ* 6:1–8.

Uozaki, H., Ishida, T., Kakiuchi, C., Horiuchi, H., Gotoh, T., Iijima, T., Imamura, T., and Machinami, R. (2000) Expression of heat shock proteins in osteosarcoma and its relationship to prognosis. *Pathol Res Pract* 196:665–73.

van 't Veer, L. J., Dai, H., van de Vijver, M. J., He, Y. D., Hart, A. A., Mao, M., Peterse, H. L., van der Kooy, K., Marton, M. J., Witteveen, A. T., Schreiber, G. J., Kerkhoven, R. M., Roberts, C., Linsley, P. S., Bernards, R., and Friend, S. H. (2002) Gene expression profiling predicts clinical outcome of breast cancer. *Nature* 415:530–6.

van Gent, D. C., Hoeijmakers, J. H., and Kanaar, R. (2001) Chromosomal stability and the DNA double-stranded break connection. *Nat Rev Genet* 2:196–206.

Vargas-Roig, L. M., Gago, F. E., Tello, O., Aznar, J. C., and Ciocca, D. R. (1998) Heat shock protein expression and drug resistance in breast cancer patients treated with induction chemotherapy. *Int J Cancer* 79:468–75.

Viatour, P., Merville, M. P., Bours, V., and Chariot, A. (2005) Phosphorylation of NF-kappaB and IkappaB proteins: Implications in cancer and inflammation. *Trends Biochem Sci* 30:43–52.

Vogelstein, B., and Kinzler, K. W. (1993) The multistep nature of cancer. *Trends Genet* 9:138–41.

Wadhwa, R., Taira, K., and Kaul, S. C. (2002) An Hsp70 family chaperone, mortalin/mthsp70/PBP74/Grp75: What, when, and where?*Cell Stress Chap* 7:309–16.

Wang, Q., An, L., Chen, Y., and Yue, S. (2002) Expression of endoplasmic reticulum molecular chaperon GRP94 in human lung cancer tissues and its clinical significance. *Chin Med J (Engl)* 115:1615–9.

Wang, Y., Theriault, J. R., He, H., Gong, J., and Calderwood, S. K. (2004) Expression of a dominant negative heat shock factor-1 construct inhibits aneuploidy in prostate carcinoma cells. *J Biol Chem* 279:32651–9.

Workman, P. (2004) Altered states: Selectively drugging the Hsp90 cancer chaperone. *Trends Mol Med* 10:47–51.

Wu, C. (1995) Heat shock transcription factors: structure and regulation. *Ann Rev Cell Dev Biol* 11:441–469.

Wu, G., Osada, M., Guo, Z., Fomenkov, A., Begum, S., Zhao, M., Upadhyay, S., Xing, M., Wu, F., Moon, C., Westra, W. H., Koch, W. M., Mantovani, R., Califano, J. A., Ratovitski, E., Sidransky, D., and Trink, B. (2005) DeltaNp63alpha up-regulates the Hsp70 gene in human cancer. *Cancer Res* 65:758–66.

Xiao, K., Liu, W., Qu, S., Sun, H., and Tang, J. (1996) Study of heat shock protein HSP90 alpha, HSP70, HSP27 mRNA expression in human acute leukemia cells. *J Tongji Med Univ* 16:212–6.

Zhang, F., Hackett, N. R., Lam, G., Cheng, J., Pergolizzi, R., Luo, L., Shmelkov, S. V., Edelberg, J., Crystal, R. G., and Rafii, S. (2003) Green fluorescent protein selectively induces HSP70-mediated up-regulation of COX-2 expression in endothelial cells. *Blood* 102:2115–21.

Zhou, V., Han, S., Brinker, A., Klock, H., Caldwell, J., and Gu, X. J. (2004) A time-resolved fluorescence resonance energy transfer-based HTS assay and a surface plasmon resonance-based binding assay for heat shock protein 90 inhibitors. *Anal Biochem* 331:349–57.

Zou, J., Guo, Y., Guettouche, T., Smith, D. F., and Voellmy, R. (1998) Repression of heat shock transcription factor HSF1 activation by HSP90 (HSP90 complex) that forms a stress-sensitive complex with HSF1. *Cell* 94:471–80.

Index

σ^E, 41
σ^S, 48
σ^W, 50
5' untranslated region (5' UTR), 39, 43–44

A

ABL mutant, 188
acid stress, 48–49
acid-resistance (AR) systems, 48
acquired thermotolerance and Hsp70 abundance, 11–12
actin cytoskeleton in rat cardiac myoblasts, 147–48
actin polymerization-Hsp27, 147
activating transcription factor 4 (ATF4), 60–61
acute ischemia-rat kidney models, 15–19, 25–27
adenine nucleotide translocase 1 (ANT1), 106
adeno-associated virus (AAV) DNA, 298–99
ADP-bound Hsc70, 230–32
aging. *See also* reactive oxygen species (ROS) sensors
   elevation of HSF1 and role of Hsp, 404–405
   role of sHsps, 158–60
Aha1, binding of, 331
Akt kinase, 153–54
alkaline stress, 49–50
α-crystallin domain, 143
α-crystallins HspB4 & HspB5,
   post-translational modifications, 159–60
αB-crystallin protein, 145–48
   mitochondria-mediated apoptosis, 152
α-crystallin/sHSP superfamily, 12, 143–44, 164
ALS. *See* amytrophic lateral sclerosis
ALS-associated mutant SOD1, 397–98, 402
Alzheimer's disease (AD), 75, 396
amyloid precursor protein (APP), 396
amytrophic lateral sclerosis, 396–97
   GluR2, abnormal editing of, 408

androgen insensitivity syndrome (AIS), 293–94
androgen receptor in prostate cancer, 346–47
angiogenesis-Hsp as promoter, 435–37
angiogenesis-UPR activation, 71
antigen-presenting cells (APC), 191–96, 367
   MHC class I antigen processing pathway(s), 370–72
   stress proteins influencing phenotypes, 382–84
APCMHCclass I molecules, 367
Apg-1, 180
Apg-2, 181
aquarium fish-experimental model, 8
arimoclomol, 406
   disease progression delays in ALS mice, 406–408
AR-P723S mutant, 301
Aryl-hydrocarbon receptor-interacting protein (AIP), 263
ATF6, 61–62
ATP binding to gp96, 367
ATP hydrolysis for chaperone activity, 210–12
ATP-binding domain of hsp110, 187
ATP-binding proteins, 184
ATP-dependent proteases, 40
auditory brainstem response (ABR), 105
auxilin 2, 238

B

B16 melanoma tumors, 191–93
bacterial cells, 36
BAG. *See* Bcl-2–associated athanogene
BAG-1
   and GrpE, 215
   association with CHIP, 319–20
   HSP70-based chaperone system, 261–62
BAG-1/Raf-1 complex, 215

BAG-1/RAP46/HAP46, 214–16
  ATPase domain of Hsc70, 215
BAG-2, association with CHIP, 320
Bcl-2–associated athanogene, 214–16, 261, 320
BCR-ABL (p210Bcr-Abl), 345
β islet cells, 66–67
β-amyloid precursor protein (APP), 75
β-sheet domain, 180, 185–86, 188
β-sheet domain in DnaK, 186
β-sheet sandwich structure, 143, 147, 164
benzoquinoid ansamycin antibiotics, 343–44
BiP binding, 62–63
bone-marrow-derived dendritic cells (DCs), 191
B-Raf, 346
brain abnormalities- Hsf1-deficient mouse, 107–108
brain development-role of Hsf2, 112
brain microvasculature, 15
breast and uterine cancer, role of FKBP, 301–302
"Brownian-ratchet" model for translocation, 243
BRX-220, 406

C
C. elegans
  role of sHsps in aging, 160
C58J-coats, 240
cancer
  and its genetic instability, 341–44
  role of HSP's
    HSP gene transcription, 2, 426–27
cancer cell surface-targeting Hsp90 on, 350–51
  chaperoning efficiency for oncogenic proteins, 429–30
cancer cells-protective role of Hsp90, 342–44
cancer therapy-role of HSP's, 439–40
cancer-UPR activation, 69–71
carotid endarterectomy, 15, 366
casein kinase 2 (CK2) phosphorylation, 286
caspase-3, 152
caspases, 151
CD4+ T cells, 385
CD4+CD25+ or CD4+CD25− phenotypes, 385
CD45RA+RO− and CD45RA−RO+T cells, 379
CD8+ T cells, 365
CD95-mediated apoptosis, 434
Cdc37, 326–27
  as a co-chaperone of Hsp90, 330–31
  role of overexpression, 333–34
CEA. *See* carotid endarterectomy
celastrol-neuroprotective effects in animal models, 411
cell life and death processes-role of sHsps, 149–58

cell lines, 10
cell-cycle-related proteins, ectopic expression of, 401–402
cellular homeostasis, 36
cellular localization of sHsps, 146
cellular-stress response, 25
chaperone-associated peptides, 191
"chaperone complexes" with gp100, 194
chaperone-receptor heterocomplex, 283–84
chaperoning and cytoprotection-role of *hsp110*, 181–82
chaperonins, 326
chimeric J-protein, 214
chimeric protein NPM-ALK, 344–45
CHIP (COOH-terminal Hsc70 interacting protein), 125–25, 216–18
  role in regulation of stress responses, 317–19
CHIP activity regulators, 319–21
CHIP as an HSF1 activator, 318
CHIP homodimer, 316–17
  HSP70 and HSP90 chaperone systems, 262
CHIP/chaperone complex, 316
CHIP's E3 ligase activity, 217
CHOP, 64
  effect in tumor cells, 70
chorioallantoic placenta-role of Hsf1, 101–103
cisplatin, 71
class II Hsp40s, 213
class III Hsp40s, 213
clathrin cage, 238
clinical implications for FKBP, 300–303
co-chaperone ubiquitin ligase, 314–17
co-chaperones-TPR domain, 315
cold shock proteins, 42–44
cold shock sensors, 42
cold-shock mRNA, 44
cold-shock translation, 43–44
colon 26 tumors-hsp110 or gp170, 190
complete AIS (CAIS), 300–301
conformational changes in Hsp70, 230–32
CSPs. *See* cold shock proteins
CT26-hsp110 cells, 191
C-terminal TPR/Hsp90-binding domain, 287–88
cultured cells, 7
cycloheximide, 13
CyP PPIase, 285
cytochrome C, 151–52
  interaction with Hsp27, 156
  release, 64
cytokine cardiotrophin-1 (CT-1)- induced expression of FKBP52, 300
cytoplasmic membrane-based proteins, 45
cytoplasmic osmolality, 44

cytoprotection, 9
  of zinc ions, 13
cytoprotective vs. proapoptotic response aspects, 64–65
cytosolic HSP70 family protein, 257–58. *See also* mammalian hsp70 family proteins

D

D. melanogaster dmhsp22mRNA, 159
damage control laparotomy, 26
Daxx, 130
  receptor-mediated apoptosis (extrinsic), 153
death-inducing signaling complex (DISC), 152
denaturation of cellular protein, 123
desaturases, 43
desmin-related myopathy-role of sHsps, 161–64
DHR domain (downstream of HR-C), 94
diabetes-UPR activation, 73–74
dimers of RepA-Hsp40/DnaJ-Hsp70/DnaK, 237–38
disaggregating chaperones, 40
dissociation constant (Kd) value, 330
DjA1-null mice, 266–67
DjA3 (hTid1), 267–68
DjA3, 259
DjA4, 259
DjB1 (hsp40/hdj-1), 259, 268
DjB11 (HEDJ/Erdj3), 268
DjB2 (Hsj1), 268
DjB4 (Hlj1), 268
DjB6 (Mrj), 268
DjB9 (MDG1/ERdj4), 268
DjC1 (Mtj1/ERdj1), 268–69
DjC19 (DNADJ1), 269
DjC2 (MPP11/MIDA1), 269
DjC3 (P58IPK), 269
DjC5 (CSPα), 269
DjC5 (cysteine string proteins), 259
DjC6 (auxilin 1), 269
DjC6 (auxilin), 259
DNA binding of *Hsf2*, 109
DNA damage-response pathways, 430, 434
DnaK chaperone, 40
domain deletion mutants of grp170, 194–95
domain I of Ydj1, 235
Drosophila HSP mRNA, 1
Drosophila melanogaster, 11
  role of sHsps in aging, 160
Drosophila salivary gland, 1
Drosophila stress genes, 3

E

E. coli OxyR, 46–47
E. coli-lon protease, ClpA and ClpX chaperones, 240–41
E1 (ubiquitin-activating enzyme) hydrolyzes ATP, 314
E2 (ubiquitin-conjugating enzyme), 314
E3 (ubiquitin ligase), 314
E6AP carboxy-terminus (HECT), 314
EGFP-*neo*, 109
eIF-2α kinase (PEK), 66
eIF2α$^{S51 \to A}$, 66
eIF-2α^{S51A} mice, 67
embryonic lethality-*Hsf2* gene, 109–12
endogenous gp96, 383
endogenous Hsp70, 383
endogenous stress proteins-anti-inflammatory nature, 379–80
endoplasmic reticulum (ER) stress-role of *sHsps*, 148–49
endoplasmic reticulum (ER), 57–58
enHsp60-reactive CD30+ T cells, 380
epidermal growth factor receptor (EGFR), 346
ER chaperones, 58–59, 67–68, 72–73, 183–84
ER luminal Hsp70, BiP, 215–16
ER stress-response elements, 61, 181
ER-associated degradation (ERAD), 58, 149
  Hsp70 and regulators of Hsp70, role of, 218–21
Erk1/2-mediated pathway, 154
ERSE. *See* ER stress-response elements.
eukaryotic microorganisms, 7
eukaryotic translation initiation factor-2α(eIF-2α), 60
exposed mesentery tissue, microcirculation of, 14
external pH stress sensors, 48–50

F

F_0F_1 ATPase operon, 49–50
FADD (Fas associated death domain protein), 152–53
fatty acids of phospholipids, 42–43
fever-like temperature, 398, 400, 403, 404
fibrosarcoma cell line, 70
5α-reductase, mutations in, 294
51KO and 52KO+51KO mice, phenotypes of, 296
52KO male and female mice, reproductive phenotype of, 293–96
52KO MEF, 293

Index

FK506 binding proteins (FKBP), 284
  gene duplication event, 299–300
  pharmacological prospects and toxicological concerns, 303
FK506 enhanced GR-mediated reporter gene expression, 290
FK506, 284–85
FKBP action, mammalian cellular models, 292–93
FKBP binding in receptor complexes, 289
FKBP chaperone activity, 287–88
FKBP gene knockout mice, phenotypes of, 293–96
FKBP12, 286–87
FKBP51
  interactions with cellular factors, 296–97
  transcriptional regulation of, 299
FKBP51 and FKBP52, 285–87
FKBP52
  in GR complexes, 289
  interactions with cellular factors, 297–99
  male sexual development and fertility, 294
FKBP52 by FKBP51, antagonism of, 292
FLT3, 345
folder chaperones, 40
folding/assembly of secretory proteins, 183–84
FtsH, 241
full CHOP induction, 73–74
full-length mutant huntingtin neuronal cell phenotype, 410

G

GadE/*gadE*, 48–49
GAL4-HSF1 chimera, 128
γ-crystallin family, 115
gastrointestinal stromal tumors (GIST), 345–46
GC-inducible down-modulator of GR activity, 302–303
gene knock-in, *Hsf2* null mice, 108–109
GFP, 66
glial cells in the mammalian brain, 398
glial to neuron transfer system, 399
glucocorticoid receptor, 244
glucose-6-phosphate dehydrogenase (G6PD), 106
glucose-regulated proteins, 178
glutathione, 106. *See also* oxidative stress protection by Hsp27
Gly/Phe-rich region, 233–34
glycogen synthase kinase-3β (GSK3β), 64
GM1-ganglioside in ER, 63–64
gp100-grp170 complexes-chaperoning functional mutants, 194

gp96-bound peptide, 367
GroEL, 40–41
Grp/grps. *See* glucose-regulated proteins
grp170 antisense-transfected C6 glioma cells, 184
GRP170, 71
Grp170/orp150, 180–81
  role in chaperoning, cytoprotection and polypeptide transportation, 183–85
grp170-H, chaperoning ability of, 188
grp78/ BiP, 178. *See* immunoglobulin binding protein
  acquired drug resistance, 436–37
  anti-inflammatory properties, 381, 384
grp94/gp96, 178, 195
  antigenic repertoire binding, 370–72
  anti-inflammatory properties, 381–82
GSH. *See* glutathione
GSH:GSSG ratio, 106

H

HAC1 mRNA, 60
Hanks' balanced salt solution (HBSS), 21
Harc, middle domain of, 327
heat shock elements
  definition of, 122
  HSP expression in cancer, 427
heat shock factors, 91
  definition of, 122
heat shock gene expression, 16
heat shock proteins, 1, 38, 178
  and replicative senescence, 435
  protection of Synapses, 399
heat shock response, 38
  capacitor of stress, 317–18
  celastrol as a novel inducer, 409–10
  neural cell death, 400–401
heat-denatured luciferase, 186–87
  CHIP-mediated ubiquitination, 217
hepatitis C virus, 74
high-dose gp96 administration, 382
high-molecular-weight protein complex (HMW), 159–60
high-temperature proteins, 38
high-temperature response-sensors of, 38–42
H-NS, DNA-binding protein, 38
holder chaperones, 40
homodimeric CtsR repressor, 41
Hop, 262, 283, 288
hormone-bound receptor, 282–83
HPLC/MALDI-TOF, 369
Hsc70 interacting protein (HIP), 212
Hsc70, 210

Hsc70–Bag-1 complex, structure of, 261
Hsc70-binding surface of the J-domain, 240
Hsc70t, 257
HSE DNA-binding activity, 122–23. *See also* phosphorylation of *HSF1*
HSE sequence, 2
HSE. *See* heat shock elements
Hsf1-knockout models of, 94–108
HSF. *See* heat shock factors
Hsf1 activation, 93
Hsf1 in female reproduction, 103–104
Hsf1 knockout mice, 94–101
HSF1 oligomerization-repression of, 123–25
HSF1 to DNA-binding form, 403–404
HSF1 transcriptional competence, 125–31
HSF1 trimers, 130, 317
Hsf1 wild-type female, 104
Hsf1, 406. *See also* Hsp90 inhibitors
   compositional design, 122–23
   heat shock response, regulation of, 403–404
   pathway of stress activation, 132f
Hsf1-deficient animals, 101–103
Hsf1-deficient normal adult hearts, 106
HSF1-FKBP52 interaction, 125–26
HSF1-Hsp90 interaction, 125
Hsf2 null mice brain-experiment (s), 112–13
Hsf2, 108–14
Hsf2-β-geo, 109
Hsf4 expression, 114–15
Hsf4-/- lens epithelial cells, 116
hsp gene expression, 1
hsp gene induction in eukaryotes, 180
HSP genes, 2
HSP levels, 3
Hsp mRNAs, 131
Hsp/HSP. *See* heat shock proteins
Hsp10, 377
   anti-inflammatory activities, 384
   anti-inflammatory properties, 380–81
hsp110 and Hsc70/Hdj-1
   chaperoning capability of, 192–94
   interaction between, 189
hsp110 family, 179–80
hsp110 protein/hsp105, 180
   human papilloma virus oncoprotein, E7, 182
   secondary structure prediction, 185
hsp110 Sse1, 189–90
hsp110-hsc70 complex, 189
hsp110-ICD (intracellular domain) complex, 193
Hsp26, chaperone activity of, 147
HSP27 and HSP70, 431–34
Hsp27 and αB-crystallin
   multiple levels of apoptotic pathways, 156–58

Hsp27/25 expression in the uterine cervix, 423–24
hsp27/Hsp27
   anti-inflammatory properties, 384
   cytoprotective effect against apoptosis, 156
   in astrocytes, 108
   proteasome-mediated protein degradation, 149
Hsp27-IKKβ, *154*
Hsp30, 12
Hsp40 Chaperones/DnaJ proteins/J proteins, 212–14
Hsp40, 232–36
   chaperone-receptor heterocomplex, 283–84
Hsp40/DnaJ, 235, 255
   in vitro folding, 263–65
   in vivo functions, 266–70
   members of family, 259–61
Hsp40/DnaJ-Hsp70/DnaK
   defects in, 240–41
   disassembler, 237–38
Hsp40/Sis1-Hsp70/Ssa1, 235
Hsp40–Hsp70 chaperone machines, 228
   functions of co-chaperones, 261–63
   minimal model (s), 235–36
   protein translocation across membranes, 242–44
Hsp40–Hsp70 disassembler, 236–40
Hsp60
   antigen-presenting capacity, 382–83
   immunoregulatory properties of, 380
HSP60 and HSP70-anti-inflammatory properties, 379–80
Hsp70 (DnaK) hydrolytic cycle, 212
Hsp70 ATPase cycle, 211f
Hsp70 ATPase domain (AD), 229
   of HSP70 protein, 257–58
Hsp70 gene family, 27
Hsp70 mRNA, 14
hsp70 superfamily members-molecular chaperoning, 194–95
HSP70, 1–2, 210–21
   effect of CHIP on, 318–19
   P. lucida, 9
   Poeciliopsis hybrids, 11
   structural domain, 228
HSP70.2-senescence pathways, 435
Hsp70/BiP, 242–43
Hsp70/DnaK, 235
Hsp70/DnaKPBD (peptide-binding domain), 229
   on Cis/Trans-side of membrane, 242–44
Hsp70/Ssc1 mutant, 243
Hsp70's ATPase activity by Hsp40, 214

Hsp70-2, 257
HSP70B′ isoform, 9
HSP70-based chaperone system, 255–57
Hsp70-binding-protein 1 (HspBP1), 216
    association with CHIP, 320–21
HSP70–DjA1, 263–65
HSP70–DjA2, 263–65
HSP70–DjA4, 263–65
Hsp70-Hsp40 complexes, 213–14
    ERAD of polytopic membrane proteins, 219
Hsp90 cochaperone p23, 288
    nucleotide exchange and ATP hydrolysis, 344
Hsp90 inhibitors, 341. *See also Hsf1*.
    androgen receptor in prostate cancer, 346–47
    cell death pathways, 431–35
    clinical trials, 353
    containing multichaperone complex, 124–26
    nucleotide-dependent cycling, 344
    sensitizing effects of 17-AAG, 350
HSP90 proteins, 2, 262–63
    proteasome-mediated degradation, 349
    role in regulating cancer signaling nodes, 342–44
        self-sufficiency in growth signals, 428–31
    telomerase stability, 435
Hsp90-p23-immunophilin multichaperone complexes, 125–26
HspB1 phosphorylation, 159
hspR gene, 40
Hsps-based vaccines, 155–56
HT-1080 cells, 350–51
HTPs. *See* high-temperature proteins
human papilloma virus oncoprotein, E7, 182
human telomerase reverse transcriptase (hTER), 366
Huntington's disease (HD), 354–55, 402
hydrophobic heptad repeats (HR-C), 92
hypoxia-inducible factor1 α (Hif1α), 70
    VEGF expression, inhibition of, 347–48

I

iatrogenic stressors, 15
immunological escape, 365
immunophilins, 284
immunoprecipitation, 190
immunosuppressive action of FKBP, 303
in vivo function of gp96, 368
induced thermotolerance, 398
insulin release-role for grp170, 184
interaction of Cdc37 and Hsp90 with protein kinases, 332
interaction of MyoD with Cdc37, 332
intracellular expression, 3

intracellular transport of receptor complexes-role of FKBP52, 298
intraperitoneal temperature, 16
invasive/metastatic capacity of tumors correlation with *HSPs, HSF1*, 437–38
invasive cardiovascular surgical procedures, 15
Ire1p, 62–63
ischemia-UPR activation, 72–73
isoform of HSP70, (iHSP70), 21
    heat shock protein, 24

J

J-domain interactions with proteins, 232–33
J-domain-AD interface, 230
J-domains of polyomavirus T/t antigens, 238
JNK pathway
    receptor-mediated apoptosis (extrinsic), 153

K

K562 erythroleukemia cell line, 93–94
KdpD sensor kinase, 45
kinase and non-kinase interaction with Cdc37, 331–34
knockdown of FKBP51, 297
(knockout) KO models, 108
    glucocorticoid signaling, 296

L

lens development-role of *Hsf4, 115–16*
LexA-HSF1 chimera, 126
ligand binding domain (LBD), 281
lipid rafts, 399–400
loss of function (LOF), 94
LPS (CD14, CD40, TLR)-stress protein activity, 383
Lys27, 217
lysosomal storage disease, 74–75

M

male and female fertility, role of FKBP, 302
male Sprague-Dawley rats-survival rates and renal vascular resistance, 20
mammalian cells-UPR pathway, 60–62
mammalian hsp70 family proteins, 179–85. *See also* cytosolic HSP70 family protein.
mammalian tissue culture systems, 403–404
MAPK-activating protein (MAPKAP) 2/3, 145–46
meiotic M phase spermatocytes-apoptosis of, 113
met receptor tyrosine kinase, 348
methylcholanthrene-induced (Meth A) fibrosarcoma, 381

MHC class I epitopes, 371
MHC class I molecules, 367–71
MHC class I-peptide complex, 371
minimum functional unit of Hsp70, 230
misfolded proteins, 397
mitochondrial permeability transition pore, 106, 411
mitochondrial protein precursors-role for cytosolic Hsp70, 242
molecular chaperones, 2, 209. *See also* Steroid hormone actions.
    anti-apoptotic mechanisms of sHsps, 155
    pharmacologic induction of, 353–55
mouse embryonic fibroblast (MEF), 293
MPTP. *See* mitochondrial permeability transition pore
mRNA, 37
mutated or chimeric kinases, 344–46
mutator phenotype, 430

N

NAD(P)H:Quinone Oxidoreductase I (NQO1), 352
nascent protein folding, 58
    interaction with Hsp40–Hsp70, 237
National Cancer Institute (NCI), 344
NCI 60-cell line screen, 344
NEF-Hsp70 complexes, 215
NEFs. *See* nucleotide exchange factors.
neurodegenerative diseases, 74
    characteristics of, 396–97
    drugs for experimental assays, 408–409
    HSP90 inhibitors, 353–55
    protein folding quality control mechanisms, 405
    role of sHsps, 160–64
        animal models, 402
Neurodegenerative Drug Screening Consortium, 408–11
neurologic damages, 15
neurologic dysfunction, 15
neuronal cell death, 401–402
neuronal cell populations, 398
neurons, 400–401
New World primates, 291
NFκB, 154
NK cells-Hsp70 as cell-surface immune mediator, 191
noise stimulation- role of *Hsf1* stress pathways, 105
non-classical J-domain proteins, 236
N-terminal domain, role in kinase binding, 333

nucleotide exchange factors, 209, 212, 215, 216, 220, 261
null mice-role of PERK, 70–74

O

off-pathway folding intermediates, 405
OhrR (organic hydroperoxide resistance), 47
oligomerization of *sHsps, 146–47*
organ preservation-translational model, 16
osmolality, 44
osmolyte concentration gradient, 44
osmoregulation, 45
osmotic upshift, 44–45
osteoblasts, normal physiology of, 68
oxidative stress in L929 cells, 148
oxidative stress protection by Hsp27, 148. *See also glutathione.*
oxygen regulated protein (Orp), 178

P

P. lucida Hsp27, 12
P. lucida, 8–10
P170 (*mdr*-1 gene), 433
p53-tumor suppressor, 70–71, 426
p63-head and neck cancer, 426
Pachytene spermatocytes, 104
PAM16, 243
PAM18, 243
pancreatic islet cells, 73
Parkinson disease (PD), 354
    Drosophila model, 402
PARP (poly(ADP-ribose)polymerase), 151–52
PCD. *See* programmed cell death
PDZ domain of DegS protease, 41
peptide antigens of gp96, 367–70
peptidylprolyl *cis/trans* isomerase (PPIase) immunophilins, 285
peripheral T cells, 379
PerR, 47
PhoP-PhoQ, 49
phosphorylation of *sHsps*, 144–46
phosphorylation of Cdc37, 328–29, 333
phosphorylation of *HSF1, 126–30. See also* HSE DNA-binding activity
PKR-like endoplasmic reticulum (ER) kinase (PERK), 60–63
plasma cell differentiation, role of UPR, 67–68
plasma cell differentiation-UPR activation, 64
Poeciliopsis Hsp27, 12
Poeciliopsis HSPs, 11
Poeciliopsis, 8–10
polo-like kinase 1, 129
polyglutamine proteins (poly Q), 161–64

post-transcriptional regulation of *FKBP52*, 300
pp60[v−src], 330
PPIase Co-Chaperone Structure, 285–87
PR gene (PRKO) female mice, 295–96
PR. *See* progesterone receptor
preinfarction angina, 26
prfA mRNA, 38–39
PR-Hsp90 complexes, 289
progesterone receptor, 288–90, 295–96, 299–303
programmed cell death, 150
  Hsp gene (s), 423
  Hsp inhibitors as *cell death pathways*, 431–33
prostate cancer, role of FKBP, 301
protein 9 or AIP2 (AhR-interacting protein 2)-association with CHIP, 320
protein aggregates, 353
protein gel electrophoresis, 21, 24–25

R

R120G mutated αB-crystallin, 161
radicicol and geldanamycin, 406
Raf MAPKKK, 154
ratios of BAG-1/CHIP to HIP or HOP, 218
reactive oxygen species (ROS) sensors, 46–47. *See also* aging
recombinant chaperone vaccine approach-advantages of, 193
redox homeostasis-role of *Hsf1*, 107
refolding of proteins, 397
release of Hsp into blood stream, 400
renal epithelium, 13
RET receptor tyrosine kinase, 349
retention of VEGF antigen, 184
retro-translocation of ERAD substrates, 218
revascularization during wound healing, 184
RING and U-box domain E3s, 314–15
RNA chaperone, 43, 44, 181
RNA polymerase, 38
rodent kidney, cold storage time of, 16
rodent renal transplant, 19, 22
rodent spermatogenesis-role for *Hsf2*, 113
rpoH transcript coding, 38–39
RseA, 41–42
RseP, 42
RT-PCR, 108
  expression of fibroblast growth factors (FGFs), 116

S

S. cerevisiae grp170 (i.e., Lhs1p), 184
S. cerevisiae, 234
Saccharomyces cerevisiae, 291
salubrinal molecules, 76

Sandoff disease, 75
satellite III transcripts, 130–31
SBMA. *See* spinal and bulbar muscular atrophy.
(SDS-resistant) complexes, 367–68
Sec63, 236, 243
secondary structure of hsp110, 185
secondary structure predictions of grp170, 185
sensitivity to chemotherapy-UPR activation, 71
Ser^{121} phosphorylation, 127, 130
Ser^{230}, 127–28
Ser^{303} and Ser^{307}, phosphorylation of, 128–30
Ser^{326}, phosphorylation of, 129
Ser^{363}, phosphorylation of, 129–30
Ser^{418}, phosphorylation of, 129–30
17-AAG, 341, 346
  clinical trial in malignant melanoma, 351–53
17-AAG dihydroquinone, 352
17-aminogeldanamycin (17-AG), 351
17-DMAG, 343–44, 351–54
sHSPs. *See* small heat shock proteins.
Sigma-32, proteolysis of, 241
SIINFEKL, 369–70
Skp1/cullin/F-box E3 complex (SCF), 316
small heat shock proteins
  anti-apoptotic activity, 151
  as Molecular Chaperones, 146–47
  beneficial effects on aging, 160
small heat shock proteins, 12
SoxR, 47
spinal and bulbar muscular atrophy, 182, 354
sponge matrix heterograft, 21, 24
squirrel monkey (SM) cells, endogenous GR in, 291
Ssa1p cytoplasmic Hsp70 chaperone, 218
Sse1 function in vivo-ATP binding, 187
stable cell lines-ovalbumin as antigen, 371–72
steroid hormone actions, 281–83. *See also* molecular chaperones.
steroid hormones, 281
steroid receptor activity, effects of FK506 and Cyclosporin CsA, 290
steroid receptor signaling, role of FKBPs, 287–96
Sti1 and Cdc37, physical interaction between, 331
stress and immunology, role of *FKBP*, 302–303
stress conditioning, 13–14, 16, 25–27
stress factors, 36–37
stress granules, 130–31
stress protein-reactive T cells, 385
stress proteins, 378
  anti-inflammatory properties, 382–85
  cell-to-cell transfer, 399

stress response biology, 7
stress restitution mediator, 318–19
stress sensors, 37
stress-conditioning principles, 26–27
stress-conditioning protocols, 25–26
stress-induced apoptosis-Hsp27, 150
structural domains of hsp110 and grp170, 186–88
structure of Cdc37, 327–28
superoxide dismutase (SOD1) gene, 397
SV40 capsid, 244

T
T47D breast cancer cells, 290
targeted neuronal overexpression of grp170, 185
temporary threshold shift (TTS), 105
10kDa α-helical subdomain (PBDα), 229–30
testis
　role of *Hsf1, 104–105*
　role of *Hsf2, 113–14*
tetratricopeptide repeat (TPR), 235
　FKBP51 and FKBP52 domain, 286–87
Th2-type CD4$^+$T cells, 380
thermal maxima, 10
thermal resistance, 10–11
thermotolerant state, 10–11
3T3 fibroblasts, 349
33-residue C-terminal tail, 229
TIM44, 243
TlpA, 39
Tom70, 262–63
TPR domain of FKBP cochaperones-Hsp90 binding, 288–89
TRADD (TNF receptor death domain protein), 152–53
transcriptional regulation of FKBP52, *300*
transcriptional regulators-HSF family, 93–94
transcriptionally active HSF, 92
transient cold shock response, 42
transplant rejection, 380
Trp 7 of Cdc37, role in kinase binding, 332–33
TSTA of MethA sarcoma, 367, 373
tumor cell lines, 7
tumor progression, 424–26, 432–3
tumorigenicity-role of Hsps, 155
tumor-specific CTL response, 191
tumor-specific transplantation antigens (TSTAs), 364–66
20-kDaβ-sandwich (PBDβ), 229–30

26S proteasome, 149, 314
　association with BAG-1, 319
type I/II/III *HSP40/DnaJ* proteins, 266–70

U
ubiquitin/proteasome system, 313–14
U-box and TPR domain- CHIP-induced degradation 216–17
U-Box E3, 315–16
unfolded protein response
　antiapoptotic and proapoptotic elements, 65f
　characteristics of, 57–59
UPR activation-conventional means, 62–63
UPR. *See* unfolded protein response
UPR-inducible reporter gene, 66
UPR-transducers, 62–63
US Food and Drug Authority (FDA), 408–409

V
vascular endothelium, 13–14
ventriculomegaly, 107
viral infections-UPR activation, 69, 74
virF gene, 38
virF promoter, 38
VSV8 with gp96, interaction of, 368–69

W
warm ischemia-reperfusion, 19–21, 22–24
water stress sensors, 44–45
wild-type littermates (WT), 293
winter flounder renal tubules, 13
Wolcott–Rallison syndrome, 73
wound healing, 14
WT-AR activity, role of FKBP52, 301

X
XAP2 (hepatitis B virus X-associated protein)/ Ara9 (AhR-associated
X-box protein 1 (XBP-1) transcript, 60, 62
XBP-1 null ES cells, 67
XBP-1 protein, 68
XBP-1(S), 64
X-linked AR gene, mutation in, 293

Y
yeast two-hybrid screens, 298
yeast type I HSP40 Ydj1, 259–61
yeast UPR, 59–60

Z
Zn-binding and C-terminal domains, 234–35

Printed in the USA